READ IN CANADIAN GEOGRAPHY

READINGS IN CANADIAN GEOGRAPHY

THIRD EDITION

EDITED BY ROBERT M. IRVING

Holt, Rinehart and Winston of Canada, Limited
Toronto

Editor:
Robert M. Irving, Ph.D.
Department of Geography
University of Waterloo

Canadian Cataloguing in Publication Data

Main entry under title:
Readings in Canadian geography
ISBN 0-03-925494-1

1. Canada — Description and travel — 1950-   *
I. Irving, Robert M., 1930-

FC75.R42 1978   917.1   C77-001804-1
F1016.R42 1978

Library of Congress Catalogue Card Number: 78-50890

In this series of readings, the views expressed are those of the individual authors, and do not necessarily reflect the opinions of the editor or the publisher.

Photo Credits

Cover and Page 1: Department of Mines, Quebec.
Page 17: Convention and Tourist Bureau of Metropolitan Toronto.
Page 139: British Columbia Government Photo.
Page 195: B-A Oil Photo.
Page 269: (Top) National Film Board; (Bottom) Central Mortgage and Housing Photo.
Page 341: Miller Services.

Printed in Canada
1 2 3 4 5   82 81 80 79 78

To the Young Canadians

Jane
Stephen
Jennifer
Lisa

and

Donald Fulton Putnam
1903-1977

teacher
scholar
mentor

Preface

Since the publication of the first edition of *Readings in Canadian Geography* in 1968, a decade has passed — a short time geologically, but a significant time in human terms. Change is rapid.

In the third edition of *Readings*, I have tried to capture some of this change. In the process, and as I review the contents of the previous editions, the emphasis has shifted from the "case study" to broader, nationwide interpretations. In part, this shift reflects my own shifting perspectives, and also the availability of a much richer field of geographic literature.

I have elected to arrange the readings on a systematic rather than a regional basis — with a few exceptions. As emphasized in the past, this selection of readings is not comprehensive; certain areas are untouched because other volumes cover the field. Likewise, in the selection of papers, I have exercised judgement and hence some potential contributions have not been included which, given extended space limits, are worthy of wider distribution. In this process, of course, I assume the responsibility. On the other hand, I have received helpful comments from a number of colleagues, and specifically I would like to express my appreciation to Ralph Krueger, Lorne Russwurm, and Richard Preston, of the University of Waterloo who have been most helpful. The map and other illustrative materials have been prepared by Gary Brannon, chief cartographer in the Faculty of Environmental Studies, University of Waterloo, with the assistance of Norm Adam and Barry Levely. I express my appreciation to Jean Baker of Kearney, Ontario, who typed most of the manuscript, and to Rosemary Ambrose in the Department of Geography at Waterloo, who also contributed in the typing and general correspondence associated with this project.

Bob Irving

Waterloo, Ontario.
November, 1977.

Table of Contents

IDENTITY CANADA

URBAN CANADA

RURAL CANADA

RESOURCES CANADA: SOME ISSUES, DEVELOPMENTS, AND QUESTIONS

DISPARITIES CANADA

FUTURE CANADA

IDENTITY CANADA

Introduction

Ten years ago, in the First Edition of *Readings in Canadian Geography*, I wrote:

> Canadians have long been digging up their ancestral roots to inspect the nature and direction of their country's growth. Size, physical diversity, sparseness and unevenness of population distribution, the presence of two founding cultures, and proximity to the United States have contributed to create many problems. One hundred years after Confederation, Canadians are asking still the questions, "What are we?" and "Does Canada have a national identity?".

In the intervening decade new crises, events, and ongoing processes have disturbed and aroused many Canadians: the La Porte-Cross tragedy in Quebec; the continued presence of foreign ownership of resources; the energy crisis; native land claims and northern economic development; the election of the Parti Québecois in Quebec; and the current Depression-like economic and unemployment conditions, reminiscent of the Thirties. Chief among this selection of events, the election of the Parti Québecois on November 15, 1976, has motivated Canadians to examine once again the form and structure of the ancestral roots — as manifested by the growing number of unity debates, symposia, articles, and books, and capped most recently by the creation of the Federal Task Force on Canadian Unity.

It is not possible to reflect all of the diverse views and interpretations in this anthology. Two selections are presented. In the first paper Wallace Stegner draws parallels and contrasts between American and Canadian attitudes at different periods in their development, as reflected in the writings of novelists and scholars in both countries. In the process, Stegner raises a number of questions and challenges — pertaining to regional and cultural identity, the melting pot versus the mosaic, and similarities between Canada and the United States — that can form the basis for thoughtful discussion and debate.

The second selection, by Andrew Burghardt, focusses on the key issue of the unity debate — the question of Quebec separatism. Not infrequently one can learn, if not resolve, basic differences by examining similar events elsewhere. This Burghardt does by tracing briefly the history of Hungary within the Austrian Empire, indicating many parallels to the Canada-Quebec situation. Although Burghardt does not speculate whether Quebec will separate from Canada, he does emphasize some of the consequences of such an event and concludes that the great Canadian corridor (the St. Lawrence) would likely be destroyed in the process and "that without the intactness of that corridor there is no viable land called Canada".

The Provincial Consciousness

WALLACE STEGNER

The last thing I would try to do is to instruct Canadians in their own culture, rebuke them for not having one, or tell them how to handle either their internal cultural problems or their relations with the powerful neighbour who provides them about equally with things they don't want and things they can't do without, examples they deplore and temptations they can't resist.

All I will try to do is tell you how the present state of Canadian culture appears to a visitor of barely two months, and to suggest some American parallels, both historic and contemporary. They may provide either horrible examples or the sort of comfort and encouragement harried parents feel when they read Spock or Gesell and find that almost *all* fifteen-year-olds act that way. At worst, it may infuriate some Canadians by suggesting that even in the matter of cultural inferiority-complexes the United States has to insist that it got there first, and that Canada, as usual, is compulsively repeating American experience two generations or more behind.

I don't want to infuriate anyone; and although I will defend the United States on most counts, even in the deplorable year 1973, I don't come as a Yankee freebooter bent upon annexing Canada. If there was ever a Manifest Destinarian in me, he exhausted himself on the way West, and doesn't operate north-south. Moreover, my loyalties are mixed, for I spent a half dozen of my most impressionable years on a Saskatchewan homestead, and missed becoming a Canadian by only about 25 mm of rain. So if I say something detestable, please remember that I am trying to speak simultaneously from within and without the family. I am an American visitor but also I am a kind of knot-headed country cousin, more to be pitied than censured, and more to be loved than reviled.

Let me summarize how the Canadian literary scene strikes the visitor. None of this will be new to you, but you ought to know the conceptions from which I begin.

First, there is an extraordinary stir, great activity, great interest, and a frequent and often assertive use of the word "Canadian". One finds all kinds of little magazines, many of them with a nationalist or ethnic cast, and even the larger-circulation magazines like *Maclean's,* even the newspapers, reflect patriotic cultural ideas. The legislature wants Ontario's colleges to get their faculties up to at least 80 percent Canadian, there is a drive to get more Canadian books taught in the schools. The bulletin boards announce poetry readings, lectures, symposia, debates, seminars. I had hardly arrived on the University of Toronto campus and been handed my free New Testament before someone topped it with a book entitled *Read Canadian.* The bookstores and libraries feature whole sections of Canadian books, the book review pages are alert to cover new ones. There seems to be an extensive series of grants, prizes, and awards to encourage Canadian writers and Canadian publishers, and one whole paperback library lines up all the Canadians together like garden gods. It is exhilarating to be in a place where literature is taken seriously by so many kinds of people.

Second, one is struck by the identity crisis Canada seems to be going through – seems to have been going through ever since World War II when the bond with Britain finally wore thin. I have talked to those who are determined to discover a Canadian identity (and assume it or invent it if necessary) and those who doubt there is any such thing. National characteristics, always elusive, have been rounded up like cattle on the range, and are kept milling in a cloud of dust, but there are skeptics who insist that they're rustled stock, that none of them wears a clear Canadian brand; they all have burns or earmarks that declare them British, or American, or ethnic.

Nobody seems to think any more that a Canadian is a colonial Englishman, and almost everybody seems determined that he shall not be a provincial American, but any positive attempt to define him runs head-on

● WALLACE STEGNER is Professor of English, Stanford University. Reprinted from *University of Toronto Quarterly*, Vol. 43, 1974, by permission of the author and University of Toronto Press.

into the abiding division between the French and British Canadas, so that if there is an identity it is a divided one. I don't suppose it helps the state of mind of English-speaking Canada to be aware that of the two Canadas the French has the clearer character. It has a language, called debased but still its own, that marks it off from both English Canada and France; and it has a culture that has been developing in relative isolation ever since Louis XV abandoned Quebec to its fate. Culturally, that isolation may have been good for it, and it could be on the brink of brilliant productiveness.

But defining English Canada is like wrapping four watermelons. The differences, even the antipathies, between Ontario and British Columbia, the Prairie Provinces, and the Maritimes may not be as acute as those between Ontario and Quebec, but they are real; and the mosiac theory, Canada's answer to the melting pot, actively promotes not unity but diversity of culture. As Northrop Frye has pointed out, those who assert a unitary identity risk being absurd, while those who assume that true identity is either regional or ethnic are on their way to balkanization and — he seems to fear — separatism.

There are those who believe that the physical environment and climate exert an irresistible torque upon people over a period of time, shaping both their way of life and their character. I tend to think so myself. Yet the other night I had a stirring argument with an economist from Saskatchewan, where I would have thought the environmental determinism most demonstrable, who insisted that culture creates geography, not the other way around. Furthermore, there is a large contingent of Canadians, resigned or bitter, who feel that American mass culture has already so infiltrated Canadian society that it can never be Canadian, but always some variant of the global village culture described by Marshall McLuhan. It is said that Canada comes too late in communications history to develop an indigenous culture.

When discussing the identity drive one should not forget those who espouse the rather desperate faith that the Canadian character has been shaped throughout its history by contact with the wilderness. The first book I read when I began to read Canadian was an anthology of Canadian writing called *Marked by the Wild*. It contained many things that I was glad to read, but I don't think it proved its thesis, which seemed to be that the wilderness has set its seal upon Canadian hearts, and that if one looks and listens and reads carefully one will discover that all Canadians are mystically one, and that the sound of canoe paddles echoes through books as various as those by Morley Callaghan, W. O. Mitchell, and Marie-Claire Blais. I exaggerate, of course; but even if the thesis could be demonstrated, it is hard to see how the Canadian identity could be differentiated by that test from the American; for Americans too, especially in the West, still try to define their character as an inheritance from the frontier.

I seem to find in Toronto a good deal not only of literary patriotism but also of what a nationalist would probably call cultural defeatism — the belief that, never having had a revolution, nor even a frontier in the stripped-down American sense, English Canada (or at least Ontario) has been a "garrison society", dedicated to remaining British in a far land, and conservative and anachronistic British at that; and that when it did begin to grow away from England it was already in the stranglehold of the United States. I hear that, perhaps as a consequence, English Canada has produced no heroes or archetypal figures. I hear that it has no language of its own, only residual British, now fading, and borrowed American; and that even its one claim to specialness, its pronunciation of the sound "out", may be heard in the Virginia tidewater. I hear that Canadian literature, despite its patriotic fervours, does not compete in the big leagues. I hear Mordecai Richler, one of its better novelists, joking sourly about being "world famous all over Canada". I read one of Canada's distinguished critics to the effect that "if evaluation were one's guiding principle . . . criticism of Canadian literature would become only a debunking project".

Along with the patriotism and the defeatism, the overpraise and underpraise of Canadian writing, the American visitor can't miss the widespread, sometimes savage hostility to the United States. It surprises him, not that he wants to defend Watergate, Vietnam, economic and cultural bullying, and exploitation of another nation's resources, or applaud the perpetuation in Canada of a branch-plant economy, or approve the domination of Canadian publishing by American firms which add insult to injury by overcharging Canadians for their imported books. And not because he likes American TV programs any better than his Canadian brother, or the advertising that in Bill Mitchell's words seems so often to present professional models confiding more than one cares to know about their armpits. He is only dismayed to find himself attacked for things he dislikes just as much as Canadians do. And he keeps running into ambivalent responses like the Letter to the Editor I saw some weeks ago, asking angrily why the detestable programs from below the border are always so much more professional and entertaining that Canadian attempts to compete with them. That attitude reminds me irresistibly of the distinguished Massachusetts Republican with whom I listened to radio returns of the 1940 presidential election. When it was clear that Roosevelt had won again, he poured himself a drink and said, "The son of a bitch may be our greatest President."

Gnashings of teeth, beatings of breasts, skepticism and aspiration, ambition and impotence, cultural dependence bitter toward what enslaves it or protesting too much its independence, a double preoccupation with who one is and what others think of one, an abiding desire to make it in the American world and a reiterated indifference to that world, these are the stigmata of the provincial state of mind. Toronto is an authentic capital, a world city, and Ontario is a province which within the Canadian sphere is more potent than any half dozen states within the Americas. But many of its artists and intellectuals obviously feel that it is an outlier of the English-speaking world, culturally a province, and it responds to its anxieties in the way provinces always have and always will. Provincialism and nationalist fervour are aspects or stages of the same thing.

At the end of its first war with Great Britain, and even more frenetically after its second, the United States was busy hunting down its native identity and asserting its cultural independence. It reacted furiously to English snobbery and condescension, it went around like the Chinese farmer pulling up its rice plants to see how they were growing. Pick up Robert Spiller's anthology *The American Literary Revolution, 1783-1837,* and read in Royall Tyler, Joel Barlow, Timothy Dwight, Noah Webster, and the group who created *The North American Review* sentiments very like those I heard in a conference on Problems in Canadian Literature. Read Crèvecoeur's *Letters from an American Farmer,* with its reverberating question, "Who is this new man, the American?" and with the change of a name you might think yourself in contemporary Canada. "Who is this other new man, the Canadian, and how has he evolved away from the earlier New Man?" Read Tocqueville's *Democracy in America,* and among much sympathetic observation of American institutions and customs you will find expressed the fear of democratic levelling, the fear of cultural and intellectual mediocrity, the fear of a "homogenized" society, to take a word from Canadian critics, that makes many Canadians uneasy about following in American tracks.

The North American Review believed that the future belonged to America; the brash young giant of the New World; I hear Canadians saying that the twenty-first century belongs to Canada. All through the nineteenth century the notion of American innocence as contrasted with European corruption is an abiding theme in American writing and a constant in Americans' notion of themselves (it crops up again in regional terms as late as *The Great Gatsby*). It crops up also in Canada. Canadians, it is said, have never been guilty of the violence, disorder, and corruption prevalent below the line; Canadians are simply more *moral* than Americans, a decenter and less rapacious breed.

The debates about language that agitated the United States in the late eighteenth century have been spared English Canada, and in a way that is Canada's loss. They don't have a language to rally around, they don't need a Noah Webster to legitimize their speech. They have been speaking North American for a long time, and I hear few Anglophile or purist Canadians rebuking their countrymen in the way James Fenimore Cooper rebuked his: for such barbarities as saying "cucumber" instead of the mellifluous and correct "cowcumber". But I hear all that going on in Quebec, in the disputes between the *joual* group who accept a humiliating and debased language and resolve to lift it into legitimacy and eloquence, and those who cling to "good"

5

French. On the strength of historical precedent I would put my money on *joual*, and bet that before too long some Chaucer, Rabelais, Mark Twain, or Bjornstjerne Bjornson will come along to show *joual* capable of every effect a writer can ask of it, a vernacular matured from below into a "national" language.

Writers of the early nationalistic period in the United States had cause to lament their cultural poverty. They tried to transplant elegances and they tried to transplant forms developed in the old country to American subject matters. Hence the epics about Niagara Falls and Daniel Boone, hence the blank verse, the Addisonian prose, the togas on the statues of statesmen. Nevertheless the cultural cupboard was pretty bare, and both Hawthorne and James, in successive generations, lamented it. Hawthorne tried what seems to me the better alternative when he made an effort to create a usable past; James took the route of expatriation. Not all of this is apposite to modern Canada. Canadians don't feel *culturally* inferior to the United States, as Americans felt culturally inferior to England. Some look down on the corrupt, vulgar, plastic culture to the south with as much disdain as British travelers in the 1840's did. It could be that their attitude is residual-British rather than Canadian, a survival of the garrison mind, and it could be that it is more prevalent in Ontario than in western Canada, and it could be it is guilty of looking on the United States as a monolith. No matter. It is somewhat different from the provincial inferiority feeling of the young American republic. And yet Hugh MacLennan spent half the space of his first novel, *Barometer Rising*, establishing Halifax as a legitimate setting for fiction, and I think I detect some of the same impulse, less obtrusively handled, in the novels of Robertson Davies about Ontario cities. And many Canadians have elected expatriation, either to Europe or the United States. Perhaps the brain drain shows signs of reversing itself, as the drain from the United States to England has reversed itself in the last quarter century, but it still seems at worst or best a two-way flow, about a standoff.

In 1837 Emerson delivered in Cambridge his celebrated address "The American Scholar", with its declaration of cultural independence: "We have listened too long to the courtly muses of Europe." That was fifty years after the first wasp-nest stirrings of American cultural nationalism, and Emerson's declaration did not mean that the battle was won, by any means. Irving had transplanted some German legends to the Hudson Valley, and Cooper (the American Scott, we called him proudly) had for lack of any American society turned to the sea and the forest, and within the American wilderness had created one of the great archetypal figures of American identity, Leatherstocking. And Hawthorne and Poe had begun to fashion the form of the modern short story. But by comparison with England, American literature in 1837 could show only what one critic called "a few dim, blinking lights". Cultural identity, the cultural independence of a new country, is not won in a day, or in fifty years. It could be provocatively asserted, and plausibly defended, that Canada in 1974 is somewhere around the stage of self-definition and literary accomplishment reached by the USA about 1837.

But again, if Canadians are anxious, there is instruction in history. America in 1837 had little to be confidently proud of in its literature, and only foreshadowings of a national identity. Wait twenty-five years and there are *The Scarlet Letter*, *Leaves of Grass*, *Moby Dick*, the books of the great romantic historians. Wait another twenty-five and there are *Huckleberry Finn*, much of James and Howells, a whole broad pyramid base of secondary writers. There is applause from abroad, the sweetest music that cultural nationalism knows. And there is a whole gallery of figures who in one way or other might represent Crèvecoeur's "new man, the American". They run all the way from Uncle Sam and the Yankee types who begin with Royall Tyler's play *The Contrast*, through Leatherstocking and all the Boone-Crockett-Carson-Bridger avatars of Leatherstocking, and on to Lincoln, Mark Twain, Howells, and still on to the literary creation Henry James called Christopher Newman, in *The American*.

I would like to emphasize that it did take a hundred years and the efforts of four generations of writers, some of them writers of the highest genius, to answer Crèvecoeur's question. And by the time the ink was dry on *The American*, the trial synthesis represented by Newman — a Wasp with democratic, frontier, and business overlays and with a plentiful supply of American naïveté, American innocence, and American integrity

— was already obsolete. There would have to be new and larger syntheses, taking account of vast non-Wasp immigrations, the closing of the frontier, the rapid industrialization and urbanization of the country, the emergence of new regions, and a considerable loss of innocence. And following that still another synthesis allowing for the recalcitrant elements that the melting pot never melted down, the stubborn and persistent subcultures of blacks, chicanos, Indians, orientals, to some extent the Boston Irish, to some extent the Jews.

Those syntheses have not yet come about, and I doubt that either will, short of the millennium. The nation is made up of too many kinds. The archetypal figures of the American have been limited by region, or ethnic background, or by chronology, or by all three. And just about when people gave up trying to synthetize an archetypal American, they gave up calling for the Great American Novel. It began to be clear that our Yankees, Leatherstockings, Pikes, Rednecks, Hoosiers, our Webfeet and Mormons and hispanos and cowboys and steamboatmen and railroaders, derived from a region or an occupation or a stage of the frontier, not from the nation as a whole. Folklore and local colour exploited them all, both romantically and realistically, and it emphasized not the Americanness that made them vaguely alike but the picturesque differences that divided them.

Local colour was the literary fashion from the early 1860's to the turn of the century. Its romantic and picturesque elements then went somewhat out of fashion, but American literature in the twentieth century has been dominated by a series of sectionalisms either geographical or ethnic — the midwesterners in the twenties, the southerners on their heels, the New Jewish school on the heels of the southerners, the black on the heels of the New Yorkers. Regional or ethnic, they have had things in common. Check the writings of the Fugitive Group that wrote the script for the South, especially such a book as the symposium *I'll Take My Stand*, or Donald Davidson's unreconstructed blast at New York, *The Fight Against Leviathan*, and you see writers making a hate object out of New York as the early American nationalists made one out of England and Canadians tend to make one out of the United States, monolithically conceived,

and the Quebeckers make one out of English Canada. The Midwest was similarly suspicious of New York, and still is; so are the Far West and the other regions. As for Black writers, they fill that space marked "Hate Me" with the vague malevolent face of Whitey, and tend to rally around a language, Black English, in a society of soul brothers.

Every successive group, pulling itself out of the regional mud or out of the anonymity of the ghettos, has defined itself in opposition to something. Every such group that I know anything about has been guilty of backscratching, praising its own crowd, forming a counter-coterie against the reigning one. There is a lot of that today among writers in the West, who feel, not without provocation, that the reigning New York clique ignores both them and their region. Because I am a westerner, and was once a sort of midwesterner, I have shared those resentments. The other night, when a somewhat heated Canadian nationalist told me he'd rather live in Saudi Arabia, which he described as a benevolent dictatorship, than in the United States, I had to disagree with him; but if he'd said New York I might have agreed. I take some pleasure in the statistics showing that only 13 per cent of Americans living in the largest American cities *want* to live there. Zoos, full of hostile, over-sexed, and dangerous zoo animals; the dismal sewers into which the American Dream has finally run. But there are other Americas than the great cities, newer and better ones than New York, which is hardly an American city at all. The twenty-first century may belong to Canada, but it will have to share it with the western United States, which are about as young as Canada, just as conscious of superior health and morals, just as provincial, and just as ambitious to avoid those sewers.

I feel completely at home in the Canadian literary climate, even when its winds get shrill. I could be enlisted as a b.b. calibre cultural Lafayette. Nothing in New York or Detroit, or for that matter in Mississippi, tells me anything about my identity or my roots. As I said in *Wolf Willow*, I came back to Saskatchewan looking for those roots; and what Saskatchewan did not put into me, Montana and Utah and California did. I am content to have my world citizenship rooted in the West, though I don't want to be just a regionalist — and I will now risk outraging some of you by saying that it doesn't much matter whether the West means Canadian or American. I don't see much difference.

Identity, the truest sense of self and tribe, the deepest loyalty to place and way of life, is inescapably local, and it is my faith that all the most serious art and literature come out of that seedbed even though the writer's experience goes far beyond it. Much of the felt life and the observed character and place that give a novel body and authenticity, much of the unconsciously absorbed store of images and ideas, comes ultimately from the shared experience of a community or region. There is a kind of provincialism, minus the aggressiveness and self-consciousness, that encompasses the most profound things that a writer has to say. Like civilizations, communities and regions have a youth, a maturity, an old age, perhaps a senile old age. The ones I am talking about are the nascent ones of the New World. It is the sense that they are so far unsaid, that something personal to one's self and one's tribe remains uncommunicated, that drives some of us to write. And I submit that that has little to do with nationality, and certainly with *nationalism,* when we are talking about the United States or Canada.

I know about the Innis theory of east-west force in Canada, and about Professor Creighton's metropolitan circles of influence, and I grant validity to both. I know how indignant Canadians can get when one brings up, as they say, that old "north-south boundary lines business". Nevertheless I am sure that the logger in British Columbia *does* have more in common with his counterpart in Washington than he does with somebody from Ontario. I am even more sure that the Saskatchewan wheat farmer shares more, in experience, climate, geography, occupational habit, folklore, language, vision, with the other farmers down the long sweep of short-grass country that runs from the North Saskatchewan to the Staked Plains, than he does with an Ontario farmer. I have heard, and I tend to believe it, that the Maritimes flow more naturally into New England, the Boston States, than into other parts of Canada, no matter how strongly the fur trade and the railroads have directed the consolidation of Canada east and west. I am not now speaking of either economics or politics. I will applaud when Canada frees itself from

too much American domination of its resource development, and I will be very depressed if Canada or any segment of it breaks off to try secession or annexation with the United States. I am talking about culture, way of life, shared experience and recognitions. No cordon sanitaire, no nationalist fervour, is going to prevent that kind of rapport between Canadians and Americans where great geographical and climatic regions happen to slop over the forty-ninth parallel. Whether patriots of either side like it or not, we are one heterogeneous people sharing a continent. Historically, the sections have been far more important than social and economic class in the United States. I think in the long haul they may be more important than nationality too, at least in places, and sometimes more important even than race and colour.

And that, I know, calls for very hasty explanation, and the explanation involves the observed or probable results of the melting pot on the one hand and the cultural mosiac on the other.

Officially and unofficially, the United States has tried to Americanize all its various peoples for two hundred years. In theory an egalitarian democracy, it aroused in Tocqueville and others those fears of stereotype and mediocrity of which I spoke earlier. But is it a truly homogenized society, as George Grant and some others declare? On the contrary. Except briefly during two wars, it has never been a nation. It is the wildest mixture of races, colours, political and religious and economic faiths, regions and sub-regions, Establishment and counter-culture, highbrow, middlebrow, and lowbrow, conservative and revolutionary, in the entire round world. The melting pot has not melted Americans down, the mass media have not homogenized more than their surface.

Working against the forces of homogenization has been a certain American genius for refractoriness, rebellion, lawlessness, and change, which uncorrected can lead and has led to near anarchy. The American way, which invented mass production with interchangeable parts, also invented the principle of annual retooling. Traditions, including the tradition of Americanism, get trampled. Americans are as close to an ethnic slumgullion, and their society as close to an experimental one, as the world can show. There is,

I suppose, a sort of standard American language, but it is subverted by pockets of stubbornly preserved dialect as well as by dialects that are in process of development out of the standard tongue, and this in spite of radio and television. Black English, to take one example, is close to unintelligible except to those who speak it.

Similarly, the regions, which ought to be flattened out into uniformity according to the theory, remain differentiated, and even grow farther apart, and they even put their mark on people whom race and religion would seem to dominate. Ralph Ellison, for one example, is a very different type, with different speech and different reactions, from the usual black writer. He grew up neither in the South nor in a northern ghetto, but in the West.

Despite the flattening and levelling tendencies of the mass culture, and despite the temptations to ethnic or cultural militancy and separatism, and despite modern communication, the regions do within broad limits produce cultures and types that develop and maintain their own integrity. Between that sort of individuation and the more hectic sort promoted by ethnic groups, America has become something very different from the vulgar monolith its critics fear or despise. In the deeper levels of culture, the great monolithic neon-lighted barbarity is a bugaboo, a myth. The melting pot has been a bust.

But I wonder if the cultural mosaic won't be a partial bust too. Quebec, again, I exempt; it is already culturally distinct and beyond much doubt will remain so indefinitely. But I wonder if the so-called ethnic cultures will persist for more than a generation or two in the free society of Canada. The way to make immigrant or native cultures persist is to persecute and oppress them, as the United States has oppressed the blacks, chicanos, Indians, orientals, and others who now revive secessionism in ethnic and cultural terms a hundred years after the Civil War disposed of secessionism as a regional and political philosophy. Leave them free, in a predominantly English-speaking country, and they will pretty surely become Canadians, English-speaking, with perhaps residual cultural attachments such as religion and some trimmings like Fastnachtkuchen or fattigmand at Christmas, plus some contributions that have gone toward the alteration, or creation, of the Canadian image.

It seems to me a significant irony that the melting pot, which was meant to produce uniformity and a national character, produced instead diversity and half a hundred regional or ethnic characters with only broad, vague resemblances. It would be a comparable irony if the mosaic theory, meant to retain diversity, should produce more uniformity than most people expect. I think it will. I also think, and have said, that the uniformities will exist ultimately, if they don't already, within a regional diversity. And that is the pattern which the United States seems to me to be reaching by the opposite route.

When William Kurelek painted the pictures that illustrate his *Prairie Boy's Winter* I doubt that he was being either self-consciously Canadian or self-consciously Ukrainian. He was painting out of his memory of indelible childhood experience, and his memory is beautifully and evocatively regional. Those pictures speak to me like trumpets, they are full of instant recognitions, they remind me at every turn of the page of things I have not forgotten but only mislaid. They do in paint what I tried only half successfully to do in words in *Wolf Willow*. I guarantee they will speak with great force to anyone, of any ethnic background, who grew up in short-grass country, on the plains, whether those plains are north of the 49th parallel or south of it. I'd like to try them on George McGovern and see if they don't say to him, like church bells, SOUTH DAKOTA! But I know they are not going to speak with quite that force to someone from British Columbia, where the physical bases of life are different.

A hundred years and more after the search for an American identity began, William Dean Howells advised his country's writers to be as American as they unconsciously could. That advice comes close to Robinson Jeffers's dream of being passionately at peace, but it will probably have to do. A Canadian literature as Canadian as it can unconsciously be will depend on what such things always depend on: on the perceptiveness, passion, and integrity of individual writers, and the eloquence with which they make their own lives and the lives of their tribe meaningful to the world. There is nothing wrong with Canadian literature that one or two world-class writers wouldn't cure. They will happen. The nationalistic debates of the Connecticut Wits led by a crooked but sure path to *Huckleberry Finn*.

9

Canada and Secession: Some Consequences of Separatism

ANDREW F. BURGHARDT

A land as attenuated and as strongly regionalized as is Canada must expect to feel severe sectional strains. For Canadians beset by separatism in Quebec, as well as complaints from many other areas, it is important to recognize this fact, and to realize too that such concerns are not new. The concept of "two nations" warring within one bosom goes back at least to Lord Durham's report of 1839, which concerned itself only with parts of what are now Quebec and Ontario. The particular viewpoints of the west coast, the interior plains, the great north, the Atlantic continental coasts, and offshore Newfoundland have since become articulated to add amplifications and overtones to every report emanating from Quebec.

The recent separatist activity is remarkable not so much for its newness as for its vigour. This increase in intensity may be seen, partly, as one of the prices Canada has had to pay for the attainment of independence. Like many other post-colonial states, Canada has had to face strong internal strains once the external political authority was removed.

Until recently the *habitants* accepted "the conquest" as the unloved but obvious base of all political life. Quebec had by force of arms become a part of British North America, later the Dominion of Canada, which was in turn a part of the British Empire. There was no way of avoiding this fact, hence the strivings of the more nationalistic Francophones were limited to the maintenance of the ancestral culture and to attempts to gain a greater measure of political power within the system. As was true in Poland, Ireland, Greece, Serbia, and Croatia, it was the Church above all which nurtured the culture and assured the continuance of the people as a distinct group, a "nation".

The achievement of independence by Canada did not occur dramatically, with displays of ceremony, as it did in the new African states. Even after the Statute of Westminster (1931), Canada remained a part of the British world and entered World War II alongside Britain. Not until the total collapse of the imperial structure after 1945, the increasing manifestation of British weakness, and the development of a distinctively Canadian presence in world affairs, did the true independence of Canada become obvious to French Canadians. Thus it was 1960 before Quebec intellectuals recognized their province as being a large and powerful part of a "middle power", rather than the small French-speaking segment of an overwhelming British structure.

Every vital nationalism requires a territorial base. Without a specific extent of land, nationalistic activity remains amorphous and unanchored. Few people are able to sustain or communicate the fervour of such a cause without calling to mind the images of a special countryside, a sacred soil sanctified by generations of forefathers, and of enshrined cities. The matrix of the national feeling is formed from the events of the shared past which have as much a locational and areal as a historical character. Thus every strong nationalism has two interwoven dimensions, culture and territory, and the latter serves just as much as the former as a bearer of the common heritage.

Hence, French-Canadian activity has been fitted very closely to the bounds of Quebec. In fact the most extreme positions are usually summarized under the term "separatism", a territorial concept, and the *habitants* often seem willing to ignore the existence of substantial numbers of Francophones in neighbouring Ontario, New Brunswick, and elsewhere. Political activity, even more than nationalism, requires sharp territorial limits.

The legal aspects of separatism are finally beginning to be aired, albeit slowly and indistinctly. Does any province have the right to

● ANDREW BURGHARDT is in the Department of Geography, McMaster University. This article is based upon a previous article "Quebec Separatism and the Future of Canada", in R. Louis Gentilcore (ed.) *Geographical Approaches to Canadian Problems* (Scarborough: Prentice Hall of Canada, Ltd., 1971). The paper has been revised, updated, and retitled for this volume.

alter unilaterally the structure of Confederation? Obviously, Canada's constitution, the BNA act, makes no allowances for an opting-out by one of the provinces, nor has, to my knowledge, the constitution of any other federation. Yet one does hear the catchy phrase "the right of self-determination" proclaimed, particularly by university students. Although obviously attractive, this slogan seems to beg the question of whose "self-determination", and on what level. Most federations have felt the self-determination (integrity, viability) of the whole to be more important than the self-determination of only a portion. The essence of a federation is that the various parts join together for the sake of a greater common good, and that once this union has been achieved, all belongs to all. The port of Montreal belongs to the manufacturers of Ontario and the wheat growers of Saskatchewan as much as to the residents of the city or of its province. It is no surprise then that in most states of the world the unilateral secession of one part of a federation has been answered with a resounding "No".

The current separatist agitation has, however, been argued on what appear to be reasonable, unemotional grounds. Political separation but economic cooperation — two independent states linked in a Common Market — this is the essence of the proposal put forward by the *Parti Québecois*.

The Austro-Hungarian Parallel

To a student of Central European history this sounds depressingly familiar. The proposed solution has been tried before, although within a somewhat different context. The history of Hungary within the Austrian Empire provides an example strikingly similar to the situation of Quebec within Canada. It may prove profitable to note what happened when such a proposal was actually implemented.

The Kingdom of Hungary was a strong, flourishing state before the rise of the Habsburgs. When absorbed by the Austrian Empire after 1683, Hungary had its own language and culture distinct from those of its conqueror. Like Quebec after its conquest by Britain, Hungary strove incessantly to maintain its identity. In both cases religion

was a potent force. In Quebec, Catholicism was the spiritual bulwark against English-Scottish domination; in Hungary, Calvinism served as the spiritual force opposing the Habsburg forms of the Catholic Counter-Reformation. In both Hungary and Quebec, abortive attempts were made to impose the official external language: Lord Durham's aim was to submerge the French and assimilate them; the Emperor Joseph II (1780-90) attempted to impose German as the one official language throughout Hungary. In both areas commercial life was controlled by the external group — in Quebec by British merchants, in Hungary by German merchants. (In Hungary, most guilds required that apprentices knew German.)

Until the Napoleonic wars Austria was a part of the greater German-speaking Holy Roman Empire, of which the Habsburgs were the emperors, much as Canada was a part of a greater English-speaking British Empire. Following the dissolution of the Holy Roman Empire in 1806, the Hungarians found themselves within only the Habsburg lands, free of the external imperial ties, much as the Québecois now see Canada free of such ties. Among the Hungarians a time of great nationalistic activity followed, culminating in the fight for independence in 1848-49. The Habsburgs attempted to treat Hungary as if it were a province among provinces, but the Hungarians claimed that Hungary deserved some special status. It seemed as out of scale then to compare Hungary with Styria as it does now to compare Quebec with Prince Edward Island. Following the defeat of Austria by Prussia in 1866 — fortunately, there is no Canadian counterpart to this — the Austrian monarchs capitulated to the Hungarian "separatists". The result was the Dual Monarchy, or the Austro-Hungarian Empire which was formed, interestingly, in the same year as Canadian Confederation, 1867.

Within the new political structure Hungary and Austria (which consisted of the remaining provinces) functioned as two distinct states. They were united only by the monarchy and hence in foreign affairs, and by the relative absence of internal trade barriers. But each had its parliament and civil service, and each handled its economic planning independently of the other.[1] Now, a century later, it is of great interest to observe the outcome of this arrangement.

First it should be noted that the most ardent Hungarian nationalists were not satisfied, and pressed for complete independence. They regarded the emperor-king Franz Joseph as a foreign monarch, and not as a rightful king of Hungary.

Second, the new political structure did not appreciably lessen the degree to which German financiers, industrialists, and merchants controlled the economic life of Hungary. Despite the rise of Budapest, Vienna continued to dominate the economy of the entire Monarchy, and Hungarian complaints against the Austrian merchants were as frequent and loud after, as before, 1867.

Third, on the other hand, Hungary enjoyed a far greater degree of decision-making power than previously, since control of the overall development of Hungary lay largely in the hands of the Hungarian government. This resulted in a tremendous concentration of growth in the capital city, at the relative expense of the district towns. It resulted also in a railroad network focussed entirely on Budapest, a net whose local lines often ended just short of the Austrian boundary.

Fourth, the Hungarian government embarked on a massive effort to make Hungary a unilingual state, despite the fact that over half of the population was linguistically non-Magyar. By 1907 all schooling had to be conducted in Magyar. These actions encouraged a blossoming of Magyar cultural life, but also helped cause that deep dissatisfaction among minority groups which led to the partition of Hungary in 1919.

Fifth, the other provinces of Austria pressed for a status similar to Hungary's. This was particularly true of Bohemia, which, like Hungary, possessed both a cultural and a territorial nationalism. Bohemia had been a medieval kingdom with its own legendary kings and religious heroes, and since 1848 the Czechs had enjoyed a cultural revival. Further, Bohemia's economy was highly developed, so its "separatism" had to be taken seriously (perhaps Alberta and British Columbia would be the closest Canadian counterparts). The desire for equal status felt by all the other provinces only accentuated an already intense regionalism. When the Monarchy collapsed in 1918, even the truly Austrian provinces of Salzburg, Tirol, and Vorarlberg voted in local referenda to join other states.[2]

Finally, the remaining ties between Austria and Hungary deteriorated under the frictions of financing. Because there were still a number of expenses to be shared by the two partner states, a formula for their financing had to be worked out. Under the terms of the *Ausgleich*, the two partners met every ten years to renegotiate the share of the overall costs to be borne by each state. These negotiating sessions became more prolonged and arduous with each decade.

The fact of the situation was that, although Austria had some 70 percent of the empire's wealth and population, Hungary held all the locational trump cards. Austria was but a crescent of territory, fragmented into a number of distinct regions, around the central bulk of Hungary. There were portions of "Austria" (eastern Galicia and Bukovina) which could be reached effectively only through Hungary. If Hungary had elected to sever the last remaining ties, Austria would have fallen apart. Hungary in turn might have suffered some decline in its "standard of living", but then, as now, that argument carried little weight with the nationalists.

By the early twentieth century it was clear that the *Ausgleich* of 1867 had not solved the problems of the Habsburg Monarchy, but rather had intensified old problems and added a few new ones. Austria itself often seemed to have reached the point where it could scarcely function politically. It was this situation that was captured in that cynical Viennese comment, "the situation is hopeless, but not serious".

What kept the Dual Monarchy together? Perhaps inertia, perhaps economic considerations. However, the strongest bond appears to have been widespread loyalty to that ultimate father-figure, the emperor Franz Joseph, who ruled for sixty-eight years, from 1848 to 1916. The death of the emperor almost coincided with the loss of the disastrous war; the component pieces of the Monarchy fell apart, to be absorbed in time by the nearby powers.

Judged by its more obvious results, this attempt to associate states within a loose union must be deemed to have been unsuccessful. Independent states, especially those motivated by a strong nationalism, must be expected to place their own interests above some common good to be shared with some other state or states. Perhaps old established states, which have grown accustomed to independence, may be able to attune their priorities to those of their neighbours, but new states, striving to solidify their independence, must be expected to be very sensitive on matters relating to sovereignty.

The Consequences for Canada

Turning now to the Canadian setting, we should note what the geographical consequences would be if Quebec were, in fact, to secede from Canada. The basic fact is that the Canadian corridor might well be destroyed.

Neither the history nor the present economic life of this country can be understood without the recognition that Canada has developed along one east-west corridor of movement, communications, and population. From the first days of French settlement, the colony along the St. Lawrence extended its trade into the interior of the continent. Within a century, the waterway corridor to the west had been explored and exploited: the St. Lawrence, the Ottawa and French rivers, the upper Lakes and the Sault, the Grand Portage or Kaministiquia, the Rainy River to Lake Winnipeg, and the giant rivers of the prairie West. Even the Atlantic Provinces, although not entirely astride the corridor, were often seen as vital flanking positions along the estuarian entry to the corridor.

The route of the *coureurs de bois* has become the St. Lawrence-Great Lakes Seaway, and despite all efforts to improve the Hudson's Bay route, this Seaway remains the principal water entry into the Canadian interior. The continued unity of this corridor is also attested to by the fact that there is only one Trans-Canada Highway, and that both major railways and both major airlines focus on Montreal, Toronto, Winnipeg, and Vancouver.

All but the three Maritime provinces extend to the northern seas or to the 60th parallel; thus the belt of transport and population is cut across completely by every province from Quebec westward. Even Edmonton, the northernmost major city, is only 54° north. All trade, all movements of people, all communications within Canada and with Europe and Asia flow east-west across these six provinces. The only other movement of any consequence is that north and south between the individual segments of this corridor and the adjacent United States. The loss of any of these six provinces would destroy that elongated web of personal, business, and administrative ties, and without the existence of the entire corridor it is hard to conceive of a viable Canada. It would indeed be ironic if the St. Lawrence portal, which instigated the development of the corridor, were to destroy it through secession.

Should Quebec secede, Canada would be divided not into two but rather into three parts. An independent Quebec would separate the Atlantic Provinces from Ontario and the West. Canada would resemble the initial Pakistan (1947-1971), whose eastern and western portions were separated by India. If the two "English" parts were to remain united politically, some transportational connection would have to be established. Obviously this would have to pass either through Quebec, as now, or through the United States. In either case, there would be a depressing dependence on the transport network of a foreign state.

The separation would also have a psychological impact. The Maritimers would find themselves separated by several hundred kilometres of foreign territory from their centres of national administration. One may safely assume that they would feel neglected, unrepresented, and remote — and in truth, they probably would be. Certainly there would arise the temptations to seek other solutions: perhaps independence for the four (or three) provinces together, or some kind of union with Britain or the United States. The history of East Pakistan (now Bangla Desh) does not promise well for the continued unity of a fragmented state.

The western portion of "English" Canada would be severed from its established ties to the Atlantic world. These would have to be negotiated either through Quebec or through the United States. The temptation to turn away from Montreal towards New York would be strong indeed, given the greater facilities and financial resources of the American port, and the possible resentment felt against the separating province.

Nor would it be certain that this part of English Canada would remain united. In a country as highly regionalized as Canada, the independence of one portion could well lead to the separation of others, on the theory that the federation is finished. The agitation in Quebec has already been countered by suggestions for an independent combination of British Columbia and the Prairie Provinces. Further, were the east-west continuum of transportation and communication broken, the centrifugal or southward tendencies felt by the various regions could well be strengthened.

The new boundaries would cause difficulties. Even though Canada is at present a strongly regionalized state, the boundaries of Quebec pass not through empty areas but rather through zones of population and development. This is most marked in the federal capital region. Of all the metropolitan areas of Canada, only Ottawa straddles a provincial boundary. One may easily imagine the local confusion that would result if an international boundary were demarcated between Ottawa and Hull and the Gatineau! At the very least Hull could expect to lose the many federal offices which have recently been moved across the river.

Further north the boundary runs through the mining-Clay Belt area. The similarity in interests between Noranda-Rouyn and Kirkland Lake and Timmins is well known; locally there have even been suggestions made for the creation of a new province comprised of present northeastern Ontario and northwestern Quebec.

Even the barren uplands of the Gaspé and northern New Brunswick do not form a clear divide. Coastal Gaspé is, in its economy and language, much like the neighboring eastern shore of New Brunswick, even though there is a strong sense of difference between the Québecois and the Acadians.

The most curious example of the divisive tendencies of the boundary would occur along the border with innermost Labrador. It is surely a remarkable coincidence that in all those hundreds of thousands of barren square kilometres the only marked development has occurred directly on the boundary. In short, the transformation of a provincial into an international boundary would create innumerable possibilities for border friction.

No city would be more severely affected by the imposition of international boundaries than would Montreal. As the headquarters of the fur trade, the base for exploration of the West, the disembarkation point for immigrants, and the exporting port for the lumber and wheat of the interior, this great city has long served as the Atlantic portal for most of Canada. It is the headquarters for the major national transportational and communicational networks. It commands all ties between the Maritimes and Ontario, that is, between the two segments of English Canada. Perhaps more than any other large Canadian city, Montreal is dependent upon the entire national market. A Toronto left only with Ontario could possibly continue to prosper, but a Montreal without all of Canada is hard to envisage. A clear separation of Montreal from non-Quebec Canada could only be a disaster for the city. Like the Vienna of 1919, it would find a new international boundary within 80 km. The port might well decline to marginal enterprise, since the overseas trade of Ontario would be channeled through either the nearby American ports, or Saint John and Halifax.

Metropolitan Montreal contains half the population of Quebec. A severely depressed Montreal would place intense strains on any Quebec government. Very few administrations have been able to face crowds of the unemployed without reverting to some kind of authoritarian rule. The case of Vienna, after 1919, is a reminder of what can happen to a great city once it is forced to endure these strains.[3]

Finally, it is doubtful that an independent Quebec could supply its cultured, sophisticated people with the manner of living they have come to expect. It is misleading to conclude the opposite from the very special cases of Switzerland and Sweden. Those states have developed their economies and their unique skills over long periods of time; they have not had to face sudden major economic retrenchments. Any province or region of Canada which opted to exist alone would be faced by just such a major retrenchment, as the present national market is replaced suddenly by a provincial market only one-quarter its size. Thus, it is not Sweden, but rather post-1918 Austria which should be viewed as having undergone an approximation of the experience of a post-secession province. The balance of payments has always been a problem for Canada; for a Quebec alone this problem would threaten to be of gigantic proportions.

Clearly the separatist planners do not envisage the black economic picture drawn above. They express the hope of having an independent Quebec and yet still maintaining the existing economic linkages. They look for the cultural advantages of independence and the economic advantages of confederation co-existing somehow in some political solution.

However, it is mischievous to speak confidently of some future common market arrangement. We would not have here a counterpart to the Western European states joining hands in the EEC, for there we witnessed the coming together of long-established states. Here, on the contrary, we would have the breaking-up of an established state into its component parts – just the opposite process. One must look instead at examples of fragmentation, such as the Balkanization of the Habsburg Empire, the splitting up of former Spanish America, or the division of British India into India and Pakistan. In none of these cases did the successor states reverse the process of fragmentation to re-establish the pre-existing economic unities. Quite the contrary! South America, the Indian subcontinent, and Central Europe have all suffered numerous wars.

Perhaps more than any other factor, the separatists seem to ignore the importance of emotions. They almost seem to believe that *les Anglais* would accept the break-up of Canada with complete equanimity. The truth is that fragmentation is always accompanied by bitterness, and bitterness leads states to seek solutions which will not help the mistrusted neighbour.

It is not even certain that the St. Lawrence Seaway could lead to continued close cooperation between Quebec and Ontario. The Seaway is of much greater concern to Ontario than to Quebec. The port of Montreal lies below the entrance to the Seaway and thus feels no dependence on it. In contast, the ports of Ontario can only be reached from the Atlantic along this waterway. Yet Quebec controls the exit of the system at St. Lambert. The possibilites for disagreement about maintenance, enlargement, tolls, etc, are limitless. We must expect that each state would approach the questions concerning the Seaway from the vantage point of its own narrowly-defined self-interest, and that these viewpoints would not always coincide.

To complete this quick look at the major consequences of secession, one must take note not only of the internal Canadian situation, but also of the possible reactions of the powerful neighbour to the south. The United States would inevitably find itself very closely involved in any attempted major reorganization of the structure of Canada. The two countries are linked by a great number of treaties, covering everything from defense to fisheries. It is obvious too that Canada is of unique importance to the strategic and economic security of the U.S.A., for these reasons: Canada's location between the U.S.A. and the Soviet Union, its proximity to the industrial heart of America, its flanking position along the North Atlantic and North Pacific sea and air routes, its control of sea access to the Great Lakes, its bridging function between Alaska and the conterminus U.S., and between Detroit and the north-eastern states, and because of the inter-meshing of the resource and manufacturing activities of the two countries.

Therefore, even if the United States were unable to prevent the act of secession, she would probably feel forced to see to it that any possible effects of such a move detrimental to U.S. interests would be prevented. In either case it is difficult to conceive of a truly independent Quebec (or Canada) emerging from the necessary deliberations.

The justice of many of the Québecois demands is obvious. Certainly no Canadian should be penalized in any way – economically, legally, or culturally – because his fluency is in French rather than in English. The point of this article is not that the Québecois aspirations are wrong, but rather that secession is not the way to achieve them. Further, a pulling-out by Quebec would amount to an abandonment of the Francophone communities in the rest of Canada.

English and French are among the closest approximations to universal languages we have in the world today. It would be a tragedy if the one country which has the opportunity to develop those two together were to be shattered because of the felt inability of the Québecois to find their cultural fulfillment within Canada.

Because of historic and physical factors, Canada has grown as an east-west corridor of settlement. With the exception of the Atlantic Provinces, no province can opt out of

Confederation without destroying that corridor, and without the intactness of that corridor there is no viable land called Canada.

NOTES

[1] The terms of the *Ausgleich* of 1867 may be found in Charles and Barbara Jelavich, *The Habsburg Monarchy. Towards a Multinational Empire or National States* (New York: Holt, Rinehart and Winston, 1959).

[2] Salzburg and Tirol voted to join Germany, Vorarlberg to join Switzerland. The Swiss were not interested, and the Allies would not allow any enlargement of Germany.

[3] The loss of the empire it had served brought on a severe depression in Vienna. Financial problems were severe and unemployment high even during the relatively prosperous 1920's. The city population divided between militant rightist and leftist factions; a three day riot in 1927 cost eighty-nine lives. The arming of the two principal political parties culminated in 1934 in a brief civil war, which was fought with artillery in the streets of Vienna. This was followed by one-party rule and, four years later, by the Nazi occupation. At the present time, despite the current prosperity of Austria, the population of Vienna is still 500 000 less than it was in 1918.

Introduction

Canada is an urban country and the distribution of urban places and population is uneven. According to the preliminary figures from the 1976 Census, just over three-quarters of the population is classified urban and of this total 55 percent is resident in the twenty-three Census Metropolitan Areas. Rather than focus on specific urban phenomena and problems which are dealt with in other anthologies, this section examines some of the broader processes, patterns, and characteristics of the Canadian urban system.

In the introductory paper, Richard Preston traces the spatial growth and development of the urban system in post-Confederation Canada. In the process he reveals some of the relationships between national economic policy and the development of the urban system, and offers some principles to describe and explain those relationships. Cities do not exist in a vacuum and Canadians are among the most mobile people in the world. However, it was not until census data became available after 1971 that the patterns of spatial mobility between cities could be identified. In his paper, Jim Simmons examines the migration patterns and interrelationships in the Canadian urban system from 1966 to 1971. The national significance of the Quebec City-Windsor axis is the subject of Maurice Yeates' contribution. He portrays cartographically the spatial growth of the urban and semi-urban population in the axis for 1961 and 1971, and projects these patterns to 2001. Should this growth forecast be realized, it is projected that 20 percent of the Canadian population would reside in the Toronto area. The desirability of permitting this to occur, Yeates suggests, should be *the* priority item on the national urban agenda.

The growth and the problems associated with the urban-rural fringe have been the subject of many studies for individual cities in Canada. Lorne Russwurm's paper goes beyond this and presents a national overview of urban fringe problems based on extensive field examination around the major urban centres in Canada in the early and mid-1970's. He identifies and discusses eight major urban fringe problems: conflicting land use activities; land conversion problems; impact on agriculture; impact of the natural environment; taxation; impact on surrounding settlements; social issues; and planning and administrative problems.

Cities can be classified in a number of ways but one of the most useful classifications for geographers is that based on economic function. In their paper, Li, Scorrar, and Williams identify eight dominant economic functions for the 137 urban areas in Canada — extraction, manufacturing, construction, transportation, retail trade, community service, personal service, and public administration. From this analysis a clear distinction emerges between the heartland cities, which account for two-thirds of the manufacturing cities, and the more diversified hinterland cities. In the second part of the paper some of the trends and characteristics of manufacturing activities are examined.

Canadian cities exhibit not only a wide range of economic functions, but also a diversity in ethnic composition, although the diversity has marked regional differences when examined in detail. This analysis is the subject of Fred Hill's contribution in which he discusses the ethnic and cultural diversity of Canadian urban areas under six headings: source of immigrants; immigration, urban size, and growth; birthplace of immigrants; mother tongue; ethnic origin; and religion.

Over the years certain concepts and beliefs have been built into our thinking regarding urban and metropolitan growth in Canada. Whereas in fact some of these beliefs, e.g. the assumption of continued national and urban growth, are no longer valid, they still persist. In his paper Larry Bourne's observations, based on research, hindsight, and 1976 census data, lay to rest some of these myths and suggest some possible future directions for urban research in Canada.

The Evolution of Urban Canada: The Post-1867 Period

A national urban system did not exist in Canada at the time of Confederation because both intercolonial trade and interurban linkages were insignificant.[1] The nearest thing to an integrated urban system was in what was becoming the national urban-industrial heartland, where despite political separation between 1791 and 1841, Upper and Lower Canada functioned as a single economic system focused on the St. Lawrence River. A fragmented urban system existed in the Atlantic region, and although Halifax, Saint John, and St. John's were growing as focal points for their respective hinterlands, prospects for an integrated economic system within the region were dim. Urbanization in Prairie and Pacific Canada was only beginning.

Shaping Forces

Confederation, in 1867, brought a *National Policy* designed to create a viable economic unity out of the political transcontinental union just realized. The intent was to promote an east-west flow, and this was to be facilitated by construction of a transcontinental railroad, by imposition of tariff barriers against imported manufactured products, and by settling the West.[2]

In 1885 the Canadian Pacific Railway was completed from Montreal to Vancouver, and was extended to Saint John in 1890. The hoped-for immediate rush of settlers to the West did not follow completion of the transcontinental railroad, since conditions were still too attractive for settlers in the western United States.

Protectionist tariffs were first instituted in Canada in 1879. The intent was not only to generate government revenue but also to facilitate the development of domestic manufacturing by creating a protected home market for those products which could be produced in Canada at a reasonable cost.[3] Only the industries of central Canada were sufficiently developed at that time to benefit from protection.[4] In contrast, in the Atlantic region the tariff increased the cost of goods to the point that, with the exception of the Sydney steel industry, the development of manufacturing was inhibited. There seems little question that the impact of the 1879 tariffs on the nation's economic situation in the latter part of the nineteenth century was important in confirming Ontario and Quebec as the industrial heartland. As in the case of the transcontinental railroad, the impact of the 1879 tariff system was not immediate. Evidence for this is conclusive, since in the 1880's and 1890's the nation's rate of population growth declined. Thus, despite the transcontinental railway, the tariff system of 1879, and a vigorous government-sponsored immigration policy, there was little progress toward integration of the nation's regional economies before 1895.[5]

By the mid-1890's the environment for economic growth became more attractive owing to: (a) a steep rise in the price of wheat; (b) a decline in ocean freight rates; (c) the exhaustion of the more accessible free land in the United States (a fact which made Canada a high choice for settlers who would previously have gone to the United States); and (d) a reduction in rail freight rates for wheat and flour moving eastward toward the head of navigation on Lake Superior (the Crow's Nest Pass agreement of 1897). All these events benefitted settlement and agriculture in the Prairies, and encouraged massive immigration, attracting more than three million new settlers by 1914, including a record 400 000 in 1913.

While the above events were taking place, railroad track distance was increasing from 21 158 km in 1890, to 28 410 km in 1900, to

● RICHARD PRESTON is in the Department of Geography, University of Waterloo. This paper is a revised version of Parts V and VI of R. E. Preston, *The Canadian Urban Pattern*, unpublished manuscript prepared for a CAG/Canada Studies Foundation Project (Waterloo: Department of Geography, February, 1977), and is published by permission of the author.
(Wherever possible, 1976 Census figures have been used. In some cases they are preliminary counts and may differ somewhat from final figures.)

61 737 km in 1917. Two more transcontinental railways were completed: the Canadian Northern was opened from Quebec to Vancouver in 1915, and the Grand Trunk Pacific connected Winnipeg and Prince Rupert in 1914, thus linking with Quebec (and later Moncton) via the National Transcontinental Railway. Because of financial problems, both the Grand Trunk Pacific and the Canadian Northern were taken over by the Federal Government between 1917 and 1923, and merged with the National Transcontinental, the Grand Trunk, and the Intercolonial to form the Canadian National Railways.

Prairie railroad distance alone accounted for about 50 percent of the national increase; with 6436 km in 1901, the Prairie total grew to 9654 km in 1906, 12 872 km in 1911, and 22 526 km in 1916. Construction of branch lines was the main goal of Prairie railwaymen during the 1920's, and by 1929 railroad distance in the region had increased to 28 962 km.[6] Much of the heavy post-1895 immigration was destined for the Prairie Provinces, where population increased from 420 000 in 1901 to 2 359 000 in 1931, 23 percent of the national total at that time. The railways completed the major features of Canada's settlement pattern and established Vancouver as the main western terminus. Moreover, west of Ontario, the railway was *the* transportation route, and towns were seldom located anywhere but next to it.

The role of wheat production during this period deserves special consideration. Prairie wheat production stimulated the first important interregional movement of commodities in Canada, and thus formed the base for an economic integration of the various regions and colonies which had been united politically by Confederation. Prairie wheat production was also the main force behind national economic development between 1896 and 1930. While population in the Prairie Provinces increased by almost 2 000 000 between 1901 and 1931, the area of occupied land in these provinces increased from 6.2 x 10^6 ha to 44.4 x 10^6 ha. Wheat rapidly increased its share of total Canadian exports, by value, from less than 5 percent in 1900 to 16 percent in 1911, and became the nation's most important export commodity by the end of the First World War. During the 1920's wheat consistently represented over 25 percent of total exports, and by 1928 wheat accounted for 40 percent of exports.

The relation between increasing wheat production and development, not only in the Prairies but also in the nation as a whole, was remarkable. The wheat boom stimulated rapid railway network extension, and together they were the main supports of national economic development in the decades just prior to World War I.[7] The flow of wheat from the Prairies was complemented by flows of people, machinery, building materials, and capital. This integrated and enlarged Canadian economy made it possible for manufacturers to take advantage of a new domestic market for nonprimary products, and thus created an opportunity for import substitution of such goods and related services on a scale heretofore unknown. Entrepreneurs were taking advantage of scale economies in production and were using a national market for the first time. The events of this period set in motion a cumulative causation development process, which in turn produced new minimum thresholds on the market side. Expansion of both population and economic opportunities further enhanced the march of industrialization. These events allowed the basic conditions for urban development to emerge: rising incomes, expanding markets, capital and labour concentrations, and accumulations of people seeking opportunities in cities. With these events came a transformation of the labour force from primary industry towards secondary and tertiary occupations. This transformation continues to this day, and is further considered below. The transformation itself essentially represents the economy's attempt both to satisfy new demands for goods and services and to adopt technology to achieve greater worker productivity. Such conditions were cumulative, and beyond some minimum threshold were capable of sustaining the growth process. Lithwick pointed out that "the very success of this urban transformation entailed the transformation of the economy itself, so that Canada was launched on its current growth path".[8]

These events are revealed by Table 1, which shows that over one-half of the national economic activity was primary and nonurban in 1870, but by 1920 that proportion had fallen to one-quarter and by 1965 to just one-tenth.

RUCTURE OF FINAL DEMAND
CANADA 1870–1965

Year	Share in Output		
	Primary %	Secondary %	Tertiary %
1870	51	25	24
1920	29	32	39
1926	23	26	51
1965	14	32	54

Note: The measures of output are not exactly comparable. The first set (1870-1920) are based on GNP and are derived from M. C. Urquhart and K.A.H. Buckley, *Historical Statistics of Canada*, Macmillan, Toronto, 1965, Series E214-E244: The latter (1926-1965) are based on GNP and are from DBS *National Accounts, Income and Expenditures*. Primary includes all resource industries; Secondary, manufacturing and construction, and Tertiary, the rest.

Source: N. H. Lithwick and G. Paquet, *Urban Studies: A Canadian Perspective* (Toronto: Methuen, 1968), p. 29.

Table 1

Prior to considering the role of manufacturing industry in Canadian development, it is useful to examine some implications of the foregoing for the national space economy in general and for the national urban pattern in particular. The above events were especially significant in central Canada, where external economies, like transportation advantages, tariff advantages, market accessibility, and a rapidly diversifying industrial base all contributed to this area's emergence as the national urban-industrial heartland. Moreover, the financial and commercial firms that developed earlier in Montreal and Toronto placed branches in Winnipeg and other western cities, and thereby organized and directed western development in a way that integrated the West into the central Canadian financial centres, and further reinforced heartland dominance.[9]

Three scholars of Canadian urban-industrial development see this era around the turn of the century as being of pivotal importance in the country's emergence as a modern nation. Nader states "The National Policy, which was designed to forge a transcontinental economy and to create interregional interdependence or complementarity, was finally achieved, based mainly on the strength of wheat as an export staple."[10] Stone noted that the combination of immigration, western settlement, the emergence of the wheat staple, and the associated integration of Canadian regions was related to the upsurge of urbanization throughout the nation between 1891 and 1911. While James Simmons commented that:

The great surge of urban growth in Canada took place around the turn of the century, specifically between 1891 and 1911, when the Canadian urban system as we now know it effectively took shape. The rapid development of the western Canada urban system and its implications for the growth of central Canada preceded by a decade the fullest expansion of ecumene . . .[11]

Compared to this surge of urban growth, the postwar urban boom of the late 1940's and 1950's was much less dramatic and certainly less structure-changing, at least at the national level (Figure 1).

By 1900 the continued prominence of central Canada in the nation's urban-economic system was apparent. Largely as a result of manufacturing expansion, the urban population increased its proportion of the national total from 18 percent in 1871 to 35 percent in 1901. The development of a substantial domestic industry was one of the foremost objectives of national policy, and such expansion played a special role in both the pattern of concentration and the overall urbanization process.

When the first manufacturing census was taken in 1870, there was little industry in the country, and sawmilling, flourmilling, shoemaking, and clothing accounted for almost one-half of the value of manufactured goods.[12] But between 1870 and 1900, national manufacturing expansion was steady and marked by increasing specialization by firm, by an increase in average firm size, and by an ever greater concentration of production in cities. These events were especially important in central Canada, where the concentration of manufacturing in cities was a major cause of the rapid increase in urban population. Over the same period, the share of national manufacturing accounted for by the Atlantic Provinces declined.

The settlement of the Prairies and the Pacific coast, the railway boom, and massive growth of the domestic market stimulated development of secondary industry generally, but particularly reinforced its concentration in the Ontario-Quebec urban-industrial heartland. Thus, between 1900 and 1911 one of the most outstanding periods of urban-industrial growth in national history took place, and of special importance in this upsurge was an increase in the number of large corporations and a decline in the number of partnerships and individual ownerships. Two of the rapidly growing industries were iron and steel and transportation

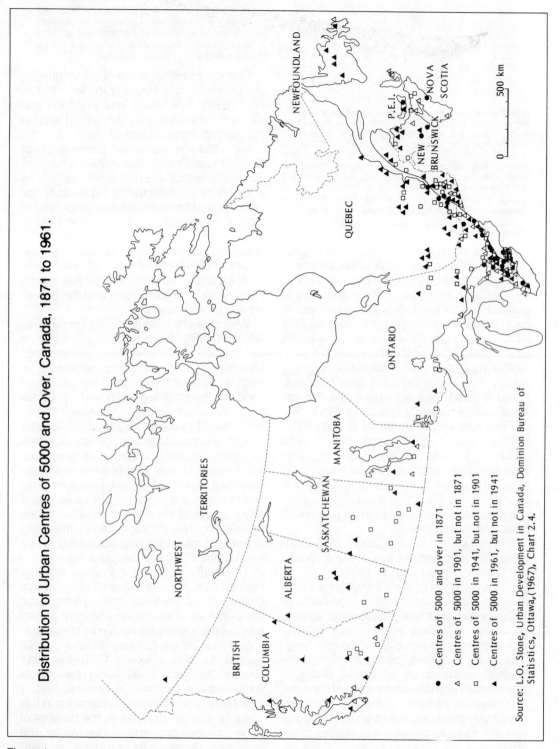

Distribution of Urban Centres of 5000 and Over, Canada, 1871 to 1961.

● Centres of 5000 and over in 1871
△ Centres of 5000 in 1901, but not in 1871
□ Centres of 5000 in 1941, but not in 1901
▲ Centres of 5000 in 1961, but not in 1941

Source: L.O. Stone, Urban Development in Canada, Dominion Bureau of Statistics, Ottawa, (1967), Chart 2.4.

Figure 1

equipment, both associated with railway expansion. Spurred on by an expanding American market, pulp and paper production also expanded, as did industries linked with post-1900 mineral discoveries and hydroelectric developments in the Shield. The region benefitting most was central Canada, where American capital, technology, and markets were most accessible. After 1900 such capital and technology were transferred increasingly to Canada as branch plants.[13]

World War I provided the next great impetus to urban-industrial development in the country by creating an unprecedented demand for manufactured products. The result was a further concentration of economic opportunities in the cities of central Canada. This was also a time of radical technological change in agriculture. Such change intensified in the 1920's, and resulted in an expanded use of farm machines and an associated reduction in demand for farm labour. The migration of rural people to cities helped to keep the rate of urbanization in central Canada at its record pace of 1901-1911 through 1911-1921. In contrast, the rate of urbanization declined markedly in the agricultural Prairies.[14]

After World War I, the United States rapidly replaced Great Britain as the principal source of foreign capital, and by 1922 was the more important of the two. Also, by 1920 manufacturing industry had displaced agriculture in terms of value of production. Without question the economy was changing from primary forms of production toward secondary and tertiary activities. Moreover, the areal concentration of these activities was clear.

Since the First World War, just over 80 percent of the national manufacturing industry, in terms of employment and value of production, has been accounted for by Ontario and Quebec, and in 1971 the nine metropolitan centres along the Quebec-Windsor industrial axis (Quebec, Montreal, Ottawa-Hull, Toronto, Hamilton, St. Catharines-Niagara, Kitchener, London, and Windsor) together accounted for 55 percent of the national manufacturing employment.[15]

Canada, like other countries in the North Atlantic Community was hit hard by the "Great Depression" of the 1930's, and Stone summarizes the impact of that event on the trends just described.

The Great Depression, which began generally in 1929, but was evident in Saskatchewan as early as 1928, was marked by an enormous dampening of the factors promoting urbanization. Immigration and population growth decelerated markedly, the demand for the products of nonprimary activities fell off considerably, and the rate of investments in technological changes declined greatly. Accompanying this matrix of economic contraction was a marked downturn in the pace of Canadian urbanization.[16]

Since the outset of World War II, Canada has experienced a period of unprecedented urban-industrial development. The war was the "mother of necessity" for impressive technological changes and for new perspectives on the mobilization of resources. Wilson, Gordon, and Judek have pointed out that:

Not only had industrial research begun on a large scale, but many entirely new industries had been established (for example, synthetic rubber, roller bearings, diesel engines, antibiotics, high octane gasoline, aircraft manufacturing, and ship building). Further processing of some manufactured goods, hitherto unimportant, likewise gave the Canadian economy a taste of new manufacturing capabilities . . . In many industries (for example, steel) basic capacity was permanently enlarged.[17]

Accompanying the events of the 1940's was a further marked decline in the number of agricultural workers. The above events all contributed to the growth of cities, a phenomenon further accentuated by the unprecedented national prosperity of the postwar period.[18]

Viewing Canadian urbanization in the post-World War II period produces still another combination of shaping forces. On the one hand there are the forces just described, but to these must be added new sources of economic opportunity in growth industries like oil, natural gas, paper and pulp, the automobile associated industries, as well as the great (and ongoing) expansion of opportunity in transportation and communication related activities. On the other hand, urbanization has been further buoyed by a postwar immigration wave which has focused its impact on urban areas – particularly Toronto, Montreal, and Vancouver. Stone argues that the above are the key forces explaining metropolitan growth in this country in the 1950's and 1960's.

THE CHANGING OCCUPATIONAL STRUC-TURE OF THE CANADIAN LABOUR FORCE

Occupational Categories	1881	1921	1951	1971
Primary	51.3	36.6	19.8	8.3
Manufacturing	24.3	20.8	25.1	22.2
Construction	4.5	5.8	6.2	6.0
Transportation	2.9	7.8	9.5	8.8
Trade & Finance	5.3	9.4	10.1	16.9
Service	11.6	19.2	28.2	37.6
TOTAL	100.0	100.0	100.0	100.0

Note: The labour force is composed of noninstitutional population, ten years of age and over in 1881 and 1921, but fourteen years of age and over in 1951 and 1971.

Source: D. Michael Ray, "Canada: The Urban Challenge of Growth and Change", Discussion Paper B.74.3 (Ottawa: The Ministry of State for Urban Affairs, 1974), pp. 4-11. Reproduced by permission of Minister of Supply and Services Canada.

Table 2

Recent decades, then, have also witnessed a continuing decline in the proportion of the labour force in primary activities. At the same time, employment in secondary industries has remained remarkably stable (Table 2).

This means that mechanization has simply kept pace with a high rate of expansion in secondary industry. The big change in employment has come neither in the primary activities, so basic to the nation's development, nor in the secondary industries, basic to the rapid urban-industrial expansion in the later nineteenth and early twentieth centuries. Rather, great expansion in employment came in tertiary activities (Table 2). In

Table 3

1881 16.9 percent of the labour force was engaged in tertiary (trade, finance, and services) activities, but by 1971 the figure was 54.5 percent. Details of contemporary changes in employment are shown in Table 3. Weir summarizes this structural transformation of the labour force as follows:

The increasing degree of specialized skills needed by labor, together with the voracious appetite of an affluent society for all kinds of services, especially professional, personal, and clerical, has been the chief contributor to the rate of urban growth. Half the number of those working in towns and cities are busy "servicing" the other half employed in manufacturing, construction, transportation, and trade.[19]

The preceding discussion appears to confirm Lithwick's position that "the role of staples as the engine of growth has been continually reduced, and endogenous urban growth has taken over".[20] While probably an overstatement of the actual case, the direction of change is clearly identified. This position does not mean that foreign markets are unimportant, since such trade accounted for 25 percent of Canada's GNP in 1969. And, although staple exports remain of great importance, foreign trade "increasingly is based on idustrial specialization on a continental scale rather than on primary products exclusively".[21] (Table 4.)

LABOUR FORCE 1961-1971: CANADA

	1961	1971	% Change	% Distribution of Change
Agriculture	640 786	481 190	-24.91	-7.41
Forestry	108 560	74 380	-31.48	-1.59
Fishing and Trapping	36 263	25 435	-29.86	-0.50
Mining	121 702	139 035	14.24	0.80
Manufacturing	1 404 865	1 707 335	21.53	14.04
Construction	431 093	538 225	24.85	4.97
Transportation	385 031	403 735	4.86	0.87
Storage	17 677	16 295	-7.82	-0.06
Communications	130 074	163 190	25.46	1.54
Utilities	70 504	87 845	24.60	0.80
Wholesale Trade	289 884	348 815	20.33	2.73
Retail Trade	701 606	920 475	31.20	10.16
Finance	110 936	185 170	66.92	3.44
Insurance & Real Estate	117 969	172 890	46.56	2.55
Education Services	266 901	569 485	113.37	14.04
Health & Welfare Services	307 433	513 095	66.90	9.54
Religious Organizations	53 130	47 210	-11.14	-0.27
Amusement and Recreation	39 837	75 065	88.43	1.63
Personal Services	437 518	509 990	16.56	3.36
Services to Business Management	98 987	208 760	110.90	5.09
Miscellaneous Services	59 556	117 785	97.77	2.70
Public Administration	309 896	499 840	61.29	8.81
Defence	173 029	139 745	-19.24	-1.54
Unspecified	158 593	681 940	329.99	24.28

Source: C. D. Burke, *The Parasites Outnumber the Hosts* (Ottawa: Ministry of State for Urban Affairs, 1975), p. 27.

COMMODITY COMPOSITION OF TRADE, CANADA, 1957—1968

| | 1957 | | 1968 | |
	Exports %	Imports %	Exports %	Imports %
Grain	8	—	5	—
Industrial Materials	53	29	43	21
Highly Manufactured Products	6	31	25	44
Services	25	31	21	29
Other	8	8	7	7

Source: Economic Council of Canada, *Sixth Annual Review,* September 1969, Table 5-5, p. 89, Lithwick, 1970, p. 74.

Table 4

Figure 2

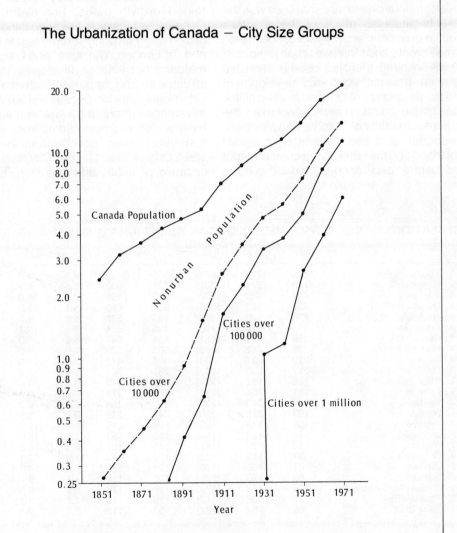

The Urbanization of Canada — City Size Groups

Source: James W. Simmons, Canada as an Urban System: A Conceptual Framework,
<u>Ekistics</u>, Vol. 41, (February 1976), pp.68—75.

The structural transformation of the labour force described above is intimately connected with urbanization because increasingly the source and destination of Canadian exports are the urban system. Throughout the continuing reorientation of the national economy toward endogenous production, the large urban centres have continued to grow (Figure 2). Outstanding in this respect is that by 1900 the larger centres, especially Toronto and Montreal, had become sufficiently large and their economies so well developed and diversified that they appear to have achieved a state of self-sustained growth. Lithwick argues that at the turn of the century the national economy was on the threshold of a new developmental phase, and that such was the case because so much of the country's modern economic activity was concentrated in the Toronto and Montreal urban regions.[22] What transpired, then, is a close relationship between national economic development and the development of those leading cities. Today, little doubt remains concerning the economic maturity of Toronto and Montreal, the former as a service centre for a vast hinterland and the latter as the great entrepôt and centre of French-Canadian culture.

Moreover, these urban regions are marked by self-sustaining growth, pushed by both high rates of growth and large shares in key indicators of national wealth like market, income, and capital.[23] The third emerging giant in the national urban system is Vancouver, where recent growth has been rapid and perhaps brought the metropolis over the threshold of self-generating development by the early 1960's.

Accompanying the rise of the country's three great cities has been the development of smaller urban systems, each focusing on a substantial metropolitan area. A number of these subsystems, as evidenced by the growth of their primate city, exhibit particular features indicating that they may be on sustained growth paths. For example: (a) Oshawa and Oakville are flourishing as part of the rapidly growing Toronto region market; and (b) London, Winnipeg, and Calgary are maturing because of developing industrial structures and important service roles. Others are growing because of exceptional advantages afforded by their immediate locations; for example, Edmonton, with oil, distribution, and government; Hamilton, steel; Ottawa-Hull, Quebec City, and Halifax because of public administration. Lithwick

Table 5

POPULATION CHANGE CENSUS METROPOLITAN AREAS (CMA's), CANADA, 1961–76

Rank (1976)	CMA	Population (in 000's)					% Change		
		1961	1966	(1971)a	1971b	1976	1961-66	1966-71	1971-76
1	Toronto	1825	2290	(2628)	2602	2803	18.3	14.8	7.7
2	Montreal	2110	2571	(2731)	2731	2802	15.5	6.7	2.7
3	Vancouver	790	933	(1082)	1082	1166	12.9	16.0	7.8
4	Ottawa-Hull	430	529	(603)	620	693	15.1	13.9	11.8
5	Winnipeg	476	509	(540)	535	578	6.8	6.2	5.2
6	Edmonton	337	425	(496)	496	554	18.9	16.4	11.7
7	Quebec City	358	437	(481)	501	542	15.6	10.0	8.1
8	Hamilton	395	457	(499)	503	529	13.6	9.0	5.2
9	Calgary	279	331	(403)	403	470	18.5	21.6	16.5
10	St. Catharines-Niagara	217	285	(302)	285	302	5.8	6.3	5.6
11	Kitchener-Waterloo	155	192	(227)	239	272	24.3	18.0	14.1
12	London	181	254	(286)	252	270	14.4	12.9	6.9
13	Halifax	184	210	(223)	250	268	7.7	6.2	6.9
14	Windsor	193	238	(259)	249	248	9.5	8.5	-0.5
15	Victoria	154	175	(196)	196	218	12.5	11.7	11.5
16	Sudbury	111	137	(155)	158	157	5.7	16.0	-0.4
17	Regina	112	132	(141)	141	151	16.9	6.3	7.4
18	St. John's	91	118	(132)	132	143	11.0	12.1	8.8
19	Oshawa	81	106	(120)	120	135	30.8	13.0	12.4
20	Saskatoon	96	116	(126)	126	134	21.3	9.1	5.8
21	Chicoutimi-Jonquière	105	133	(134)	126	129	5.9	0.9	1.6
22	Thunder Bay	92	108	(112)	115	119	17.6	3.2	4.0
23	Saint John	96	104	(107)	107	113	5.2	2.4	5.8

Note: For the percentage change 1961-66, 1966-71, 1971-76 respective areal definitions used are those for 1966, 1971, and 1976.
a Based on 1971 census area definitions
b Based on 1976 census area definitions
Source: Statistics Canada, 1976.

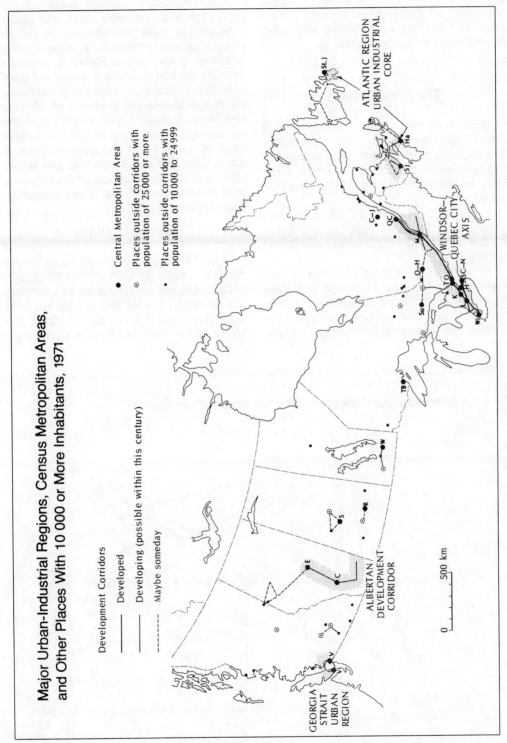

Major Urban-Industrial Regions, Census Metropolitan Areas, and Other Places With 10 000 or More Inhabitants, 1971

Development Corridors
—— Developed
—— Developing (possible within this century)
-- -- Maybe someday

● Central Metropolitan Area
◉ Places outside corridors with population of 25 000 or more
⊙ Places outside corridors with population of 10 000 to 24 999

ATLANTIC REGION URBAN INDUSTRIAL CORE

WINDSOR–QUEBEC CITY AXIS

ALBERTAN DEVELOPMENT CORRIDOR

GEORGIA STRAIT URBAN REGION

0   500 km

Figure 3

27

feels the above conditions are key inputs into a national economic system and that they will be even more so in years to come. Moreover, all these urban regions should respond positively to sustained economic development in the heartland.

The Areal Pattern

Such were the forces shaping national urban development since Confederation. The evolving areal pattern is summarized in Figure 1, and will be considered here. First, however, the contemporary pattern is established.

In 1976, 76 percent of the Canadian people lived in urban centres, and these people were anything but evenly distributed (Figure 3).

About 30 percent of the population live in metropolitan Montreal, Toronto, and Vancouver. An additional 25 percent live in twenty other medium-sized metropolitan centres like Winnipeg, Edmonton, Calgary,

Victoria, Thunder Bay, Quebec City, Halifax, and St. John's (Table 5).

The remaining 21 percent of the nation's urban folk live in smaller cities and towns with 1000 or more inhabitants. The hierarchical structure of the national urban system is shown graphically in Figure 4, while its overall territorial structure is shown in Figure 5. Both the manner in which the urban population was distributed in places of different size in each major region in 1971 and the names and populations of the five largest cities in each region are given in Table 6 (page 30). The maps, graphs, and tables provide reference materials for the following brief descriptions of the urban pattern in each region.

Central Canada

Today central Canada embraces the national heartland. In 1971 almost 60 percent of Canada's people lived in an area between the American Border and a 1050 km east-west line from Quebec City to Sault Ste. Marie. Within that area there is continuity of

Figure 4

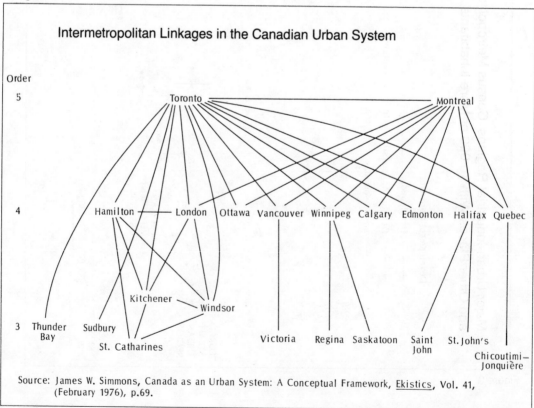

Source: James W. Simmons, Canada as an Urban System: A Conceptual Framework, Ekistics, Vol. 41, (February 1976), p.69.

The Territorial Structure of the Canadian Urban System

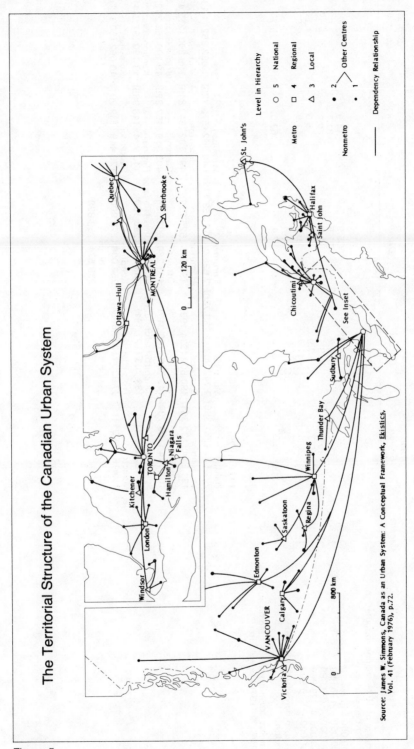

Source: James W. Simmons, Canada as an Urban System: A Conceptual Framework, Ekistics, Vol. 41 (February 1976), p.72.

Figure 5

Table 6

URBAN POPULATION DISTRIBUTION IN CANADA IN THE 1970's

	Canada	Atlantic Region	Quebec	Ontario	Prairies	British Columbia
Percentage of 1971 regional population living in urban centres of:						
a) 500 000 and over	31.9	–	44.2	35.9	14.9	42.4
b) 100 000 – 499 999	15.6	9.2	8.7	17.6	31.7	7.5
c) 30 000 – 99 999	9.0	15.8	8.1	11.6	2.9	5.2
d) 10 000 – 29 999	8.1	10.3	7.9	7.9	5.6	11.1
e) 1 000 – 9 999	11.5	20.5	11.7	9.3	11.9	9.5
Total urban percentage of regional population	76.1	55.9	80.6	82.4	67.0	75.7

Five largest urban centres per region and their 1971 & (1976) population

1971 (1976)

Atlantic Region	Quebec	Ontario	Prairies	British Columbia
Halifax 222 637 (261 366)	Montreal 2 743 208 (2 758 780)	Toronto 2 628 043 (2 753 112)	Winnipeg 540 262 (570 725)	Vancouver 1 082 352 (1 135 774)
St. John's 131 814 (132 391)	Quebec City 480 502 (534 193)	Hamilton 498 523 (525 222)	Edmonton 495 702 (542 845)	Victoria 195 800 (212 466)
Saint John 106 744 (109 700)	Hull 149 230 (166 317)	Ottawa 453 280 (502 536)	Calgary 403 319 (457 825)	Prince George 49 100 (58 292)
Sydney 91 162 (87 210)	Chicoutimi-Jonquière 133 703 (127 181)	St. Catharines-Niagara 283 852 (298 129)	Regina 140 734 (148 965)	Kamloops 43 790 (57 241)
Moncton 71 416 (74 911)	Trois Rivières 97 930 (102 368)	London 252 000 (270 000)	Saskatoon 126 449 (132 291)	Kelowna 40 092 (50 111)
		Kitchener 238 574 (269 828)		

Source: Adapted from Statistics Canada, 1971 *Census of Canada, Population*, and 1976 *Census of Canada, Population* (preliminary).

settlement, but there also exists large unsettled tracts. This area, whose greatest north-south reach is 435 km, comprises only about 2.2 percent of the nation's land area.

The settlement pattern in central Canada is most constricted in Quebec between Quebec City and Trois-Rivières, where there is but a narrow fringe of settlement on either shore of the St. Lawrence, with mainly empty land behind. The nine largest Census Metropolitan Areas (CMA's) in central Canada are here and account for over one-third of the national population: Toronto, Montreal, Ottawa-Hull, Quebec City, Hamilton, St. Catharines-Niagara, Kitchener-Waterloo, London, and Windsor.

In terms of population size, political power, and economic wealth, the dominance of central Canada over the rest of the nation was well established by Confederation. Much of this importance was attributable to and centred in the major cities. In 1871 the region included the country's three largest cities, Montreal (107 000), Quebec City (60 000), and Toronto (56 000). Decision-making in the nation had also localized in these cities plus Ottawa, and, as shown above, national governmental policies continued to confer advantages on this already favoured region.

In Quebec, Montreal became a primate city in every sense of the word. By 1921 its metropolitan population was more than five times that of its old rival Quebec City. There are 6 141 496 people in Quebec today (1976), and it is the second most highly urbanized region in Canada, with 80 percent of its population in urban places. As was the case in 1921, when only 50 percent of the provincial population was urban, about one-half of Quebec's city people lived in Montreal in 1976. Montreal had approximately 2 802 000 inhabitants in 1976, but was surpassed in population by Toronto during the first half of the 1970's. Nevertheless, Montreal is both the core of one of the two major poles of influence in the national heartland, and the unrivaled economic and social heart of French Canada. The second city of the province is Quebec City with 534 193 inhabitants – an administrative centre with an important port function. Next is Hull with 166 317 people, but this is misleading as Hull is functionally part of the Ottawa-Hull metropolitan area, and the population of both cities totals over 693 000.

Next in size are Chicoutimi-Jonquière, Sherbrooke, and Trois-Rivières, all with populations of less than 150 000 and all focal points for outlying urban subsystems. The remainder of Quebec's urban dwellers currently occupy a large number of smaller cities, most of which are in the St. Lawrence Lowland and Eastern Townships.

Montreal has a diverse economic base, serving as a national financial centre, a great entrepôt, a provider of high order goods and services, and an important manufacturing centre. Moreover, within the heartland, interaction between Montreal's metropolitan region and southern Ontario contributes greatly to the development and maintenance of the modern integrated urban system along the axis. Quebec City is the provincial capital, but in comparison with Montreal, it has a much narrower economic base. Secondary centres are located in the Eastern Townships, where manufacturing and central-place services are important and where Sherbrooke provides a focus for the most important subsystem. Centres outside the major urban areas just described are small and are supported by primary activities; for example, by mining in Wabush-Labrador City, by farming in the Lake St. John Lowland, by hydroelectric power production along the Manicouagan River, and by forestry, especially the pulp and paper industry, in numerous places peripheral to the St. Lawrence Lowland.[24]

Ontario, with a 1976 population of 8 131 618, is presently the most highly urbanized of the nation's five major regions, with approximately 82 percent of its population urban. This situation is dramatized by the location of seven of the fourteen largest metropolitan areas in the country within the province. Toronto became the nation's largest city between 1971 and 1976, dominating the urban-industrial region of southwestern Ontario, and is the main national link with the American manufacturing belt. There are at least six secondary metropolitan areas in Toronto's urban system, focusing on Hamilton, St. Catharines-Niagara Falls, Kitchener-Waterloo, London, and Windsor, and each is the coordinator of a well developed urban region.

Most current economic growth in Ontario is concentrated in a crescent of land around the western end of Lake Ontario. This is the "Golden Horseshoe", embracing an area from Oshawa to Kitchener to Niagara Falls. The core is the continuous built-up area of the Toronto agglomeration, stretching along the lake front from Oshawa to Aurora to Oakville and including a population, in 1971, of over two and a half million. Most of the cities in the Golden Horseshoe are diversified manufacturing centres and are at once both the core and most rapidly growing component of the Canadian heartland and an integral part of the North American Manufacturing Belt.[25]

Northern Ontario towns are mostly dependent on primary activities like mining, forestry, and hydroelectric power, but some, like Sudbury, Thunder Bay, and Timmins, also offer extensive administrative and institutional services. For the most part the smaller cities along lakes Huron and Superior specialize in primary activities, tourism, and depot functions.

A useful view of the state of urbanization in Ontario and Quebec can be gained by comparing their respective urban rank-size distributions. There are sharp contrasts in their patterns of urban development, and one of the main differences is in their respective ranges of alternative growth complexes to the Toronto and Montreal metropolitan areas. Outside of metropolitan Toronto (2 803 000) there are at least four alternative locations available to investors. Included here are the Kitchener-Guelph cluster, the Niagara regional cluster, and the London and Windsor urban regions. The key point is that each of these clusters of urban places represents a viable locational alternative to the Oshawa-Toronto-Hamilton concentration. In fact, Oshawa and Hamilton offer alternatives to Toronto, though in a more limited sense. All of these urban complexes are located within the heartland and all have attained adequate size to offer sufficient infrastructure and external economies to likely assure self-sustaining growth.[26]

In comparison, in Quebec similar alternative locations to Montreal for accepting and generating new urban-industrial growth do not exist. The Montreal metropolitan area (2 802 000) contains nearly 50 percent of the provincial population, and beyond this region, growing middle-sized cities are few. Next to Montreal in size is Quebec City, but it is locationally isolated from the core of the provincial space economy. The next group of

centres are Saguenay, Trois-Rivières-Cap-de-la-Madeline, Shawinigan-Grande Mère, and Sherbrooke-Lennoxville. These are much smaller than the four urban clusters in Ontario mentioned above. They are also more peripheral in location and have been growing at a slower rate. Additionally, all four lack a strong and dominating central city as exists in the Windsor, London, Kitchener-Guelph, and Niagara (Hamilton) urban regions of Ontario. The result of such differences is that the growth potential for Quebec urban centres beyond the Montreal region appears substantially less than that for cities in Ontario beyond metropolitan Toronto.[27] These may well be the most substantial observations that can be offered regarding the comparative structure and prospects of the urban systems in the two provinces.

The Atlantic Region

In 1976 2 141 692 people lived in the Atlantic Provinces, and this population is most unevenly distributed, with only the smallest province, Prince Edward Island, completely occupied. Large parts of interior Nova Scotia, New Brunswick, and the Gaspé Peninsula are vacant, as is most of the interior of Newfoundland where settlement is confined mainly to a broken fringe along the coast.

About 56 percent of the region's people were classed as urban in 1917; however, only 9 percent lived in places of over 100 000 population, by far the lowest percentage for any major region. Moreover, Halifax, with 261 366 inhabitants, is the smallest of the nation's regional capitals. The second and third cities of the Atlantic Provinces – St. John's and Saint John – both have less than 150 000 people and are thus smaller than all but Regina and Saskatoon among cities in the country that command regional subsystems. The outstanding trait of the region's population pattern, however, is the large concentration of people in the "small urban places" (less than 10 000) which, at approximately 20 percent, is twice that of both the nation and of any other major region. Likewise, the region has an excessive concentration of people in the rural nonfarm class. The essential traits of the settlement pattern in the Atlantic region are: (a) three tenuously linked urban hierarchies; (b) small nondiversified urban places; and (c) rurality.[28]

The state of rurality in the region is summarized by Macpherson:

The total rural population in all of the Atlantic Provinces constituted at least 40 to 50 percent, and in Prince Edward Island's case 63.4 percent, of the total population in 1966. The regional percentage (46.5) was almost double the national figure of 26.4. Moreover, although decline in relative strength since 1951 was common to the country, the region, and the individual Atlantic Provinces, the percentage decline in the contribution of the rural sector in the Atlantic Provinces only slightly exceeded the national decline (eleven points) in the cases of Newfoundland and Prince Edward Island; New Brunswick fell rather less (eight points) that the country as a whole; and Nova Scotia's relative decline was less than three percentage points. During the period 1951-66 it would appear that the forces of urbanization were acting less efficiently upon the rural sector of the population in the Atlantic Provinces, particularly in Nova Scotia, than in Canada as a whole. The rural tradition, as expressed in current structures of settlement hierarchy, economy, and social structure, would appear to have retained much of its ability to resist change, presumably by a retention of reinforcement of conservative values in the way of life.[29]

Right after Confederation, the Atlantic Provinces entered a period of economic depression, a situation in bitter contrast to the period 1847 to 1867, when the region enjoyed unprecedented prosperity based on forest industries, shipbuilding, fishing, and trade.[30] Nader feels national attempts at economic integration have had a generally negative effect in Atlantic Canada, for example:

National economic policies, such as the protectionist tariffs of 1879, largely favoured the commercial manufacturing and financial interests of central Canada, reinforcing its position as the national heartland and relegating the peripheral regions to subordinate roles of primary production.[31]

Neither did the Maritimes benefit from the waves of immigration that swept the country after Confederation. In fact, until recently (1971-1976) the region has suffered from chronic out-migration. In reality the effect of Confederation was simply to accelerate a decline caused mainly by technological change. The area's economic difficulties stem from the facts that: (a) it does not possess significant natural resources besides the sea; (b) it is isolated by distance and topography from growth in central Canada and on the western frontier; (c) it has a small and dispersed population; and (d) it is increasingly separated from national Maritime trade by improvements in the navigability of the St. Lawrence, development of routes through the United States, and reorientation of considerable Prairie trade to Vancouver. All these factors limited development of a self-generating consumer market or a regionally-supported industrial structure.

As far as industry is concerned, only shipbuilding and sawmilling have ever been very important, but even these declined rapidly after Confederation. The tariff system of 1879 hit Maritime industry hard because it militated against small scale firms, which were the dominant type in the region. Moreover, the impact of the railroad was also negative because not only did the rails not traverse the agriculturally-oriented and heavily populated Saint John Valley, but they also tended to bring the area increasingly under central Canadian economic dominance.[32]

Use of iron steamships assisted the metropolitan position of Halifax and Saint John by concentrating what shipping there was in fewer ports. Halifax developed a more diversified economy than Saint John and became the largest city in the region. The New Brunswick city's extreme dependence on lumbering and shipbuilding hampered its development. Aside from Saint John, the only centre of any size in New Brunswick was Moncton (10 000 in 1921). Moncton grew rapidly after completion of the Intercolonial Railway in 1876, and its role as a distribution centre for the region was further enhanced by completion of the National Transcontinental Railway. In Nova Scotia, Sydney's urban area grew after 1890 with the development of coal mining and an iron and steel industry. On Prince Edward Island, Charlottetown became the main city, performing central-place and administrative functions for the island. The only other city of importance was Summerside, much smaller than Charlottetown but also basically a central place.

On Newfoundland, St. John's developed early as the main mercantile centre and has maintained that position ever since. The trans-island railway became operational in 1900 and gave rise to the first inland communities. The lumber industry was the chief beneficiary of the railway, and communities based on forest industries have grown up at Grand Falls, Windsor, and Exploits River. Over the years, St. John's has increased domination of Newfoundland's economy by adding senior governmental functions to its traditional activities, and in 1976 it accounted for 24.1 percent of the provincial population. Other towns with over 10 000 inhabitants are the pulp and paper towns of Corner Brook (24 798), Grand Falls-Windsor (15 069), and Stephenville (10 120), and the mining town of Labrador City (15 600).

As noted above, the Atlantic Provinces are a land of small towns, and for the most part such places exist because they either perform a special function or are located in relation to a particular resource. Dalhousie, Newcastle, and Liverpool are paper and pulp mill towns; Yarmouth, Carbonear, Lunenburg, Grand Bank, Caraquet, Harbour Grace, and Shelburne are home ports for the dragger fleets, and have fish processing plants; Springhill, Wabush, Labrador City, and Newcastle are mining towns; Bay Roberts is a wholesale centre; Gander, Yarmouth, Deer Lake, Port-aux-Basques, and Botwood are service centres occupying break-of-bulk locations. Gander and Deer Lake are also important central places and are joined in this function by Kentville and Windsor in Nova Scotia; Bridgewater is a logging centre.[33]

In summary, the Atlantic Provinces are presently the least urbanized part of the country. The individual provinces are small in size and population relative to those in central and western Canada. Additionally, both urban population and functions are distributed among several medium-sized cities rather than concentrated in any one centre. The regional economy continues to be based mainly on resource-related activities — for example, fishing, agriculture, forestry, and tourism — while strong economic links of an essentially colonial nature are maintained with the national heartland. The urban-industrial system in the Atlantic Provinces remains noncompetitive with central or western Canada. The cities are comparatively small, and no city exercises any significant degree of influence over the

entire region. Rather the region is fragmented into a number of jurisdictions, each served by a small to medium-sized city, while metropolitan functions, if available in the Atlantic region at all, are provided by either New England or central Canadian cities.[34] Charles Forward writes:

> The geographical, political, and social fragmentation of the Atlantic Provinces region sets it apart as unique among the major regions of Canada. The islands and peninsulas are like pieces from different jigsaw puzzles that do not fit together. Transportation systems have been ineffective in drawing the subregions together to focus on one major urban centre. Hence, a number of separate hinterland areas that are tributary to several urban centres of similar size have developed independently.[35]

The Prairie Region

Approximately 3 700 000 people inhabited the Prairie Region in 1976. Of these, 67 percent were classed as urban. While there are five large cities — Winnipeg, Edmonton, Calgary, Regina, and Saskatoon — about half of the region's population is scattered in hundreds of small agriculturally-based towns and villages.

The degree of metropolitan development in the region is indicated by the fact that 46 percent of the population lived in urban places with over 100 000 people in 1971. This percentage was lower than that for central Canada (58 percent) and British Columbia (50 percent), but higher than the Atlantic Provinces. Another aspect of the region's rank-size distribution is the low percentage of its population in places in the 10 000 to 30 000 and 30 000 to 100 000 ranges, especially in the latter. Thus, the region is one in which emergent metropolitan areas are definitely lacking.

The areal structure of the Prairie urban system is straightforward. The Winnipeg metropolitan area, with 570 725 people, dominates the eastern approach. At present this core region is showing some tendencies toward corridor formation in the direction of Portage la Prairie and Brandon. Otherwise long-distance connections extend Winnipeg's influence over the remainder of the province and into northwestern Ontario. On the west, the emerging Albertan development corridor dominates the Rocky Mountain front. The northern part of Alberta, much of the Northwest Territories, and parts of northeastern British Columbia and northwestern Saskatchewan all lie within Edmonton's area of dominance. The Peace River country, with Grande Prairie at the top of its urban subsystem, is clearly within Edmonton's (542 845) sphere, while Red Deer, on the main highway between Edmonton and Calgary, marks the zone of territorial competition between Edmonton and Calgary. Calgary, with 457 828 people, is the corridor's second city, Edmonton's principal competitor, and major centre of southern Alberta. The urban subsystems centring on Lethbridge and Medicine Hat are within Calgary's area of influence. Between the metropolitan concentrations in Manitoba and Alberta are two smaller urban systems in Saskatchewan, focussing on Regina and Saskatoon, both cities with less than 150 000 people. Regina dominates the Moose Jaw subsystem while Saskatoon dominates the Prince Albert subsystem.

The situation described above was not always the case. In the second half of the nineteenth centry, western Canada awaited forces capable of overcoming its isolation from central Canada. Prior to the 1880's the fur trade and gold mining dominated economic life in the West, and aside from the trading posts, economic activities were located peripherally. Victoria had been selected colonial capital and dominated west coast trade and commerce from its location on the tip of Vancouver Island. On the Prairies, Winnipeg was the focal point for all traffic into and out of the region, and was in every sense a classic gateway city.[36] Edmonton and Calgary were only small trading posts.

Winnipeg played a special role in the urban-economic development of the Prairies. The agriculturally productive section of the region is shaped like a fan bounded by climate and terrain on the north, by the Rockies on the west, by the Pre-Cambrian Shield on the east, and by the United States boundary on the south. Settlement entered the region from the east and progressed towards the northwest. Moreover, until recently most products exported from the Prairies flowed eastwards, so Winnipeg's location at the point of the fan was well suited to serve as the gateway to the West. Such was Winnipeg's (then Fort Garry) function even before the railway era.

Winnipeg was the only city of importance in the Prairies prior to the construction of the Canadian Pacific Railway in 1885, and its development was due primarily to its transportation connections with the United States,[37] and only secondarily to its connection with western Canada.

With completion of the nation's first transcontinental railway in 1885, the western part of the continent was opened up for development. Towns and villages emerged along the railway and cities grew at divisional points where car-repair and other maintenance facilities were located. Examples are Winnipeg, Portage la Prairie, and Brandon in Manitoba; Regina and Moose Jaw in Saskatchewan; Medicine Hat and Calgary in Alberta; and Vancouver in British Columbia. All of these cities received their greatest stimulus for early growth from construction of the Canadian Pacific. The significance of the railroad to Prairie settlement is illustrated by the fact that in 1931 approximately 80 percent of the farmers in Saskatchewan lived within 16 km of a railway station.[38]

Mineral discoveries also played an important role in the development of western Canada. For example, in southwestern British Columbia settlements were brought into existence by mineral discoveries in the Kootenay region between 1890 and 1895. They were literally "Boom Towns", and places like Nelson and Trail, as well as smaller settlements, have persisted while at the same time adding supplementary functions. McCann summarizes settlement at the turn of the century by stating "Influenced by the route of the railroads, agricultural opportunities, and mineral discoveries, the basic outline of today's urban pattern was beginning to emerge."[39]

The story of the Prairies in the early twentieth century is highlighted by massive immigration, agricultural development, and rapid urbanization. The urban proportion of the region's population increased from 19 percent to 28 percent between 1901 and 1911 as the wave of settlement spread from east to west. Most new communities were directly associated with growing wheat for export, and the Prairies became dotted with hamlets, villages, and towns. Most of these places functioned as small agricultural service centres, and had populations between 1000 and 2500.

35

As the rail lines spread westward and the wheat economy expanded, Winnipeg developed rapidly as the primary wholesale centre for the region. In fact, "During the first decade of this century all of present Manitoba, Saskatchewan, and Alberta seemed securely within the dominance of "Peg" wholesalers."[40] At that time, the Rocky Mountains formed a kind of buffer zone between the trade area of Winnipeg and those of Pacific coast towns. But,

As the tide of occupance reached the extremities of the Prairie "fan", two new gateway cities developed, on a smaller scale (than Winnipeg), at its outer margins. Calgary, also on the Canadian Pacific Railway, had begun as a Mounted Police post but did not begin its rapid growth until around 1910. It was the entrance of the main routes into the Rockies and served as a gateway to British Columbia and the Pacific. Edmonton had been a portage site on the North Saskatchewan River, but soon developed its special roles as capital of Alberta and gateway to the far northwest of Canada. The rapid growth of these two cities (aided by the subsequent discovery of oil) rolled back eastward the borders of Winnipeg's area of dominance.[41]

And, in the same vein,

More recent yet has been the steady, if unspectacular growth of the cities in the middle of the Prairies triangle, Regina and Saskatoon. These have never aspired to be gateways, but are emerging as true central places in control of the Saskatchewan portions of the wheatlands.[42]

Recent expansion of the trade areas of the two Saskatchewan cities has also been at Winnipeg's expense.

Edmonton and Calgary mushroomed from small service centres in 1901 to cities of about 80 000 in 1921. At that time Calgary appeared to have a slight edge on Edmonton because of oil and natural gas strikes in its hinterland in 1913. Lethbridge and Medicine Hat emerged as subregional service centres in southern Alberta, while a number of agricultural service centres grew up in the fertile soil areas south and east of Edmonton. In Saskatchewan, Saskatoon became a focal point for urban growth in a rich agricultural belt in the central part of the province, while Regina had the dual advantage of an early start as a service centre and designation as provincial capital. Moose Jaw, Prince Albert, and Yorkton were all growing as regional subcentres.

In Manitoba the situation was more complex. Winnipeg solidified its hold as the province's primate city, but, at the same time, was losing its position as gateway to the Prairies. The opening of the Panama Canal in 1914, the loss of much of Alberta's trade to the port of Vancouver, and the expiry of rail privileges and preferential freight rates held by Winnipeg, all led to Edmonton, Calgary, Regina, and Saskatoon taking over some of the functions previously handled by Winnipeg.[43] Andrew Burghardt summarizes Winnipeg's situation.

This severe retrenchment in wholesaling marked the transition of Winnipeg from gateway to central place status. However, the city remains the major rail center of the Prairies, and is in control of a high proportion of Canadian wheat sales and shipments.

As a central place, Winnipeg is hampered by its position at the point of the fan; the city has been thrown back onto a particularly limited portion of its former domain. Since, unlike the fertile Prairies, the Pre-Cambrian Shield east of Winnipeg witnessed the rise of no large centres, Winnipeg's service area now extends, without competition, somewhat more towards the east than towards the west . . . Winnipeg is now as much a gateway from the West to the vast empty areas of westernmost Ontario as it is to the Prairie West it once dominated.[44]

Although the degree of urban development is greater in the Prairies than in the Atlantic Provinces, the region has not been dominated by metropolitan centres to the same extent as central Canada or British Columbia. For example, the proportion of Prairie population concentrated in the five metropolitan centres (Winnipeg, Saskatoon, Regina, Edmonton, and Calgary) was 13 percent in 1901, 24 percent in 1941, 41 percent in 1961, and 49 percent in 1971. For comparison with other major regions see Table 6.

The low level of metropolitan dominance in the Prairies is, of course, related to the structure of the regional economy and to the position of the region in the national economy. Although the region has urbanized rapidly since 1941, the urban proportion of its population is still below the national average of 76 percent. About three-fifths of the Prairie population live in urban places, with Alberta having the highest percentage at 73.5 and Saskatchewan the lowest with 53 percent. Barr notes that the:

distribution of Prairie population in urban places reflects the relative importance of small agricultural and resource-extractive towns (under 10 000 population) and the dominant administrative, distributional, and service functions of the five largest regional centres (over 100 000 population). Centres of intermediate size (30 000-99 999) with a strong secondary manufacturing base are not yet significant in the system of Prairie cities. A low density of population in the region encourages major manufacturing enterprises to serve the Prairie market from plants in central Canada, the United States, or overseas countries.[45]

From the outset the Prairie Provinces have occupied a colonial status within the national economy, in that they were producers of primary materials for sale in international markets and purchased the majority of their fabricated products from tariff-protected industries in central Canada.[46] Although this situation would probably have evolved under free market conditions, it unquestionably has been encouraged by national economic policies with respect to railway freight rates and tariffs. A final point relating to the role of the Prairie region in the national economy is that the nature of the links with the heartland encouraged the specialization of the larger Prairie cities in regional service functions.

The role of manufacturing industry has varied greatly in Prairie urban development. With the decline of its gateway function, manufacturing replaced commerce as the chief economic activity in Winnipeg during the interwar period. However, this point should not play down the dominant position of Winnipeg in the Prairies in the financial sphere. Bercuson summarizes this role:

> Winnipeg formed part of the central Canadian hinterland as it acted out a metropolitan role for the Prairies while other, smaller centres, which were tied to Winnipeg, captured their own hinterland areas throughout the West. The pattern was repeated as large cities attracted smaller cities which in turn attracted nearby towns – the metropolitan impulse was transmitted into every corner of the Prairies. The pull of the Winnipeg Grain Exchange was felt in every city, town, elevator, and farm, just as the effects of decisions made in banking boardrooms on Bay or St. James Streets were acutely felt in Winnipeg.[47]

By comparison with Edmonton and Calgary, Winnipeg has developed a diversified industrial structure that is related to its regional role. Agricultural products are produced (for example, meat-packing and flour milling), and railway equipment is manufactured, as well as a wide range of consumer goods and farm machinery. Edmonton and Calgary have had little success in decreasing their dependence on regional resources. They process primary products and perform central place functions. These traditional activities have not been of diminishing proportions. Recent growth of manufacturing employment in the two cities has been at a rate less than proportionate to the growth of the overall labour force. Edmonton has benefitted more than Calgary from manufacturing growth, and has attracted capital-intensive plants, particularly those associated with the petrochemical industry. Secondary manufacturing continues to be of minor importance to the economies of both cities.[48] Likewise, Regina and Saskatoon both suffer from Saskatchewan's lack of economic diversification.

In summary, early urbanization in western Canada produced the rise of Winnipeg as gateway to the West, and the subsequent rise of Vancouver, Calgary, and Edmonton, all wresting away some of Winnipeg's realm. Later still, and primarily based on central-place functions, came the emergence of Regina and Saskatoon. Thus, early in this century these cities and their supporting hierarchies formed the main elements in the urban system in western Canada. Today the Prairies are less urbanized than other regions in the country, except the Atlantic region. Economic ties among the urban subsystems in the region are not well developed. Alberta's cities are more closely tied to British Columbia and the North, while Winnipeg is tied to central Canada.

Historically, Winnipeg was the main urban place in the region. As noted above, this role has faded. However, Winnipeg remains supreme within Manitoba. Over 50 percent of the province's population is located in Winnipeg, and the city is the centre of the grain trade as well as an important distribution and financial centre for the eastern Prairies. Winnipeg's prospects are limited, however, because the city's urban system is weak. There are few secondary centres in the province (none really beyond Brandon and Portage la Prairie), and thus the overall small size of Winnipeg's urban system is a barrier to economic development.

Edmonton and Calgary are growing faster than Winnipeg. The economies of these two cities are benefitting from expanding oil and beef industries, as well as from their traditional functions as the province's main centres for collection, distribution, manufacturing, and in the case of Edmonton, for government services. Saskatchewan is mostly rural, with most of its people residing in small urban centres. Regina and Saskatoon are the two major cities and, in addition to Regina's governmental role, they are the collection and distribution service centres for the province. In the Prairies there is a strong trend towards a gradual disappearance of smaller settlements, which cannot compete with the attractions of large cities, on the one hand, and whose trade areas have been depopulated by technological advances in agriculture, on the other hand.

British Columbia

There were 2 406 212 people living in British Columbia in 1976. Of these, 76 percent were classed as urban. Only central Canada, where over 80 percent of the inhabitants of both Ontario and Quebec are urban, is more urbanized than British Columbia.

> There is continuity of settlement throughout the southern half of British Columbia, but in the form of narrow interconnecting strips following mountain valleys and coastal plains. Between the valleys, large areas are empty of permanent settlement. The settled strips occupy about 0.7 percent of the area of Canada and contain over 10 percent of its population. More than 5 percent of the national population is located in the Lower Fraser Valley.[49]

Thus the provincial settlement pattern is dominated by the Georgia Strait Urban Region, the core of which is lower mainland British Columbia, which in turn focusses on Vancouver with its 1 135 744 inhabitants.[50] Walter Hardwick described this region in 1972 ". . . as a dispersed city, a galaxy of subcommunities, of which the inner city of Vancouver is clearly the focus".[51] He also notes that

> Most of the urban population of British Columbia is clustered around Georgia Strait in the southwestern corner of the province. Victoria, founded in 1843, Nanaimo (1852), New Westminster (1858), and Vancouver (1886) are the historic cores and present nuclei around which the region

is organized. Each city has its unique character and distinctive origin, but continued growth in population, associated with economic expansion and improved accessibility within the region, is contributing to increased functional interdependence.[52]

The second city of both the urban region and of the province is Victoria with 212 466 people. Below these two cities there is a large gap in the province's rank-size distribution, with Prince George, 58 292, Kamloops, 57 241, Kelowna, 50 111, and Nanaimo, 39 655, next in size. In terms of territorial dominance, in addition to the lower mainland Vancouver dominates directly the subsystems centring on Prince Rupert, Prince George, Kamloops, Kelowna, and Trail-Nelson, as well as in the Yukon, where Whitehorse is the main focus. However, Vancouver meets competition from Spokane throughout the southern Okanagan and Kootenays, from Calgary in the Cranbrook-Kimberley area, and from Edmonton in the northwest section of the province. The Port Alberni and Nanaimo subsystems are dominated by Vancouver indirectly through Victoria.

British Columbia was part of San Francisco's economic realm until the completion of the transcontinental railway in 1886, and commercial contacts were maintained through Victoria.[53] However, when Vancouver was chosen as the western terminus of the Canadian Pacific Railway, the new city quickly replaced Victoria as the commercial centre of the province. Both the first train from the East and the first Canadian Pacific Railway ship from Yokohama arrived in Vancouver in 1887, and these events emphasized the city's function as a meeting place of transcontinental and trans-Pacific routes, and foretold its evolution as the major trading city of the province.[54] Vancouver was destined to become the western gateway to the nation and the hub of the Canadian west coast.

Numerous factors contributed to Vancouver's growth after 1900 and after its selection as western terminus for the Canadian Pacific Railway.[55] Just as in the other western cities, the speculative economic bubble that rose soon after the turn of the century collapsed in Vancouver around 1913; however, it appeared that in the decade prior to World War I the foundation for the new city's eventual rise to national metropolitan status was already in place. For example, freight rate schedules, formerly discriminating in favour of Winnipeg, were removed, and this allowed Vancouver to become the wholesale and distribution centre for British Columbia. The Vancouver Stock Exchange was founded in 1907, and became a control point for development of the province's resource-based industries. The impact of the completion of the Panama Canal, in 1913, hit the city around 1920. The Canal reduced by half the distance to Europe by sea, and at the same time, as a result of a decrease in the amount of shipping at the end of World War I, ocean freight rates fell sharply. These events allowed Vancouver to compete for the first time with eastern Canadian ports in the export of Prairie wheat. Moreover, low ocean rates by way of the Panama Canal caused transcontinental rail rates to be reduced, and further enhanced Vancouver's importance as a wholesale centre. Vancouver had overtaken Winnipeg in terms of population by 1921, and by 1930 (excluding bulk commodities such as wheat) Vancouver's wholesale trade was greater than Winnipeg's. Vancouver was a metropolitan centre in every sense of the word and was the gateway to the Prairies from the West.[56]

Several features of the 1921 map of urban development in British Columbia are of note. Smaller regional centres serving the specialized agricultural region of the Okanagan Valley had sprung up at Penticton, Vernon, and Kelowna. The Grand Trunk Pacific Railroad (now the Canadian National Railway) reached the west coast in 1915, and along the way aided the rapid growth of Saskatoon and Edmonton, as well as smaller places on the Prairies. The railway also put Prince George and Prince Rupert on the map for the first time in British Columbia, and represented new accessibility to central and northern British Columbia. This led to the exploitation of forest, mineral, and agricultural resources that had heretofore been out of reach of men and machines.[57] It appears that outside lower mainland British Columbia, the railroads did not fundamentally alter the character of the provincial economy, which has remained largely dependent on exploitation of agricultural, forest, mineral, and fishing resources.

Vancouver has continued to dominate British Columbia. By 1971 census definitions, the city's metropolitan area increased its proportion of provincial population continuously, from 44 percent in 1921 to 50 percent in 1931, a higher proportion than that of any other major city in the country at that time. Vancouver's proportion of provincial population presently stands at 47 percent.

Today Vancouver is the prime city of the province. In addition to its province-wide metropolitan function and its regional financial role, much of Vancouver's contemporary growth is based on the city's break-of-bulk and transportation activities. It is a wood products and grain exporting centre to both the Far East and to the United States west coast, and an import centre for the whole of Canada for goods from the Pacific region. Moreover, Vancouver and Victoria perform functional roles for the province that are increasingly complementary. Victoria is the provincial government's legislative-executive centre and thus the decision-making centre for the public sector, while Vancouver is the decision-making centre for the private sphere of the economy.

Notwithstanding Victoria's role, it is clear that, beyond the Georgia Strait Urban Region, British Columbia's urban subsystems lack economic diversity.[58] In the interior valleys there are small and scattered communities that have grown rapidly in recent years. However, in each of them livelihood tends to be based primarily on primary economic activities. The Fraser Valley, the Peace River country, and the Okanagan all have economies dominated by agriculture, with forestry playing an important secondary role. Service-centre activities are also significant, particularly in places like Vernon, Kelowna, and Penticton in the Okanagan. In the Peace River region, mining and oil and gas extraction are also important economic activities. The east and west Kootenay areas display similar economies in that both emphasize mining and service functions, while agriculture plays a minor role. Trail and Nelson are the main service centres. Forestry is the economic base of Vancouver Island beyond Victoria and of the coast north of Vancouver, while fishing complements forestry around Prince Rupert. The multiple-resource-based economy of the Prince George and Kamloops areas seems to be generating well-developed urban subsystems and seems to have particularly strong growth potential.[59] Growth in the importance of tourism is apparent in some places, for example in the case of Kelowna, which appears to have emerged as the dominant place in the Okanagan.

Some Conclusions and Some Thoughts about The Pre-1976 Pattern of Urbanization

Emerging Themes

This study had two goals: first, to reveal some of the relations between national economic development and the evolution of the Canadian urban system, and second, to offer some principles useful in describing and explaining those relations. Numerous themes have emerged, and some of them are summarized here.

1. *Heartland-hinterland*

As the nation's space economy has been transformed from a resource-based rural economy toward an urban-industrial one, disparities in material levels of life have arisen between the national heartland and most of the remainder of the country (hinterland). This situation also appears at the regional scale, where disparities exist between regional heartlands and the remainder of each region.[60] In this situation lie both some of the causes and answers to regional problems associated both with excessive concentration and underdevelopment.

2. *Staples and Spatial-Temporal Regional Differentiation*

Early in the nation's history, externally-determined national development led to a highly specific patterning of urban growth. Such development was sustained through a fortunate sequence of export expansion in which each stage provided basic conditions for urban development, i.e., incomes rose, markets expanded, capital and labour accumulated, expected structural changes were forthcoming, and these were capable of launching a sustained growth process. The very success of this pattern of urban-industrial development entailed the transformation of the economy itself. Thus, Canada was launched on its own development path, with the direct role of staples as the engine of growth becoming increasingly masked behind development of the industrial-service sector, and with foreign trade remaining important.

For example, the fur trade gave rise to a single large entrepôt, Montreal, and to a far-flung system of functionally similar trading posts and forts. Mining, in comparison, led to a virtually unconnected urban subsystem featuring clusters of towns varying greatly in size and concentrated around a single mineral complex, for example, Sudbury and Asbestos. Agriculture created subsystems of highly integrated trading centres, each tied to a hierarchy focusing on a regional centre, for example, the systems of Regina or Saskatoon. Industrial and tertiary concentrations have created the integrated urban-industrial complexes in the heartland.

Finally, exploitation of a particular staple can give a great boost to a particular city by conducting its financial affairs there,[61] Montreal benefitted from the fur trade, Toronto from mining and indirectly, through Winnipeg, from the wheat economy, Calgary was spurred by oil production, and today Vancouver is increasing its share of financial control over western oil, mining, lumbering and other resource-development activities in the West. The indirect urban growth induced by staples can also be seen in historic confrontations between cities, e.g., Toronto vs. Montreal, Calgary vs. Edmonton, and Calgary and Edmonton vs. Winnipeg. Today, Vancouver and Toronto are striving for control over the Prairie economy, and Vancouver and Edmonton are competing to be the gateway to the North.

The result of the nation's staple-based pattern of regional development is what Simmons[62] calls a "spatial-temporal" variation in Canada's national development. This has produced a clear regional differentiation in the process of settlement and urbanization: Quebec in the eighteenth century, the Atlantic Provinces in the nineteenth, Ontario in the late nineteenth century, and western Canada in the early twentieth. Each of these regions reflects quite different patterns of urban development, and this situation makes it difficult to generalize about urbanization in Canada. Accordingly, Simmons argues that the study of evolving patterns of intercity interaction should lead to discussions of functional subsystems, e.g., the settlement ports, the agricultural towns, the resource towns.[63] Also he argues that it

THE GROWTH OF MAJOR URBAN CENTRES
Historical Increase in Population of Metropolitan Areas (thousands)

Region	City	1650-1750[a]	1750-1850[b]	1850-1900	1900-1950
East	Charlottetown		7	5	6
	Saint John		27	23	46
	Halifax		25	26	130
Quebec	Quebec City	3	32	54	266
	Montreal	8	69	296	1534
Ontario	Kingston[c]		8	10	36
	Ottawa		15	81	334
	Toronto		31	241	1553
	Hamilton		14	65	304
Prairies	Winnipeg			48	428
	Regina			2	110
	Edmonton			3	324
	Calgary			4	273
West	Vancouver			27	763
	TOTAL			858	6107
	CANADA			2893	8299
	% of Increase in Urban Areas			30	74

a Dates are approximate.
b Population as of 1850.
c In 1831, Kingston was actually larger than Toronto (then called York).

Source: Lithwick, N. H., *Urban Canada: Problems and Prospects* (Ottawa: Central Mortgage and Housing Corporation, 1970), p. 72.

Table 7

should be possible to view the nation in specific periods as a series of unconnected elements – the widely separated settlements whose only links are maintained through European ports and institutions. At other times it should be viewed as an integrated system of centres, each with a particular function determined by the lords of the fur trade or the directors of the railroad. The national urban system should be treated ultimately as a complex set of entities in which each city performs several roles, each of which link the city to a different subset of urban places.[64]

3. *The Size of the Nation and the Distance between Cities*

Understanding the national urban system is complicated by the sheer size of the nation and by the distances between cities. Urban subsystems are separated by vast distances, and there are few large cities, 124 over 10 000 in 1976. Physiography, climate, and resources also vary greatly, so relative accessibility in the system is of paramount importance. Given such conditions, the role of transportation comes to the fore in any consideration of the structure and development of the Canadian urban system. These conditions all contribute to the spatio-temporal variation in the regional urban system. When viewing the Canadian urban system, however, Simmons offers three ideas, expressed in terms of comparative advantage, useful in explaining regional differences in urban economies within the system: first, natural advantage (the resource base); second, accessibility to the national market; and third, initial advantage (scale). Each of these have exerted varying influences on primary, secondary, and tertiary activities of each region.[65]

4. *Cities as Places where National Development has Occurred*

The critical role of urban places in national development is revealed by the last line of Table 7, which shows that over the last fifty years growth was almost exclusively in the largest cities. Likewise the spread of urban growth from east to west is shown by the same Table. Without regions, there is the same tendency (Table 8).

In this regard, Simmons points out that by 1971 urban-economic concentration had proceeded to the point where the ten largest metropolitan nodes contained 40 percent of the population and 60 percent of the industry and received over 50 percent of the income, while occupying less than 1 percent of the land area of the country. In some critical areas they play an even more decisive role, as corporation headquarters and media decisions are almost wholly concentrated in these cities.[66]

CENSUS YEARS IN WHICH CANADA AND
THE MAJOR REGIONS HAD REACHED OR
SURPASSED SELECTED LEVELS OF
URBANIZATION

	Levels of Urbanization[a]				
	25%	35%	50%	65%	75%
Canada					
Maritimes	1891	1901	1931	1961	—
Quebec	1901	1921	1961	—[b]	—
Ontario	1881	1891	1911	1941	1961
Prairies	1911	1951	1961	—	—
British Columbia	1891	1891	1911	1951	—

a The level of urbanization is measured by the percentage of population classified as urban.

b In this table, the dash indicates that the area in question had not attained the pertinent level of urbanization as of 1961, according to the source data.

Source: D. M. Ray, "Urban Growth and the Concept of Functional Regions", in H. Lithwick & G. Paquet, *Urban Studies: A Canadian Perspective* (Toronto: Methuen, 1968,) p. 62.

Table 8

5. *Temporal Stability in the National Urban System*

Since the Canadian urban system was established around the turn of the century, it has been highly stable (Table 9). So, despite distance, the spatio-temporal settlement sequence, and the systems exceptional openness to foreign influence, there has been considerable stability in the relative positions of its component cities. By this is meant that once the relative size and importance of urban places are established these relations do not change greatly.[67] Such is particularly the case within regions. Thus, once established, urban subsystems or hierarchies appear to functionally reinforce the existing urban pattern. On this subject of stability in the national urban system, Simmons observes, first, that the canals, railroads, freeways, and airports are all basically system reinforcing. They fortify existing interurban relationships. Moreover, decisions about such routes and the sources of capital for their formation flow from the financial and political powers in the larger cities.[68]

Secondly he argues that the network changes that have occurred are mainly the result of system expansion, which has been biased towards the frontier — first westward then north (Figure 1). The most dramatic changes in the network occurred when a formerly tributary city becomes a competitor, e.g., Montreal and Quebec before 1851, Toronto and Montreal between 1891 and 1901,

and Edmonton and Winnipeg between 1951 and 1961. The older city suddenly loses a whole set of linkages and the younger one moves ahead. Initial growth in the hinterland is always welcomed by the core, but at some point it appears to grow beyond the older city's control.[69]

6. *The Changing Structure of the Top Level of the National Urban Hierarchy*

It also appears that the structure of the top level of the national urban hierarchy is changing.

Not only has Toronto surpassed Montreal in terms of population, but it is also suggested by Yeates[70] that overall development in Toronto's urban system is steadily outstripping that of Montreal's. Thus Toronto is increasingly dominating all of Canada directly via high-order nation-serving functions, while Montreal dominates only Quebec and the Atlantic Provinces directly. At the same time Vancouver's role in the far west is growing, and by the end of the century Vancouver may dominate directly British Columbia, Alberta, and part of Saskatchewan. Thus, instead of the present national hierarchy topped by two principal cities of approximately equal importance, a hierarchy will exist with Toronto as the national centre and with Montreal and Vancouver as major regional centres. The zone of active competition between Toronto and Montreal will likely be around the Quebec-Ontario border, while the zone of active competition between Toronto and Vancouver could be a north-south zone in the vicinity of Saskatoon and Regina.

7. Finally, all of Lithwick's major conclusions regarding the development of the nation's urban system appear to be confirmed here:[71] namely, (a) that the early history of Canada played a key role in the development of the urban system; (b) that mature urban units have come to dominate the national economy; and (c) that this dominance has shaped the structure of the urban system in terms of hierarchy and connectedness.

8. *The Canadian Urban Pattern between 1971 and 1976: Some Possible Deviations from the Trend?*

During the 1950's and 1960's the dominant areal development trend shaping the Canadian urban pattern was one of population concentration in metropolitan areas, particularly around the "Big Three" — Toronto, Montreal, and Vancouver — and elsewhere

Table 9

POPULATION STABILITY OF THE CANADIAN URBAN SYSTEM

1851	1901	1941	1971	Rank by Population
MONTREAL (79.7)	Montreal (392.1)	Montreal (1193.2)	Montreal (2587.3)	1
QUEBEC CITY (45.5)	Toronto (270.9)	Toronto (865.7)	Toronto (2465.1)	2
TORONTO (30.8)	Quebec City (88.6)	Vancouver (338.3)	Vancouver (1026.9)	3
ST. JOHN'S (30.5)	OTTAWA (85.3)	Winnipeg (302.0)	Ottawa (555.0)	4
SAINT JOHN (23.7)	Hamilton (83.3)	Hamilton (224.7)	Winnipeg (540.3)	5
HALIFAX (20.7)	LONDON (51.6)	Ottawa (208.9)	Hamilton (488.9)	6
HAMILTON (17.6)	Saint John (51.2)	Quebec City (196.7)	Edmonton (486.7)	7
KINGSTON (11.6)	Halifax (51.0)	Windsor (128.6)	Quebec City (463.3)	8
	WINNIPEG (48.5)	EDMONTON (124.9)	Calgary (403.3)	9
	VANCOUVER (43.4)	CALGARY (111.6)	Windsor (237.6)	10
	St. John's (40.0)	Halifax (98.6)	London (235.8)	11
	KITCHENER (36.0)	London (97.2)	Kitchener (226.8)	12
	SYDNEY-	Sydney-	Halifax (204.8)	13
	GLACE BAY (35.7)	Glace Bay (96.7)	Victoria (195.8)	14
	VICTORIA (23.5)	Kitchener (82.8)	SUDBURY (145.0)	15
	WINDSOR (22.4)	Victoria (81.0)	Regina (139.5)	16
		Saint John (70.9)	SASKATOON (126.4)	17
		St. John's (60.9)	ST. CATHARINES (124.8)	18
		REGINA (58.8)	OSHAWA (119.4)	19
		THUNDER BAY (56.3)	St. John's (112.4)	20
		TROIS-RIVIÈRES (56.3)	CHICOUTIMI-	21
			JONQUIÈRE (109.1)	
			Thunder Bay (108.4)	22
			Sydney-Glace Bay (106.0)	23
			Saint John (102.1)	24
			Trois-Rivières (96.9)	25
	Kingston (19.8) (rank 16)	Kingston (34.8) (rank 30)	Kingston (76.8) (rank 28)	

Note: Each time a new city is introduced to the chart, it appears in capital letters. Note the difference in the growth of Quebec City and Saint John. Data obtained from Statistics Canada, Census of Canada, for various years, using the 1961 spatial definition of a Census Metropolitan Area.

Source: Simmons, J. W. "The Growth of the Canadian Urban System", (Toronto: University of Toronto Centre for Urban and Community Studies, November, 1974), Research Paper No. 65.

in the Windsor-Quebec City Axis and Alberta Development Corridor. Expected results were that rapid urban growth would continue, that an excessive proportion of the country's total population would be concentrated in the three largest cities, and that the present gap between social and economic opportunity in major urban regions and the remainder of the nation would continue to widen. Today, however, urban experts are less comfortable than they were a few years ago with some of their predictions about the nation's population growth, migration flows, and urban trends. Preliminary returns from the 1976 Census of Population have given cause for reappraising popular views of the Canadian urban future. At the broadest level it appears that, while the direction of change identified in the 1950's and 1960's remains the same, rates of change are lessening, and there seem to be signs at least of some territorial redistribution of population. One of the early notices that national urban trends were shifting came from Burke and Ireland who found that,

> The total population estimate for Canada (for 2001) is still valid, but the provincial estimates seem extremely dubious. Since 1971, there have been major shifts, with the Atlantic region and Saskatchewan showing much greater strength, and British Columbia and Ontario losing some of their dominance in terms of in-migration.[72]

Other trend variations were soon recorded, and they are summarized in Bourne's paper, in this volume, and by Preston and Russwurm[73]. Some of the points raised that might alter the present view of the country's urban future are as follows:

(a) Accepting substantial and rapid urban growth as an inevitable and ongoing condition in Canada is no longer an accurate perspective.

(b) The trend to population loss by the Prairies and Atlantic region to central Canada may be stabilizing or reversing itself.

(c) The proportion of the nation's total population living in Census Metropolitan Areas is levelling off at about 55 percent.

(d) Medium-sized CMA's like Victoria, Edmonton, Calgary, and Oshawa are growing faster than the Big Three.

(e) Absolute population losses are being experienced by the core's of Metropolitan Toronto, Montreal, Saint John, and Halifax, and very low growth rates by the cores of Hamilton and Vancouver. At the same time, these and most other CMA's are exhibiting a pattern of high growth rates at the periphery of their metropolitan areas.

These are but a few of the deviations from the pre-1971 trend revealed by the 1976 Census. Both the forces creating these shifts and their implications for the future are far from understood, but the presence of such shifts and our slowness in discovering them clearly show the uncertainty surrounding attempts to describe, explain, and predict which path the Canadian urban system is likely to follow.

REFERENCES

[1] Nader, G.A., *Cities of Canada* Vol. I (Toronto: MacMillan, 1975), p. 201.

[2] *Ibid.*

[3] *Ibid.*, p. 204.

[4] Ministry of State for Urban Affairs, *Human Settlement in Canada* (Ottawa, 1976), pp. 7-9.

[5] Nader, p. 204.

[6] *Ibid.*, p. 203.

[7] *Ibid.*, p. 205.

[8] Lithwick, N. H., *Urban Canada: Problems and Prospects* (Ottawa: Central Mortgage and Housing Corporation, 1970), p. 73.

[9] Nader, pp. 205-207, and D. J. Bercuson, "Urbanization in Prairie Canada: Problems from the Historical Perspective." Unpublished paper (Toronto: Canadian Studies Foundation, March 1972).

[10] Nader, p. 205.

[11] Simmons, J.W., "The Growth of the Canadian Urban System" (Toronto: University of Toronto Centre for Urban and Community Studies, November, 1974), Research Paper No. 65, p. 3.

[12] Nader, p. 207.

[13] *Ibid.,* pp. 209-210.

[14] Stone, L.O., *Urban Development in Canada* (Ottawa: Dominion Bureau of Statistics, 1967), p. 21.

[15] Nader, p. 210.

[16] Stone, p. 21.

[17] Wilson, G., S. Gordon and S. Judek, *Canada: An Appraisal of Its Needs and Resources* (Toronto: Twentieth Century Fund New York, University of Toronto Press, 1965), pp. 44-45.

[18] Stone, pp. 21-22.

[19] Weir, T. R., "The People", in J. Warkentin (ed.), *Canada: A Geographical Interpretation* (Toronto: Methuen, 1968), p. 155.

[20] Lithwick, p. 73.

[21] *Ibid.*

[22] *Ibid.,* p. 21.

[23] *Ibid.,* p. 52.

[24] Putnam, D.F., and R. G. Putnam, *Canada: A Regional Analysis* (Toronto: J. M. Dent and Sons, 1970), pp. 123-178.

[25] Leman, A.B. and I.A. Leman (eds.), *Great Lakes Megalopolis: From Civilization to Ecumenization* (Ottawa: Ministry of State for Urban Affairs, 1975), pp. 114-119.

[26] Bourne, L.S. and G. Gad, "Urbanization and Urban Growth in Ontario and Quebec: An Overview", in L. S. Bourne and R. D. MacKinnon (eds.), *Urban Systems Development in Central Canada* (Toronto: University of Toronto, Department of Geography Research Publications, 1972), p. 32.

[27] *Ibid.,* pp. 32-33.

[28] Forward, C.N., "Cities: Function, Form and Future", in A. G. Macpherson (ed.), *The Atlantic Provinces* (Toronto: University of Toronto Press, 1972), pp. 137-176.

[29] Macpherson, A.G., "People in Transition", in A. G. Macpherson (ed.), *The Atlantic Provinces* (Toronto: University of Toronto Press, 1972), p. 50.

[30] Erskine, D., "The Atlantic Region", in J. Warkentin (ed.), *Canada: A Georgraphical Interpretation* (Toronto: Methuen, 1968), pp. 253-259.

[31] Nader, p. 235.

[32] *Ibid.,* p. 236.

[33] Macpherson, pp. 48-49.

[34] Nader, p. 241.

[35] Forward, p. 137.

[36] Burghardt, A.G., "A Hypothesis about Gateway Cities", *Annals*: A.A.G. Vol. 61, 1971, pp. 269-285.

[37] Nader, pp. 246-247.

[38] *Ibid.,* pp. 244-245.

[39] McCann, L.D., "Urban Growth in Western Canada 1881-1961", *The Albertan Geographer* Vol. 5, 1968-69, p. 69.

[40] Burghardt, p. 274.

[41] *Ibid.*

[42] *Ibid.*

[43] McCann, p. 72.

[44] Burghardt, pp. 275-276.

[45] Barr, B.M., "Reorganization of the Economy Since 1945", in P. J. Smith (ed.), *The Prairie Provinces* (Toronto: University of Toronto Press, 1972), p. 68.

[46] Nader, p. 249.

[47] Bercuson, p. 6.

[48] Nader, p. 255.

[49] Ministry of State for Urban Affairs, p. 4.

[50] Gibson, E.M., *The Urbanization of the Strait of Georgia Region* (Ottawa: Lands Directorate, Environment Canada, Geographical paper No. 57, 1976).

[51] Hardwick, W.G., "The Georgia Strait Urban Region", in J. L. Robinson (ed.), *British Columbia* (Toronto: University of Toronto Press, 1972), p. 119.

[52] *Ibid.*

[53] Nader, p. 255.

[54] *Ibid.*, p. 256.

[55] Hardwick.

[56] Nader, pp. 256-257.

[57] McCann

[58] Denike, K.G. and R. Leigh, "Economic Geography 1960-70", in J. L. Robinson (ed.), *British Columbia* (Toronto: University of Toronto Press, 1972), pp. 69-86.

[59] *Ibid.*

[60] Berry, B.J.L., E. C. Conkling and D. M. Ray, *The Geography of Economic Systems* (Englewood Cliffs, N.J.: Prentice Hall, 1976), p. 271, and D. M. Ray and T. N. Brewis, "The Geography of Income and its Correlates", *The Canadian Geographer,* Vol. 20, 1976, pp. 41-71.

[61] Simmons, J.W., *The Canadian Urban System* (unpublished book length manuscript, 1975).

[62] _____, "The Evolution of the Canadian Urban System," Unpublished paper (Toronto: University of Toronto, Department of Geography, December, 1972), p. 1.

[63] _____, "The Evolution of the Canadian Urban System," Unpublished research proposal (Toronto: Department of Geography, 1971), p. 3.

[64] *Ibid.*, pp. 2-3.

[65] Simmons, *The Canadian Urban System*, p. 21.

[66] _____, "Canada as an Urban System: A Conceptual Framework" (Toronto: University of Toronto Centre for Urban and Community Studies, May, 1974). Research Paper No. 62, p. 3.

[67] _____, "The Evolution of the Canadian Urban System," p. 1.

[68] *Ibid.*, p. 9.

[69] *Ibid.*

[70] Yeates, M., *Main Street: Windsor to Quebec City* (Toronto: MacMillan Company of Canada, 1975), pp. 353-354.

[71] Lithwick, p. 107.

[72] Burke, C. D. and D. J. Ireland, *Holding the Line — A Strategy for Canada's Development* (Ottawa: Ministry of State for Urban Affairs, 1976), p. 18.

[73] Preston, R.E. and L. H. Russwurm, "The Developing Canadian Urban Pattern: An Analysis of Population Change, 1971-1976", in Russwurm and Preston (eds.), *Essays on Canadian Urban Process and Form* (Waterloo: University of Waterloo, Department of Geography Publication Series, No. 10, 1977).

Migration in the Canadian Urban System

JAMES W. SIMMONS

The 1971 Census provides the first opportunity to examine flows of migrants within Canada at a spatial scale finer than the provincial level. In particular we are able to study migration among various components of the Canadian urban system – the set of urban regions which together make up the country. These data tell us a great deal about the urban system and how it operates. Migration is important not only for its effect on local growth or decline, and on the demographic process, but also for what it tells us about the social and economic contacts among cities. An intricate but regular pattern of intercity migration emerges from this analysis – a pattern which severely restricts the potential for rapid change within the urban system.

The first part of this paper briefly describes the Canadian urban system as it is defined here, with its 124 urban-centred regions, a five level hierarchy, and eleven urban subsystems. The second part describes the various migration ratios compiled for the 124 urban regions, including in, out, gross, and net migration and their correlates. The third part, the heart of the paper, looks at the actual relationships among cities. Where do people go from Brandon, or Timmins, or Trois-Rivières? How do they get to Thompson or Labrador City? Finally these intercity relationships are aggregated into net flows among the various urban system components. Do cities of one size systematically gain from cities that are larger or smaller? What kind of broad regional interchanges exist?

The Canadian Urban System

The most meaningful spatial units for analysis at this scale are the various elements of the urban system. Over two-thirds of Canada's population, and a large proportion of its economic activities, are found within the 137 urban places of greater than 10 000 persons.[1] It is the characteristics of these places, the interactions among them, and the dynamics of their relationships which dominate the Canadian geography and its change over time.

These urban nodes can be extended in two ways in order to obtain an operational definition of an urban system.[2] We can extend the urban nodes in space until they exhaust the area of the nation (Figure 1). Each urban node then represents an extended area for which it provides services and, in turn, depends upon as an economic base. Brandon, for instance, is closely linked to an extensive agricultural area which determines Brandon's prosperity. The operational units for this spatial extension are the counties and census divisions used by Statistics Canada. When the 260 census divisions are allocated to the 137 census metropolitan areas, urban agglomerations, and cities of greater than 10 000 population, 124 extended urban regions result. In some cases two cities occur within the same census division (e.g., Chatham and Wallaceburg) and one of them disappears. In other cases an insolated census division, served by a town or city slightly smaller than the 10 000 threshold, is awarded urban status (e.g., Yellowknife).

The other embellishment, essential to the urban system concept, is the imposition of a pattern of organization or linkages among cities, which provides a basis for grouping and aggregating urban regions in a systematic manner. The simplest assumption about these relationships (and the one used here) is that of a nested hierarchy. Each urban centre is linked to a single larger place, which is assumed to provide high order services. Five levels or orders of centres are identified. Thus Summerside (1) to Charlottetown (2) to Halifax (4) to Toronto (5) (no level three

● JAMES SIMMONS is in the Department of Geography, University of Toronto. This article is Research Paper No. 85, Centre for Urban and Community Studies, University of Toronto, February, 1977, and is reprinted with permission. It has been revised by the author.

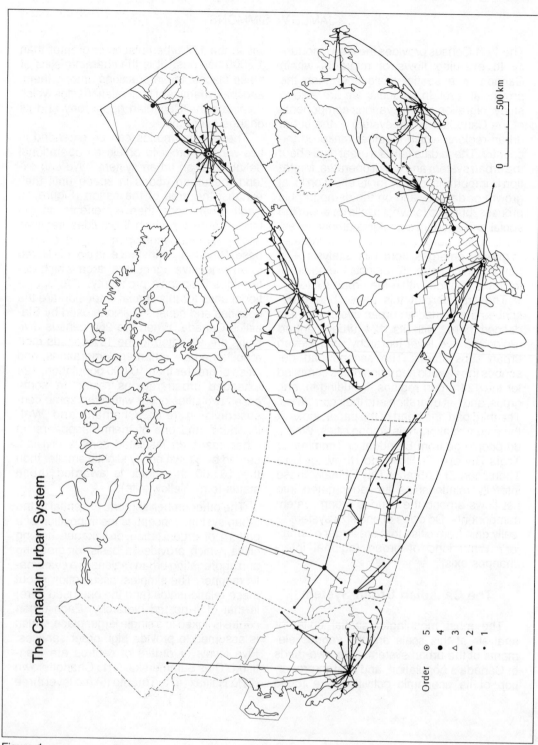

The Canadian Urban System

Order
⊙ 5
● 4
△ 3
▲ 2
· 1

500 km

Figure 1

centre in this sequence); or Chatham (2) to Windsor (3) to London (4) to Toronto (5) (no level one centre). This enables us, in a later section, to examine relationships among cities of different levels across Canada and to define regional subsystems according to linkages to high order places. The assumptions of the hierarchy and the particular linkages involved are quite arbitrary, reflecting the inadequate theoretical discussions of these contact patterns, and the lack of data on intercity contacts in Canada. In part this study tests the validity of the urban system concept itself.

The migration data were obtained from unpublished files of the 1971 Census. Statistics Canada prepared a tape containing an estimate of all moves among 260 census divisions, based on the census question:

Where did you live five years ago on June 1, 1966?

☐ Same dwelling

☐ Same city, town, village, or municipality

☐ Outside of Canada

☐ Different city, town, village, or municipality in Canada
give its name _____

The data relate to the migration status of persons five years and over. The migration status of persons five to fourteen years was assumed to be the same as that of the family head. One household in three was asked these questions, and the expanded sample was random rounded to 0 or 5. About 1.4 percent of the sample were unable or unwilling to locate their residence five years earlier.

Table 1 provides an overview of the national migration patterns. The population which moves is much younger and better educated than the population as a whole, and this is particularly true for those persons who migrate to different urban regions.[3] These latter moves are the primary concern in this study, and although they involve only 11.3 percent of the population over five years old, they include almost one-quarter (23.9 percent) of the movers. Immigration from abroad makes up only 4.2 percent of the population, but accounts for a considerable proportion of the variation in population growth.

Mobility Rates

Rates of movement were compiled for each of the 124 urban regions, for moves at three different spatial scales: all moves, interregional moves, and international

Table 1

MIGRATION STATUS

Total Population (5 years and over) in 1971
19 717 200 (100.0%)

Nonmovers (Same Dwelling)
10 371 300 (52.6%)

Movers (Different Dwelling)
9 345 900 (47.4%)

Same Urban Region
6 280 300 (31.9%)

Different Urban Region
3 065 600 (15.5%)

Intercity
2 242 000
(11.3%)

Abroad
823 600
(4.2%)

CORRELATIONS AMONG MOBILITY RATES

Variable	1	2	3	4	5	6	7	8	9	10
1. Mobility Rate	1.000									
2. In-Migration	0.864	1.000								
3. Out-Migration	0.417	0.553	1.000							
4. Gross Migration	0.775	0.929	0.823	1.000						
5. Net Migration (Turnover)	0.705	0.761	-0.119	0.466	1.000					
6. Migration Efficiency	0.708	0.718	-0.107	0.442	0.940	1.000				
7. Natural Increase	0.322	0.138	0.241	0.201	-0.023	-0.045	1.000			
8. Immigration	0.661	0.584	0.252	0.511	0.500	0.558	-0.005	1.000		
9. Log (Population, 1966)	-0.047	-0.387	-0.480	-0.478	-0.088	-0.029	-0.023	0.058	1.000	
10. Growth Rate	0.853	0.818	0.107	0.605	0.892	0.850	0.287	0.729	-0.062	1.000
Mean	44.1%	14.2	14.4	28.6	4.5*	15.0*	5.6	0.7	—	6.3
Standard Deviation	11.6%	7.5	4.9	11.0	6.5	18.4	2.9	3.5	—	9.5
Coefficient of Variation	0.26	0.53	0.34	0.39	1.44	1.23	0.52	5.00	—	1.70

*Absolute values

Table 2

moves. In each case the population of the place in question is the denominator (in fact, the denominator = the average of the total population recorded in 1966 and the population over 5 years of age in 1971). The most fundamental mobility rate is a simple measure of transience: the proportion of 1971 residents who live in a different dwelling than they did five years before. As Table 2 indicates, the average value for all urban regions is 44 percent with a standard deviation of 13 percent. The highest values occur in northern resource centres like Labrador City, the lowest rates in rural areas of eastern Canada. In fact, the high levels of correlation between this simple measure of mobility and other measures of migration and growth suggest a single dominant dimension of transience which is maintained throughout the country.

This measure, "mobility", includes moves of all kinds, both down the block and across the ocean, as well as from city to city. The Statistics Canada tape permits us to identify both out-moves and in-moves to and from other urban regions, and we can generate a variety of other mobility rates, including net migration (in-migration minus out-migration), gross migration or turnover (in-migration plus out-migration), and migration efficiency (net migration/gross migration).

Turnover, the sum of in-migration and out-migration, indicates the degree to which each urban area participates in the process of intercity migration. The most striking aspect of the patterns shown in Figures 2 and 3 and Table 3, is the existence of two separate migration regions in the country. East of Montreal, every place but Labrador City has both in-migration and out-migration under the average rate of 14.2 percent for all cities. West of Montreal, the majority have gross turnover rates of over 30 percent. The poorest, most disadvantaged areas of the country, then, appear to be less able to adapt quickly to economic change. Whether this is a cause or a result of economic deprivation is a question to be examined later. It may well be that entrenched regional cultures and low levels of education retard migration response as much as the lack of nearby opportunities.

Other notable patterns are the relatively low turnover rates of larger cities (Toronto: 17.5, Montreal: 13.2, Vancouver: 22.4) and the regular increases in turnover as one goes down the urban hierarchy. Both in-migration and out-migration contribute to the latter effect, although the pattern is more regular for out-migration. Small places h.. fewer migration opportunities within their borders, but this is only part of the explanation. We also know that the growth rates of small centres fluctuate greatly over time. British Columbia exhibits very high migration rates. Obviously, a location peripheral to the

Table 3

INTERURBAN MOBILITY RATES BY ORDER AND REGION, 1966–71

		Region[a]					
		B.C.	Prairies	Ontario	Quebec	Atlantic	Total
In-migration							
Order 4th and 5th	(11)[b]	14.0	12.1	9.5	8.0	10.9	9.9
3rd	(14)	19.3	14.3	11.0	5.9	5.9	10.5
2nd	(36)	28.4	10.5	14.7	11.5	7.2	13.2
1st	(63)	26.0	12.7	15.4	8.2	7.1	13.4
Total	(124)	20.1	12.2	11.3	8.6	8.8	12.2
		(19)[b]	(21)	(31)	(36)	(17)	(124)
Out-migration							
Order 4th and 5th		8.4 (1)	11.6 (3)	9.7 (3)	7.7 (3)	11.5 (1)	9.2
3rd		12.2 (1)	19.5 (2)	9.7 (7)	9.8 (2)	8.2 (2)	11.0
2nd		21.2 (3)	16.1 (3)	12.8 (12)	11.3 (13)	10.0 (5)	12.6
1st		22.1 (14)	17.9 (13)	15.2 (9)	12.9 (18)	11.8 (9)	15.6
Total		13.7	14.5	10.7	9.4	10.3	12.2
Gross migration (or Turnover)							
Order 4th and 5th		22.4	23.7	19.2	15.7	22.4	19.1
3rd		31.5	33.8	20.7	15.7	14.1	21.5
2nd		49.5	26.6	27.5	22.8	17.2	25.8
1st		48.1	30.6	30.6	21.1	18.9	29.0
Total		33.8	26.7	22.0	18.0	19.1	24.4

a Regions are groups of urban subsystems: Atlantic = Halifax
Quebec = Montreal, Quebec City, Ottawa
Ontario = Toronto, Hamilton, London
Prairies = Winnipeg, Calgary, Edmonton
B.C. = Vancouver

b Number of cities in group.

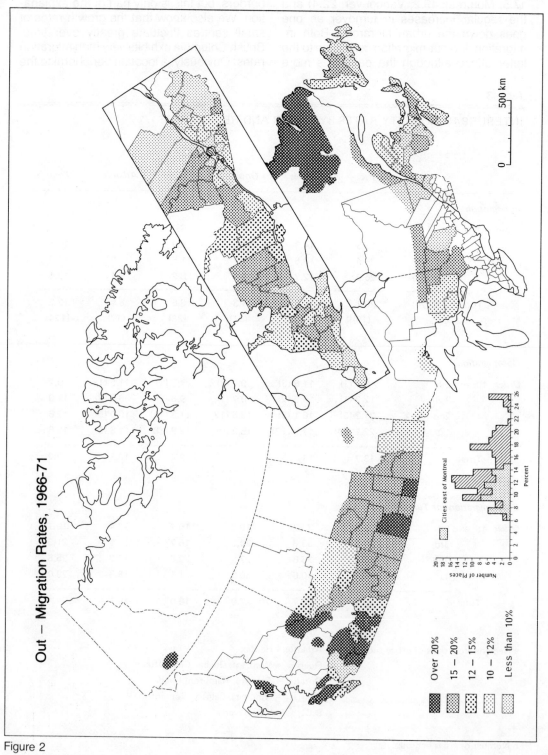

Out – Migration Rates, 1966-71

Over 20%

15 – 20%

12 – 15%

10 – 12%

Less than 10%

Cities east of Montreal

Number of Places

Percent

500 km

Figure 2

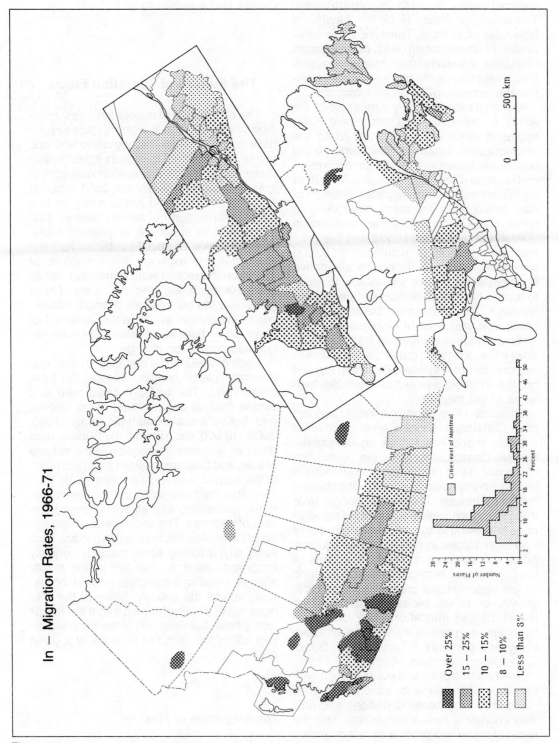

In – Migration Rates, 1966-71

Over 25%

15 – 25%

10 – 15%

8 – 10%

Less than 3%

Cities east of Montreal

500 km

Figure 3

53

rest of the urban system need not have a lower rate of turnover. In fact, isolated resource towns – such as Whitehorse, Thompson, or Prince Rupert – have very high rates of change. Turnover is the composite of in-migration and out-migration rates, but the correlation matrix suggests that in-migration is the more significant component, because of its greater variance.

Out-migration is largely a measure of the ability to cut one's ties and leave. Out-migration rates are closely linked to the demographic structure, because young people age fifteen to thirty are so much more mobile than other age groups. The other significant factor in out-migration is length of stay; newcomers are much more likely to move on. The result is the relationship pointed out by Cordey-Hayes and Gleave,[4] that out-migration is correlated (+0.553) with in-migration, overriding the alternative hypothesis that people are unlikely to move to a city where large numbers of people are leaving.[5] The differences between urban areas east and west of Montreal are quite striking. Eastern Quebec, in particular, despite weak economic growth, is very slow to respond by out-migration. Declining regions on the Prairies, in contrast, display high levels of out-migration.

Patterns of in-migration rates suggest how Canadians choose among different economic and residential environments. Eastern Canada reflects the low overall rate of mobility in that region, but the variations in levels of in-migration to western cities outline the spatial pattern of economic change. Note that the frequency distribution of in-migration rates is skewed to the left, with a number of very high values in rapidly growing towns. Again the evidence supports the ideas of Cordey-Hayes.[6] In-migrants tend to respond to job opportunities created by economic growth, or to vacancies created by high levels of out-migration. We can also hypothesize a positive response to attractive living environments – particularly British Columbia and western Alberta.[7]

The patterns of in- and out-migration indicate how Canadians in various locations respond to local social conditions and how they choose a new environment. The net result of these two processes is a migration component of population redistribution, but we will leave that discussion until later in the paper. At this point we turn to an examination

of the range of choice, or the pattern of preference. Where did the people who left Corner Brook decide to go?

The Pattern of Migration Flows

The distribution of population sizes, modified by the variation in mobility rates that we have noted, defines the beginning and end points of the actual migration flows. Within these constraints, innumerable variations of linkages can exist. A city can send migrants to all other cities or to just one city, or to a subset (subsystem) of nearby centres. Distance may be important, or perhaps transportation links, or social contacts. To trace these patterns explicitly, three matrices of migration interaction were generated: actual flows, gross flows, and net flows. Gross flows give the best picture of social interaction or integration available for Canada. Net flows tell us how these relationships cumulate into population change.

Each of these three matrices, in turn, can generate a large number of maps. We have chosen two. The first one used here is a simple map of actual flows, using arrows, and based on a logarithmic scale: 1000, 3000, 10 000 etc. (Figure 4). Such a map gives an accurate impression of the volume of flow, and thus emphasizes the importance of the largest metropolitan centres. By their very size they dominate the migration patterns, generating and attracting large numbers of migrants. The second map (Figure 5) simply identifies the main migration link from each city, treating each place as equally important. Such a map fills in the empty spaces, creating a complete network of linkages across the country. Many of the linkages will be relatively small within the matrix as a whole, but each one is the most important within the vector of contacts of a given city.

The Magnitude of Flow

Figure 4 presents the main features of the entire migration matrix, using a logarithmic scale which de-emphasizes the enormous

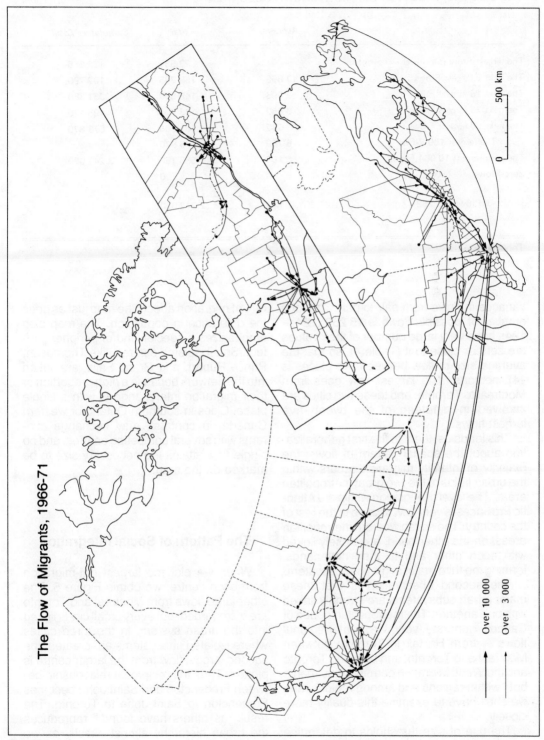

The Flow of Migrants, 1966-71

Over 10 000

Over 3 000

500 km

Figure 4

55

THE SIZE DISTRIBUTION OF FLOW MAGNITUDES, 1966—71

	Average	*Total*	*Cumulative Total*
The largest Flow (Montreal — Toronto)	—	23 555	23 555
The next 9 largest Flows	17 693	159 235	182 790
The next 15 largest Flows	9 906	148 590	331 380
The next 25 largest Flows	6 552	163 790	495 170
The next 50 largest Flows	3 632	181 600	676 870
The largest 100 Flows	6 769	676 870	—
The next largest 10 064 Flows	155	1 565 150	2 242 020
Zero Flows (5088)	0	0	—
All Flows 15 252	147	2 242 020	—

Table 4

variance in flows from pair to pair. The total number of possible flows is 15 252 (124^2 — 124), but a very large number of their values are zero or close to it (Table 4), so that the average size of flow between two places is 147 persons. The largest flow goes from Montreal to Toronto, and these two cities are involved in nineteen of the twenty-five largest flows.

This leads us, then, to the first generalization about the spatial pattern of flows: the majority of intercity migration occurs within the urban fields of the two great metropolitan areas. The level of migration within the Atlantic Provinces is very low, relative to the rest of the country. The seven western metropolitan areas, on the other hand, are closely linked with each other and to their hinterlands, forming the third major migration subsystem.

The second observation is that these major urban subsystems are strongly linked with one another. The metropolitan nodes of Canada generate very large interregional flows — from Halifax to Toronto, between Montreal and Toronto, and between Toronto and the West. Migration permits adjustments both within regions and among regions, and we shall have to examine this duality more closely.

The use of size thresholds to determine which flows are plotted emphasizes the links among the largest places rather than the small, and plays up the relationships among high turnover regions, since the variance in

rate of migration affects the map just as does the difference in population. The map also emphasizes a special kind of peripheral effect. Small isolated centres, like Thompson, Prince Rupert, and Grand Falls, are linked into the network because a high proportion of their migration interchange is with a single place. Cities in southern Ontario or western Canada, in contrast, may exchange migrants with several competing centres, and no single flow attains the necessary size to be marked on the map.

The Pattern of Social Integration

When we plot the largest out-migration from each centre, we obtain Figure 5. The other large flows from Montreal and Toronto are disregarded, but every small city is linked into the urban system. In the three cases where isolated subsystems are created, the second largest flow from the larger centre is plotted. Thus the reciprocal relationship between Fredericton and Saint John becomes Fredericton to Saint John to Toronto. The result, as others have found,[8] reproduces the urban hierarchy almost exactly as we defined it earlier on the basis of service relationships.

Migration exchanges, then, tend to follow hierarchic structures. The main migration

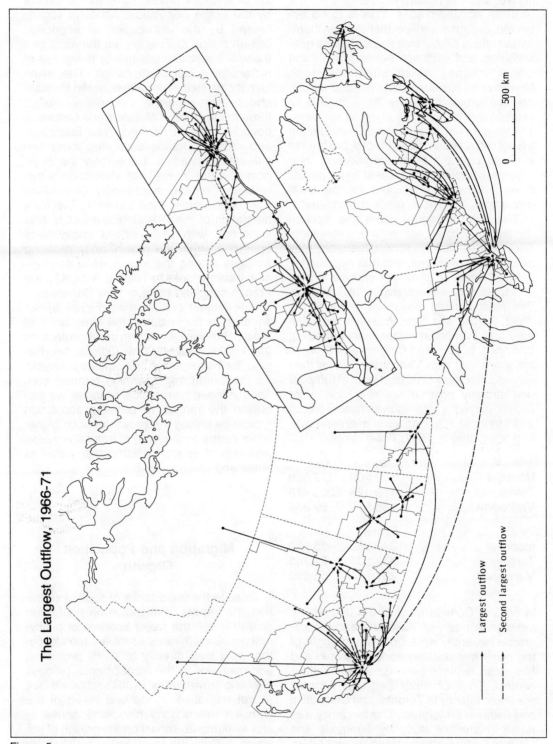

The Largest Outflow, 1966-71

Largest outflow

Second largest outflow

Figure 5

destination is the next highest service centre, and this pattern is strongest for the spatially isolated places. This is to be expected because service hierarchies themselves reflect simple size and distances relationships, and establish transportation and social linkages which affect migrants. Nonetheless it is important because we can use the urban hierarchy as a means of establishing the choices open to migrants. The latter do not, in most cases, consider the full set of Canadian urban places, but tend to evaluate only the other members of their urban subsystem. The highest order places in each subsystem account for the great proportions of flows to other subsystems.

The few deviations from the service hierarchy defined earlier are interesting. They suggest that the service hierarchy may be changing over time, and that migration — itself an agent of growth or change — may anticipate future hierarchical relationships. The most notable deviations occur in the West. Regina and Saskatoon link with Calgary and Edmonton rather than Winnipeg; Wininpeg, in turn, and Calgary and Edmonton are linked with Vancouver, rather than Toronto. Because of its high level of turnover and strongly positive net migration, Vancouver played a role equivalent to Toronto and Montreal in the national migration system from 1966 to 1971. For example:

	In	Out
Montreal	162 200	177 880
Toronto	179 125	203 410
Vancouver	138 545	82 955

	Gross	Net
Montreal	340 080	−15 680
Toronto	382 535	−24 285
Vancouver	221 500	55 590

In eastern Canada most of the differences between the service hierarchy and the migration hierarchy arise from the strength of the major metropolitan nodes, Toronto and Montreal, which override intermediate centres. Thus Charlottetown and Windsor are linked directly to Toronto, and Chicoutimi and Bathurst to Montreal. This tendency was noted in another study by Simmons and Lindsay,[9] which attempted to project future interaction patterns.

The two maps, Figures 4 and 5, are the best measures of social integration that we have for Canada, in that they measure the actual linkages among cities as carried out by the entire population. Although slightly biased by the distribution of economic growth during this period, for the most part these linkages are essentially measures of reciprocal social relationships. The maps identify the lack of contact between English- and French-speaking Canadians, except through the cities of Montreal and Ottawa. A boundary across northern New Brunswick and eastern Ontario separates these two cultural subsystems. Conversely the maps point out the intensity of interaction within western Canada, increasingly centred on Vancouver rather than the east. The fragmentation of the Atlantic Provinces is also apparent, with four distinct subsystems operating within the region, and no strong linkages among the four. Instead each regional centre looks to Toronto, except for the French-speaking parts of New Brunswick.

The largest outflow map (Figure 5) delineates, in some detail, the hinterlands of most of the third and fourth order centres, so that Halifax, Quebec City, Ottawa, and Regina, for instance, can be bounded precisely. To the extent that migration remains confined to these hierarchical linkages, we can outline the ranges of possible population growth by linking the growth of each higher order centre to the level of natural increase and rate of spatial redistribution within its hinterland.

Migration and Population Growth

Despite the importance of social integration, the impact of migration on population growth is still the major concern of policymakers. Growth rates which are too slow or too rapid lead directly to social problems, and migration is the most obvious process affecting growth rates. In this section we look at net migration — the end result of the complex interactions discussed above — and evaluate its impact on the growth of the urban system, by level and subsystem. In addition, two other sources of population growth are discussed: immigration from abroad and natural increase.

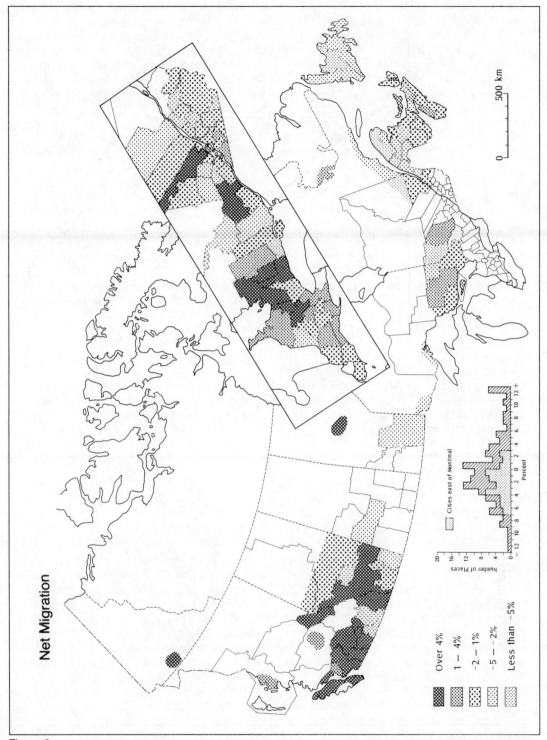

Net Migration

Over 4%
1 — 4%
-2 — 1%
-5 — -2%
Less than -5%

Cities east of Montreal

Number of Places

Percent

500 km

0

Figure 6

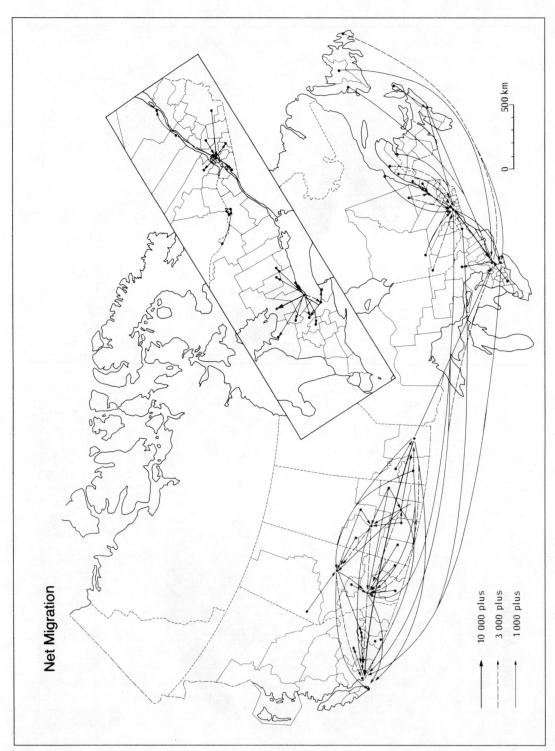

Net Migration

10 000 plus
3 000 plus
1 000 plus

0   500 km

Figure 7

Net Migration

The difference between in-migration and out-migration is defined as net migration, and its distribution, over space, is more complex than either one of them (Figure 6). At the same time the magnitudes of net migration are relatively small, with the majority of places altered by less than 5 percent of their population. Net migration rates are highest in the counties around Toronto and Montreal, which absorb the spatial decentralization of those cities, and in British Columbia, which serves the same role for the country as a whole. In addition, some northern resource towns — Labrador City, Thompson, and Yellowknife (the Northwest Territories) — are also growing through net migration. Negative rates of net migration are found throughout the Maritimes, in rural Quebec, and in most of the Prairies, with the lowest rates in declining mining towns such as Rouyn, Timmins, and Trail. These patterns support what we know about the distribution of population growth rates. Growth rate variance is highest for the smallest places, and for peripheral single industry resource towns.

Figure 7 identifies the main sources of these migration gains or losses. Each arrow represents the net exchange between two places. The map tends to emphasize the single dramatic relationship between two large centres, such as the extraordinary flow from Montreal to Toronto. The cumulative negative relationships, which a place like Sydney has with almost all other nodes, are disregarded.

Although some major interregional flows exist, particularly from the Atlantic Provinces to Montreal to Toronto, the bulk of net migration occurs within the three subsystems centred on Montreal, Toronto, and among the seven cities of western Canada. Many of these flows are hierarchical, from a city to its larger or smaller contact within an urban subsystem. Other net migration flows transfer population from one subsystem to another — from Halifax to Toronto or Toronto to Vancouver. For the map as a whole, the predominant trend is westward. Only a few localized hierarchic shifts go in the opposite direction (with the single exception of Winnipeg to Toronto).

The main differences from the pattern of total migrant flow in Figure 4 are the westward trend, and the much higher level of interregional transfers, with places like Grand Falls and Corner Brook contributing directly to Toronto's growth. Keep in mind that the sum of positive net migration moves throughout the system is only 300 345 persons or about 13.4 percent of all moves. The population growth effect of migration flows, hence the response to economic change, is only a small part of the total migration picture.

Given these patterns of gain or loss due to net migration, and the specific complementarity of gaining and losing cities, we can evaluate the role of net migration in the population change of various components of the urban system.

Population Growth Components

In Table 5 the various components of population growth are ordered in the same fashion, so that we can compare net migration to immigration and natural increase. Net migration shows the greatest variation from place to place, and the pattern is most consistent for interregional (i.e., intersubsystem) differences. No regular variation with order of centre occurs. Immigration, in contrast, varies markedly with city size — supporting the very weak positive correlation between growth and size that we observe. Interregional variations in net immigration are substantial, but smaller, and less likely to be strongly negative than internal migration rates. Net immigration is the essential element of Ontario's continued growth. Natural increase shows no consistent variation with size of centre and only modest regional variations. It contributes very little to differential rates of growth.

Before exploring the further implications of Table 5, let us take a close look at immigration and natural increase. The net immigration from abroad is computed as a residual after net internal migration and natural increases have been subtracted from population growth. As a result it contains all errors in measurement and definition. The pattern is of interest, however, because of the skewed distribution. A few large centres account for

61

COMPONENTS OF POPULATION GROWTH, BY ORDER AND REGION

		B.C.	Prairies	Ontario	Quebec	Atlantic	Total
				Region[a]			
Net migration							
Order	4th and 5th	5.5	0.5	-0.2	0.3	-0.6	0.7
	3rd	7.1	-5.2	1.3	-3.9	-2.3	-0.5
	2nd	7.2	-5.6	1.9	0.2	-2.8	0.6
	1st	3.9	-5.2	0.2	-4.7	-4.7	-2.2
Total		6.4	-3.3	0.6	-0.8	-1.5	0.0
Net immigration							
Order	4th and 5th	7.7	2.8	8.1	1.6	1.1	4.4
	3rd	1.9	-1.8	2.8	-1.0	0.0	1.2
	2nd	6.5	-1.1	1.3	-1.0	-1.4	0.1
	1st	3.8	-1.5	1.7	-1.9	-1.4	-0.4
Total		5.9	0.8	4.9	0.3	-0.6	2.3
Natural increase							
Order	4th and 5th	3.9	6.4	6.0	4.8	6.0	5.4
	3rd	3.5	7.3	5.3	5.4	7.6	5.8
	2nd	7.6	4.8	4.5	6.1	5.7	5.4
	1st	5.9	6.0	3.0	5.7	6.0	5.5
Total		4.7	6.2	5.3	5.3	6.3	5.5

a Regions are groups of urban subsystems:

Atlantic	= Halifax
Quebec	= Montreal, Quebec City, Ottawa
Ontario	= Toronto, Hamilton, London
Prairies	= Winnipeg, Calgary, Edmonton
B.C.	= Vancouver

Table 5

virtually all of the net immigrants, with most small places (except in parts of Ontario and B.C.) showing negative values. Toronto, at 9.3 percent net immigration, is far and away the highest value in the system. There appears to be a widespread international out-migration rate of about 1 percent, regardless of the in-migration level. This out-migration is higher for the border towns of New Brunswick and the Eastern Townships.

The nature of net immigration can be further expanded by looking at the statistics on actual immigration. The importance of immigration in population growth, then, lies in its high level of migration efficiency, so that a smaller number of individuals lead to growth rate differentials of the same magnitude as internal migrants.

Natural increase, on the other hand, although widely used to forecast population growth, plays a very minor role in differentiating growth rates. The data are provided by Statistics Canada (1973). The pattern of natural increase can be compared to the fertility arch natural increase pattern mapped for 1970 by Hill.[10] Newfoundland still shows high rates of fertility, but for the rest of the country most variations simply reflect the age structure of the overall population.

Net Migration and the Urban System

We have seen that net migration among cities, although relatively small in numbers, accounts for a substantial variation in growth rates from one part of the country to another. The origins and destinations of these transfers are outlined more precisely in Tables 6 and 7, indicating the flow of net migrants among different levels in the hierarchy and different regional subsystems respectively. Figure 8 outlines the main patterns of net migration movement from both internal and external sources.

Many of the major net migration flows take place within the urban subsystem — from Toronto and Montreal, for instance, to their surrounding cities. Surprisingly, the gain or loss of net migration is not consistent with city size. The smallest (order 1) and largest (order 5) places lose population, while orders 2 and 4 gain in the exchange. The two fifth order centres, Toronto and Montreal, both lose migrants to surrounding industrial centres, particularly those of order 2, due to the spread effect of metropolitan growth. Peripheral order 4 centres do tend to absorb growth from smaller places, however, as is shown in Figure 8 (see Quebec and Winnipeg). The growth of British Columbia cities is so dramatic that it often overwhelms the net changes in the rest of the urban system, reducing, for instance, the net migration loss of level 1 centres.

Immigration from abroad conforms more closely to expectations, increasing regularly with city size and accounting for over 30 percent of the growth in the two highest orders. The urban system appears to be

Table 6

NET MIGRATION AMONG ORDERS IN THE URBAN HIERARCHY (1966–1971)

Between	1	2	3	4	5
Order 1	(112 500)				
2	-17 600	(78 200)			
3	-13 300	900	(40 000)		
4	-32 300	-20 000	-29 500	(143 100)	
5	-10 700	22 800	3 205	24 600	(32 200)
Net Migrants	-73 900	21 300	-13 800	106 400	-40 000
From Abroad**	-34 100	2 800	46 900	176 000	264 100
Natural Increase	182 800	198 800	183 300	276 100	256 900
Population Growth	74 800	222 900	216 400	558 500	481 000
1966 Population	3 327 900	3 681 200	3 165 500	4 969 300	4 750 000

Diagonal elements in parentheses are gross flows within a given order. A negative flow indicates that net flow is *from* the column order *to* the row order.

**As a residual

Table 7

NET MIGRATION AMONG URBAN SUBSYSTEMS

	1	2	3	4	5	6	7	8	9	10	11
1 Halifax	(89 140)										
2 Quebec City	+250	(49 430)									
3 Montreal	-2 720	-28 315	(187 275)								
4 Ottawa	-4 320	-2 900	-11 690	(23 410)							
5 Toronto	-14 720	-4 720	-27 515	+2 755	(234 125)						
6 Hamilton	-2 045	-1 085	-4 835	-355	-10 260	(18 885)					
7 London	-1 565	-1 010	-3 320	+20	-2 615	-1 350	(37 790)				
8 Winnipeg	+110	-80	-295	+4 225	+6 925	+1 030	+465	(135 510)			
9 Calgary	-1 735	-370	-2 070	-350	-2 715	-460	-390	-22 880	(14,680)		
10 Edmonton	-1 055	-720	-1 715	-100	-1 295	-170	-160	-19 280	+2 980	(26 890)	
11 Vancouver	-6 310	-2 085	-10 455	-1 920	-16 745	-3 185	-2 475	-46 000	-11 930	-19 855	(186 135)
Sum	34 110	-41 035	-30 860	23 185	17 495	14 445	7 280	-100 540	22 030	-1 650	120 960
From Abroad**	-11 300	-30 600	35 100	16 800	257 300	29 200	24 800	-9 000	20 500	16 100	111 300
Natural Increase	116 100	96 500	203 900	41 100	237 300	51 400	467 00	104 900	40 400	70 100	89 500
Population Growth	70 700	24 900	208 100	81 100	512 100	95 000	78 800	-4 600	82 900	87 900	321 800
1966 Population	1 855 300	1 643 200	4 089 700	774 900	4 302 000	1 031 500	987 600	1 958 500	588 700	903 300	1 888 200

**As a Residual

Net Migration Among Urban Subsystems

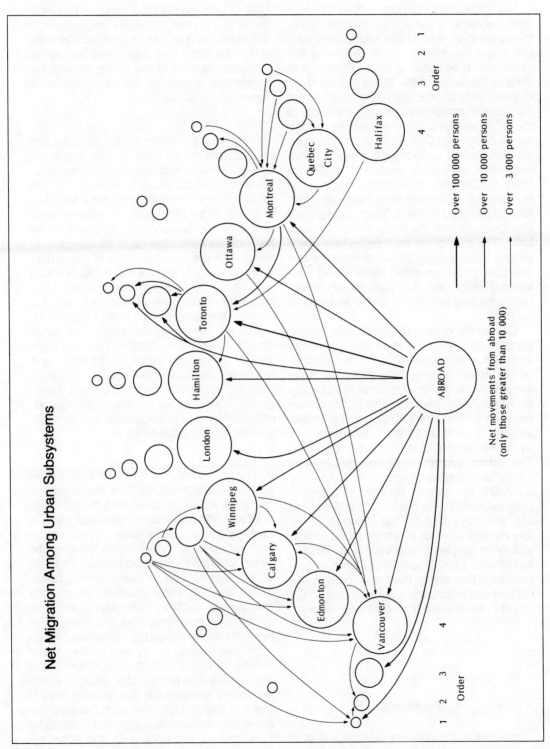

Over 100 000 persons

Over 10 000 persons

Over 3 000 persons

Net movements from abroad
(only those greater than 10 000)

Order
1 2 3 4

Order
4 3 2 1

Figure 8

65

rather sensitive to the effects of immigration regulation.

Net movements among urban subsystems are rather small and are proportional to the population size of the subsystem, with two major exceptions – the decline of the Winnipeg subsystem (i.e., Manitoba and Saskatchewan) and the growth of the Vancouver subsystem (British Columbia). About half of the population change in each of these subsystems is accounted for by the single entry of 46 000 net migrants between the two. Except for Winnipeg, the flow of net migrants is inexorably westward – from the East to Ontario, from Ontario and the Prairies to B.C. Distinct subgroupings occur as well, linking Halifax to Toronto, Montreal and Quebec, and the three Prairie subsystems. The bulk of the exchanges take place among the highest order centres, but again Winnipeg is an exception. Regina and Saskatoon, which are assigned to Winnipeg, obviously look to Calgary and Edmonton as well.

Overall the intercity migration data is more important for what it tells us about contacts and relationships than for its impact on population growth. Immigration tends to have almost as large and diverse effects on growth *per se*. The weakness in correlation between the two forms of migration growth is striking (r = 0.500) and contradicts any simplistic models of labour force response. The urban system dampens the internal migration response to economic change by restricting the options available to a native-born migrant. The opportunities in the home city are compared to half-a-dozen or so alternatives – most of which share similar economic problems or advantages. The immobility of French-speaking Canadians is evident: Montreal is their only outlet. The fluctuations in labour supply, then, within the urban system, have their own internal logic.

Summary

Inherent in the concept of the urban system is a concern with the interdependence of cities in the sense of regular ongoing linkages, as well as the diffusion of growth and change from one city to the next. The study of migration flows illuminates relationships at both of these temporal scales, and provides some of the strongest evidence, to date, on the functioning and change of the Canadian urban system. This paper has simply described some of these patterns and has suggested some of the implications as summarized below.

1. Patterns of mobility rates are complex and do not lend themselves to any simple interpretations. In-migration and out-migration rates are correlated at a level of 0.6; net migration and immigration from abroad at 0.5. East of Montreal, a slow-growth economy and entrenched regional cultures produce uniformily low rates of migration. In the rest of the country, rates of movement vary more widely and are more clearly linked with local economic conditions. Out-migration rates are highest for smaller places, and places with high levels of in-migration. In-migration rates are more closely linked to job creation. They are highest in the urban fields of Toronto and Montreal, and in the developing resource-based cities of the North and West. Immigrants from abroad are most strongly attracted to the largest centres and to rapid growing resouce-based cities.

2. The pattern of migrant flows displays two patterns simultaneously. The migration matrix as a whole is dominated by the large number of flows between Montreal and Toronto and their nearby cities. A third important complex of flows exists among the seven western metropolitan areas. The bulk of all migration takes place within, rather than between, these clusters. When only the largest outflow from each centre is examined, however, the importance of hierarchical relationships emerges. In almost every case a city sends the largest number of migrants to the place above it in the service hierarchy. The map of these relationships suggests the growing importance of Vancouver in western Canada; the increasing polarization of central Canada by Toronto and Montreal, with the latter restricted to French-speaking cities; and the fragmentation of the Atlantic Provinces subsystem.

3. The patterns of migration response to social and economic change and of destination options for migrants, affect the patterns of population growth during the study period, but to a lesser extent than one might have anticipated. Net migration is not clearly related to order in the hierarchy, and the only major movement among urban systems is from Winnipeg to Vancouver. International immigration is equally important to the growth of most cities and subsystems.

Each of these groups of generalizations generates further research questions. Why is the level of migration so low in the depressed areas of eastern Canada? Are the patterns of migration flow sufficiently stable, over time, to incorporate into demographic projections? Can the discussion of net migration be tackled directly, using economic models, or is it simply the fall-out of essentially independent demographic processes of out-migration and in-migration? Why is the immigration pattern different from the effects of internal migration?

REFERENCES

[1] Ray, D. Michael, *et al.* (eds.), *Canadian Urban Trends* Vol 1. (Toronto: Copp Clark, 1976).

[2] Simmons, J. W., "The Canadian Urban System: A Conceptual Framework", University of Toronto Centre for Urban and Community Studies, Research Paper No. 62, Toronto, 1974; and J. W. Simmons (forthcoming), "Short Term Income Growth in the Canadian Urban System", *The Canadian Geographer.*

[3] Stone, Leroy O., *Geographical Mobility in Canada: A Study of its Frequency and Socio-Economic Composition.* 1971 Census Monograph. Ottawa: Statistics Canada.

[4] Cordey-Hayes, M. and D. Gleave, "Migration Movements and the Differential Growth of City & Regions in England and Wales", *Papers, Regional Science Association,* Vol. XXXIII, 1974, pp. 99-123.

[5] Renshaw, Vernon, "The Relationship of Gross Migration to Net Migration: A Short Run, Long Run Distinction", *Regional Science Perspectives,* Vol. V, 1975, pp. 109-124.

[6] Cordey-Hayes, Martin, "Migration and the Dynamics of Multi-regional Population Systems," *Environment and Planning,* Vol. VII, (November) 1975, pp. 793-814.

[7] Svart, Larry M., "Environmental Preference Migration: A Review", *Geographical Review,* Vol. LXVI (July), 1976, pp. 314-330.

[8] Nystnen, J. and M. Dacey, "A Graph Theory Interpretation of Nodal Regions", *Papers and Proceedings of the Regional Science Association, Vol. VII, 1961, pp. 29-42.*

[9] Simmons, J. W., and I. Lindsay, "Intercity Linkage Patterns" in L. S. Bourne *et al.* (eds.) *Urban Futures for Central Canada* (Toronto: University of Toronto Press, 1974), pp. 140-157.

[10] Hill, Frederick I., "Age Structure and the Family Cycle", in Ray *et al.* (eds.) *Canadian Urban Trends* (Toronto: Copp Clark, 1976), pp. 203-228.

The Windsor-Quebec City Urban Axis

MAURICE YEATES

This paper focusses on the most significant population concentration in Canada – the Windsor-Quebec City Axis.[1] Specifically, the focus is on:
- identifying the limits of urban development within the axis in 1961 and 1971;
- outlining the importance of the urban axis to Canada;
- projecting the spatial extent of urban development in the axis to the year 2001;
- the implications of these growth trends for Montreal and Toronto.

Urban Development in the Axis Area

The outer limits of the Windsor-Quebec City axis are indicated in Figure 1. This large area (174 825 km²) contains a vast area of agricultural and waste land, as well as the majority of Canadian cities.

Russwurm[2] has provided a fairly good argument for using particular critical limits to distinguish between urban, semi-urban, and rural census subdivisions. In his empirical survey of population densities in the "corridor" between Stratford and Toronto, he determined that those census subdivisions in which the population density was less than 20 persons per square kilometre had a population that was usually less than half nonfarm, while those with a population density of greater than 20 persons per square kilometre usually had a population of greater than half nonfarm. Thus, a density of 20 persons per square kilometre can be used to distinguish between rural and semi-urban (including urban) census subdivisions.

The argument supporting the density distinguishing between semi-urban and urban census subdivisions (46 persons per square kilometre) is rather more complex. Briefly, it is based on an estimation of the density of nonfarm residences that are usually found in

an area when preliminary urban services seem first required. This is estimated to be about one residential parcel per 10 ha, to which is added the average rural farm density in southern Ontario of 8 persons per square kilometre.

These critical limits (20 and 46 persons per square kilometre) are interesting because they are based on some empirical investigations of densities, and they are similar to the densities used in the ekistic literature to define the limits of megalopolitan areas.

The Axis in 1961 and 1971

The maps of distribution of population density for 1961 and 1971 (Figure 1) indicate an increase in both urban and semi-urban areas between the two periods. The semi-urban and urban areas are quite continuous along the whole axis in 1971, though the middle section in eastern Ontario is mostly rural. The area embraced by the census subdivisions defined as urban in 1971 is quite extensive (28 676 km²) and has an average population density of 345 persons per square kilometre. To place these figures in perspective, we should note that the area defined as urban is about the size of Belgium, which has an average population density of about 317 persons per square kilometre.

The increase in extent of urbanization between 1961 and 1971 has been quite dramatic. The area defined as urban has increased by almost 19 percent, while the population contained within it has increased by more than 28 percent. The total area defined as semi-urban and the population in this area have increased slightly. There is, therefore, a continuous strip of urban and semi-urban development between Windsor and Quebec City. The area can be divided into thirds, in which urbanization is fairly

● MAURICE YEATES is Professor and Head, Department of Georgraphy, Queen's University. This paper is reprinted from *Ekistics*, Vol. 41., No. 243, 1976, pp. 120-122, published by the Athens Centre of Ekistics, Athens, Greece, with permission.

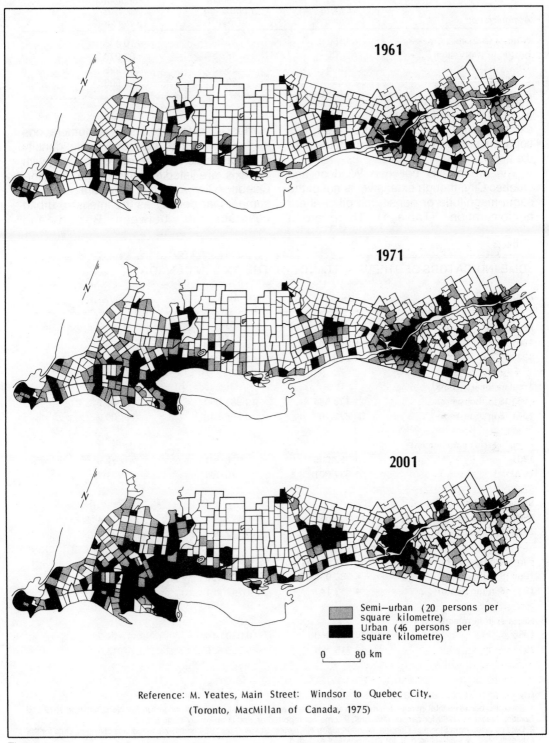

1961

1971

2001

Semi–urban (20 persons per
square kilometre)

Urban (46 persons per
square kilometre)

0   80 km

Reference: M. Yeates, Main Street: Windsor to Quebec City.
(Toronto, MacMillan of Canada, 1975)

Figure 1

69

THE AXIS COMPARED WITH OTHER ELONGATED URBAN AGGLOMERATIONS

Agglomeration	Length (kilometres)	Estimated population (millions)	Population per linear kilometre
Windsor to Quebec City	1150	12	10 435
Boston to Washington	725	45	62 070
Tokyo to Osaka	595	35	58 825
London to Manchester	320	20	62 500

Table 1

extensive in the western and eastern sections, but somewhat fragmentary in the middle zone.

The urban axis between Windsor and Quebec City, though extensive, is not of the same magnitude or density as other linear agglomerations (Table 1). There are a number of extensive linear agglomerations in the world. The size and crude density characteristics of three of these, along with the axis, are listed in Table 1. The Windsor-Quebec City axis has nowhere near the same linear density as the intensive urban developments between Boston and

Table 2

SOME INDICATORS OF THE IMPORTANCE OF THE AXIS IN CANADA

Indicators	Canada	Axis	Percent
Population			
1961	18 328 247	9 730 950	53.1
1971	21 568 311	11 917 655	55.3
Total national income[1]			
1961 ($ in thousands)	21 479 544	12 987 050	60.5
1971 ($ in thousands)	50 825 409	30 858 380	60.7
Manufacturing employment[2]			
1961	1 352 605	986 391	72.9
1970	1 637 001	1 173 094	71.7
Valued added[2]			
1961 ($ in thousands)	10 931 561	8 128 813	73.4
1971 ($ in thousands)	21 417 784	16 132 072	75.3
Farm cash receipts[3]			
1966 ($ in thousands)	4 280 033	1 533 490	35.8
1971 ($ in thousands)	4 513 147	1 747 281	38.7
Hectares in farmland[3]			
1966	70 520 960	10 030 310	14.2
1971	68 715 789	8 956 535	13.0

Notes:
1. Estimates based on total income derived from an analysis of all taxable and nontaxable returns in *Taxation Statistics, 1972* and *Taxation Statistics, 1963* for census divisions. Income not reported or filed is obviously excluded.
2. Data obtained from Statistics Canada, *Manufacturing Industries: Series G* for 1964 and 1973, for census divisions. Data for the axis estimated by subtracting data for municipalities excluded.
3. Data and estimates for the axis obtained from Statistics Canada, *Number and Area of Census Farms, 1971;* Advance Bulletin and Farm Cash Receipts, 1971.

Washington, Tokyo and Osaka, and London and Manchester. While these three major agglomerations have about 62 150 persons per linear kilometre, the axis area has a linear density of less than one-fifth of this. Thus, the special problems and remedies touted for these major urban agglomerations should not be transferred automatically to the Windsor-Quebec axis area.

But, even though the axis is not of major proportions on an international scale, it is extremely important to Canada (Table 2). The area defined as urban in 1971 contained a far greater proportion of Canada's population (46 percent) than that of any other major linear agglomeration for its respective country. Furthermore, the axis area contains over 70 percent of the manufacturing employment in the country, and the average income in the area is over 10 percent greater than in the rest of the nation.

The Axis in the Year 2001

The methodology used in this study involved forecasting in large regions in Canada, then allocating the forecasts for the axis to subregions which are subsequently allocated to the census subdivisions (and aggregations in Quebec) using the method of cascading, averaging, and differencing as described by Curry and Bannister.[3] The advantage of this procedure is that it preserves the relative stability of large region forecasts, while still permitting variations in growth between census subdivisions to be expressed.

The product of this hierarchical method of forecasting yields estimates of the population size of Canada for the year 2001, for four major regions in the country (one of which is the axis area), and the probable distribution of urban densities in the axis. Figure 1 indicates the distribution of urban and semi-urban areas in the year 2001, and demonstrates a considerable increase in both the area defined as urban and the number of people residing in the urbanized area between Windsor and Quebec City. At the turn of the century, 16.6 million people, or 50 percent of the forecast population of the country, will be residing in those census subdivisions classified as urban in Figure 1. The area occupied by these census subdivisions has increased from 28 676 km^2 in 1971 to 41 862 km^2 by the year 2001, with a concomitant increase in average density to almost 397 persons per square kilometre.

Concentration on Montreal and Toronto

One aspect of the 2001 map in Figure 1 which is quite noticeable is the apparent increase of population in the areas immediately adjacent to Montreal and Toronto. The forecasts suggest that whereas 30 percent of the population of Canada resided within 64 km of either Toronto or Montreal in 1971, by the year 2001 about 34 percent will be concentrated around these two metropolises (Table 3). The trend data, supported by a close examination of some economic indicators, suggest that the increase in population within one hour's commute of Toronto is likely to be much greater than that for Montreal. In fact, it is forecast that by 2001 one out of every five residents of Canada will be located in the Toronto area.

Conclusion

These forecasts are based on national, regional, and local (township) trends in population growth. The assumption is that we will

Table 3

THE EXPECTED CONCENTRATION OF POPULATION (IN THOUSANDS) AROUND TORONTO, MONTREAL, OTTAWA, AND QUEBEC CITY IN 2001, COMPARED WITH 1971

City	1971		2001	
	Population	Percent	Population	Percent
Montreal (64 km)	3173	14.7	4558	13.8
Toronto (64 km)	3397	15.8	6645	20.2
Ottawa (20 km)	573	2.7	1018	3.1
Quebec City (24 km)	484	2.2	713	2.2

Note: The percentages relate to the total and forecast population of Canada.

continue allowing growth to occur in the same way as it has in the immediate past. Growth is directly related to employment opportunities, and the only way that growth can be affected is by influencing the location of jobs. The map for 2001 presents a scenario of what might happen if we permit the concentration of employment opportunities in the urbanized portion of the axis to continue. The desirability of the continuance of this trend should be "item one" for debate on our national urban agenda.

REFERENCES

[1] M. Yeates, *Main Street: Windsor to Quebec City* (Toronto: MacMillan of Canada, 1975).

[2] L. Russwurm, *The Development of an Urban Corridor System: Toronto to Stratford Area, 1941-1966* (Toronto: Government of Ontario, 1970).

[3] L. Curry and G. Bannister, "Forecasting township populations of Ontario, from time-space covariances", in L. Bourne *et al.*, *Urban Futures for Central Canada* (Toronto: University of Toronto Press, 1974), pp. 34-59.

Urban Fringe Problems in Canada

LORNE H. RUSSWURM

Our living experience clearly reveals that Canadian and American cities have sizable urban fringe zones surrounding them. Why do these belts of land, with their transitory, somewhat unstable, mixed, and often conflicting activities, occur around our cities? Five major reasons stand out:

(a) the population growth of our cities;
(b) the high level of transportation and communication facilities available;
(c) the almost unrestricted competition for land that is still so readily accepted in our society[1];
(d) the general acceptance of four individual rights – life style preferences, space and territory, property, accessibility and services;
(e) the lack of unitary and consistent planning controls over both the built-up city and its urban fringe.[2]

A growing population provides the initial thrust; our technology makes possible accessibility to the built-up city from outside locations; competition for land accelerates the thrust; our insistence on individual rights ensures the thrust; and a lack of unitary planning means that limited controls exist to control the processes creating continuing change. It is a system dominated mainly by positive feedback mechanisms which work to change the city's surroundings.

In this paper eight classes of land problems are identified. Four relate directly to land and four to the activities of people and only indirectly to the land. The direct land problems are land use activity conflicts, land conversion difficulties, impact on the environment, and impact on agricultural land. The indirect land problems are the equity of services versus taxation, governing and planning difficulties, impact on surrounding settlements, and social issues. Most frequently identified are land use conflicts, land conversion difficulties, governing and planning difficulties and the equity of services versus taxation. Clearly, these eight problem areas overlap each other. Separate discussion, however, can help clarify the particular issues involved in each problem area.

Conflicting Land Use Activities

By their very nature, some conflicts over land use activities are inevitable in the urban

● LORNE RUSSWURM is in the Department of Geography, University of Waterloo. This reading, which forms Ch. 4 and a part of Ch. 5 in L. H. Russwurm, *The Surroundings of our Cities* (Ottawa: Community Planning Press, 1977), is reprinted with permission.

72

fringe. As a regional environment, the surroundings of our cities serve three broad land functions, namely, as a natural environment, as a production environment, and as a lived-in environment.[3] And given many different land owners with different purposes for owning land, it is easy to conceptualize a variety of potential land use activity conflicts. The gravel pit operator, the farmer, the non-farm household, the industry unwanted in the built-up city, the highway-oriented business, the space-using industry; the wildlife habitat, the outdoor recreation facility, all exist and often side by side. Little wonder that land use activities can seem scattered and haphazard in appearance.

Three broad areas of conflict over land use activities are identifiable. One is conflict between the actions and desires of individuals and municipalities. Such conflicts lead to what Friedmann[4] calls the collective phenomenon. For instance, scattered country residences (individual actions) and the high costs of providing needed services to them (collective actions) is a collective phenomenon. A second conflict arises between present uses and potential future uses. A scenic area containing several country residences may later be needed for a regional park or for its sand and gravel resources. Or by the year 2050, our descendants may bemoan our lack of wisdom in building on class one agricultural land which comprises only 0.5 percent of Canada's land area or 5 percent of the land usable for improved farmland.[5] The third conflict is between economics, with its highest bidder and associated highest use notions, and aesthetic and economic concepts. Lessening of any of these conflicts demands regional government units having a regional planning authority based on urban regions as the administrative unit. Such regional units imply provincial recognition in the complexity of the land use problems and their viewing from a wider perspective. An excellent case study illustrating these conflicts is that by Halkett[6] on the Saanich Peninsula in the urban fringe of Victoria.

What are some of the questions that interested citizens should be examining in their own area? One is the question of who owns what land. Another is how land use activities are changing and whether or not these changes seem to relate sensibly to the longer run potential of the land. Do we really need that dam or reservoir on the river or does it smack of being an engineering project? Could that highway be better located so that forest, swamp, and farmland are less affected? Despite a considerable human and financial investment in planning since the 1960's, it is still often difficult to find out for your own area the trends in land use over a ten-year period. Can you say that the pattern of changes is for the better, for the worse, or is it inevitable progress?

We also need to ask what are the unique environmental potentials and how do they jibe with the locational qualities of the land space resource of a particular city. Must the relocated or expanded airport, or the proposed new town, be located on good agricultural land? Can the gravel be removed without damage to the stream flowing nearby? Why should the farmer, who was there first, be blamed by the country resident for the smells his manure creates or the dust he stirs up which settles on the wash hanging in the sunshine?

Moreover, the concept of sequential land use activities needs to be more seriously considered. Can the physical resource site of sand and gravel deposits be exploited, be restored to the biological use of agriculture or the amenity use of a small fishing lake, or become the location for a new subdivision or a waste disposal site?

The Canada Land Inventory, Lands Directorate, Environment Canada, provides land capability maps for agriculture, recreation, forestry, and wildlife at a national scale. But what is needed for the surroundings of individual cities are more detailed local capabilities. A puny lake like Chestermere in the Calgary urban fringe is more important in that setting than one of Dartmouth's many lakes. Water supplies may be a critical problem in one urban fringe (Edmonton, Thunder Bay); buildable residential land in another (Halifax, Sudbury); recreational land in a third (Swift Current, Toronto); the need to protect specific agricultural land in a fourth (St. Catharines, Vancouver). For a given activity, what is rated as low-potential in a national scale inventory in one urban fringe may indeed be of high potential in another urban fringe, given its lesser resources potential for a given activity.

Two specific land space resource use activities, country residences and sand and gravel extraction, need to be emphasized.

Often, especially in southern Ontario, they conflict directly with each other and to a lesser degree with other urban fringe land use activities. One's dream home in the country does not include fantasies of gravel trucks rumbling by! For economic reasons, the extractive industries of sand and gravel occur dominantly in the surroundings of our cities.[7] The costs of moving aggregate material about 120 km exceed the current value of the material itself unless it can be transported in bulk by boat or train.

As a collective benefit, country residences, especially estate types, can provide an amenity resource for others as well as for the owner. In the surroundings of our cities, large and small, the country estate or gentleman's farm of the wealthy is often part of the Sunday afternoon drive. But too many country residences soon threaten to consume significant amounts of land of agricultural, recreational, and scenic potential needed by the community at large. This is an example of the collective phenomenon of Friedmann or the economic problem of externalities which is so prevalent in urban fringe problems.

Another collective phenomenon, part of the spatial form and pattern of all our cities, is the peripheral shopping centre, usually located in what was part of the inner urban fringe surrounding our cities. In the 1970's such developments are primarily problems for smaller cities of 50 000 or fewer people. They can compete and thus conflict directly with and contribute to the deterioration of the downtown businesses of smaller cities. They may reorient the development direction of a small city. They may be located outside the city boundaries thereby resulting in local municipal conflict. Recent examples include Goderich and Timmins, Ontario. And since they demand flat land, they are usually located on good agricultural land.

All seven other problem categories interface directly with land space resource uses and their associated conflicts. Services and the taxation needed to provide them relate directly to particular land use activities and their spatial patterns. Conversion of land to urban uses, impact on agricultural land, on environment, and on surrounding settlements are all part of land space resource use activities. All the problem areas interrelate as critical factors of total environmental quality of our cities and their surroundings.

Great need exists to evaluate the various potentials for different land space resource use activities and the various types of possible conflict between present and potential activities at national, provincial, and local scales. The underlying processes affecting land in the urban fringes of our cities appear to operate in much the same way across Canada and for all sizes of cities. But unique local qualities, such as those of the Niagara or Okanagan Fruit Belts, the Niagara Escarpment in Ontario, or the river valleys of many cities, must be identified for individual urban fringes. These can have national significance like the Niagara fruit lands, regional significance like the agricultural land of the Lower Fraser Valley, or local significance like the lakes north of Regina.

Only lately, with the development of regional planning areas which reflect provincial concern, have real possibilities existed for better utilizing the land space resource potential of the urban fringe. Only lately has the will to apply the legal tools available become a significant force as indicated by legislation in various provinces. A 1975 discussion paper by the Canadian Council of Resource and Environment Ministers summarizes land issues facing Canadians and outlines the various provincial approaches. The issues identified coincide to a considerable degree with the problems outlined in this paper. They are seen as reaching their zenith in the urban fringe for land use planners, politicians, and citizens. "The urban fringe of today lies within the city of tomorrow, and the ultimate selection from the many alternatives must be made by informed citizens."[8]

Land Conversion Difficulties

Land speculation, rising land values, public land banking, and increasing land fragmentation are all part and parcel of the problems involved in land conversion from other uses to urban uses. And land conversion difficulties relate closely to land use conflicts through the mechanisms of competition for land. In this section the emphasis will be primarily on factors resulting largely from land ownership attitudes through which land is seen as a commodity rather than a resource.

Since 1950 the nature of land conversion from rural to urban uses has altered fundamentally.[9] It still occurs primarily in the inner urban fringe, and will do so for the rest of this century. The actors and the structure of the process described by Martin[10] have remained fundamentally similar since 1950. But the scale, the number of actors involved, and their decision-making power have changed drastically. In the 1950's, every city had many small builder-developers and only minimal services were provided by municipalities at the lot or community level. By the mid-1960's, the land conversion process was dominated by limited numbers of large developers with governments and planning departments providing services and enforcing many regulations.

What is the magnitude of the actual demand for land? Given Canada's current population increment of about 250 000 people per year, at present densities about 12 000 ha per year have to be converted to urban uses. Most of this conversion uses land space of the inner urban fringe. Conversion for urban-associated uses like recreational areas, country residences, sand-gravel extraction, and transportation and other infrastructure needs, demands approximately another 12 000 ha.

Rising Land Values

The past few years were marked by land values that rose more rapidly than housing costs.[11] This is partly a result of an increase in average lot size. Overall, however, the dominant factor is an increased demand combined with a restricted supply which partly reflects a lag effect resulting from the development of better controls to ensure better quality developments and better environmental protection. Recognizing that the demand for land is a derived demand for housing, one economist states, "high house prices are a function of rising incomes, general inflation, tax inducements to home purchasers, and long term government intervention to improve housing and environmental standards".[12] Under government intervention, Baxter includes the considerable increase in red tape and thus aggravating delays for developers. Important as all these factors are, two other factors must be added: land speculation and control

in some cities of much of the developable land by a few developers. In the larger cities of Ontario, Manitoba, and Alberta, a few large firms produce about three-quarters of the detached housing.[13]

In 1973 the Hon. Ron Basford, then Minister of State for Urban Affairs, noted three objectives of a possible national urban land policy as: (a) to provide an adequate supply of serviced land to accommodate urban populations; (b) equity in the distribution of land between various needs; and (c) ensuring that land markets operate effectively in the public interest. All three of these objectives relate to land value, and their effect is to a considerable degree expressed in the land space of the surroundings of our cities.

Residential Land Banking

Increasingly, advance public land acquisition for residential purposes is suggested as a means of dealing with rising land prices. Other public objectives include recouping land value increments created by the growth of the collective, the community at large, and improved long range planning where planners guide rather than react. Price, profit, and planning are the triune god of public land banking.

Major land development firms recognized this triune god earlier than did federal, provincial, and local governments. As good businessmen, they recognized during the 1950's that good planning for them meant having land available for subdivision under their control. Rapidly rising land prices subsequently spelled good profits. Thus major developers have assembled private residential land banks as income property and as a hedge against the uncertainty of when to develop and about where the development would go, given increasing public intervention. Some become public companies. Increasing consolidation has marked the industry with large vertically integrated companies like Trizec, Cadillac-Fairview, Ladco, BACM, Block, Campeau, and others becoming national scale organizations. In 1974, by way of illustration, forty-seven firms averaging six subsidiary companies held 48 600 ha in the surroundings of twenty-one cities.[14] Sixteen of these forty-seven firms propose to house one million people by

1990. At the same time, Spurr notes that the total public land inventory held by all levels of government stood at only 24 300 ha.

Public land acquisition for residential purposes has been successful in retarding price increases in Saskatoon, Regina, and Red Deer.[15] It seems evident from the few successful experiences that for public residential land banking to be effective, control over a ten-year supply of land, including at least one-quarter of the land likely to be developed, is essential. Moreover, a close relationship to official plans is also necessary so that the public land will in fact be *leading* the development thrust.

Land Fragmentation

Fragmentation of land in the surroundings of our cities is of particular concern for farming operations and for future expansion of the built-up urban area into the surrounding inner urban fringe.

Five specific disadvantages of land fragmentation are identifiable. It helps drive up land values generally since smaller parcels are almost always worth more per hectare or square metre. It increases the number of owners who have to be dealt with in daily municipal administration. It makes future large scale land assemblies difficult. Recently, in the large Calgary annexation proposal defeated by public referendum in late 1974, an area of many small holdings in the Bow River Valley was excluded from the annexation proposal for just this reason.

It has considerable effect on adjacent farmland values for two reasons. One is the higher prices which nonfarm residents willingly pay. The other reason is the uncertainty effect on farmers and their operations when a scattered pattern of small holdings develops.[16] The increase in land fragmentation in parts of southern Ontario is documented by Russwurm and Found and Morley.[17]

And fifth, in the context of country residential development, it places in passive use considerable farmland areas which may become difficult to reconvert to other more active uses. Remember that minimum lot sizes of 4 to 10 ha are common requirements for country residences in Canada

from Ontario westwards. Hoffman[18] states this disadvantage as follows:

> Not only is land used up to provide building space for the rural nonfarm dweller but it is also used to provide services. Increased numbers of people in the country necessitate wider and more numerous roads and highways, more overhead telephone poles and hydro lines are needed and more land is used for recreation centres. Worst of all, the subdivision of the land into small lots makes rural planning impossible.

Impact on Agricultural Land and Activities*

Despite the vast literature that exists about this problem area and its long range importance, many people are still uneasy about our lack of knowledge and understanding about urbanization and its interactions with agriculture. Two studies on these interactions stand out in the Canadian context. At a national scale, C. R. Bryant[19] has analyzed these interactions for the urban fringe-urban shadow zones surrounding Canada's twenty-two largest cities. At the provincial scale, we have the study, *Planning for Agriculture in Southern Ontario.*[20]

Five general urbanization-agriculture processes affect farmers operating in the urban fringe.[21] They are: (1) the land conversion mechanisms and their ramifications; (2) the increasing mobility and urban employment possibilities for farmers, and consequently, the growth of part-time farming; (3) the use of the countryside for a wide range of recreational purposes; (4) the move of an increasing number of urbanites to homes in the countryside; and (5) the use of other countryside resources like sand-gravel and water for urban purposes.

Land Conversion Ramifications

In Canada, except for considerable areas in the Prairie Provinces, much of our agriculture takes place in the urban fringe-urban shadow areas surrounding our cities. Hence land conversion processes are closely linked with agricultural land. Farmland suitability depends on two key factors: suitable soils and suitable climate. General knowledge

---

*Ed. Note. For further discussion see, in this volume, papers by Williams, Pocock and Russwurm, Rodd, and Pearson.

tells us that most of our cities are located on the best land in their region. Even in northern Ontario, Sudbury, Sault Ste. Marie, and Thunder Bay, in the heart of the Canadian Shield, are located in pockets of farmland. In a recent quantitative study the climatic heat resource (suitable climate) and census occupied farmland (suitable soils) were combined and then compared with the location of urban population.[22] The results showed that 50 percent of Canada's urban population lives in areas having the best 5 percent of our farmland.

Moreover, three-quarters of the urban population growth took place on this best 5 percent of our farmland of 1961-1971. This trend will continue into the future unless we adopt planning policies to direct the urban growth onto land other than the best farmland. Regional plans being developed across Canada from St. John's to Victoria are trying to preserve the best agricultural land. Whether the implementation will be successful remains to be seen. To date only British Columbia has firmly established Agricultural Land Reserves through firm provincial legislation.*

Given the clearly finite limits of the land resource, it seems common sense that the irreversible conversion of good farmland (Classes 1-3) should be severely minimized. G. Runka, Chairman of the B.C. Land Commission, expresses it this way:[23]

> Only a very small percentage of our vast land area is capable of food production. Unfortunately, this is also the area where other resource capabilities are high and where pressures for urban expansion are the greatest. But what animal destroys the food habitat next to its shelter – and survives?

Given our legal framework concerning land, what methods do we have to keep good farmland from being converted irreversibly or sterilized for agriculture? Hoffman[24] identifies five approaches. We can institute a land freeze by establishing Agricultural Land Reserves as British Columbia has done. We can establish provincial farmland banks as Saskatchewan has done. The land is then leased to farmers at a constant percentage of the market value for a ten-year period with the option of purchase after five years. We can apply various zoning and development controls as proposed in the general plans in Ontario and St. John's, Newfoundland. We can preserve rural conditions via planning controls as proposed in regional plans in Alberta and Ontario. This approach involves some combination of identifying places where new growth will be concentrated, establishing criteria for and numbers of farmland parcels that can be severed, and applying regulations concerning the environment, waste, water, and general servicing needs. And we could insist that land be maintained at some minimum level as proposed but not implemented in Prince Edward Island. The success of many of these approaches for the farm community, however, depends upon national and provincial agricultural policies which will allow farming to remain an economically viable enterprise.

Part-time Farming

Whether or not the increase in part-time farming is a collective phenomenon with attached negative external costs is uncertain. That it occurs with increasing frequency in the surroundings of our cities across Canada is certain. Mage,[25] in a detailed study of the Regional Municipality of Waterloo surrounding the Kitchener Census Metropolitan Area, noted that a third category of permanent part-time farmer now exists, who prefers to combine farming with a full-time other job, usually urban. The two more traditional categories of part-time farming also continue to occur: farmers gradually moving out of farming and younger farmers raising capital to move into full-time farming. Both agricultural and regional planning bodies are having difficulties in dealing with the part-time farm phenomenon. Does it lower productivity? Does it encourage land speculation? Is it a beneficial part of the urban occupation of the countryside? What does it mean as far as a land ethic is concerned? Does it reflect inadequate agricultural policies? Our knowledge at present is far from sufficient to answer these questions.

*Ed. Note. For discussion of this, see the paper by G. G. Pearson in this volume.

The Countryside and Recreation

Already in his early paper, Wehrwein[26] noted recreation aspects as a critical factor in urban fringe development. Today two broad problem areas exist. They are the loss of needed recreational land, and conflicts between recreationists and farmers.

It appears axiomatic that the urban fringe land space, in some places more from accessibility than real potential, will have to meet the increasing demands for the play function of land. As noted earlier, a local resource, though not of high quality on a national scale, will be peculiarly valuable if no other resources exist. Rajotte,[27] based on her study of the recreational hinterland of Quebec City, has conceptualized the need to view recreational activities in the surroundings of our cities in national, provincial, and local contexts. Provincial governments, by increasingly establishing parks near cities, have recognized this demand and its relation to accessibility. Recent examples include parks in the urban fringes of Calgary, Winnipeg, London, Toronto, Drummondville, and St. John's.

Conflicts between recreationists and farmers occur in at least three ways. One type involves urbanites carelessly and without permission using farmers' property. The visual result is the increasing number of "no trespassing" signs. Another type relates to service demands made by resort settlements when a farm-dominated rural council may be happy to collect taxes but not to provide services, or may find that the cost of services exceeds the tax return. A third type arises from increasing demands placed on country roads and other facilities, again with costs falling inequitably on the farm owner.

On balance it is not clear whether outdoor and leisure pursuits benefit the rural community. But it is clear that increasing demands are being made on the urban fringe countryside. Some authors urge farmers to take advantage of latent recreational opportunities presented by the increased influx of urbanites.[28] A current development is the spread of national chain campgrounds like KOA, Safari, Jellystone Park, etc., which usually locate on the inner urban fringe in country settings that commonly involve travel over country roads. Does the farm community benefit at all or do they only get dust on their crops and houses, or pay to keep the roads dustless or to pave them?

Country Residential Development

The urban demand for residential homes in the countryside is nationwide. It is also part of our emerging form of city development which combines built-up areas and dispersed areas described by concepts like urban regions and urban fields. A large zone of urban fringe is part of our form of city development. No longer can the urban fringe be viewed primarily as a transition zone ultimately destined to be converted to more or less continuous urban uses. Only in limited parts of the inner urban fringe is continuous conversion expected. Rather our cities are surrounded by a countryside interspersed in various ways by urban activities which take up about 5 to 10 percent of the land space.

An increasing proportion of this land space is occupied by scattered country residences located outside the villages and towns. The population of these scattered country residences is about 5 to 10 percent of that of the built-up city. Put another way, a city of 100 000 can be expected to have about 1500 country homes within 40 km of its built-up edge. Specific data exist for the 5440 km² Toronto-Stratford axis. There the total population grew from 255 000 (excluding Toronto) in 1941 to 862 000 in 1971. The farm population declined from 17.6 percent of the total to 4.5 percent, while the scattered nonfarm population living outside villages and towns increased from 2.7 percent to 8.2 percent of the total.[29]

The processes involved with increased country residential development affect farmers in three ways. One is the problem of land fragmentation and its ramifications discussed earlier. A second is who pays and who gets — the services versus equity problem. The third is the variety of conflicts that can be generated between people.

These conflicts occur at the daily existence level and at a political level. Country residents complain about dust, crop spraying, farm odours, and machinery movement on roads. Farmers complain about trespass, fruit pilfering, road usage, and the like. At the political level, at some point in time if numbers of country residents exceed those of farmers, then they may end up as a significant force on local councils and a battle may ensue over what services are needed and taxes required for them.

What must be recognized today is that in the surroundings of our cities where agricultural activities are prevalent, two major groups — farmers and urbanites — are living in and using the same geographic space in overlapping ways. While rural and urban overlap in many more ways than in the past, differences in goals, motives, values, and attitudes still exist. Some conflicts are thus inevitable. What we need to strive for is an understanding of the possible interactions as part of the system that is our emerging form of city.

Other Urban Uses of Countryside Resources

Four land use activities, using the land as a physical resource or site, are of special significance. These are: the extraction of sand, gravel, and stone; the removal of water and the building of reservoirs; the use of land for urban waste disposal; and the use of land for transportation and communications facilities. All four hinge on the acquisition of farmland.

All four are generating considerable citizen patricipation as planners attempt to develop policy to reconcile the land use conflicts which arise. The problems of the aggregate industry in the urban fringe are well described in three short articles by McLellan.[30] Farmers seem less perturbed by gravel pits and waste disposal sites than are country residents. Farmers are more perturbed by water removal and transportation and communications facilities than are country residents. Where gravel acquifers are tapped for city wells, as around London and Kitchener, lowering water tables have affected farm wells and springs. For example, compensation has been paid to farmers west of Kitchener to ease the conflict.

Routes of hydro lines, gas and oil lines, and highway expansions are equally contested by farmers who don't want their land involved, and by others over environmental issues, aesthetic aspects, and preservation of good farmland. Expropriation rights applied in the past against farmers, with less than full explanation in some areas, have left a bitter residue of feelings. Now environmentalists and country residents are often allied with farmers over routeway conflicts.

Route problems for such facilities are more severe in the surroundings of cities. There the routes focus and concentrate. There the range of possible sites becomes more constricted. Consequently, the real possibility of eliminating conflict is greatly reduced. Hard decisions have to be made. But they need to be made with full participation of citizens and with full availability of facts.

Impact on the Natural Environment

Lumped under this problem area is the whole range of environmental concerns involving protection of our life support systems. The concern is both with natural ecosystems *per se* and with their resources for recreational and amenity purposes. Also included is concern with deterioration of scenic qualities resulting from the impact of man's careless activities. Interactions between natural, economic, and cultural environments are the crux of this problem area.

Four areas of impact on the natural environment are identifiable in the urban fringe and elsewhere. These involve ecological protection, pollution impacts, landscape amenities, and potential for future uses.

Ecological Protection

Ecological protection includes all land use and land cover necessary for airshed, watershed, landform, soil, vegetation, and wildlife habitat maintenance. In some places, ecological protection may mean leaving natural systems alone. In other places it may mean providing special protection, e.g., the habitat of a rare plant or animal. In still other places, it may mean improving on or even replacing natural systems, e.g., in the plant species used for the rehabilitation of mine sites.

The surroundings of our cities have many areas where carelessness in man's activities has left degraded natural environments. All terrain vehicles and snowmobiles damage tree seedlings and delicate swamp lands, critical water recharge areas are built upon, marshes which filter water, store water, and provide nesting places for wildlife are drained. The list of specific examples is lengthy.

Twelve aspects of ecological protection are singled out. They are: areas subject to temperature inversions and thus air pollution, watershed supply sources, ground water recharge areas, estuaries, marshes, shorelines, floodplains, easily erodible slopes, areas subject to landslides, areas subject to soil pollution, unusual natural ecosystems, and wildlife breeding areas.

Environmentally sensitive areas should be established and protected in regional plans. Recognition of this need has gradually evolved, e.g., outlining areas of unstable clay in the Ottawa-Carleton official plan. Even more has been done in the pending official plan of the Regional Municipality of Waterloo. Over fifty sensitive areas, mostly on privately-owned land including floodplain land, forest strips, glacial lakes, and special ecosystems, were legally designated for no or minimal development.

Areas needing ecological protection will vary from city to city. For instance, the Qu'appelle River Valley and its associated lakes in the urban fringe-urban shadow of Regina are under increasing recreational and residential demand. Its physical environment is sensitive in six aspects: erosion of slopes, stability of slopes, flooding, water supply, waste disposal (compact clays), and aesthetics of scenery.[31]

Pollution Impacts

A close interrelationship exists between pollution and ecological protection. Often it is man-made pollution that needs protecting against. A resource like small streams and rivers adjacent to cities and in cities is much more valuable to people than a wilderness stream far away. The daily pleasure of a child being able to fish — the kind or size of fish is not important — or sail a toy boat downstream is often forgotten. Is it because we have freely permitted pollution of streams near cities that wilderness streams seem more important to us? Let us emphasize restoring the natural environment where most people can use it and appreciate it — adjacent to or in our cities!

Outputs of pollution from cities are primarily absorbed in urban fringe areas. Downwind areas may suffer odours and fallout from industry. The natural environment of the urban fringes of Sudbury and Trail-Rossland still show the devastation caused by smelters before pollution controls were put into force. Transitory temperature inversions sometimes make living unpleasant around parts of Sarnia, Montreal, Edmonton, and Saint John.

But not all pollution is from cities! Cattle feeder lots, heavy farm fertilization, pesticide residues, hog and poultry enterprises, and natural erosion are part of the pollution impact in urban fringes as elsewhere. But as a general rule, agriculture tends to be more intensive near cities and thus poses increased pollution possibilities.

Given that pollution flows across many municipal boundaries, regulations and policies to solve problems need to be tailored to meaningful administrative areas. Whether this is best done by regional municipalities, groupings of municipalities, or special areas is still uncertain.

The problem as usual demands understanding interrelationships of natural, economic, and cultural environments. The need must be recognized, the cost must be accepted, and the physical, chemical, and biological processes must be understood.

Landscape Amenities

Both natural and man-made amenities must be considered. The white picket fence of the gentleman farmer, the old bridge, the escarpment view, the oil pump, the golden wheat field, the Holsteins grazing amidst lush greenery, the country estate, the lake, the river, the hill, the woodland are all part of landscape amenities surrounding our cities. Many disamenities exist as we all know, but so do many amenities.

In this area, we are dealing mainly with the cultural environment, with what people perceive as desirable in the urban fringe landscape. Historical buildings, dams, and fences all contribute to what urbanites appreciate as rural New Brunswick or rural Alberta. Since about four-fifths of our population live in urban places, what is rural to most of them is what is easily seen in the urban fringe. Slowly we are moving in the direction of preventing slovenly things like abandoned cars or stark hydro towers, and are at last attempting to improve the landscape. More and more official plans call for maintaining the rural character. Increasingly too, more often than not it is the country resident rather than the farmer who is really concerned about improving landscape amenities.

There is abundant evidence of the importance of natural scenic attributes as magnets for country residential location. The woodland, the stream, the hilltop, and the view into the valley mark the site of most country residences. But too many such residences scattered about without careful planning control leads to deteriorating landscape amenities. Then the problem of environmental impact becomes one of how many country residences we can permit and where they should go.

Potential for Future Uses

Closely related to the problems of ecological protection, pollution, and landscape amenities is the need to assess the potential of the land space of the urban fringe for various mixes of land use activities, for different densities of population, and for different intensities of use. What farmland should be preserved? What areas need carefully guided activities so that ecological protection occurs and scenic qualities are maintained? What areas should we designate for country residential? Existing needs must be met, but not at unnecessary restrictions on future choice.

We must also strive for better understanding and recognition of the resource potential of the urban fringe for all kinds of recreation – public, private, passive, active, collective, individual. The recreation potential largely depends on the biophysical resource base. But biophysical sites having good potential for recreational use are also attractive for private residential use and sometimes for extractive minerals.

Recreational demands are increasing but so is the demand for other uses of the urban fringe land space resource. Hence, active competition for land occurs. Inventories and interpretative classification of the biophysical resources should be a part of the development of official plans. Only when detailed and expert analysis of biophysical

resources is available can protection of environmentally sensitive areas be justified against competing demands. Such was the case in developing the official plan for the Regional Municipality of Waterloo.[32]

The three basic jurisdictional means of public acquisition, land use controls, and taxation devices can be used in various mixes to ensure as wise a use as possible, given current knowledge of the biophysical resource in the urban fringe. In addition, as shown in a study of the Niagara region, three principles must be accepted.[33] The urban-centred region should be recognized as the dominant recreation environment; the local policy and planning body should be regional in scale and should coincide with the governing unit; provincial and federal responsibilities need to be more definitely articulated and coordinated.

Assessment of the potential of the land space for future land use activities in the surroundings of our cities merits high priority. Land in the urban fringe is not yet as fixed in use as is much of our built-up urban area. It can meet most competing demands of urbanites and ruralites if it is treated as the crucial resource that is. An ecological planning approach should be used to provide guidance.[34]

Services Versus Taxation: Who Benefits, Who Pays?

Who benefits from services as opposed to whose taxes pay for them became an urban fringe problem with the post World War II influx of urbanites into the countryside. Water, sewers, garbage collection, roads, education, fire protection, health, welfare, all have led to problems in urban fringes across the country. Generally, urbanites living in the country receive more than they pay for, while the farmer pays more than he receives. Within the built-up city, commercial and industrial land uses take up the slack; in the urban fringe the farmers, as the prime property-owners, tend to bear the burden.

The problem continues to exist across the country despite various attempts at solution. A recent study in Parkland County, west of

Edmonton, again concluded that the 1973 tax returns on country residents did not cover the costs of services provided. In another Alberta study it was concluded that both farmers and townspeople help subsidize the country residents — the latter because with centralized schools, municipalities still pay part of costs above provincial grants.[35] In New Brunswick and Prince Edward Island, where major services are borne by the province except in a limited number of incorporated municipalities, the problem takes a different wrinkle. A lack of strong planning controls has led people to locate outside the boundaries of the incorporated municipalities, both for cheaper land and to avoid taxes for major services borne by the province. Somebody benefits, somebody pays; in the urban fringe the twain too often do not meet.

Amelioration efforts have taken two general approaches. One is to ease the taxation burden on the farmer either by preferential assessment or by provincial tax rebates. The latter is the main Canadian approach; the former is more common in the United States. The other is by large lot zoning. This can attain the objective of individual country residences paying for themselves as long as individual severances for smaller parcels are largely impossible. Such was not the case in Ontario, but large lot zoning has lessened the gap in costs between services and taxation. Help was further provided in Ontario by provincially applied subdivision control where official plans were missing or weakly applied.

The inevitable result is largely exclusionary zoning so that only the wealthier can afford a home in the country. Obviously an ethical question is at issue: should country residential development be permitted only for those who can pay their way, should it be available to all, or should it be available to none? The other alternative of letting them do with only minimal services seems to founder in a participatory democracy. What is the crux of this problem? The equity of costs and services in the urban fringe is closely associated with noncontiguous and low density residential uses. People demand services regardless. But economic and equitable provision of services generally requires compact, continuous development, not the irregular development frequently associated with urban fringes. Costs of services, like roads, ditches, water mains, sewers, power and gas lines, are basically proportional to length and the number of users located along that length. It is easy to see why costs per household, if the household were charged, would be two or three times as high as for compact development. When costs of such services are spread over the entire municipality, some people must be subsidizing uneconomic services for others. On the other hand, where existing services, e.g., school bus routes, already exist, any additional users may help reduce per capita costs.

Obviously regional government units cannot ease this problem unless accompanied by planning controls which work to lessen the equity issue. Given that large lot zoning, which really is a form of fiscal zoning, raises the spectre of exclusionary zoning, what can be done? If minimum urban services and sound use of land are to result, then country residential development must be seen as part of a system of closely interrelated parts, all of which must be planned for simultaneously. Controls such as large lot sizes or so many residences per section of land (Calgary) are not system approaches. What is needed if the right to country residences continues to be accepted is designated areas mostly for cluster development, mostly of small lot sizes but with variable lot sizes as a possibility. These designated areas would have to be part of a regional plan which tries to recognize the potential of land for various activities. The designated areas should be partly determined on the basis of road capacities, mail and school bus routes, and other more economical provision of any other agreed upon services.

In the Alberta study of the Hinton-Edson area, west of Edmonton, the following guidelines for identifying potential areas for country residential development are given: (1) restrict in areas that are biophysically sensitive; (2) restrict within 3 km of the corporate limits (an Alberta Subdivision and Transfer Regulation) to allow room for built-up urban expansion; (3) restrict where servicing problems of water and sewage can occur; (4) select accessible and scenic sites not subject to the above constraints; and (5) then design so that rural character of landscape is maintained.[36] Applying such guidelines clearly demands a carefully thought out and flexible offical plan based on a systems approach.

Impact on Surrounding Settlements

Still a rather neglected area of study in the urban fringe literature is the effect of dispersed urban expansion on hamlets, villages, and small towns (about 100 to 1500 in population). During the historical development of the system of local places that are part of today's urban regions, the place which became the major centre of the region by competition tended to eliminate or reduce to minor status places located within about 16 km of today's built-up city edge. This pattern of settlement, in varying degrees of intensity, is generally exhibited in the agriculturally settled parts of Canada. Good examples are the surroundings of London, Saskatoon, and Winnipeg. At about 24 to 32 km out from the built-up edge, where the competitive effect lessens, occurs a ring of settlements. This ring of settlements (and of course, those which survived nearer the built-up city) is today part of the urban field.

Many such settlements are unincorporated and are located in rural municipalities which have limited resources available to provide planning guidance. Others are incorporated but are usually still too small to cope with an urban influx or a large development proposal. Regional planning districts in British Columbia and New Brunswick, regional planning commissions in Alberta, and regional municipalities in Ontario are current means of dealing with such planning needs.

Regional plans increasingly also include statements about maintaining the rural character of such settlements.[37] Most such places evolved as service centres for their surrounding farm population. Most continue to provide such services even in the shadow of large cities. Given the mobility of population in the urban fields of our cities, can we so direct growth to maintain some places as primarily rural service centres (i.e., maintain rural character) while others are destined to become dormitory places? The possible pay-off in increased diversity in the landscape surrounding our cities certainly merits the attempt, limited though success may be. Indeed any success will depend on strong regional plans and controls which view the city and the surrounding settlements which make up the urban region as an integrated system which is our emerging form of City Today.

Social Issues

During the late 1940's, through the 1950's, and on into the 1960's, as urban fringes developed rapidly, considerable attention was given by a number of sociologists to the impact of the influx of urbanites into rural communities.[38] The rural-urban fringe, as most of them called it, was viewed in demographic and social terms as a separate region whose characteristics were intermediate on a continuum between urban and rural.

Thomas[39] points out that two attitudinal approaches existed concerning urban fringe research. The social attitude implies that the urban fringe is a separate community, neither urban nor rural. This view emphasized permanence. The land-based attitude implies that conflicts over land use reflect society's inability to reach agreement on criteria to allocate land. This view emphasized transition. Today, it seems clear through concepts like urban regions and urban fields, that the 24 to 32 km dispersed zone of urban fringe is a permanent part of current forms of city development.

Four social issues are briefly considered: urban-rural attitudes to landscape, the exclusionary zoning question, and the influx of newcomers both into the rural agricultural community and into pre-existing settlements.

While urban and rural are no longer clearly distinguished, attitudinal differences still exist. Urbanites are often criticized for viewing the countryside as a place to unload their psychological pains and physical needs without looking after it. This is the "beer bottle and garbage on the roadside" syndrome. Farmers are often romantically viewed as faithful stewards of the land, yet some of them dump old cars and other waste in the back woodlot probably with no more thought than the urbanite who flings the empty pop can. The reality is somewhere in between, as it so often is. There are farmers who are environmentally sloppy just as there are urbanites.

The issue of exclusionary zoning and country residences has already been discussed. Its importance as a social issue is still not fully recognized in regional planning policy. Why should only the wealthier urbanites be able to live in the countryside? Why should less expensive country residences in the urban fringe have to pay their way if similar single family residences in the built-up city don't?

A closely related issue is the question of mobile home-trailer park developments which can provide reasonable housing at a reasonable price. Many more such developments occur in the Maritimes, especially New Brunswick, and usually in urban fringe locations. Yeates[40] in his study of the Windsor-Quebec City Axis reports that while the wealthier Axis has 55 percent of Canada's housing units, it has only 13 percent of Canada's mobile homes. This difference between provinces probably reflects three things: a less exclusionary zoning approach, less stringent planning controls, and a greater need for lower income housing in the Maritimes.

In the Canadian setting, Punter[41] identified evolving spatially segregated communities dominated by managerial-professional groups in the outer urban fringe of Toronto. But in a study of the Winnipeg urban fringe, Carvalho[42] noted that two groups, managerial-professional and craftsmen-production workers, were over represented in comparison with Winnipeg's population. However, three other studies on Calgary[43], Brandon[44], and Edmonton[45] suggested similar socio-economic profiles for country residents and city residents. None of these three studies suggested any serious conflict with the rural community.

Recent research on three townships in south Simcoe County, 80 km north of Toronto in the outer urban fringe and urban shadow zone, provides a meaningful perspective.[46] Walker suggests that two major groups, farmers and urbanite country residents, are pursuing different objectives though living in the same geographic space. They come into contact but operate largely in different spheres. The social networks of the urbanites are dominantly with Toronto and secondarily with other country residents. Farmers are still the centre of the rural community social structure and politically still control the three townships, though making up only about one-quarter of the population.

In discussing urban fringe problems, Thomas[47] concluded that only the problem associated with mixtures of different social groups seems to be abating. The specific problem of urban newcomers complaining about nuisances of pre-existing farm operations is not resolved in Canada and seems morally unjustifiable. In Ontario, a code of agricultural practices and specific regulations about country residence locations and existing farm operations in regional plans may eliminate this problem. However the evidence cited is interpreted, one should keep in mind that the effects of the urban influx must be considered in three contexts: that of the farm and farmer, that of the rural community, and that of the agricultural sector as a whole.

Just as in the open countryside, so can conflicts occur through an influx of newcomers and a subdivision or two tacked onto an existing village. When urban people from nearby cities start to build houses in these places, various problems can arise. Since families moving out from the city usually have young families, new schools or at least additional rooms for existing schools are soon needed, raising local tax rates. With larger regional school districts today, this problem is lessening since the tax burden is no longer directly felt.

With further growth of such places towards dormitory status, water and sewage systems become necessary or must be expanded. So do other services, and again property taxes rise. Disagreement may arise between the original population and the newcomers whose needs seem to be responsible for the increase in taxes. Restricting the growth of scattered country residences and directing it to pre-existing settlements may eliminate certain costs, but since everything is connected with everything else in the urban region, other problems ensue.

Evidence is still slim on many questions about population additions to pre-existing settlements. Clearly there are costs and benefits. How do newcomers integrate, or do they? How do age, sex, family, and other demographic characteristics relate to social and economic factors? In a study of twelve villages in the Waterloo urban fringe, Hancock[48] showed that among the newcomers, professional-technical, higher income classes and the better educated were overrepresented. Limited evidence suggested that though most of these newcomers kept their place of work in Kitchener-Waterloo, they also became actively involved in their village.

Somewhat similar results, but with conflicts, were shown in an in-depth study of Elora, a village of about 1500, located in the overlapping outer urban fringe of Toronto, Guelph, and Waterloo.[49] Conflicts arose between the newcomers who opposed an apartment development and the established population who saw the apartment development as progress. The newcomers were not well integrated into the village community, having social and working ties elsewhere. But they had moved to Elora because of its rural settlement character and wished it to remain that way.

The intention now usually included in regional plans of maintaining the rural character of pre-existing settlements as far as possible is laudable. This is probably only possible by strictly limiting additions to settlements so designated. How should the decision be made; which villages should become submerged by newcomers and which should retain their rural character? One could identify settlements that are still active farm service centres and largely prohibit any additional growth in them. These places would usually be the larger settlements, however. Thus, the decision would be made to develop further smaller places (usually having fewer than 1000 people).

Such decisions could only be administered as part of a regional government-regional planning unit. Even then conflicts will arise as they have in the Regional Municipality of York. If the regional government were single tier rather than double tier, conflicts might lessen. Where regional government units exist, as in Ontario, local municipalities continue to compete with each other for development and political power.

Government Planning and Administrative Difficulties

This problem permeates all other problems of the urban fringe as the potential "curse" or at least the soothing salve. A lack of unitary and consistent planning is a definite factor in why urban fringes diseconomies occur. The lack of spatial coincidence between municipal boundaries and the occurrence of urban fringe processes is a key factor in these difficulties. Cause, effect, and possible solution come together under this problem area.

Three words − governmental, planning, and administrative − are deliberately used. All three have legal and regulatory overtones. Governmental difficulties revolve around the need to have urban fringe and built-up city under one legal municipality. The processes which create the dispersed city of the urban fringe and the built-up city are similar; they are parts of the same system. Local governmental reorganizations, operational and proposed, in Ontario, Quebec, British Columbia, New Brunswick, and Nova Scotia represent changes in this desired direction. In Nova Scotia the key goal is to develop a local government structure which will provide the most effective services for the money available and will make clear who should be doing what.[50]

Unitary planning is essential to eliminate a confusion of regulations or lack of regulations between adjacent but independent municipalities. Fiscal zoning results as municipalities compete for development.

Municipal governments, because they raise much of their income from property taxes, are under powerful inducement to develop everything in sight, regardless of the consequences.[51] Any planning system can only be truly effective when it deals with the urban regions as a system. Urbanization processes operate at the scale of urban regions, not individual municipalities based on past boundaries developed for other purposes. This key fact was earlier noted in discussing annexation, and is the basis of a proposed new planning act in Manitoba.[52]

The question arises of compensation for loss of value to individual land owners from stricter application of regulations administered as part of a provincial plan. At the local scale, zoning regulations assume no need for compensation. At the provincial scale, more people seem to feel compensation is merited somehow, a view I find unconvincing. The community at large determines the value of land, not individuals. Both British Columbia (B.C. Land Commission) and Ontario (Ontario Planning and Development Act), who are moving towards province-wide regulations on certain aspects of land use activities, have refused to accept that compensation is needed.

A second difficulty is how to get away from fiscal zoning. How can the local municipality say "no" to a proposed subdivision, a lot severance, or a large recreational development which does not fully meet regulations but which means tax dollars? Halkett[53] has discussed these administrative difficulties as they occur in the Saanich Peninsula, part of the Victoria urban fringe.

Administrative difficulties occur over the consistent application of existing planning regulations. The paramount difficulty is that centralized administration should mean more consistent but possibly inflexible application, and local application, while more flexible (nearer to the people), may well be inconsistent given local pressure. One of the basic arguments for regional government units is that the best of both worlds is possible. Opponents of regional governments, however, are ready to point out that the worst of both worlds is what will happen.

Administrative difficulties also occur in providing services. Again one of the basic arguments for regional government is the need to provide services like parks, sewage, water supply, schools, and hospitals at a regional scale. The basis of this argument is that people-processes, which create the need for these services, now operate at a regional scale; therefore, consistent provision of services demands a regional administrative unit. The regional service infrastructure can be likened to a regional skeleton into which the local flesh and blood fits to the benefit of the urban region as a system and of local communities as part of the larger system.

REFERENCES

[1] Racine, J.B., "L'évolution récente du phenomene peri-urbain Nord Américain : les observations traditionnelles et les prémises d'un nouveau style de vie", Revue de Géographie de Montréal, Vol. 24, 1970, pp. 43-54 and pp. 143-63: and J. N. Jackson, The Canadian City (Toronto: McGraw-Ryerson, 1973).

[2] Coleman, A., The Planning Challenge of the Ottawa Area, Geographical Paper 42, (Ottawa: The Queen's Printer, 1969), and G. Isberg, "Controlling Growth in the Urban Fringe", in Management and Control of Growth, R. W. Scott, D. J. Bower, and D. D. Miner, eds., Vol. 111 (Washington: The Urban Land Institute, 1975), pp. 29-39

[3] Russwurm, L.H., "The Urban Fringe as a Regional Environment", in Russwurm, Lorne H., Preston, Richard E., and L.R.G. Martin (eds.), Essays on Canadian Urban Process and Form (Waterloo: Department of Geography Publication Series, No. 11, 1977).

[4] Friedmann, J., Retracking America: A Theory of Transactive Planning (Garden City, N.Y.: Anchor Press/Doubleday, 1973).

[5] Sauvé, Honourable Jeanne, Minister, Environment Canada, Canada Needs a National Land Use Policy, Statement, (Ottawa: Environment Canada, 1975), 7 pp.

[6] Halkett, I., "Residential Land Use and Attempts to Preserve Open Space on the Saanich Peninsula", in Residential and Neighbourhood Studies in Victoria, C. N. Forward, ed., Western Geographical Series, Vol. 5 (Victoria: Department of Geography, University of Victoria, 1973), pp. 178-229.

[7] Rissor, H.E., and R. L. Major, "Urban Expansion — An Opportunity and a Challenge to Industrial Mineral Producers", in Focus on Environmental Geology, R. W. Tank (ed.) (New York: Oxford, 1973), and Ontario Ministry of Natural Resources, Towards the Year 2000: A Study of Mineral Aggregates in Central Ontario, Summary of Proctor and Redfern Report (Toronto, Ontario Ministry of Natural Resources, 1974).

[8] Jackson, J. N., The Urban Future (London: George Allen and Unwin, 1971), p.

[9] Spurr, P., Land and Urban Development (Toronto: James Lorimer, 1976).

[10] Martin, L.R.G., *Problems and Policies Associated with High Land Costs on the Urban Fringe*, Discussion Paper for Conference on the Management of Land for Urban Development, Toronto, 1974 (Ottawa: Canadian Council in Urban and Regional Research, 1974), mimeo, 27 p.

[11] Dennis, M. and S. Fish, *Programme in Search of a Policy: Low Income Housing in Canada* (Toronto: Hakkert, 1972), and W. W. Hamilton, *Public Land Banking — Real or Illusionary Benefits?* (Vancouver: Faculty of Commerce and Business, University of British Columbia, 1974).

[12] Baxter, D., "Speculation in Land", in *Battle for Land Conference Report* (Ottawa: Community Planning Association of Canada, 1975), pp. 25-30.

[13] Spurr, P., *Land and Urban Development.*

[14] *Ibid.*

[15] Revis, D., *Advance Land Acquisitions by Local Government: The Saskatoon Experience* (Ottawa: Community Planning Association of Canada, 1973); A. Goracz, I. Lithwick, and L. O. Stone, *The Urban Future,* Research Monograph 5 (Ottawa: Central Mortgage and Housing Corporation, 1971); N. H. Lithwick, *Urban Canada: Problems and Prospects* (Ottawa: The Queen's Printer, 1971), and Alberta Land Use Forum, *Urban Residential Land Development,* Report No. 4 (Edmonton: Alberta Land Use Forum, 1974).

[16] University of Guelph, Centre for Resources Development, *Planning for Agriculture in Southern Ontario*, ARDA Report No. 7 (Toronto: Ontario Ministry of Agriculture and Food, 1972), 331 pp.

[17] Russwurm, L. H., *Development of an Urban Corridor System Toronto to Stratford Area 1941-1966*, Research Paper No. 3, Regional Development Branch (Toronto: Queen's Printer, 1970), and W. C. Found and C. D. Morley, *a Conceptual Approach to Rural Land Use — Transportation Modelling in the Toronto Region*, Research Report No. 8, University of Toronto-York University Joint Program in Transportation (Downsview: York University Transport Centre, 1972).

[18] Hoffman, D.W., "Disappearing Farmland", *The Bulletin of the Conservation Council of Ontario*, Oct. 1975, p. 10.

[19] Bryant, C.R., *Farm-Generated Determinants of Land Use Changes in the Rural-Urban Fringe in Canada, 1961-1975* (Ottawa: Lands Directorate, Environment Canada, 1975).

[20] University of Guelph, *Planning for Agriculture in Southern Ontario.*

[21] Munton, R.J.C., "Farming on the Urban Fringe", in *Suburban Growth: Geographical Processes at the edge of the Western Cities*, J.H. Johnson, (ed.) (London: Wiley, 1974) pp. 201-23, and L.H. Russwurm, *The Urban Fringe as a Regional Environment.*

[22] Williams, G.D.V., N. Pocock and L. H. Russwurm, *The Distribution of Agroclimatic Resources in Relation to Canada's Urban Population,* in Irving, Robert M. (ed.), *Readings in Canadian Geography* (Toronto: Holt, Rinehart and Winston), 3rd Edition, 1978.

[23] Runka, G., "Who Has the Responsibility?", *The Agrologist,* Vol. 4, Autumn, 1975, p. 3.

[24] Hoffman, D. W., "Disappearing Farmland".

[25] Mage, J. A., *Part-time Farming in Southern Ontario with Specific Reference to Waterloo County*, Ph.D. Thesis (Waterloo: Department of Geography, University of Waterloo, 1974).

[26] Wehrwein, G.S., "The Rural-Urban Fringe", *Economic Geography*, Vol. VIII, July, 1942, pp. 217-229.

[27] Rajotte, F., "The Retreating Lanscape of Leisure", *Habitat*, Vol. 18(2), 1975, pp.2-7.

[28] Munton, R.J.C., "Farming on the Urban Fringe".

[29] Russwurm, L.H., *The Urban Fringe as a Regional Environment.*

[30] McLellan, A.G., "Problems Facing a Growing Aggregate Industry, a Canadian Assessment", *Cement, Lime and Gravel*, Sept., 1971, pp. 217-19; "Derelict Land in Ontario — Environmental Crime or Economic Shortsightedness?", *The Bulletin of the Conservation Council of Ontario*, Oct. 1973 pp. 9-14; and "The Aggregate Dilemma", *The Bulletin of the Conservation Council of Ontario*, Oct. 1975, pp. 12-19.

[31] Saskatchewan Departments of the Environment and Municipal Affairs, *Urban Fringe Development Problems, Seminar Proceedings, 1972* (Regina: Queen's Printer, 1973).

[32] Theberge, J., *et al.*, *Asking the Land: Study of Potential Outdoor Recreation Sites in the Regional Municipality of Waterloo* (Waterloo: Faculty of Environmental Studies, University of Waterloo, 1974), 194.

[33] Cambray, C.T., *Outdoor Recreation and Multilevel Government — The Niagara Region*, M.A. Thesis, Waterloo: School of Urban and Regional Planning, University of Waterloo, 1973.

[34] McHarg, I.L., *Design with Nature* (Garden City, N.Y.: Doubleday, 1969); R. S. Dorney, "The Ecological Approach to Planning", in *Urban Fringe Development Problems* (Regina: Saskatchewan, Department of Environment and Municipal Affairs, 1973), pp. 50-64; D. J. Coleman, *An Ecological Input to Regional Planning*, Ph.D. Thesis, Waterloo: School of Urban and Regional Planning, University of Waterloo, 1974, and C. Kitchen, "Ecology and Urban Development in Canada", in *Canada's Natural Environment: Essays in Applied Geography*, G.R. McBoyle and E. Sommerville, (eds.), (Toronto: Methuen, 1976).

[35] Alberta Provincial Planning Branch, A.A. Preiksaitis, Researcher, *Country Residential Development Study: The Hinton-Yellowhead Corridor* (Edmonton: Provincial Planning Branch, Alberta Department of Municipal Affairs, 1974).

[36] *Ibid.*

[37] York, Regional Municipality of, *Interim Policy on the Urban Settlements in the Rural York Region* (Newmarket: Planning Department, Regional Municipality of York, 1973).

[38] For example: W. T. Martin, *The Rural Urban Fringe: A Study of Adjustment to Residence Location* (Eugene: University of Oregon Press, 1953); R. Kurtz and J. B. Eicher, "Fringe and Suburb: A Confusion of Concepts", *Social Forces,* Vol 37 Oct. 1958, pp. 32-37, and H. W. Andrews, *"The New Community 11: Adjustment to Living in the Changing Rural Fringe of a Metropolitan Area,* Research Bulletin 955 (Columbus: Ohio Agricultural Experiment Station, Ohio State University, 1963).

[39] Thomas, D., "The Urban Fringe: Approaches and Attitudes", in *Suburban Growth: Geographical Processes at the Edge of the Western City*, J. H. Johnson, ed.

[40] Yeates, M., *The Windsor-Quebec City Urban Axis* (Toronto: Macmillan, 1975).

[41] Punter, J. V., *The Impact of Exurban Development on Land and Landscape in the Toronto-Centred Region 1954-71,* Report to Central Mortgage and Housing Corporation (Ottawa: Policy Planning Division, Central Mortgage and Housing Corporation, 1974).

[42] Carvalho, M.E., Director, *The Nature of Demand for Exurbia Living*, Winnipeg Region Study (Winnipeg: Department of City Planning, University of Manitoba, 1974).

[43] Whitehead, J.C., *Country Residential Growth in the Calgary Region: A Study of Ex-Urbanization*, M.A. Thesis, Vancouver: Department of Geography, University of British Columbia, 1968.

[44] Stadel, C. and L.R. Clark, *Urban Fringe of Brandon: Land Use, Population Mobility*, Brief submitted to Brandon Boundaries Commission (Brandon: Department of Geography, University of Brandon, 1971), Mimeo, 24 pp.

[45] Alberta Land Use Forum, H.L. Diemer, *Parkland County Country Residential Survey*, Report No. 4A (Edmonton: Alberta Land Use Forum, 1974).

[46] Walker, G., "Social Perspectives on the Countryside: Reflections on Territorial Form North of Toronto", *Ontario Geography,* Vol. 10, 1976, pp. 54-63, and *Urbanization of the Countryside: Notes on the Social Geography of the Toronto Hinterland,* Paper presented to the Annual Meeting, Association of American Geographers, April, 1975, Milwaukee. Toronto: Department of Geography, York University, 1975, Mimeo, 11 pp.

[47] Thomas, D., "The Urban Fringe: Approaches and Attitudes".

[48] Hancock, J.D., *An Analysis of Expanding Urbanization in the Rural Villages of Waterloo County*, M.A. Research Paper, Toronto: Department of Geography, York University, 1973.

[49] Sinclair, P.R. and K. Westhues, *Village in Crisis* (Toronto: Holt, Rinehart and Winston, 1975).

[50] *Nova Scotia Royal Commission on Education, Public Services, and Provincial-Municipal Relations,* J. Graham, Chairman (Halifax: Department of Municipal Affairs, 1974).

[51] Hilchey, J.D., "Long-Term Planning in Resource Development and Use", *The Agrologist*, Vol. 2, July-Aug., 1973, pp. 9-11.

[52] Manitoba Division, Community Planning Association of Canada, *Land Use: Control, Management, Legislation Conference Report* (Winnipeg: Manitoba Division, CPAC, 1975).

[53] Halkett, I., "Residential Land Use and Attempts to Preserve Open Space on the Saanich Peninsula", in *Residential and Neighbourhood Studies in Victoria,* C.N. Forward, ed.

City Functions, Manufacturing Activity, and the Urban Hierarchy

SHIU-YEU LI, DOUGLAS A. SCORRAR, AND MICHAEL H. WILLIAMS

Functional Classification of Cities

Dominant Function, Distinctive Function, and Specialization Index

Social scientists have developed several approaches to classifying Canadian cities. One approach, which focuses on the economic functions that cities perform, is illustrated by the work of J. W. Maxwell[1] and J. U. Marshall.[2] Their classifications identify the predominant economic function of a city by comparing its employment profile against some national norm. Thus, Maxwell uses national data to define what he terms "minimum requirements". Following Maxwell's approach, the "average minimum" is used as the norm against which to measure the excess employment of each city in each industry. The procedure can be described simply.

Figure 1

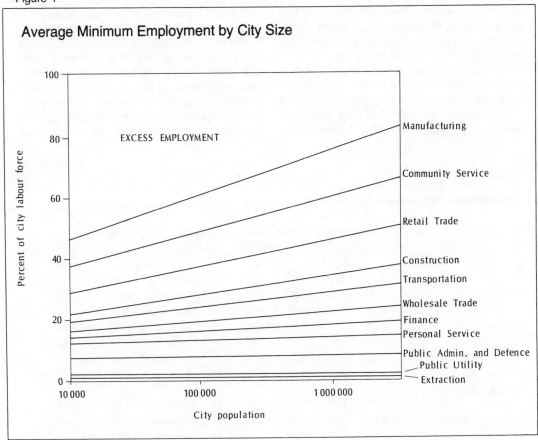

Average Minimum Employment by City Size

EXCESS EMPLOYMENT

Percent of city labour force

Manufacturing
Community Service
Retail Trade
Construction
Transportation
Wholesale Trade
Finance
Personal Service
Public Admin. and Defence
Public Utility
Extraction

City population

● SHIU-YEU LI and DOUGLAS SCORRAR are with the Ministry of State for Urban Affairs, Ottawa, and MICHAEL WILLIAMS is with Statistics Canada, Toronto. This paper has been prepared from two papers: Shiu-Yeu Li, "Labour Force Statistics and a Functional Classification of Cities," and Douglas A. Scorrar and Michael H. Williams, "Manufacturing Activity and the Urban Hierachy," in D. Michael Ray (ed.), *Canadian Urban Trends* (Toronto: Copp Clark Publishing Co. and Ministry of State for Urban Affairs, 1976). Used with permission of the authors and the Copp Clark Publishing Co.

The classification procedure builds on the *minimum average* employment calculated and graphed by city size (Figure 1). This value is the minimum proportion of employment in each industry, common to all cities of a given population. The *minimum average* is deducted from a city's total proportion in each industry, thereby measuring the *excess* employment in each industry.

The *dominant economic activity* is now defined as that industrial activity with the highest percent of excess employment. For example, in 1971 Alma had an excess employment of 19.7 percent in manufacturing. The next highest figure for Alma is an excess employment of 11.8 percent in community service. Alma's dominant function is therefore manufacturing. Similarly, the dominant function for each of the 137 cities can be identified in turn (Table 2). Since the average minimum differs with city size, two cities with an identical employment structure, but with different population totals, may have different dominant functions.

No city had finance, insurance, and real estate as its dominant function. Toronto, for example, had an excess employment of 10.85 percent in manufacturing and 2.67 percent in finance. But Table 1 shows that, on average, cities had a much higher proportion of their labour in manufacturing than in finance. And, indeed, the average excess of the labour force in manufacturing for all cities was 13.5 percent, and in finance 0.94 percent. Even given Toronto's small excess of labour force in finance, it was still 2.8 times the average excess for finance. By comparison, Toronto's excess employment in manufacturing was only 80 percent of the average. A more complete description of a city's employment profile can be provided by noting which industries are *distinctive functions*, in that they had an excess employment well above the average excess for all cities for that industry. Such a description is provided in Table 3 (pages 92-93), in which a function is classified as distinctive if excess employment for a given activity in a given city exceeds the value of the mean excess plus one standard deviation for that industry.

Manufacturing I cities are more concentrated in the heartland than Manufacturing II cities, and only six out of twenty-nine Manufacturing I cities are located in the hinterland. In contrast to the broad range of manufacturing industries found in heartland Manufacturing I cities, the hinterland cities of this type are characterized by a few large

Table 1

THE URBAN EMPLOYMENT PROFILE

Industry	Mean	Median	Montreal	Toronto	Van-couver	Minimum		Maximum	
	Average percent					Percent	City	Percent	City
Extraction	3.56	0.34	0.18	0.32	0.79	0.00	Oromocto	61.90	Labrador City
Manufacturing	23.05	19.72	28.36	27.47	18.40	1.17	Oromocto	58.06	Kitimat
Construction	6.67	6.24	5.22	6.66	7.48	0.78	Oromocto	15.18	Ste-Scholastique
Transportation	7.22	6.08	9.64	6.83	10.53	0.78	Oromocto	19.02	Moncton
Public utilities	1.24	1.01	0.91	1.31	1.15	0.00	Oromocto	8.71	Baie-Comeau
Retail trade	13.21	13.20	11.35	12.15	13.24	6.67	Petawawa	19.60	Kentville
Wholesale trade	3.90	3.58	5.21	5.94	6.77	0.65	Oromocto	7.95	Dawson Creek
Finance, insurance and real estate	3.56	3.41	6.31	7.36	6.58	1.41	Magog	7.36	Toronto
Community service	15.98	15.09	13.60	12.14	13.50	6.02	Petawawa	28.20	Kingston
Personal services	7.14	7.05	6.30	5.77	7.67	3.89	Oromocto	13.82	Penticton
Public administration and defense	9.59	6.56	5.74	5.82	5.21	1.63	Labrador City	74.58	Oromocto

Source: Canada, Statistics Canada, *1971 Census User Summary Tape* (Ottawa: Statistics Canada, 1975).

DOMINANT FUNCTIONS OF CANADIAN CITIES

	Region		City size class (in 10^3)						Total	
	Heart-land	Hinter-land	10–20	20–30	30–50	50–100	100–250	250–1000	over 1000	
Extraction	2	13	8	3	1	1	1	1	0	15
Manufacturing	47	23	27	12	14	7	3	5	2	70
Construction	0	2	0	0	2	0	0	0	0	2
Transportation	0	9	4	0	1	1	2	0	1	9
Retail trade	0	5	4	0	1	0	0	0	0	5
Community service	4	14	9	3	2	2	2	0	0	18
Personal service	0	1	1	0	0	0	0	0	0	1
Public administration	5	12	5	3	3	0	3	3	0	17
Total	58	79	58	21	24	11	11	9	3	137

Source: Canada, Statistics Canada, *1971 Census User Summary Tape* (Ottawa: Statistics Canada, 1975)

Table 2

industrial establishments concerned with processing of primary materials. Examples include ore smelting at Kitimat and Trail and pulp and paper mills at Powell River, Kapuskasing, and Port Alberni.

A final index of the employment profile is provided by weighting the excess employment in each industry in a city and summing the values to obtain a *specialization index*. The values are weighted in such a way that the index has a value of one when the distribution of excess employment in each activity in a city is proportional to the average minimum in each industry for a city of that size. The value is a maximum when the excess employment is concentrated in the activity with the lowest minimum average value. The index is calculated by the formula:

$$S = \Sigma_i \left\{ \frac{(P_i - M_i)^2}{M_i} \right\} \div \frac{(\Sigma_i P_i - \Sigma M_i)^2}{\Sigma_i M_i}$$

where:

S is the index of specialization

i refers to each of the industries in turn

P is the percentage of a city's labour force employed in each "i" industry

M is the average minimum labour force for each industry in a city of the given population, and

Σ_i means the sum of the calculations for each "i" industry.

City Types in Canada

Cities are recorded according to their dominant function in Figure 2 (page 94). A clear distinction emerges between the heartland cities, which form a manufacturing belt stretching from Windsor to Quebec City, and the hinterland cities. Hinterland cities, with the notable exception of mining towns, tend to have a more diversified employment structure as shown by their lower specialization indexes. All functions, with the exception of manufacturing, occur as dominant functions primarily in the hinterland, although the few cities with administration and defence as their dominant functions are fairly equally distributed across the heartland and hinterland alike.

Given the preponderance of cities with manufacturing as their dominant function (70 out of 137), it has been considered necessary to subdivide manufacturing cities into two classes, *Manufacturing I* cities with an excess employment in manufacturing above 25 percent of their total labour force, and *Manufacturing II* cities with an excess manufacturing employment equal to less than 25 percent of their total labour force.

The Manufacturing II cities differ from the Manufacturing I cities not only in the degree of dominance of manufacturing activity but also in their functional profile. This is indicated by their distinctive functions and specialization indexes (Table 3). Instead of having only one or two distinctive functions and a high index of specialization, as do the Manufacturing I cities, the Manufacturing II

Table 3

DOMINANT FUNCTION, SPECIALIZATION INDEX AND SELECTED DISTINCTIVE FUNCTIONS OF CANADIAN CITIES, 1971

No.	Urban Area	Dominant Function	Specialization Index	Extraction	Manufacturing	Finance Insurance & Real Estate	Administration & Defence	Transportation	Public Utilities	Wholesale Trade
1	Alma	Manufacturing II	1.64		+	+			+	
2	*Arnprior CA	Manufacturing I	2.82		+				+	
3	*Asbestos	Etraction	35.73	+	+					
4	Baie-Comeau CA	Manufacturing II	3.76		+				+	
5	*Barrie CA	Manufacturing II	1.24		+	+	+		+	+
6	Bathurst	Extraction	4.87	+	+	+		+		+
7	*Belleville	Manufacturing II	1.34		+	+	+	+	+	+
8	Brandon	Community service	1.48			+	+	+	+	+
9	*Brantford CA	Manufacturing I	4.25		+					+
10	*Brockville	Manufacturing I	2.10		+	+				+
11	Calgary CMA	Extraction	9.49	+		+		+		+
12	Campbellton CA	Community service	1.61			+		+		+
13	Charlottetown CA	Community service	1.71			+	+	+	+	+
14	*Chatham	Manufacturing II	1.87		+	+			+	+
15	Chicoutimi-Jonquière CMA	Manufacturing II	1.77		+					
16	Chilliwack CA	Pub. adm. & def.	2.13			+	+			+
17	*Cobourg CA	Manufacturing I	2.03		+	+				
18	Corner Brook	Manufacturing II	1.50		+			+		
19	*Cornwall	Manufacturing II	2.09		+					
20	Courtenay CA	Pub. adm. & def.	3.64			+	+	+		+
21	*Cowansville	Manufacturing I	3.52		+	+				+
22	Cranbrook	Tran. stor. & comm.	2.14			+		+	+	+
23	Dawson Creek	Tran. stor. & comm.	2.44			+		+	+	+
24	Dolbeau CA	Manufacturing II	1.66		+	+				
25	*Drummondville CA	Manufacturing I	3.30		+					+
26	Edmonton CMA	Pub. adm. & def.	2.21			+	+	+		+
27	Edmundston	Manufacturing II	1.44		+	+		+		
28	Flin Flon CA	Extraction	34.75	+						
29	Fredericton CA	Pub. adm. & def.	2.38			+	+		+	+
30	Gaspé	Community service	1.55		+			+		
31	*Granby CA	Manufacturing I	3.89		+	+				
32	Grand Falls CA	Manufacturing II	1.61		+	+		+	+	+
33	Grande Prairie	Retail trade	1.47			+	+	+	+	+
34	*Guelph CA	Manufacturing II	2.27		+					
35	Haileybury CA	Extraction	3.87	+		+		+	+	+
36	Halifax CMA	Pub. adm. & def.	5.04			+	+	+		
37	*Hamilton CMA	Manufacturing II	5.08		+					
38	*Hawkesbury CA	Manufacturing I	2.77		+	+				+
39	*Joliette CA	Manufacturing II	1.63		+	+			+	+
40	Kamloops CA	Construction	1.93			+		+		+
41	Kapuskasing	Manufacturing I	2.41		+					
42	Kelowna CA	Construction	2.05			+		+		+
43	Kenora CA	Manufacturing II	1.24		+	+		+	+	+
44	Kentville CA	Retail trade	1.45			+		+		+
45	*Kingston CA	Community service	2.33			+	+		+	
46	Kirkland Lake (Teck Twp.)	Extraction	7.08	+			+	+	+	+
47	*Kitchener CMA	Manufacturing I	5.08		+	+				
48	Kitimat	Manufacturing I	5.21		+					
49	Labrador City CA	Extraction	72.85	+						
50	*Lachute CA	Manufacturing I	2.71		+	+			+	+
51	*La Tuque	Manufacturing I	2.68		+			+	+	
52	*Leamington	Manufacturing I	2.37		+					+
53	Lethbridge	Retail trade	1.21			+		+	+	+
54	*Lincoln	Manufacturing II	2.02		+	+				+
55	*Lindsay	Manufacturing II	1.57		+	+				+
56	*London CMA	Manufacturing II	1.65		+	+				+
57	*Magog CA	Manufacturing I	3.42		+					+
58	Matane	Community service	1.37			+		+		+
59	Medicine Hat CA	Manufacturing II	1.13		+	+		+		+
60	*Midland CA	Manufacturing I	2.21		+			+		+
61	Moncton CA	Tran. stor. & comm.	2.69			+		+		+
62	Montmagny	Manufacturing II	1.67		+				+	+
63	*Montreal CMA	Manufacturing II	7.37		+	+		+		
64	Moose Jaw	Community service	1.58			+	+	+		+
65	Nanaimo CA	Manufacturing II	1.36		+	+	+	+	+	+
66	Newcastle CA	Pub. adm. & def.	2.21				+	+	+	+
67	New Glasgow CA	Manufacturing II	1.66		+	+		+	+	+
68	*New Hamburg CA	Manufacturing I	2.84		+	+		+		+
69	North Battleford CA	Community service	1.75			+		+		+
70	North Bay	Pub. adm. & def.	1.87				+		+	+
71	*Orillia	Manufacturing II	1.43		+	+				+
72	Oromocto	Pub. adm. & def.	14.75				+			
73	*Oshawa CA	Manufacturing I	4.00		+					
74	*Ottawa-Hull CMA	Pub. adm. & def.	18.66			+	+			
75	*Owen Sound	Manufacturing II	1.63		+	+		+		+

DOMINANT FUNCTION, SPECIALIZATION INDEX AND SELECTED DISTINCTIVE FUNCTIONS OF CANADIAN CITIES, 1971   (Continued)

No.	Urban Area	Dominant Function	Specialization Index	Extraction	Manufacturing	Finance Insurance & Real Estate	Administration & Defence	Transportation	Public Utilities	Wholesale Trade
76	*Pembroke CA	Pub. adm. & def.	1.50				+			+
77	Penticton	Personal service	1.56			+		+		+
78	*Petawawa CA	Pub. adm. & def.	12.80				+			
79	*Peterborough CA	Manufacturing II	2.16		+	+				+
80	Portage la Prairie	Community service	1.76			+	+	+		
81	Port Alberni CA	Manufacturing I	3.48		+			+		
82	Powell River	Manufacturing I	4.27		+					
83	Prince Albert	Community service	1.18			+	+	+		+
84	Prince George CA	Manufacturing II	1.73		+	+		+	+	+
85	Prince Rupert CA	Manufacturing II	1.96		+			+	+	+
86	*Quebec City CMA	Pub. adm. & def.	4.26			+	+			
87	Red Deer	Pub. adm. & def.	2.44				+			+
88	Regina CMA	Pub. adm. & def.	1.99			+	+	+	+	+
89	Rimouski CA	Community service	1.92			+		+	+	+
90	Rivière-du-Loup	Community service	1.63			+		+		+
91	Rouyn CA	Extraction	13.50	+				+	+	+
92	*St. Catharines-Niagara CMA	Manufacturing II	3.94		+					
93	*St-Georges CA	Pub. adm. & def.	1.59		+	+				+
94	*St-Hyacinthe CA	Manufacturing II	2.21		+					+
95	*St-Jean CA	Manufacturing II	2.43		+					+
96	*St-Jérôme CA	Manufacturing I	1.82		+	+			+	+
97	St. John's CMA	Community service	2.11				+	+	+	+
98	*Ste-Scholastique	Manufacturing II	2.72		+			+	+	+
99	Saint John CMA	Tran. stor. & comm.	1.50			+		+		+
100	*Sarnia CA	Manufacturing II	2.24		+	+		+		
101	Saskatoon CMA	Community service	1.71			+		+		+
102	Sault Ste. Marie CA	Manufacturing I	2.59		+					
103	Sept-Îles	Extraction	14.54	+		+		+	+	+
104	*Shawinigan CA	Manufacturing I	3.12		+				+	
105	*Sherbrooke CA	Community service	1.95		+	+				+
106	*Simcoe	Manufacturing II	1.85		+	+			+	+
107	*Smiths Falls CA	Community service	1.59		+			+		+
108	*Sorel CA	Manufacturing I	3.29		+				+	
109	*Stratford	Manufacturing I	3.21		+	+				
110	Sudbury CMA	Extraction	51.74	+						
111	Summerside CA	Pub. adm. & def.	2.47			+	+	+		+
112	Swift Current	Retail trade	1.73			+	+	+	+	+
113	Sydney CA	Extraction	9.36	+	+			+	+	
114	Sydney Mines CA	Tran. stor. & comm.	3.82	+	+			+		+
115	Terrace CA	Trans. stor. & comm.	1.57	+				+	+	+
116	*Thetford Mines CA	Extraction	31.71	+		+				
117	Thompson	Extraction	45.97	+						
118	Thunday Bay CMA	Tran. stor. & comm.	1.49					+	+	+
119	Timmins CA	Extraction	24.74	+				+		
120	*Toronto CMA	Manufacturing II	8.05		+	+				+
121	Trail CA	Manufacturing I	3.35	+	+				+	
122	*Trenton CA	Pub. adm. & def.	3.81		+		+		+	
123	*Trois-Rivières CA	Manufacturing II	2.03		+	+			+	
124	Truro CA	Manufacturing II	1.42		+	+		+		+
125	Val-d'Or CA	Extraction	12.77	+			+	+		+
126	*Valleyfield CA	Manufacturing I	2.88		+				+	
127	Vancouver CMA	Tran. stor. & comm.	2.34			+		+	+	+
128	Vernon	Retail trade	1.53			+		+		+
129	Victoria CMA	Pub. adm. & def.	3.48			+	+			
130	*Victoriaville CA	Manufacturing I	2.18		+				+	+
131	*Wallaceburg	Manufacturing I	4.05		+	+			+	+
132	Whitehorse	Tran. stor. & comm.	3.43	+		+	+	+		+
133	Williams Lake CA	Manufacturing II	1.68		+	+		+		+
134	*Windsor CMA	Manufacturing II	3.58		+	+				
135	Winnipeg CMA	Manufacturing II	1.58			+		+		+
136	*Woodstock	Manufacturing I	2.84		+	+				+
137	Yorkton	Community service	1.42			+	+	+		+

Note: Four distinctive functions, construction, retail trade, community service, and personal service have been deleted since most of the 137 urban areas claim these functions (construction – 81%, retail trade – 95%, community service – 74%, and personal service – 81%).

• Refers to cities in the heartland

Source: Canada, Statistics Canada, *1971 Census User Summary Tape* (Ottawa: Statistics Canada, 1975).

The Heartland – Hinterland Organization of Urban Functions.

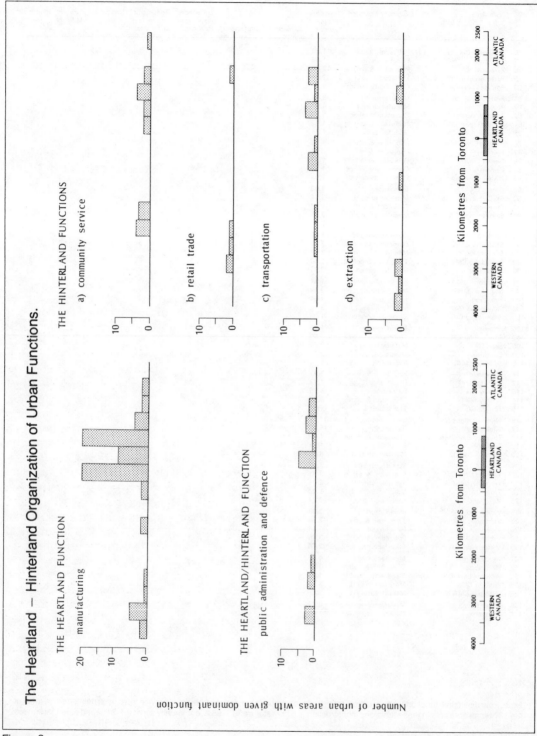

Figure 2

cities generally have several distinctive functions. Surprisingly, manufacturing may or may not be one of their distinctive functions. The type of manufacturing activity is much smaller in scale, less specialized, and may be oriented to local markets. In fact, these cities are also important service centres. The excess labour force is assigned more equitably among different industries and, consequently, Manufacturing II cities are generally among the most diversified cities in Canada. Although Winnipeg is classified as a Manufacturing II city, manufacturing is not one of its distinctive functions. In fact, it is an important transportation and trading centre of the Prairies. Kenora and Corner Brook are other similar cases. Therefore, there is a transitional group between Manufacturing I cities and the service centres.

Other City Types

The dominant function of most of the remaining cities is central place activity. The economic activity in a region is critical in determining the functions of the cities in the region. The contrast between heartland and hinterland is significant. Out of thirty-five service towns (construction, transportation, retail trade, community service, and personal service), thirty-one are found in the Prairies, Maritimes, and northern hinterland. The primary function of these towns is to serve their trade area where the dominant economic activity is agriculture. Because of the varieties of service they provide, their employment structure is diversified. They tend to have very low specialization indexes and numerous distinctive functions.

Cities with transportation as their dominant function (Figure 2c) are located at break-of-bulk points where goods are transshipped from one mode of transportation to another. Ports are an obvious example and Vancouver and Moncton are classified as the transportation centres, although Halifax is classed as an administrative centre. Thunder Bay, the major transshipment point for Prairie grain, is the point where the rail transportation of wheat from the West changes to Great Lake carriers to the East.

The administrative centres do not have a significant concentration in either the hinterland or the heartland (Figure 2). Furthermore, the degree of specialization varies. In one extreme, it is possible for a city to have 70 percent of its labour force dedicated to government or military functions and, as a result, a highly specialized employment structure emerges. The most notable examples are Oromocto and Petawawa, where military personnel comprise a large proportion of the labour force. Ottawa-Hull is an example of a high proportion of workers employed by the federal government. On the other hand, a service centre with diversified employment structure can still be classified as an administrative centre. Therefore, Regina and Edmonton are essentially service centres; yet they have the largest proportion of excess labour force in government services.

Extraction Cities

The employment structure of the last group of cities is dominated by the labour force in extraction. These cities are located adjacent to the major mines, especially on the edge of the Canadian Shield in northern Manitoba, Quebec, and Ontario (Figure 2d). Their location limits the possibility of the development of other activities. Therefore, they are the most specialized cities in Canada. For instance, the raison d'être of Labrador City is the location of iron mines. The conditions for developing manufacturing activity or farming are unfavourable and, as a result, the numbers of the labour force in service industries required to meet the demand of its hinterland are small. Extraction cities which do provide central place function to their hinterland, such as Haileybury and Rouyn, have a less specialized employment structure. Nevertheless, the degree of specialization in employment structure is still higher than that of other Canadian cities.

The spatial organization of the function structure of Canadian cities is summarized at the bottom of Figure 2, where the frequency of occurrence of cities is plotted against distance from Toronto (the centre of the heartland) in 500 km bands, by dominant fuctions. The sequence of the types of cities which are likely to be encountered with increasing distance from Toronto is: manufacturing, community service, retail trade, and transportation. There is an equal likelihood of finding an administrative centre or extraction city, irrespective of distance from Toronto.

Manufacturing Activity and the Urban Hierarchy

Manufacturing is a cornerstone of the Canadian economy and, more particularly, of its urban economy. Approximately one-half of all Canadian cities over 10 000 population have manufacturing as their dominant economic function (Table 2). Over 25 percent of Canada's real output in 1970 was attributable directly to manufacturing industries and there were also considerable indirect contributions.[3] These direct and indirect contributions of manufacturing to economic output, as well as the mobility of the manufacturing industries, make manufacturing an important component of government policies to alleviate disparities in urban growth and regional development.[4] The growth in manufacturing on the national scale has kept pace with the overall growth of the Canadian economy and, while manufacturing has maintained its relative importance over the past fifty years, some regional variations in growth have occurred. The Economic Council of Canada notes in its *Eleventh Annual Review* (p. 184) that it expects these trends to continue.

The organization of manufacturing activities has changed over the past fifty years. The data in Table 4 show that, over the period 1917 to 1971, the number of employees increased by 175 percent while the number of establishments grew by only 45 percent, indicating increases in plant size. These organizational changes, together with technological changes, resulted in the relative increase of supervisory and head-office employees from 10 percent in 1917 to about 30 percent in 1971.

A Geographic Distribution of Manufacturing

The manufacturing sector in most countries is associated with urban settlements, especially large urban settlements, and Canada is no exception. In 1971, the 137 urban areas, comprising about 70 percent of the population of Canada, accounted for 82 percent of all manufacturing employees, 85 percent of the total value of shipments and 87 percent of the total value-added. Ontario and Quebec, the two heartland provinces, with about 64 percent of the population, accounted for 80 percent of all employees and about 82 percent of value-added (Table 5). Even more striking is the concentration of population and manufacturing activity in Toronto and Montreal, which alone accounted for 25 percent of Canada's population, 26 percent of all manufacturing employees, and 35 percent of total manufacturing shipments.

The distribution of manufacturing in Canada, as shown in Table 5, has varied only slightly since 1926, the period for which

Table 4

SELECTED STATISTICS FOR MANUFACTURING, CANADA, SELECTED YEARS, 1917–71

	1917	1926	1933	1942	1951	1961	1971
Number of establishments	22 043	21 269	23 747	27 791	37 021	33 357	31 908
Total employees	585 945	558 861	468 366	1 150 616	1 258 375	1 352 605	1 628 404
Percentage administrative, office	10.7	13.5	18.5	15.4	19.7	30.5	28.3
Percentage production workers	89.3	86.5	81.5	84.6	80.3	69.5	71.7
Total salaries and wages ($10³)	479 998	625 416	435 908	1 681 150	3 276 281	5 701 651	12 129 897
Percentage paid to administrative, office	17.1	22.7	31.9	19.9	24.9	38.0	35.5
Percentage paid to production	82.9	77.3	68.1	80.1	75.1	62.0	64.5
Gross value of shipments ($10³)	2 768 046	3 090 179	1 952 904	7 548 215	16 392 187	23 438 956	50 275 917
Census value added by manufacturing ($10³)	1 170 788	1 281 021	918 923	3 305 495	6 940 947	10 434 832	21 737 514

Source: Canada, Dominion Bureau of Statistics, *Manufacturing Statistics of Canada*, Cat. No. 31-203 (Ottawa: Queen's Printer, 1964); Canada, Statistics Canada, *General Review of the Manufacturing Industries of Canada*, Cat. No. 21-203 (Ottawa: Information Canada, 1971); M. C. Urquhart and K.A.H. Buckley, *Historical Statistics of Canada* (Toronto: Macmillan Press, 1965).

REGIONAL SHARES OF SELECTED NATIONAL MANUFACTURING STATISTICS, SELECTED YEARS, 1926–71

	1926	1933	1942	1951	1961	1971	Change quotient*
Total employees (%)							
Atlantic Provinces	6.4	5.2	4.8	5.3	4.8	4.6	0.7
Quebec	31.3	33.6	34.6	33.2	33.5	31.2	1.0
Ontario	48.5	48.0	47.1	47.6	46.8	49.1	1.0
Prairie Provinces	5.9	7.1	5.7	6.5	7.2	7.1	1.2
British Columbia	8.0	6.1	7.8	7.4	7.7	7.9	1.0
Canada†	100.0	100.0	100.0	100.0	100.0	100.0	
	558 861	468 366	1 150 616	1 258 375	1 264 946	1 628 404	
Salaries and wages (%)							
Atlantic Provinces	4.3	4.4	4.1	4.3	3.9	3.7	0.9
Quebec	29.2	30.9	31.9	30.7	31.1	28.5	1.0
Ontario	51.5	50.6	50.0	51.0	49.7	52.2	1.0
Prairie Provinces	6.7	7.6	5.2	6.0	7.0	6.6	1.0
British Columbia	8.3	6.5	8.8	8.0	8.4	9.1	1.1
Canada†	100.0	100.0	100.0	100.0	100.0	100.0	
	$625 416	$435 908	$1 681 150	$3 276 281	$5 231 447	$12 129 897	
Census value added (%)							
Atlantic Provinces	4.3	4.3	3.6	4.3	3.7	3.3	0.8
Quebec	29.5	31.3	32.0	30.0	30.0	27.6	0.9
Ontario	51.9	50.6	50.5	51.4	50.8	54.1	1.0
Prairie Provinces	7.0	7.4	5.6	5.7	7.3	6.7	1.0
British Columbia	7.2	6.4	8.3	8.6	8.1	8.2	1.1
Canada†	100.0	100.0	100.0	100.0	100.0	100.0	
	$1 281 021	$918 923	$3 305 495	$6 940 947	$10 682 138	$23 187 881	

* Change quotient is the percentage in 1971 divided by the percentage in 1926.
† Canada shown total percent and absolute number.

Source: Canada, Dominion Bureau of Statistics, *Manufacturing Industries of Canada,* Cat. No. 31-204 (Ottawa: Queen's Printer, July 1964); Canada, Statistics Canada, *Manufacturing Industries of Canada: Atlantic Provinces,* Cat. No. 31-204 (Ottawa: Information Canada, August 1974); *Quebec,* Cat. No. 31-205 (Ottawa: Information Canada, July 1974); *Ontario,* Cat. No. 31-206 (Ottawa: Information Canada, July 1974); *Prairie Provinces,* Cat. No. 31-207 (Ottawa: Information Canada, July 1974); *British Columbia,* Cat. No. 31-208 (Ottawa: Information Canada, July 1974). M. C. Urquhart and K.A.H. Buckley, *Historical Statistics of Canada* (Toronto: Macmillan Press, 1965).

Table 5

there is consistent regional data. Ontario has always accounted for about 50 percent of all manufacturing employees, Quebec about 30 percent, and the rest of Canada the remaining 20 percent. Nevertheless, there has been an east-to-west growth gradient in manufacturing resulting in a regional shift of manufacturing emphasis from the Atlantic Provinces to the Prairie Provinces, as is evident in the change quotients shown in Table 5.

Change quotients for the regions are the simple ratios of their proportions of the particular variable in the most recent data to that proportion in the initial data. Over the period 1926 to 1971, the proportion of national manufacturing employment in the Atlantic Provinces declined from 6.4 percent to 4.6 percent, a change quotient of 0.7. The Prairie Provinces increased their share over the same period from 5.9 to 7.1 percent; this increase is described by the change quotient of 1.2. The other major regions retained their shares of national totals over the period 1926 to 1971, as shown by their change quotients of 1.0.

Manufacturing Threshold, Concentration, and Incidence

Three industrial characteristics – threshold, concentration, and incidence – serve to differentiate industries and explain their variations in geographic distribution. Threshold, or "the condition of market entry" for an industry, is the portion of the total national market for any specific product that an industrial plant must be able to secure in order to achieve economic operation. The higher the threshold, the more difficult it is for a new plant to begin economic operation because it must start at a high scale of production relative to the industry. Examples of high-threshold industries are vegetable oil mills (1.463) and umbrella manufacturers (4.045). This means that a new umbrella plant would, theoretically, need to capture about 4 percent of Canada's umbrella market to be competitive with the existing small plants. Conversely, new establishments in low-threshold industries such as slaughter houses and meat processors (0.002) would need to capture only two-one-thousandths of the national market to operate at the same relative level of activity as existing small plants.

The actual range of thresholds for industries in Canada, based on 1970 data, is plotted in Figure 3. Those industries which have very low thresholds, such as sawmills and planing mills, commercial printing and plastics fabricators, tend to have operations which are very general in nature and are labour- and raw-material intensive. By contrast, those industries which have very high thresholds, vegetable oil mills and umbrella manufacturers, are high-technology industries with very specific products.

Industry concentration measures the extent to which total production in any given industry is unevenly distributed among its various establishments. A raw index for each industry is computed by calculating the proportion of total value of shipments for that industry accounted for by each individual establishment, and then summing the squares of these proportions over all establishments. Where these proportions are of approximately equal size the index will be lower than where the proportions of the largest plants dominate the industry. However, inter-industry comparisons of this index are not possible because the number of establishments varies.

The adjusted concentration indexes for industries reflect the different number of establishments by expressing the raw index as a percentage of the "range" of concentration from the lowest possible concentration, where all establishments are equal in output and the index equal to 1.0 divided by the number of establishments, and the maximum concentration, where one establishment has all the output and an index of 1.0. Thus, an index of 2.5 means that the concentration level of an industry is only 2.5 percent of its maximum. The office and store

machinery and umbrella industries are examples of high-concentration industries. Their indexes are 41.06 and 36.63 percent of their respective maximums. The women's clothing industry, and machine shops, with indexes of only 0.25 and 0.28 percent of their respective maximums, are industries of low concentration.

In addition to the threshold and concentration measures of industrial structure, a further means of describing the geographic distribution is offered. The simplest measure of geographic distribution is the incidence of an industry, that is, the number of cities in which that industry is represented. For example, bakeries and commercial printing houses occur in most cities and have respective incidences of 129 and 117 out of the maximum of 137 cities. Lime manufacturers and clock and watch manufacturers are examples of sporadic industries with incidences of 2 and 4.

The interrelationships of threshold, concentration, and incidence show that Canadian industries which are ubiquitous tend to have lower threshold and concentration indexes. Conversely, those industries which are sporadic have higher market thresholds and also tend to be dominated by a few large plants. Not all these industries with high indexes of concentration and threshold are manufacturing giants; nor does a high value on one index necessarily imply a high score on the other. For example, two of the highest market thresholds are for rubber footwear manufacturers (the actual value for which is confidential) and for umbrella manufacturers (4.045). In both these cases there are only a few establishments in Canada, but the plants belonging to the former are of more equal size so that its concentration index (2.712) is considerably lower than that of umbrella manufacturers (36.628).

All three indexes of threshold, concentration, and incidence are needed to describe manufacturing industries and their geographic distribution. Industries which have high thresholds or high concentrations are likely to have most of their plants centrally located to optimize access to markets. Sporadic industries mean that the demands for the commodities are being served from a few points. If the corresponding thresholds are low, then new plants could conceivably be located in noncentral regions and hence

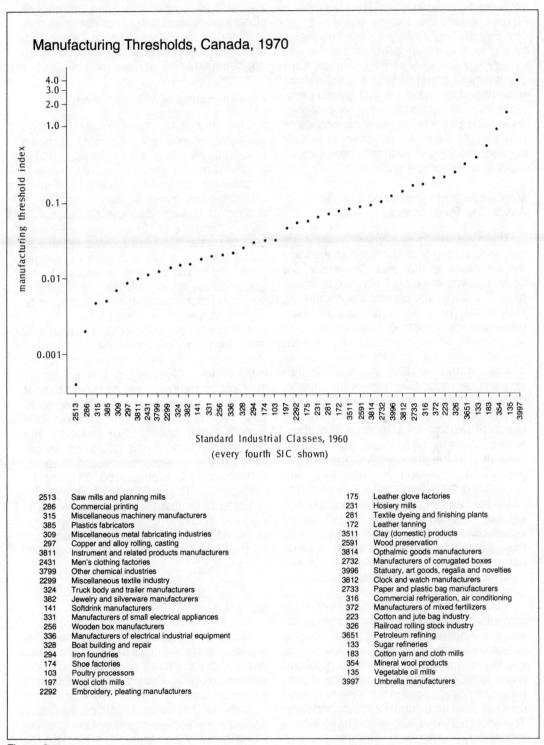

Manufacturing Thresholds, Canada, 1970

Standard Industrial Classes, 1960
(every fourth SIC shown)

2513	Saw mills and planning mills	175	Leather glove factories
286	Commercial printing	231	Hosiery mills
315	Miscellaneous machinery manufacturers	281	Textile dyeing and finishing plants
385	Plastics fabricators	172	Leather tanning
309	Miscellaneous metal fabricating industries	3511	Clay (domestic) products
297	Copper and alloy rolling, casting	2591	Wood preservation
3811	Instrument and related products manufacturers	3814	Opthalmic goods manufacturers
2431	Men's clothing factories	2732	Manufacturers of corrugated boxes
3799	Other chemical industries	3996	Statuary, art goods, regalia and novelties
2299	Miscellaneous textile industry	3812	Clock and watch manufacturers
324	Truck body and trailer manufacturers	2733	Paper and plastic bag manufacturers
382	Jewelry and silverware manufacturers	316	Commercial refrigeration, air conditioning
141	Softdrink manufacturers	372	Manufacturers of mixed fertilizers
331	Manufacturers of small electrical appliances	223	Cotton and jute bag industry
256	Wooden box manufacturers	326	Railroad rolling stock industry
336	Manufacturers of electrical industrial equipment	3651	Petroleum refining
328	Boat building and repair	133	Sugar refineries
294	Iron foundries	183	Cotton yarn and cloth mills
174	Shoe factories	354	Mineral wool products
103	Poultry processors	135	Vegetable oil mills
197	Wool cloth mills	3997	Umbrella manufacturers
2292	Embroidery, pleating manufacturers		

Figure 3

boost local employment. There are industries, such as breakfast cereal manufacturers, whose thresholds (0.057) and incidence (9) are relatively low, indicating perhaps that new establishments may be fairly easily initiated. In this case, concentration is extremely high at 36.5 percent of maximum, which suggests that there is a great deal of industry control and, thus, that growth prospects are uncertain for new operations. Knowledge of these three characteristics of industries is, therefore, important in selecting manufacturing industries to complement regional development plans.

Manufacturing Industries and Urban Areas: The Rank-orders

There is a strong relationship between the type and number of industries in an area and the population of that area. In general, the smaller towns have only ubiquitous industries. The larger the centre, the greater the likelihood of many industrial types, including some sporadic industries, being present. For instance, Montreal, Canada's largest metropolitan area, has the largest number of different industries (129), whereas Thompson and Labrador City, two of the smaller cities in the tabulations (excluding Oromocto which had no industry in 1970), have the fewest at two industries each. Cities, thus, can be ranked by the size and variety of their industries. In the same way, industries can be ranked according to their size and frequency of occurrence in the various cities.

On the basis of simultaneous rank ordering, the 137 urban areas by their manufacturing employment, and the 171 industries by their employment in urban areas, several patterns emerge. For example, at the industry level, the uneven distribution of manufacturing employment is apparent.

It is also possible to see the interrelationship between industry threshold, concentration, and incidence and industry location. The industries which are most important in total employment in Canada, such as pulp and paper mills, bakeries, and machine shops, have very low indexes of threshold and concentration and high incidence indexes. The sporadic industries which have a high technology component and require immediate access to large markets, such as the venetian-blind industry and typewriter and supplies manufacturers, are associated with the larger urban areas and tend to have higher thresholds and concentrations. There are also sporadic industries which are closely associated with a local raw-materials industry, such as leaf tobacco processing, and which have plants in locations which are not determined by city and market size.

Foreign Control of Manufacturing

Over the years the amount of foreign capital and investment in Canada has grown in importance. On a national basis, in 1970, establishments which were foreign-controlled employed 44 percent of all manufacturing employees and accounted for 52 percent of value-added and 54 percent of value of shipments. Since roughly 90 percent of this activity was located in the 137 urban areas, a major drawback of foreign-controlled manufacturing is apparent: the geographic concentration of foreign-controlled manufacturing has contributed to regional disparities in Canada.

The pattern of foreign-controlled activity is much less dispersed than that of all manufacturing activity. Also, the heartland cities in the Windsor-Quebec City axis, especially the Ontario centres, have by far the bulk of foreign-controlled activity and these represent very high proportions of their total manufacturing activity. For instance, the percentages of local manufacturing value-added that is foreign-controlled in Toronto, Windsor, and Guelph are 58, 71, and 75 respectively. This reflects high foreign involvement in industries that tend to locate in this area, such as transportation equipment, as well as an "economic shadow" effect from the proximity to the United States' manufacturing belt.[5]

Nonlocal Canadian-Controlled Manufacturing

Some of the problems associated with foreign control of a city's manufacturing also apply where the head office of an industrial establishment is located outside the area where the plant is located. There are some indications that these situations are manifested in the social problems of the communities involved. For example, a report in the *Ottawa Citizen* (October 28, 1975) claimed there is a higher incidence of medical problems resulting from stress and job anxiety in

CANADIAN AND FOREIGN CONTROL OF MANUFACTURING EMPLOYMENT AND VALUE ADDED, BY URBAN SIZE CLASS, 1970

Urban Size Class	Employment Canada	Employment Foreign	Value added Canada	Value added Foreign
10 000—19 999	52.7	47.3	43.1	56.9
20 000—29 999	49.7	50.3	42.6	57.4
30 000—49 999	52.6	47.4	44.2	55.8
50 000—99 999	51.0	49.0	41.3	58.7
100 000—249 999	53.0	47.0	44.3	55.7
250 000—999 999	54.4	45.6	48.6	51.4
1 000 000 +	58.0	42.0	47.4	52.6

Note: Excludes head office establishments.

Source: Tabulations provided by Manufacturing and Primary Industries Division of Statistics Canada, August 1975.

Table 6

DISTRIBUTION OF TOTAL MANUFACTURING EMPLOYMENT AND TOTAL MANUFACTURING VALUE ADDED BY CANADIAN NONLOCAL CONTROL, BY URBAN SIZE CLASS, 1970

Urban Size Class	Employment Local	Employment Nonlocal	Value added Local	Value added Nonlocal
10 000—19 999	72.1	27.9	74.0	26.0
20 000—29 999	79.6	20.4	77.2	22.8
30 000—49 999	76.2	23.8	76.4	23.6
50 000—99 999	81.2	18.8	85.1	14.9
100 000—249 999	81.2	18.8	80.1	19.9
250 000—999 999	80.6	19.4	80.1	19.9
1 000 000 +	94.1	5.9	93.7	6.3

Source: Tabulations provided by Manufacturing and Primary Industries Division of Statistics Canada, August 1975.

Table 7

communities where the industries are subsidiaries or branch plants of larger corporations.

To examine the pattern of nonlocal control, a sample of Canadian-controlled establishments was cross-tabulated by the location of the head office. The sample firms accounted for over 60 percent of all Canadian-controlled manufacturing activity. Toronto head offices have plants in more urban areas than Montreal and Vancouver,

but still Toronto has a higher proportion of locally-owned manufacturing (49 percent) than does Montreal (42 percent) or Vancouver (37 percent). Both Toronto and Vancouver have large proportions of their corporation control in Montreal than does Montreal in the other two cities. There is also a difference in the locational pattern of control between Toronto and Montreal. Toronto controls almost no activity in Quebec east of Montreal, whereas Montreal head offices have considerable activity in southwestern Ontario.

Foreign and Nonlocal Control by Urban Size

The incidence of foreign and nonlocal control is tabulated by urban size classes in Table 6. The degree of foreign control does not appear to be related to urban size. This lack of a relationship suggests that geographic location and industry mix are more important considerations in the location of foreign-controlled branch plants than the size of the urban place, in spite of the associated urbanization economies of larger cities.

In contrast, the tabulations by urban size class of nonlocal Canadian-controlled activity displayed in Table 7 reveal that a clear relationship exists between the location of the head offices of Canadian firms and the urban hierarchy. The smaller the centre, the more likely it is that the head office of a local industry will be located in another, probably larger, urban area. Therefore, the larger the centre in which an activity is located, the less likely that control of that activity is situated elsewhere.

This pattern is reflected for all establishments, regardless of country of control, in the

Table 8

DISTRIBUTION OF EMPLOYMENT TYPE, BY URBAN SIZE CLASS, 1961, 1971

Urban Size Class	1961		1971	
	Production	Supervisory/Head office	Production	Supervisory/Head office
10 000 to 19 999	73.4	26.6	77.8	22.2
20 000 to 29 999	77.9	22.1	77.5	22.5
30 000 to 49 999	75.0	25.0	76.7	23.7
50 000 to 99 999	71.8	28.2	72.0	28.0
100 000 to 249 999	73.3	26.7	73.2	26.8
250 000 to 999 999	69.1	30.9	72.0	28.0
1 000 000 +	64.2	35.8	66.4	33.6

Source: Data tabulations provided by Manufacturing and Primary Industries Division of Statistics Canada, August 1975.

ratio between production and non-production employees listed in Table 8. Since the normal head office functions of administration and research occur more often in larger centres, there are lower proportions of production employees in the larger urban size classes. This pattern may also be a function of the location of labour-intensive industries in small centres as compared to the concentration of complex, capital-intensive industries in major centres.

Conclusion

The interrelated problems of manufacturing are likely to be of continuing concern to Canadians because of the importance of this sector in the economy. Regional growth rates in manufacturing activity have been uneven and there has been a multiple-scale trend toward a concentration of manufacturing in the larger metropolitan centres, particularly Toronto and Montreal, in the heartland corridor from Windsor to Quebec City and from eastern Canada to western Canada. These geographic trends in manufacturing growth mirror the trends observed in population growth.

In addition to these trends, which are evident in manufacturing employment and value-added data, are the less evident but equally important shifts in the geographic location of control. The documentation of these shifts is less complete, but it is known that a major proportion of manufacturing activity in Canada is owned and controlled by foreign residents. Therefore, decisions relating to this portion of the national economy are, potentially at least, made outside Canada.

It is also evident that the control of Canadian-owned industry, as indicated by the location of head offices, is more concentrated in larger urban areas than is production activity. The location of head offices influences the location of the manufacturing activity they control and reveals a relationship between the spatial concentration of decision-making and the spatial distribution of manufacturing activity. More documentation is needed, both of this relationship between structure and process and of the interrelationships between manufacturing activity and other urban characteristics such as population growth, income levels and unemployment rates.

REFERENCES

[1] Maxwell, J.W., Grieg, J.A., and H.G. Meyer, "The Functional Structure of Canadian Cities. A Classification of Cities". (Ottawa: Department of the Environment, 1973). mimeo.

[2] Marshall, John U., "City Size, Economic Diversity and Functional Type: The Canadian Case", *Economic Geography*, Vol. 51, No. 1, January 1975.

[3] Economic Council of Canada, *Eleventh Annual Review* (Ottawa: Information Canada, 1974), p. 184.

[4] For example, the Department of Regional Economic Expansion.

[5] Ray, D.M., *Market Potential and Economic Shadow* (Department of Geography, University of Chicago, Research Paper No. 101, 1965).

Ethnicity and the Cultural Mosaic

FREDERICK I. HILL

Issues relating to cultural change in Canada have always occupied a prominent place in Canadian politics. The Royal Commission on Bilingualism and Biculturalism, the more recent review of Canada's immigration policy, and the unceasing debate over the implementation of Canada's Official Languages Act are only three of innumerable

● FREDERICK HILL is with the Research Directorate, Department of Environment, Housing, and Community Development, Canberra, Australia. This article is abridged from D. Michael Ray, *et al* (eds.), *Canadian Urban Trends: National Perspective*, Vol. 1 (Ottawa: Ministry of Supply and Services Canada, and Copp Clark Publishing Co., 1976), pp. 231-284. Used with permission of the author, the Ministry of Supply and Services, Canada, and the Copp Clark Publishing Co.

examples of the importance of the cultural differences in the Canadian population. The fact that Canada's many ethnic and language groups are not spread evenly across the country makes the challenges of coping with cultural diversity all the greater.

A person's ethnic origin or spoken language is important in an individual context, and is of significance nationally because of the relationships between the ethnocultural characteristics of the population and other dimensions of social differentiation. Ethnic origin has long been related to education levels, occupation structure, fertility rates, religious denomination, and the power structure in Canada. Studies by Porter and Clement have demonstrated the continued significance of ethnic origin in the definition of Canada's vertical, social mosaic.[1]

Cultural differences in Canada are largely the result of the history of immigration. Admittedly, ethnic groups settling in one part of the country have occasionally relocated to other regions and provinces, but the initial location of an immigrant group has an enduring impact on the ethnic map of Canada. Immigrants of particular cultural characteristics are still differentiated in terms of their choice of destination.

Even apart from the cultural differences among immigrants, however, the demographic significance of immigration in Canada is great indeed. Canada's population growth rate has been, and will continue to be, very dependent on the rate of immigration. Furthermore, if birth rates remain at their present low levels, the choice of immigrant destinations will become an even greater determinant of differences among cities and regions in terms of population growth rates. To the extent that immigrant destinations may be more easily manipulated than the relocation of persons already living in Canada, the potential for a coordinated immigration-population distribution policy in Canada is enhanced.

In this paper, cultural variations among cities in different regions, and of different sizes, are explored using several of the cultural measures which the censuses provide. National trends in the birthplace, linguistic, ethnic, and religious characteristics of the Canadian population are presented, followed in each case by a discussion of provincial and urban size variations in city characteristics. Particular attention is directed towards change between 1961 and 1971, and towards the role which immigration plays in effecting change.

Figure 1

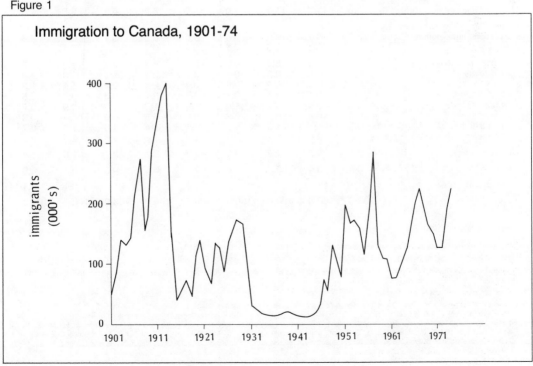

Immigration to Canada, 1901-74

A Nation of Immigrants

Annual and Decadel Levels

By land, sea, and air they have come — the oppressed and the opportunists, the refugees and the fortune-seekers, the relatives of earlier immigrants and the young adventurers. Canada is, by world standards, a nation of immigrants. As many as 400 000 immigrants have arrived in Canada in a single year. The level of immigration to Canada has been so high in relation to our population that the percentage of foreign-born in Canada is exceeded only in Australia, New Zealand, Switzerland, and Israel. In addition to the 3.3 million foreign-born living in Canada in 1971, who comprised 15.3 percent of our population, a further 4.0 million Canadian-born residents (18.5 percent of the population) had at least one foreign-born parent.[2] Thus, almost one-third of Canada's population in 1971 were first- or second-generation immigrants.

Since Confederation, Canada has had only three Immigration Acts and two collateral acts (the Chinese Immgration Act and the Immigration Appeal Board Act). Constantly changing regulations, rather than statutes, have been used to control the characteristics of immigrants allowed into Canada. The stringency of the regulations has, in turn, exercised some control over the number of immigrants. The regulations have reflected changing domestic economic conditions and attitudes towards immigration. As a result of changing regulations and world conditions, the number, source, and characteristics of immigrants have fluctuated enormously from year to year, and from decade to decade (Figure 1). In general, two world wars and economic recession in Canada retarded immigration; conversely,

Table 1

BIRTHPLACE OF CANADA'S POPULATION, 1901—71*

Birthplace	1901 %	1911 %	1921 %	1931 %	1941 %	1951 Excl. Nfld. %	1951 Incl. Nfld. %	1961 %	1971 %
Canada	87.0	78.0	77.7	77.8	82.5	84.6	85.3	84.4	84.7
United Kingdom and Ireland	7.5	11.2	11.7	11.0	8.3	6.8	6.7	5.5	4.5
Other Commonwealth and British Dependencies	0.3	0.4	0.5	0.4	0.4	0.5	0.1	0.3	0.8
Newfoundland	0.2	0.2	0.3	0.3	0.2	0.3	—†	—†	—†
European countries	2.3	5.6	5.2	6.9	5.7	5.7	5.5	7.9	7.6
Germany	0.5	0.5	0.3	0.4	0.2	0.3	0.3	1.0	1.0
Italy	0.1	0.5	0.4	0.4	0.4	0.4	0.4	1.4	1.8
Netherlands	—	0.1	0.1	0.1	0.1	0.3	0.3	0.7	0.6
Poland	—‡	0.4	0.7	1.6	1.4	1.2	1.2	0.9	0.7
Scandinavia §	0.3	0.8	0.7	0.9	0.6	0.5	0.5	0.4	0.3
U.S.S.R.‡	0.6	1.2	1.3	1.3	1.1	1.4	1.3	1.0	0.7
Other	0.8	2.0	1.7	2.2	1.9	1.6	1.6	2.3	2.5
United States	2.4	4.2	4.3	3.3	2.7	2.1	2.0	1.6	1.4
Asia‖	0.4	0.6	0.6	0.6	0.4	0.3	0.3	0.3	0.6
Other countries	—	—	—	—	—	—	—	0.1	0.4
Total	100.0	100.0	100.0	100.0	100.0	100.0	100.0	100.0	100.0

— Less than 0.05 percent.

* Exclusive of Newfoundland in censuses prior to 1951.

† Included with Canada, 1951-1971.

‡ U.S.S.R. includes Russia, Lithuania, and the Ukraine, 1901-1941, as well as Poland in 1901. Poland includes only Galicia (Austrian Poland) in 1911, the remainder being included with Russia (U.S.S.R.).

§ Denmark, Iceland, Norway, and Sweden.

‖ British Commonwealth and British dependencies in Asia are included with the former category.

Source: Canada, Statistics Canada, *1971 Census of Canada: Population: Birthplace*, Bulletin 1.3-6, Cat. No. 92-727 (Ottawa: Information Canada, 1974); Canada, Dominion Bureau of Statistics, *1961 Census of Canada: Population: Place of Birth*, Bulletin 1.2-7, Cat. No. 92-547 (Ottawa: Queen's Printer, 1963); Canada, Dominion Bureau of Statistics, *Ninth Census of Canada: Volume I: Population: General Characteristics* (Ottawa: Queen's Printer, 1953), Table 45.

the opening up of the resource frontier attracted immigrants.

The changing level of immigration into Canada is reflected in the birthplace data obtained from the decennial censuses of Canada, although it must be remembered that emigration from Canada and death rates are also important determinants of changes in Canada's birthplace data.

While the percentage of foreign-born in Canada remains high by world standards, this percentage was not as great in 1971 as it was in 1911, 1921, or 1931 (Table 1). The veritable flood of immigrants in the few years before World War I (many of whom settled on the Prairies) left Canada, in 1921, with nearly 25 percent of its population foreign-born. In 1901, by contrast, after three decades of heavy out-migration to the United States, only 13 percent had been foreign-born, the lowest percentage in any of our eleven decennial censuses since Confederation. Even the volume of immigration in the 1950's and 1960's has not increased the foreign-born proportion of the population significantly from the 15 percent figure of 1951.

Table 2 classifies Canada's immigrants by period of arrival in Canada, as enumerated in the 1961 and 1971 censuses. The foreign-born in *urban* Canada[3] in 1971 comprised 18.5 percent of the population, down slightly from 18.9 percent in 1961. By 1971, prewar immigrants comprised only 4.7 percent of urban Canada's population, a lower share than the 6.4 percent who had arrived since 1961. Three-quarters of the immigrants in urban Canada in 1971 had arrived since 1946. As early as 1961, urban Canada had more postwar immigrants than prewar immigrants. In contrast, in 1961 nonurban Canada still had more prewar than postwar immigrants. Even in 1971, nonurban Canada had almost equal numbers of prewar and postwar immigrants.

The Provincial Distribution of Immigrants in Urban Canada

Canada's provinces have not attracted immigrants in proportion to their population. While 18.5 percent of the residents of urban Canada were foreign-born in 1971, this percentage varied from only 3 percent in urban Newfoundland to 25 percent in urban Ontario (Table 2). Urban British Columbia ranked first in this respect in 1961, but fell to second place in 1971. The western provinces and Ontario had a higher proportion foreign-born than Quebec and the Atlantic Provinces. The westward spread of settlement across Canada is still evident from the distribution of immigrants, particularly those who arrived before 1946. In the urban areas of the four western provinces, in 1971, this group still outnumbered those who had arrived since 1961, and in urban Saskatchewan pre-1946 immigrants still exceeded the number of all postwar immigrants.

For urban Canada as a whole, the proportion of foreign-born declined slightly between 1961 and 1971. The four western provinces were mostly responsible for this decline. Cities in Saskatchewan, in particular, have failed to attract many immigrants since 1961, and their foreign-born population fell from 17.5 to 13.4 percent. The foreign-born proportion actually increased in urban Ontario and Quebec, and even urban Newfoundland and Prince Edward Island increased their meagre percentages.

The percentage of the population foreign-born in each of the 137 urban areas in 1971 reflects clear regional and provincial differences. Only six urban areas east of the Ontario-Quebec boundary had more than 5 percent foreign-born population, and only one of these (Montreal) was in the province of Quebec. Within Atlantic Canada, the five urban areas with over 5 percent foreign-born population were large or functionally specialized. Two are capitals (Halifax and Fredericton), Oromocto and Kentville have military bases, and Labrador City is a new iron-mining centre.

Within Ontario, the cities in the southwestern portion of the province had consistently higher percentages of foreign-born than eastern and northeastern Ontario cities. Leamington, with its demand for immigrant labour on vegetable farms and in canneries, had the second highest foreign-born percentage in Canada.

None of the urban areas of western Canada had less than 10 percent foreign-born population in 1971. The high percentage in Winnipeg placed urban Manitoba ahead of urban Saskatchewan. Most of the cities in Alberta and British Columbia had more than the national level of 15 percent foreign-born population in 1971. Kitimat ranked third in the nation in this respect.

Table 2

URBAN IMMIGRANTS, BY PROVINCE AND SIZE CLASS, 1961 AND 1971

Province and Size Class (1971)	Number of Immigrants		Percentage Immigrants		Immigrants by period of arrival in Canada, as a percentage of total population				
					Arrived Before 1946		Arrived 1946-60	Arrived 1946-61*	Arrived 1961-71†
	1971	1961	1971	1961	1971 census	1961 census	1971 census	1961 census	1971 census
Province:‡									
Newfoundland	5 775	3 229	3.1	2.2	0.4	0.7	1.1	1.5	1.6
Prince Edward Island	1 695	1 299	4.3	3.8	1.3	1.7	1.5	2.0	1.5
Nova Scotia	22 935	21 207	5.6	5.6	1.9	3.0	1.8	2.7	1.9
New Brunswick	13 695	12 919	4.7	5.0	1.7	2.5	1.5	2.5	1.6
Quebec	439 235	358 318	10.3	10.1	2.0	3.5	4.1	6.6	4.2
Ontario	1 551 095	1 191 526	24.7	24.4	5.2	9.2	10.6	15.2	8.9
Manitoba	116 535	122 011	19.0	22.9	7.6	13.0	6.4	9.9	5.0
Saskatchewan	49 855	53 428	13.4	17.5	7.5	11.9	3.5	5.6	2.5
Alberta	194 125	173 370	19.2	23.8	6.1	11.1	7.7	12.7	5.4
British Columbia	406 875	340 922	24.5	27.6	8.9	15.9	8.7	11.7	6.9
Size class (1971)									
1 000 000 +	1 585 480	1 190 035	24.6	24.0	5.0	9.1	10.0	14.9	9.6
Montreal CMA	405 680	326 165	14.8	14.7	2.7	4.9	6.0	9.8	6.0
Toronto CMA	893 315	627 685	34.0	32.7	5.8	10.9	14.3	21.8	13.9
Vancouver CMA	286 485	236 185	26.5	28.6	8.9	16.2	9.4	12.4	8.1
250 000–1 000 000	682 420	598 608	17.6	19.6	5.1	9.0	7.3	10.6	5.3
100 000–250 000	227 670	201 697	13.6	14.8	4.6	7.5	5.2	7.3	3.8
50 000–100 000	88 935	80 283	10.4	10.6	3.1	4.8	4.4	5.8	3.0
30 000–50 000	92 940	86 821	10.3	11.8	4.3	6.7	3.7	5.1	2.3
20 000–30 000	44 180	43 774	7.7	9.0	2.9	4.5	3.0	4.5	1.8
10 000–20 000	81 605	78 372	10.1	11.5	3.8	6.0	3.7	5.5	2.6
Urban Canada	2 803 230	2 279 590	18.5	18.9	4.7	8.1	7.5	10.8	6.4
Nonurban Canada	492 300	564 673	7.7	9.1	3.8	5.8	2.5	3.3	1.5
Canada	3 295 530	2 844 263	15.3	15.6	4.4	7.3	6.0	8.3	4.9

* Includes the first five months only of 1961.

† Includes the first five months only of 1971.

‡ Provincial data refer only to urban areas over 10 000 population in 1971. All of Ottawa-Hull Census Metropolitan Area and all of Hawkesbury Census Agglomeration are included in Ontario. All of Flin Flon Census Agglomeration is included in Manitoba.

Source: Canada, Statistics Canada, *1971 Census of Canada: Population: Birthplace*, Bulletin 1.3-6, Cat. No. 92-727 (Ottawa: Information Canada, 1974); Canada, Statistics Canada, *1971 Census of Canada: Population: Citizenship and Immigration*, Bulletin 1.3-7, Cat. No. 92-728 (Ottawa: Information Canada, 1974); Canada, Statistics Canada, *1971 Census of Canada: Characteristics of Census Agglomerations*, Bulletin SG-2, Cat. No. 98-702 (Ottawa: Information Canada, 1974); Canada, Dominion Bureau of Statistics, *1961 Census of Canada: Population: Citizenship and Immigration*, Bulletin 1.2-8, Cat. No. 92-548 (Ottawa: Queen's Printer, 1963); 1971 census summary tapes; 1961 census microfilm tabulations.

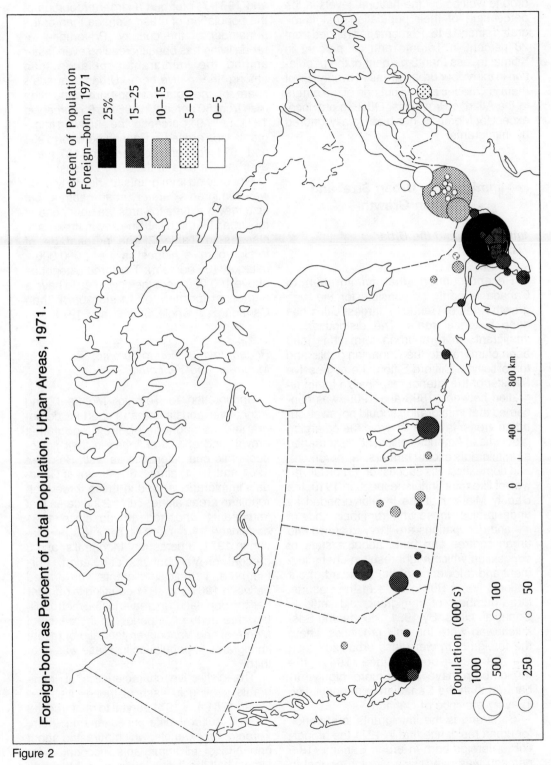

Foreign-born as Percent of Total Population, Urban Areas, 1971.

Percent of Population
Foreign-born, 1971

25%
15–25
10–15
5–10
0–5

Population (000's)

1000
500
250

100
50

800 km

400

0

Figure 2

When only postwar immigrants are considered, Saskatchewan cities, in particular, drop to well below the national levels in the percentage of their population that immigrated since 1946. Percentages ranged from 28 percent in Toronto and 31 percent in Kitimat to less than 2 percent of the population in thirty-four urban areas — twenty-six of them in Quebec and all but one of the others in the Atlantic Provinces. Quebec province, except for Montreal, has been totally ignored by immigrants.

Immigration, Urban Size, and Urban Growth

Immigration and the Urban Hierarchy

Recent concern that immigration to Canada is partly responsible for the high growth rates of Canada's largest cities has deep historical roots. The distribution of immigrants in rural-urban terms has long been of interest to the Canadian public and to politicians. Clifford Sifton, the aggressive Minister of the Interior responsible for immigration between 1896 and 1905, was concerned that immigration should not swell our urban areas. In order to avoid the emerging problems of American cities, he discouraged all immigrants except farmers, farm workers, and domestics. His successor, Frank Oliver, was of like mind in this respect. In 1910, the Deputy Minister of the Interior branded as undesirables "those who from their mode of life and occupations are likely to crowd into urban centres and bring about a state of congestion which might result in unemployment and a lowering of the standard of our national life".[4] But in spite of their efforts, large numbers of immigrants did settle in Canadian cities. By 1931, Alberta and Saskatchewan were the only provinces where the foreign-born were less urbanized than the Canadian-born. Since 1941, the foreign-born have been more highly urbanized than the Canadian-born population in every province of Canada.

So strong is the immigrants' preference for urban residence that in 1971 the proportion of foreign-born in urban Canada (18.5 percent) was over twice as high as that in nonurban Canada (7.7 percent). Immigrants who arrived in Canada in each of the three tabulated periods (before 1946, 1946-60, and 1961-71) formed a larger proportion of the population of urban Canada than of the remainder of the country. Preference for urban living has been increasing even faster among the immigrant population than among the native-born. Urban Canada's share of the total foreign-born population rose from 80 percent in 1961 to 85 percent in 1971. Fully 91 percent of the surviving immigrants who arrived in Canada between 1961 and 1971 were enumerated in the 137 areas of urban Canada in the 1971 census!

Not only do immigrants (particularly postwar immigrants) prefer urban centres, but also they gravitate towards the upper end of the urban hierarchy. Of the seven urban size classes in Table 2, the percentage of foreign-born is highest in the 1 000 000+ category (25 percent). The three categories above 100 000 were the only ones to have a higher proportion of foreign-born than Canada as a whole in 1961 and 1971.

Recent Immigration to Toronto, Montreal, and Vancouver

Immigration to Toronto, Montreal, and Vancouver constitutes an important demographic component of their population growth and effects changes in their socioeconomic characteristics as well. Between 1961 and 1971, almost 60 percent of Canada's immigrants settled in these three metropolitan areas alone. Table 2 indicates that immigrants who had arrived since 1961 comprised 14 percent of Toronto's population in 1971, 6 percent of Montreal's, and 8 percent of Vancouver's population. Furthermore, Toronto immigrants who arrived between 1961 and 1971 equalled in number half the population increase in that metropolitan area during this period. In the cases of Montreal and Vancouver, the ratio of recent immigrants to population increase was one-third.

The distinctive characteristics of immigrants settling in each of these cities between 1961 and 1971 served to reinforce the existing cultural differences among them. Toronto, for example, which attracted about one-third of all immigrants, accounted for almost half the Italian immigrants. Montreal, on the other hand, attracted nearly half the

Jewish and French immigrants. Vancouver attracted a disproportionate share of Canada's Asian, Scandinavian, and Lutheran immigrants, as well as those with no stated religious affiliation.

The Age Structure of Immigrants

By virtue of their age structure, immigrants contribute more to Canada's population growth and potential for cultural change than their numbers alone suggest. For example, in 1971, 32 percent of Canada's population was between fifteen and thirty-four years of age, while 50 percent of the immigrants who had arrived between 1961 and 1971 were in this most fertile age group. On the other hand, such a small percentage of recent immigrants are over sixty-five years of age that they place no immediate demands on Canada's old age security system and facilities associated with an aged population.

Even though Canada's immigrants tend to be in their young adult years when they arrive in Canada, one should not assume that Canada's total foreign-born population is young. In fact, quite the opposite is true. The history of immigration in Canada has been such that in 1971 almost 20 percent of Canada's foreign-born were over sixty-five years old, and a further 26 percent were between the ages of forty-five and sixty-four. The large numbers of young immigrants who arrived in Canada early in the first three decades of this century had reached the most advanced age categories by 1971. The long- and short-term impacts of high rates of immigration on the age structure of the population are thus quite different.

The Birthplace of Immigrants

The National Trend

Not only has Canada experienced changes in the proportion of foreign-born population, but also the countries of birth of the foreign-born have changed even more widely since 1901 (Table 1).

Immigrants born in the United Kingdom and Ireland accounted for only 4.5 percent of Canada's population in 1971, the lowest percentage since Confederation. Each census since 1921 has seen a decline in this percentage from a peak of nearly 12 percent in 1921, following two decades of heavy immigration from the United Kingdom.

The remainder of the British Commonwealth countries have never accounted for more than 1 percent of Canada's population by birthplace, but this group registered a significant increase between 1961 and 1971 as immigrants from the British West Indies, India, Pakistan, Hong Kong, and other Commonwealth countries took advantage of Canadian immigration policy.

As a whole, the European countries (excluding the British Isles) increased their share of Canada's population by birthplace, though somewhat irregularly, from 2 percent at the turn of the century to 8 percent in 1971 (Table 1).

Like the British-born, the percentage of American-born has declined in each census since 1921. Those born in Asia (excluding British Commonwealth countries in Asia) increased their percentage of Canada's population to 0.6 percent in 1971, regaining their level of 1911 to 1931. The rest of the world – Australia, Africa, the West Indies, South and Central America – still account for only 0.4 percent of Canada's population by birthplace in 1971, but this percentage has increased from less than 0.05 percent in 1951.

Provincial Differences in Birthplace, 1971

Where immigrants choose to live in Canada is influenced by where they were born (Table 3). Only in urban Ontario and British Columbia did the percentage of the population born in the United Kingdom exceed the level of 5.3 percent for urban Canada in 1971. Although in urban Canada as a whole immigrants from the rest of Europe accounted for nearly twice as many people as immigrants from the United Kingdom, in the four Atlantic Provinces and in British Columbia, U.K. immigrants exceeded other European immigrants. American and Asian immigrants, like all the others, preferred Ontario and western Canada. In British Columbia, the percentage of the population born in Asia was twice as high as in urban Canada as a whole. American immigrants reached their peak percentage in the cities of Alberta and British Columbia.

BIRTHPLACE OF URBAN POPULATION, BY PROVINCE AND SIZE CLASS, 1971

Province and Size Class (1971)	Same Province %	Different Province %	United Kingdom %	Other Europe %	United States %	Asia %	Other %
Province:*							
Newfoundland	92.2	4.7	1.4	0.7	0.6	0.3	0.2
Prince Edward Island	76.6	19.0	1.5	1.2	1.1	0.4	0.1
Nova Scotia	80.2	14.2	2.3	1.6	1.0	0.4	0.3
New Brunswick	78.8	16.5	1.8	1.4	1.1	0.3	0.2
Quebec	85.2	4.5	1.4	6.4	0.8	0.6	1.1
Ontario	64.1	11.2	7.4	13.4	1.3	1.1	1.5
Manitoba	65.2	15.8	5.3	10.8	1.3	0.7	0.9
Saskatchewan	73.1	13.5	4.0	6.0	2.2	0.8	0.4
Alberta	56.4	24.4	5.1	9.7	2.5	1.1	0.8
British Columbia	46.6	29.0	9.3	9.1	2.4	2.3	1.4
Size class (1971)							
1 000 000 +	64.2	11.2	6.4	13.2	1.4	1.5	2.1
Montreal CMA	79.5	5.7	2.1	9.3	1.0	0.8	1.6
Toronto CMA	55.8	10.2	9.6	18.6	1.4	1.8	2.7
Vancouver CMA	46.1	27.5	9.7	9.8	2.4	2.8	1.7
250 000–1 000 000	67.8	14.6	5.3	9.3	1.6	0.7	0.7
100 000–250 000	72.8	13.5	4.8	6.4	1.3	0.6	0.5
50 000–100 000	82.3	7.3	3.7	4.9	1.2	0.3	0.3
30 000–50 000	75.2	14.5	3.3	4.6	1.5	0.5	0.4
20 000–30 000	82.5	9.8	2.5	3.5	1.1	0.4	0.2
10 000–20 000	73.3	16.5	3.1	5.0	1.3	0.5	0.3
Urban Canada	68.9	12.5	5.3	9.7	1.4	1.0	1.2
Nonurban Canada	83.4	8.9	2.1	3.6	1.5	0.2	0.2
Canada	73.3	11.5	4.3	7.9	1.4	0.8	0.9

* Provincial data refer only to urban areas over 10 000 in 1971. All of Ottawa-Hull Census Metropolitan Area and all of Hawkesbury Census Agglomeration are included in Ontario. All of Flin Flon Census Agglomeration is included in Manitoba.

Table 3

Within each province, urban areas differ in the birthplace of their population. Victoria, with 14 percent of its population born in the United Kingdom, far outranked all other urban areas in this respect. Vancouver, Toronto, and Hamilton followed with nearly 10 percent U.K.-born. Kitimat and Leamington tied for first place according to the percentage born in the rest of Europe, but Toronto ranked third. Medicine Hat and Williams Lake were first in the percentage American-born; the Asian-born peaked in Leamington; and Toronto was first in the percentage born in the rest of the world.[5]

Provincial differences in attractiveness to interprovincial migrants are apparent from Table 3. Ninety-two percent of the residents of urban Newfoundland in 1971 were born in that province. In the cities of British Columbia, on the other hand, less than half the population was born in that province. Only in Quebec, Ontario, and Manitoba have the cities attracted more international migrants than interprovincial migrants. Quebec is extremely isolated from the rest of the country in terms of migration. Only 4.5 percent of urban Quebec's population in 1971 was born in another province — and only 2.2 percent if Montreal is excluded.

Thirty-one cities (all of them in Quebec or Newfoundland) had over 90 percent of their population born in the province of residence in 1971. In ten cities, on the other hand, over one-third of the population were born in a province other than where they lived in 1971. Seven of these were in British Columbia; the others were the military city of Oromocto and the northern frontier cities of Whitehorse and Thompson.

Urban Size Differences in Birthplace, 1971

Urban size differences based on birthplace were much less pronounced than provincial differences (Table 3). Nevertheless, there was a tendency for immigrants born in the United Kingdom, other European countries, and Asia to live in the bigger cities. Asians were the most highly urbanized of these broad nativity groups. There was little variation among the seven urban size categories in the percentage of their population born in the United States. Of the birthplace categories shown in Table 3, the American-born were the only ones to be over-represented in the nonurban population of Canada. This fact is a result of the heavy immigration of American farmers into the Canadian West early in this century. While many European immigrants at that time also settled on farms in the West, the volume of postwar immigration from Europe which has been directed to urban Canada, especially metropolitan Canada, has tipped the balance of European immigrants towards the upper end of the urban hierarchy.

Interprovincial migrants also comprise a higher percentage of the population in urban Canada than in the rest of the country, but the urban size groups show no pattern in the percentage born in a different province.

The Mother Tongue of the Population

The National Trend

The millions of immigrants who have come to Canada and the existence of two founding nations, England and France, have made this country one of the most linguistically diverse in the world. One of the ways in which our linguistic heritage has been measured is by mother tongue data, tabulated in each decennial census since 1921. Mother tongue refers to the first language learned,

Table 4

MOTHER TONGUE OF CANADA'S POPULATION, 1921–71*

Mother Tongue	1921†	1931†	1931	1941	1951 Excl. Nfld.	1951 Incl. Nfld.	1961	1971
	%	%	%	%	%	%	%	%
English	62.8	59.1	57.0	56.4	58.1	59.1	58.5	60.2
French	26.0	25.9	27.3	29.2	29.8	29.0	28.1	26.9
German	3.0	3.5	3.5	2.8	2.4	2.4	3.1	2.6
Greek	0.1	0.1	0.1	0.1	0.1	0.1	0.2	0.5
Indian & Eskimo	—‡	—‡	—‡	1.1	1.1	1.0	0.9	0.8
Italian	0.6	0.8	0.8	0.7	0.7	0.7	1.9	2.5
Netherlands	0.3	0.3	0.3	0.5	0.6	0.6	0.9	0.7
Polish	0.6	1.2	1.1	1.1	0.9	0.9	0.9	0.6
Scandinavian §	1.6	1.7	1.5	1.3	0.8	0.8	0.6	0.4
Ukrainian‖	1.3	2.4	2.4	2.7	2.6	2.5	2.0	1.4
Yiddish	1.3	1.5	1.4	1.1	0.8	0.7	0.5	0.2
Chinese & Japanese	0.7	0.7	0.7	0.5	0.3	0.3	0.4	0.5
Other	1.7	2.9	3.8¶	2.6	1.9	1.9	2.1	2.7
Total	100.0	100.0	100.0	100.0	100.0	100.0	100.0	100.0

* Exclusive of Newfoundland in censuses prior to 1951. For changes in definition of mother tongue, see text.

† Data for 1921 and the first column of data for 1931 refer only to the population 10 years of age and over, and exclude the Native Indian and Eskimo population and the entire population of the Yukon and Northwest Territories.

‡ The Indian and Eskimo population are excluded from the total in 1921 and in the first column for 1931. In the second column for 1931, persons reporting Indian and Eskimo languages as mother tongue are included in the category called "other."

§ Danish, Icelandic, Norwegian, and Swedish.

‖ Includes Bukovinian, Galician, and Ruthenian.

¶ Includes Indian and Eskimo languages.

Source: Canada, Statistics Canada, *1971 Census of Canada: Population: Mother Tongue,* Bulletin 1.3-4, Cat. No. 92-725 (Ottawa: Information Canada, 1973);
Canada, Dominion Bureau of Statistics, *Ninth Census of Canada: Volume I: Population: General Characteristics* (Ottawa: Queen's Printer, 1953), Table 54;
Canada, Dominion Bureau of Statistics, *Seventh Census of Canada, 1931: Summary,* Volume I (Ottawa: King's Printer, 1936), pp. 984-988.

which is still understood at the time of the census.

Changes in the percentage of the population of each mother tongue are affected by several factors: differential fertility rates, mortality rates, and age structures of the population by mother tongue; the mother tongue of immigrants and emigrants; the propensity of each group to transmit their mother tongue through successive generations in Canada.

Table 4 includes all mother tongues that accounted for 100 000 people, or 1 percent of the population, in any census since 1921. English and French have accounted for between 84 and 89 percent of Canada's population, by mother tongue, for the last fifty years. French has lost ground since 1941, but the decline amounts to only 3.3 percent. Persons with English mother tongue now outnumber those with French mother tongue by more than two to one, yet only 60 percent of Canadian residents claimed English

mother tongue in 1971. The third most common mother tongue in Canada in 1971 was German, closely followed by Italian, which had increased its percentage fourfold since 1921.

Provincial Differences in Mother Tongue, 1971

Table 5 reports, on a provincial and size-class basis, the percentage of the population with each of the five mother tongues accounting for at least 1 percent of urban Canada's population in 1971. Provincial differences are very pronounced. Over 90 percent of the population of urban Newfoundland, Prince Edward Island, and Nova Scotia reported English as their mother tongue. Three-quarters of urban Quebec and one-fifth of urban New Brunswick had French as their mother tongue. Although English dominated in Ontario and the West, over 15 percent of the population in all provinces

Table 5

MOTHER TONGUE OF URBAN POPULATION, BY PROVINCE AND SIZE CLASS, 1971

Province and Size Class (1971)	English %	French %	German %	Italian %	Ukrainian %	Other %
Province:*						
Newfoundland	98.3	0.8	0.2	0.1	0.0	0.6
Prince Edward Island	94.0	4.9	0.1	0.1	0.1	0.8
Nova Scotia	95.1	2.4	0.3	0.3	0.1	1.9
New Brunswick	77.7	20.8	0.2	0.2	0.0	1.0
Quebec	15.5	76.5	0.7	3.1	0.3	3.9
Ontario	74.9	7.5	2.4	5.4	1.2	8.7
Manitoba	72.2	5.4	6.0	1.1	6.7	8.5
Saskatchewan	79.9	2.4	6.6	0.4	4.9	5.7
Alberta	79.8	2.5	5.2	1.4	3.8	7.3
British Columbia	83.0	1.8	3.9	1.6	1.0	8.8
Size class (1971):						
1 000 000 +	53.0	29.2	2.1	5.7	0.9	9.1
Montreal CMA	21.7	66.3	0.9	4.8	0.4	5.9
Toronto CMA	73.8	1.7	2.6	8.4	1.3	12.1
Vancouver CMA	81.5	1.7	4.0	1.8	1.0	10.1
250 000–1 000 000	65.3	20.3	3.1	2.5	2.4	6.3
100 000–250 000	76.9	12.5	3.0	1.2	1.5	4.9
50 000–100 000	63.3	30.5	0.7	2.0	0.4	3.1
30 000–50 000	57.2	35.5	2.0	0.9	0.7	3.6
20 000–30 000	57.0	37.4	2.0	0.4	0.5	2.6
10 000–20 000	62.9	29.0	2.1	0.9	1.3	3.9
Urban Canada	60.3	25.8	2.4	3.5	1.3	6.8
Nonurban Canada	59.8	29.3	3.2	0.2	1.7	5.7
Canada	60.2	26.9	2.6	2.5	1.4	6.5

* Provincial data refer only to urban areas over 10 000 in 1971. All of Ottawa-Hull Census Metropolitan Area and all of Hawkesbury Census Agglomeration are included in Ontario. All of Flin Flon Census Agglomeration is included in Manitoba.

Source: Table 4.

west of Quebec had a mother tongue other than English or French. German-speaking people peaked in Saskatchewan, Italian-speaking in Ontario, and Ukrainian-speaking in Manitoba. The variety of mother tongues in Ontario and the West is also evident from the size of the "other" mother tongue category in these provinces.

Urban Size Differences in Mother Tongue, 1971

Most measures of cultural characteristics in urban Canada, including mother tongue, display less variation by urban size than by region or province. There are, however, some noteworthy urban size variations in mother tongue.

As far as Canada's two official languages are concerned, the main point with respect to urban size is that the French mother tongue is under-represented in the 100 000-1 000 000 range. Quebec's urban hierarchy is lacking in cities of this size, in comparison with Canada as a whole, and the mother tongue data reflect this fact. French is also a less common mother tongue in urban Canada than in the nonurban parts of the country. German and Ukrainian are both over-represented in the 100 000-1 000 000 range, since several of the Prairie cities are in this size range. These languages are also more common in nonurban areas than in urban Canada. Italian, on the other hand, is almost entirely an urban mother tongue, found mainly at the upper end of the urban hierarchy. As a group, the mother tongues other than these five are also more common in urban Canada (especially metropolitan areas) than in nonurban areas.

The Ethnic Origin of the Population

The National Trend

Ethnic origin provides an alternative measure of the cultural background of Canada's population. Country of birth and mother tongue provide a good indication of the cultural group with which many Canadians identify, particularly if they are immigrants or speak a minority language. There are many native-born Canadians with an English or French mother tongue, however, who retain some identification with a foreign cultural group through several generations of Canadian residence.

Admittedly, language is a very important means of passing cultural traditions through successive generations. Nevertheless, to maintain that the third-generation, English-speaking Canadian of Scottish ancestry is no different from his Greek, Chinese, or German counterpart is to ignore the fact that a degree of ethnic identity is often retained, even without the benefit of speaking the language of the ancestral cultural group. Although it can be argued that, for many Canadians with several generations of Canadian residence, tracing their cultural roots to their overseas origin is pointless, to Canadian immigrants, and to at least a substantial portion of the Canadian-born, ethnic origin is a meaningful measure even if they have English or French mother tongue.

A second reason for including ethnic origin is that ethnic origin data were available at the enumeration area level in 1961, while mother tongue and birthplace data were not. Enumeration area data were necessary for retabulation of 1961 data to 1971 urban boundaries. Since it is very important to ensure that identical urban boundaries are used when data are presented for two census years, ethnic origin data were chosen to provide information on cultural change in Canadian cities.

Ethnic origin refers to ethnic or cultural background traced through the father's side. The question asked to obtain this information was: "To what ethnic or cultural group did you or your ancestor (on the male side) belong before coming to this continent?". In some cases, language spoken by the respondent or his parental ancestor was used as a guide to the determination of his ethnic group.

In spite of the many problems with ethnic origin data,[6] the main thrusts of change in the ethnic composition of Canada's population since 1901 can be discerned from Table 6.

In 1971, the percentage of British Isles ethnic origin reversed its long-term decline, but at 45 percent it still stood 12 percentage points lower than at the turn of the century. The French ethnic group has fluctuated around 30 percent throughout this century. As a result of heavy immigration in the first two decades of this century, the French

ETHNIC ORIGIN OF CANADA'S POPULATION, 1901–71*

Ethnic Group	1901	1911	1921	1931	1941	1951 Excl. Nfld.	1951 Incl. Nfld.	1961	1971
	%	%	%	%	%	%	%	%	%
British Isles	57.0	55.5	55.4	51.9	49.7	46.7	47.9	43.8	44.6
French	30.7	28.6	27.9	28.2	30.3	31.6	30.8	30.4	28.7
Other European:	(8.5)	(13.1)	(14.2)	(17.6)	(17.8)	(18.7)	(18.3)	(22.6)	(23.0)
Austrian	0.2	0.6	1.2	0.5	0.3	0.2	0.2	0.6	0.2
German	5.8	5.6	3.4	4.6	4.0	4.5	4.4	5.7	6.1
Greek	—	0.1	0.1	0.1	0.1	0.1	0.1	0.3	0.6
Hungarian	—†	0.2†	0.1	0.4	0.5	0.4	0.4	0.7	0.6
Italian	0.2	0.6	0.8	0.9	1.0	1.1	1.1	2.5	3.4
Jewish	0.3	1.1	1.4	1.5	1.5	1.5	1.5‡	1.4‡	1.4
Netherlands	0.6	0.8	1.3	1.4	1.8	1.9	1.9	2.4	2.0
Polish	0.1	0.5	0.6	1.4	1.5	1.6	1.5	1.6	1.5
Russian	0.4	0.6	1.1	0.8	0.7	0.6	0.6	0.5	0.3
Scandinavian	0.6	1.6	1.9	2.2	2.1	2.1	2.0	2.1	1.8
Ukrainian	0.1	1.0	1.2	2.2	2.7	2.9	2.8	2.6	2.7
Yugoslav	—	—	—§	0.2	0.2	0.2	0.2	0.4	0.5
Other ‖	0.2	0.5	1.0	1.4	1.4	1.6	1.5	1.9	2.0
Asian	(0.4)	(0.6)	(0.8)	(0.8)	(0.6)	(0.5)	(0.5)	(0.7)	(1.3)
Chinese and Japanese	0.4	0.5	0.6	0.7	0.5	0.4	0.4	0.5	0.7
Other	—	0.1	0.1	0.1	0.1	0.1	0.1	0.2	0.6
Native Indian and Eskimo	2.4	1.5	1.3	1.2	1.4¶	1.2¶	1.2¶	1.2	1.4
Other and not stated	0.9	0.7	0.5	0.3	0.3	1.3	1.3	1.3	1.0**
Total	100.0	100.0	100.0	100.0	100.0	100.0	100.0	100.0	100.0

— Less than 0.05 percent

* Enclusive of Newfoundland in censuses prior to 1951.

† Includes Lithuanian and Moravian in 1901 and 1911.

‡ In order to increase the comparability of the Jewish ethnic group in 1951 and 1961 with other years, persons who reported a non-Jewish ethnic origin but Jewish religion were re-classified as Jewish ethnic origin in 1951 and 1961.

§ Serbo-Croatian in 1921.

‖ Includes Serbian in 1901 and 1911, and Swiss in 1901, 1911 and 1921. Persons reporting Swiss origin in 1931 and 1941 were classified as German, French, or Italian according to language spoken. In 1951, 1961 and 1971, those reporting Swiss ethnic origin were classified as German, French and Italian in the ratio of 2:1:1. In 1921, the Swiss numbered 12 837, i.e., 0.1 percent of Canada's population.

¶ Includes 35 416 Half-breeds in 1941 (0.3 percent of Canada's population).

** In 1971, "not stated" cases were computer assigned.

Source: Canada, Statistics Canada, *1971 Census of Canada: Population: Ethnic Groups,* Bulletin 1.3-2, Cat. No. 92-723 (Ottawa: Information Canada, 1973);
Canada, Dominion Bureau of Statistics, *1961 Census of Canada: Population: Ethnic Groups,* Bulletin 1.2-5, Cat. No. 92-545 (Ottawa: Queen's Printer, 1962);
Canada, Dominion Bureau of Statistics, *1961 Census of Canada: Population: Religion by Ethnic Groups,* Bulletin 1.3-8, Cat. No. 92-559 (Ottawa: Queen's Printer, 1964);
Canada, Dominion Bureau of Statistics, *Ninth Census of Canada, 1951: Volume II: Population: Cross Classifications of Characteristics* (Ottawa: Queen's Printer; 1953), Tables 34 and 35;
Canada, Dominion Bureau of Statistics, *Eighth Census of Canada, 1941: Volume I: General Review and Summary Tables* (Ottawa: King's Printer, 1950), p. 222;
Canada, Dominion Bureau of Statistics, *Sixth Census of Canada, 1921: Volume I: Population* (Ottawa: King's Printer, 1924), Table 24.

Table 6

declined to only 28 percent in 1921, but subsequently gained until 1951. Large-scale, non-French immigration since then, together with declining birth rates among the French, have again caused a decline in the French proportion.

The most spectacular gains have been registered by the other minority European ethnic groups, which in 1971 accounted for nearly one-quarter of the Canadian population. In addition to the Germans, who were the only other European ethnic group to comprise even 1 percent of the population in 1901, the Italians, Ukrainians, Dutch, Scandinavians, Poles, and Jews exceeded this level by 1971. The Austrians also exceeded 1 percent in 1921, presumably because many Germans claimed Austrian ethnic origin after World War I. The resurgence of Austrians in 1961 probably results from confusion with Ukrainians. The German ethnic group fluctuated as Germans fled to Austrian, Dutch, and Russian ethnic origins in 1921 and 1941.

The Jewish, Dutch, Polish, Russian, Scandinavian, and Ukrainian groups were established as substantial minority groups early in this century, whereas the Hungarians, Greeks, Italians, and Yugoslavs experienced most of their growth since World War II.

Table 7

ETHNIC ORIGIN OF URBAN POPULATION, BY PROVINCE AND SIZE CLASS, 1961 AND 1971

Ethnic Origin

Province and Size Class (1971)	British 1971 %	British 1961 %	French 1971 %	French 1961 %	German 1971 %	German 1961 %	Italian 1971 %	Italian 1961 %	Netherlands 1971 %	Netherlands 1961 %	Polish 1971 %	Polish 1961 %	Scandinavian 1971 %	Scandinavian 1961 %	Ukrainian 1971 %	Ukrainian 1961 %	Asian 1971 %	Asian 1961 %	Other 1971 %	Other 1961 %	Average Index of Ethnic Diversity* 1971	Average Index of Ethnic Diversity* 1961
Province:†																						
Newfoundland	94.5	94.7	2.2	2.1	0.6	0.6	0.1	0.1	0.2	0.1	0.1	0.1	0.3	0.4	0.1	0.0	0.6	0.3	1.5	1.6	0.171	0.244
Prince Edward Island	82.5	78.0	13.3	17.6	1.3	0.9	0.1	0.1	0.8	0.8	0.1	0.1	0.3	0.6	0.3	0.1	0.9	0.8	0.7	1.1	0.321	0.371
Nova Scotia	80.6	76.3	7.8	9.2	3.4	3.3	0.8	0.8	1.5	2.6	0.6	0.7	0.5	0.8	0.5	0.4	0.9	0.6	3.4	5.2	0.287	0.367
New Brunswick	67.3	65.3	26.0	26.9	1.8	1.8	0.4	0.3	1.1	1.4	0.2	0.2	0.7	0.9	0.2	0.1	0.9	0.4	2.1	2.8	0.404	0.415
Quebec	12.1	13.0	74.5	75.5	1.1	0.9	3.9	3.0	0.2	0.2	0.5	0.8	0.2	0.3	0.4	0.4	0.9	0.4	6.2	5.5	0.159	0.158
Ontario	57.0	57.9	10.6	11.1	5.6	5.8	7.2	5.4	2.2	2.5	2.0	2.6	0.8	1.0	2.3	2.3	1.7	0.8	10.6	10.6	0.527	0.540
Manitoba	44.6	46.7	8.4	8.1	11.0	10.0	1.6	1.1	2.7	3.1	4.6	5.1	3.4	3.8	11.6	11.0	1.3	0.6	10.8	10.4	0.679	0.667
Saskatchewan	46.8	48.3	5.5	5.4	18.1	15.6	0.5	0.5	2.2	3.2	2.8	2.9	5.1	5.4	9.2	7.6	1.4	0.9	8.4	10.1	0.721	0.719
Alberta	49.6	50.4	5.6	5.5	13.3	12.7	2.0	1.5	3.5	4.1	2.7	2.9	5.0	5.6	8.2	6.9	2.0	1.1	8.2	9.2	0.699	0.699
British Columbia	59.7	62.4	4.3	4.0	8.5	6.9	2.7	2.4	3.0	3.3	1.4	1.5	4.9	5.7	2.8	2.2	4.3	2.7	8.6	8.9	0.643	0.645
Size class (1971):																						
1 000 000 +	39.8	41.9	29.4	31.0	3.8	3.4	7.2	5.3	1.3	1.5	1.3	2.0	1.2	1.5	1.7	1.6	2.6	1.2	11.7	10.6	0.611	0.578
Montreal CMA	16.0	17.6	64.3	65.2	1.4	1.3	5.9	4.6	0.3	0.3	0.7	1.2	0.2	0.3	0.7	0.7	1.3	0.5	5.2	8.3	0.554	0.539
Toronto CMA	56.9	61.4	3.5	3.3	4.4	4.4	10.3	7.4	1.7	2.1	1.9	3.1	0.7	0.9	2.3	2.5	2.7	1.1	15.5	13.8	0.646	0.600
Vancouver CMA	58.6	61.9	4.0	3.9	8.3	6.6	2.8	2.3	3.0	3.2	1.4	1.6	4.8	5.8	2.9	2.3	5.4	3.1	9.0	9.2	0.635	0.597
250 000–1 000 000	45.9	46.0	23.1	23.4	7.3	6.9	3.7	2.9	2.4	2.8	2.3	2.7	1.9	2.1	4.8	4.4	1.3	0.7	7.3	8.1	0.598	0.594
100 000–250 000	57.8	55.6	15.5	17.0	9.0	8.9	2.0	1.7	1.7	2.1	1.7	1.8	1.8	2.0	3.1	2.9	1.1	0.7	6.3	7.3	0.495	0.503
50 000–100 000	52.1	50.6	33.2	34.7	2.8	2.6	3.4	2.9	1.7	1.8	1.2	1.2	0.5	0.6	1.8	0.8	0.6	0.3	3.8	4.5	0.379	0.392
30 000–50 000	42.8	40.6	38.0	40.7	5.2	4.0	1.6	1.6	2.1	2.3	1.0	1.2	2.0	1.9	1.8	1.4	1.1	0.8	4.5	5.5	0.422	0.427
20 000–30 000	44.4	44.2	39.7	38.9	6.1	5.9	0.8	0.7	1.6	1.9	0.9	0.9	1.5	1.6	1.2	1.1	0.7	0.5	3.0	4.2	0.346	0.362
10 000–20 000	46.6	44.6	32.1	34.1	6.2	5.6	1.5	1.4	1.9	2.3	1.2	1.3	2.4	2.6	2.6	2.1	0.9	0.6	4.9	5.4	0.462	0.477
Urban Canada	44.7	45.2	27.5	28.8	5.5	5.1	4.6	3.5	1.8	2.0	1.6	2.0	1.5	1.7	2.6	2.4	1.7	0.9	8.4	8.3	0.445	0.455
Nonurban Canada	44.4	41.2	31.3	33.4	7.5	7.0	0.5	0.5	2.5	3.0	1.2	1.4	2.4	2.8	2.8	3.0	0.4	0.3	6.9	7.5		
Canada‡	44.6	43.8	28.7	30.4	6.1	5.8	3.4	2.5	2.0	2.4	1.5	1.8	1.8	2.1	2.7	2.6	1.3	0.7	8.0	8.0		

* Index of ethnic diversity $= 1 - \Sigma\, p_i^2$, where $p_i =$ the proportion of an urban area's population in the ith ethnic group. Twelve ethnic groups were used: the first nine shown in this table, plus Russian, Jewish, and a residual "other" category. The higher the value of this index, the more ethnically diverse is the population. The entries in this table represent the simple (unweighted) mean of the index of ethnic diversity for the urban areas over 10 000 population in 1971 in each province and size class, and the mean for the 137 urban areas in Canada.

† Provincial data refer only to urban areas over 10 000 in 1971. All of Ottawa-Hull Census Metropolitan Area and all of Hawkesbury Census Agglomeration are included in Ontario. All of Flin Flon Census Agglomeration is included in Manitoba.

‡ The 1961 figures for Canada differ from those in Table 6 in the case of the German and Polish population because no adjustment was made to the Jewish population in this table.

Source: Table 5

The rather heterogeneous group called Asian also increased its representation to 1.3 percent in 1971. While the Chinese and Japanese were almost the only Asians in Canada until 1961, other Asian groups, especially Indians and Pakistanis, were approaching the Chinese and Japanese in numbers by 1971.

Native Indians and Eskimos comprised 1.4 percent of Canada's population in 1971. Taking into account the definitional changes in 1951[7], it appears that the Indian and Eskimo populations began to increase their share of Canada's population in the 1930's. Exceptionally high fertility rates and declining mortality rates among the native peoples account for their increasing percentage of the Canadian population.

Provincial Differences in Ethnic Origin, 1961 and 1971

The ethnic composition of urban Canada in 1961 and 1971, by province and urban size class, is presented in Table 7. All ethnic groups comprising at least 1 percent of urban Canada's population are shown, except Jewish. Because of the lack of comparability in Jewish ethnic origin data for 1961 and 1971, religious data provide a better indication of the distribution of Canada's Jewish population.

In urban Canada as a whole, the only ethnic group to increase its share of the population by more than 1 percentage point between 1961 and 1971 was the Italian, although the Asian group ran a close second. Cities in all provinces recorded an increase in the percentage of their population with Asian ethnic origin, and only the three easternmost provinces showed no increase in their percentage of Italian ethnic origin. Ontario cities remained the most Italian and British Columbia cities the most Asian.

The Germans and Ukrainians were the only other groups (of those identified in Table 7) to increase their representation in urban Canada. The only instance of a declining percentage of German or Ukrainian population was in Ontario, where the German population fell slightly. Western Canada remained the most German and the most Ukrainian. In 1971, nearly one-fifth of the population in Saskatchewan's cities was of German ethnic origin. Manitoba's cities continued to be the most Ukrainian.

The population of French ethnic origin registered the greatest decline in its share of urban Canada's population, but the decline amounted to only 1.3 percentage points. The cities in the least French provinces (Newfoundland and the four western provinces) recorded small increases in the percentage of their population with French ethnic origin, while all the others, including Quebec, became less French. Although the French are still a highly segregated ethnic group, they are slowly spreading across the country.

The British were the most numerous ethnic group in the urban population of all provinces except Quebec, but only in the Atlantic Provinces did they account for over 60 percent of the urban population. In the Prairie Provinces, less than half the urban population was of British ethnic origin in 1971. Only the three Maritime Provinces showed an increasingly British urban population. The population of Newfoundland's cities was still 95 percent British in 1971.

Canadians of Dutch, Polish, and Scandinavian ethnic origin were most numerous in western Canada. All three groups comprised a declining proportion of urban Canada's population, and a slight increase in Newfoundland's tiny percentage of Dutch ethnic origin was the only instance of an increase in any of these three groups in the cities of any province between 1961 and 1971.

Canada has some very French cities and some very British cities. Montmagny was 99 percent French ethnic origin in 1971, and St. John's was 96 percent British. In very few cities are the French and British found in even roughly the same proportions. The non-British, non-French element is strongest in the West. In no province did the non-British, non-French proportion exceed 50 percent, but Saskatchewan's cities approached this figure in 1971.

Canada's ethnic minorities occasionally reach very high levels in Canadian cities. The top-ranking urban areas in 1971, according to the percentage of their population in each ethnic minority, are as follows: New Hamburg CA (53 percent German), Yorkton (30 percent Ukrainian and 6 percent Polish), Trail CA (17 percent Italian), Lincoln (13 percent Dutch), Kenora CA (11 percent Scandinavian), and Leamington (7 percent Asian). Most of Canada's ethnic minorities have highly skewed distributions. They are virtually absent in many cities, and constitute significant minorities in other cities, occasionally exceeding 10 percent.

Less than 1 percent of the population in nearly one-third of Canada's urban areas was of German ethnic origin in 1971. All of these urban areas are east of Ontario. In Ontario, Germans were most prevalent in the southwestern and midwestern parts of the province, particularly around Kitchener, but two urban areas (Pembroke and Petawawa) in the upper Ottawa valley were of more than 10 percent German ethnic origin. Prince Albert was the only urban area in Saskatchewan or Alberta to be less than 10 percent German. Regina, Swift Current, and Medicine Hat all exceeded the 20 percent German threshold in 1971.

Like the Germans, the Italians were absent from most cities east of Ontario, except Montreal. Although the Toronto CMA had 272 000 persons of Italian ethnic origin (some 10 percent of its population in 1971), on a percentage basis, Trail and Sault Ste. Marie were both more Italian than Toronto. Only fourteen urban areas were more than 5 percent Italian ethnic origin in 1971. In sixty of the 137 urban areas, persons of Italian ethnic origin accounted for less than 1 percent of the population. Italian Canadians have highly selective locations.

East Europeans were also rare in most Canadian cities. Cities in the Prairie Provinces, especially in Manitoba and in the parkland areas of Alberta and Saskatchewan, had by far the highest percentages of East European population. Northwestern Ontario cities (Thunder Bay and Kenora) were similar to the Prairies in this respect. Within Ontario, East Europeans were most common around the Golden Horseshoe area (the west end of Lake Ontario) and in the northern Ontario resource towns.

Urban Size Differences in Ethnic Origin, 1961 and 1971

Urban size differences in ethnic origin are less important than provincial differences. As was the case with mother tongue, birthplace, and immigration data, however, some ethnic groups show a fairly strong relationship with urban size.

Consider first the broad distinction between urban and rural living. People of French, German, Dutch, Scandinavian, and Ukrainian ethnic origin formed a larger percentage of nonurban Canada than of urban Canada in 1961 and 1971. The greater preference among these groups for rural life, in comparison with the Canadian population as a whole, has also been noted in earlier censuses. Recent immigration, however, has tended to bring the proportion of urban Canada's population in each of these groups closer to their proportion in nonurban Canada. The continued urbanization of the native-born population of these ethnic origins, and of the earlier immigrants of these origins, has also accelerated the convergence of the ethnic composition of urban and nonurban Canada as far as these groups are concerned.

People of British, Italian, Polish, and Asian ethnic origins (as well as those in the residual "other" category) are more common in urban Canada than in the rest of the country. Italians and Asians show the strongest tendency in this respect, and the differences between urban and nonurban Canada in the percentage of Italian and Asian increased between 1961 and 1971. The difference between urban and nonurban Canada in the percentage of British ethnic origin all but disappeared by 1971, as urban Canada became less British and nonurban Canada more British. It is most unlikely that British immigration to nonurban Canada has been responsible for its increasingly British ethnic composition. Rather, the continued drift of the prewar non-British immigrants and their descendants from the Prairie farms and small towns to urban Canada is probably the main component of ethnic change in nonurban Canada.

Asians, Italians, and the residual "other" category have a strong preference for the upper end of the urban hierarchy. The French were most common in the 10 000-100 000 size classes, because Quebec is over-represented by cities in this size range. With all the ethnic groups, in fact, it is really necessary to examine the ethnic composition of cities by size class *within* regions or provinces because of the very strong regional dimension to the Canadian ethnic mosaic.

An Index of Ethnic Diversity

Although the "ethnic geography" of Canada is usually described in terms of the varying distribution of each ethnic group in turn, the concept of ethnic diversity provides a single summary measure of the ethnic composition of the population. The ethnic diversity of a city measures the potential for day-to-day contact of people of different ethnic origins. Ethnic diversity also influences perception of Canadian society. The residents of a city as diverse as Winnipeg must surely have a much better understanding of what cultural pluralism means in Canada than the inhabitants of Montmagny or St. John's.

Table 7 presents the average indexes[8] in the cities within each province and urban size class. In 1971, Winnipeg, with an index of 0.770, was Canada's most ethnically diverse city, and Montmagny was the most homogeneous, with an index of 0.021. As shown in Table 7, Quebec's cities were the most homogeneous, followed closely by those of Newfoundland.[9] Ethnic diversity is highest in Saskatchewan's cities, but all four western provinces have very diverse cities. Larger cities tend to be more diverse than small ones, although the relationship between size and diversity does not hold below the 100 000 mark. Size class differences in ethnic diversity are confounded by regional differences, which are by far the greater of the two.

The average Canadian city changed very little in its ethnic diversity between 1961 and 1971. The average index for the 137 urban areas declined marginally from 0.455 to 0.445. Ethnic diversity in Canada and urban Canada, however, increased slightly, while nonurban Canada became a little more homogeneous.[10] In view of the amount of immigration into Canada between 1961 and 1971, it is surprising that Canadian cities did not increase their ethnic diversity, particularly since many recent immigrants have come from nontraditional sources. Perhaps if the indexes were calculated using a much larger number of ethnic categories, more change in the index would be apparent. Too many ethnic groups may have been subsumed under the rubric, "other".

The cities of Atlantic Canada showed the most change in ethnic diversity between 1961 and 1971. Most became less diverse. The province whose cities showed the largest increase in ethnic diversity was Manitoba, but the increase was very small. The only size categories of cities to increase their

ethnic diversity were the two classes over 250 000. Toronto, Montreal, and Vancouver all became more diverse.

The Religious Denominations of the Population

The National Trend

The final cultural characteristic considered here is the religious denomination of Canada's urban population. Although the empty pews in so many of Canada's houses of worship would lead many to deny the importance of religious beliefs as a means of stratification of Canada's urban population, the continuing significance of religious activity to many millions of Canadians should not be ignored. That almost everyone in Quebec City is a Roman Catholic while Winnipeg has a sizeable Jewish minority, for example, continues to be an important difference between these two cities.

The trend in the religious denomination of Canada's population since 1901 is presented in Table 8. All denominations which have accounted for 1 percent of Canada's population, or 100 000 persons, in any year since 1901 received separate entries in the table.

Historical religious data are complicated by mergers and partial mergers of denominations. The most significant change was the creation of the United Church, in 1925, by the merger of the Methodist, Congregationalist and Presbyterian Churches. Many adherents of the Presbyterian Church, however, did not join the United Church, and Presbyterian continued to be a major Protestant denomination in Canada. Smaller numbers of Methodists and Congregationalists also continue to maintain their own identity.[11]

The religious composition of Canada has not changed a great deal since 1901, but a few trends are apparent from Table 8. The Roman Catholic Church remains the largest denomination and has increased its percentage of Canada's population in each census since 1921, from 39 percent in 1921 to 46 percent in 1971. The second largest denomination is the United Church, whose share of Canada's population fell to 18 percent in 1971 from its peak of 21 percent in 1951. Heavy immigration from the British Isles in the first twenty years of this century raised the Anglican Church's share to 16

RELIGIOUS DENOMINATION OF CANADA'S POPULATION, 1901—71*

Religious Denomination	1901	1911	1921	1931	1941	1951 Excl. Nfld.	1951 Incl. Nfld.	1961	1971
	%	%	%	%	%	%	%	%	%
Anglican	12.8	14.5	16.1	15.8	15.2	14.3	14.7	13.2	11.8
Baptist	5.9	5.3	4.8	4.3	4.2	3.8	3.7	3.3	3.1
Greek Orthodox†	0.3‡	1.2‡	1.9‡	1.0	1.2	1.3	1.2	1.3	1.5
Jehovah's Witness	—	—	0.1	0.1	0.1	0.2	0.2	0.4	0.8
Jewish	0.3	1.0	1.4	1.5	1.5	1.5	1.5	1.4	1.3
Lutheran	1.8	3.2	3.3	3.8	3.5	3.3	3.2	3.6	3.3
Mennonite§	0.6	0.6	0.7	0.9	1.0	0.9	0.9	0.8	0.8
Penecostal	—	—	0.1	0.3	0.5	0.6	0.7	0.8	1.0
Presbyterian	15.8	15.6	16.1	8.4	7.2	5.7	5.6	4.5	4.0
Roman Catholic	41.7	39.4	38.7	39.5	41.8	43.6	43.3	45.7	46.2
Salvation Army	0.2	0.3	0.3	0.3	0.3	0.3	0.5	0.5	0.6
Ukrainian Catholic	—‖	—‖	—‖	1.8	1.6	1.4	1.4	1.0	1.1
United Church	17.2¶	15.1¶	13.3¶	19.5	19.2	20.4	20.5	20.1	17.5
Other	3.3	3.3	3.1	2.7	2.6	2.3	2.2	2.8	2.7
No religion	0.1	0.4	0.2	0.2	0.2	0.4	0.4	0.5	4.3
Total	100.0	100.0	100.0	100.0	100.0	100.0	100.0	100.0	100.0

— Less than 0.05 percent

* Exclusive of Newfoundland in censuses prior to 1951.

† Includes those churches which observe the Greek Orthodox rite, e.g., Russian Orthodox, Syrian Orthodox and Ukrainian Orthodox.

‡ Includes Ukrainian Catholic in 1901, 1911 and 1921.

§ Includes Hutterite.

‖ Included with Greek Orthodox in 1901, 1911 and 1921.

¶ Includes Methodists.

Source: Canada, Statistics Canada, *1971 Census of Canada: Population: Religious Denominations,* Bulletin 1.3-3, Cat. No. 92-724 (Ottawa: Information Canada, 1973);
Canada, Dominion Bureau of Statistics, *1961 Census of Canada: Population: Religious Denominations,* Bulletin 1.2-6. Cat. No. 92-546 (Ottawa: Queen's Printer, 1962);
Canada, Dominion Bureau of Statistics, *Ninth Census of Canada, 1951: Volume I: Population: General Characteristics* (Ottawa: Queen's Printer, 1953), Table 38;
Canada, Department of Trade and Commerce, *Fifth Census of Canada, 1911: Volume II* (Ottawa: King's Printer, 1913), Table I.

Table 8

percent in 1921, but it has since fallen steadily to 12 percent in 1971. The Presbyterian Church dwindled to 4 percent in 1971, and the percentage of Baptist religion also declined in each census since 1901. Three fundamentalist churches — the Jehova's Witnesses, the Pentecostal, and the Salvation Army — have experienced steady and rapid increases, but they still do not surpass 1 percent each.[12] The Lutheran Church has fluctuated around the 3 to 4 percent level since 1911.

The churches with East European origins — Greek Orthodox and other Orthodox churches, and the Ukrainian Catholic — each accounted for just over 1 percent of Canada's population in 1971. These are relative newcomers to the religious scene in Canada, having had a combined percentage of only 0.3 percent in 1901. Since 1931, the Ukrainian Catholic Church has accounted for a declining percentage of Canada's population, whereas the Orthodox Churches have continued to increase their share.

Canada's Jewish population grew rapidly during the first two decades of this century. Since 1921, between 1.3 and 1.5 percent of Canadians have been of the Jewish faith.

Between 1961 and 1971 there was nearly a tenfold increase in the percentage reporting "no religion". This category accounted for 4 percent of Canadian residents in 1971.

*Provincial Differences in Religion,
1961 and 1971*

Provincial differences in the religious denominations of the population of urban Canada in 1961 and 1971 are presented in Table 9. Two groupings of Protestant denominations have been made on the basis of similarity of religious beliefs. One group called Principal Protestant consists of the Anglican, Lutheran, Presbyterian, and United Church denominations, the four largest Protestant churches in Canada. The second group, consisting of the Baptist, Mennonite (including Hutterite), and Pentecostal religions, has been termed the Principal Fundamentalist group. These groupings are not meant to imply that each of the component denominations of each group has a similar distribution in Canada. In fact, they do not.

Provincial differences in the religious composition of urban Canada are related to differences in birthplace, immigration, mother tongue, and ethnic origin. The Greek Orthodox and Ukrainian Catholic religions, for example, are most common in western Canada and Ontario, where East European origins are most in evidence. In the cities of the Atlantic Provinces, Protestants and Roman Catholics are found in approximately

the same numbers. The Fundamentalist group is strongest in New Brunswick cities, where this group is mainly Baptist. In the cities of Ontario and the West, the Principal Protestant group comprises approximately half the population. Jews are most common in the cities of Manitoba, Quebec, and Ontario. The residual "other" category is largest in the West, reaching 21 percent in urban British Columbia. The high percentages in the "other" category in the West are largely accounted for by the large numbers of people of no religion living there.

Provincial differences in religious denominations in urban Canada changed somewhat between 1961 and 1971, although the basic differences remain. Cities in Ontario and the West became more Roman Catholic and considerably less Protestant. The Greek Orthodox Church became more significant in Ontario and Quebec cities. The Jewish proportion declined in most provinces, and the "other" category increased in all provinces, largely because of the rapidly increasing numbers of people reporting no religion.

The maximum values reached for the various religious groupings in any of the 137 urban areas over 10 000 in 1971 were in the following cities: Matane (99 percent Roman Catholic), Cobourg CA (69 percent Principal Protestant), Kentville CA (43 percent Principal Fundamentalist, mainly Baptist), Yorkton (15 percent Ukrainian Catholic and 8 percent Greek Orthodox), Montreal CMA (4 percent Jewish), and Vancouver CMA (23 percent "other", mainly no religion).

*Urban Size Differences in Religion,
1961 and 1971*

With few exceptions, the urban size differences in religions in urban Canada are small. The most outstanding exception is the heavy concentration of Jews in the largest urban areas, and their almost complete absence from cities under 100 000 and from rural areas. In fact, the Jewish population exceeds 1 percent in only four urban areas: Montreal CMA (4.0 percent), Toronto CMA (3.9 percent), Winnipeg CMA (3.4 percent), and Ottawa-Hull CMA (1.1 percent). The Greek Orthodox also favour the upper end of the urban hierarchy.

Table 9

RELIGIOUS DENOMINATION OF URBAN POPULATION, BY PROVINCE AND SIZE CLASS, 1961 AND 1971

Province and Size Class (1971)	Religious Denomination														Average Index of Religious Diversity¶	
	Roman Catholic		Principal Protestant*		Principal Fundamentalist†		Greek Orthodox‡		Jewish		Ukrainian Catholic§		Other‖			
	1971 %	1961 %	1971 %	1961 %	1971 %	1961 %	1971 %	1961 %	1971 %	1961 %	1971 %	1961 %	1971 %	1961 %	1971 %	1961 %
Province:**																
Newfoundland	44.7	44.6	46.0	47.3	3.0	2.4	0.0	0.0	0.1	0.1	0.0	0.1	6.2	5.5	0.714	0.701
Prince Edward Island	47.7	48.8	41.0	41.5	6.2	6.3	0.1	0.1	0.0	0.0	0.1	0.0	4.8	3.3	0.692	0.680
Nova Scotia	42.5	42.5	43.8	46.5	8.0	7.8	0.3	0.2	0.5	0.5	0.2	0.2	4.7	2.2	0.698	0.694
New Brunswick	47.4	45.8	33.3	36.2	14.3	15.3	0.1	0.1	0.3	0.4	0.1	0.0	4.6	2.1	0.554	0.558
Quebec	84.0	85.0	8.1	9.7	0.9	0.5	1.4	0.9	2.6	2.9	0.5	0.2	2.6	0.7	0.113	0.099
Ontario	36.3	32.5	45.1	53.6	4.5	4.8	2.1	1.6	2.0	2.2	0.8	0.9	9.2	4.5	0.738	0.734
Manitoba	25.5	23.1	48.1	55.2	6.1	5.6	2.5	2.8	3.0	3.7	5.6	5.6	9.2	4.0	0.810	0.792
Saskatchewan	26.4	23.4	52.2	59.2	5.2	5.3	2.8	3.1	0.4	0.7	3.4	2.8	9.5	5.0	0.806	0.796
Alberta	23.8	21.7	50.6	59.4	5.2	5.0	2.8	2.9	0.6	0.8	2.3	1.9	14.7	8.3	0.816	0.810
British Columbia	18.4	16.5	53.0	66.5	5.4	5.3	1.0	1.0	0.6	0.6	0.6	0.4	21.1	9.6	0.809	0.786
Size class (1971):																
1 000 000 +	49.1	47.6	31.8	39.5	2.9	2.8	2.4	1.7	3.4	4.0	0.7	0.6	9.6	3.8	0.675	0.666
Montreal CMA	77.8	78.4	11.1	13.7	1.1	0.6	2.1	1.4	4.0	4.6	0.5	0.3	3.4	1.0	0.388	0.376
Toronto CMA	32.0	25.7	45.2	57.8	3.9	4.2	3.2	2.3	3.9	4.6	1.0	1.1	10.6	4.3	0.814	0.813
Vancouver CMA	17.9	16.2	51.4	66.1	5.3	5.3	1.2	1.2	0.8	0.9	0.6	0.5	22.8	9.9	0.825	0.810
250 000–1 000 000	42.4	40.5	39.8	45.9	4.3	4.3	1.8	1.9	1.0	1.3	1.8	1.7	8.9	4.4	0.695	0.687
100 000–250 000	39.4	38.8	44.6	49.0	5.2	5.5	1.1	1.1	0.3	0.4	1.1	1.0	8.3	4.1	0.694	0.685
50 000–100 000	53.7	53.5	34.4	37.3	5.0	5.0	0.4	0.3	0.3	0.4	0.4	0.3	5.9	3.2	0.571	0.566
30 000–50 000	51.9	53.5	33.6	36.7	4.7	4.3	0.5	0.5	0.2	0.2	0.6	0.5	8.4	4.3	0.547	0.530
20 000–30 000	53.3	51.7	35.3	38.8	4.2	4.1	0.4	0.4	0.1	0.2	0.4	0.3	6.2	4.5	0.524	0.513
10 000–20 000	47.5	47.7	37.3	40.7	5.8	5.7	0.7	0.8	0.1	0.2	0.9	0.8	7.7	4.2	0.593	0.585
Urban Canada	46.8	45.7	35.9	41.9	4.0	3.9	1.7	1.4	1.8	2.1	1.0	0.9	8.8	4.0	0.589	0.579
Nonurban Canada	45.0	45.8	38.3	40.4	7.3	6.7	0.9	1.1	0.0	0.1	1.2	1.3	7.4	4.6		
Canada	46.2	45.7	36.6	41.4	5.0	4.9	1.5	1.3	1.3	1.4	1.1	1.0	8.4	4.2		

* Anglican, Lutheran, Presbyterian, and United Church. United Church includes Congregationalists and Methodists in 1961.

‡ Includes those churches which observe the Greek Orthodox rite, such as Russian Orthodox, Ukrainian Orthodox, and Syrian Orthodox.

‖ Includes those with no religion.

† Baptist, Mennonite (including Hutterite), and Penticostal.

§ Includes "Other Greek Catholic".

¶ Index of religious diversity = $1 - \Sigma\ pi^2$, where pi = the proportion of an urban area's population of the ith religious denomination. Twelve denominations were used: Anglican, Baptist, Greek Orthodox, Jewish, Lutheran, Mennonite (including Hutterite), Pentecostal, Presbyterian, Roman Catholic, Ukrainian Catholic, United Church and a residual "other" category (including "no religion"). The higher the value of this index, the more religiously diverse is the population. The entries in this table represent the simple (unweighted) mean of the index of religious diversity for the urban areas over 10 000 population in 1971 in each province and size class, and the mean for the 137 urban areas in Canada.

** Provincial data refer only to urban areas over 10 000 in 1971. All of Ottawa-Hull Census Metropolitan Area and all of Hawkesbury Census Agglomeration are included in Ontario. All of Flin Flon Census Agglomeration is included in Manitoba.

An Index of Religions Diversity

An index of religious diversity, analogous with the index of ethnic diversity, was calculated for each urban area in 1961 and 1971. Twelve religious denominations were used: Anglican, Baptist, Greek Orthodox, Jewish, Lutheran, Mennonite (including Hutterite), Pentecostal, Presbyterian, Roman Catholic, Ukrainian Catholic, United Church, and a residual "other" category (including "no religion"). The provincial and size-class summaries are included in Table 9.

The average Canadian city became only slightly more diverse in its religious composition between 1961 and 1971.[13] The cities in all provinces, except New Brunswick, and of all size classes became marginally more diverse. No city changed its degree of religious diversity a great deal.

Quebec's cities were by far the most homogeneous in religion, being 84 percent Roman Catholic in 1971. The four western provinces were the most diverse in religious denomination, as they were in ethnic origin, and Ontario followed the West according to both measures. The Atlantic Provinces, however, were not a great deal less heterogeneous in religion than were the western provinces and Ontario, although in ethnic diversity they fell far behind. The Atlantic Provinces lacked the religions associated with the immigration streams of the twentieth century – the Ukrainian Catholics, the Greek Orthodox, the Jews, and the Lutherans – but several Protestant denominations and Roman Catholics were represented in sufficient numbers in the Atlantic Provinces to give their cities a fair degree of religious diversity.

Conclusion

Many factors account for the great cultural differences among Canada's urban areas. The single most important determinant of a city's cultural mix is the date at which settlement first occurred because, in spite of the mobility of the Canadian population and continued high immigration rates, many of the inhabitants of a city are descendants of the area's first wave of settlers. Each city acquired the ethnic, linguistic, and religious characteristics of the immigrant stream that was entering Canada at the time the area was settled. The Prairie cities, for example, acquired their peculiar ethnic mix because immigrants from Eastern and Central Europe were a much more important component of the immigration stream around the turn of the century when the Prairies were opening up than they were during the nineteenth-century settlement of Ontario.

Although the ethnic composition of the immigrant population at the time of settlement is very important, other factors determine the modern ethnic-cultural mix in each city. First, the ability of a city to attract migrants from other parts of Canada is also significant. Winnipeg, for example, derived its French component not from immigrants from France but from migrants from Quebec. Secondly, a city's continued power to attract immigrants long after the initial settlement period is a very important component of ethnic change. To some extent, immigration often reinforces cultural differences among cities. Jewish immigrants go to Toronto and Montreal where Jewish areas and institutions are to be found; Scandinavian immigrants go to Vancouver; and so on. Undoubtedly, immigrants' knowledge of city differences in Canada, often gathered through personal contact with friends or relatives who immigrated earlier, is the main cause of the continued selectivity in immigrant destinations. Thirdly, the age structure of each cultural group and its tendency to marry within, or outside, itself influence its growth rate and survival. Finally, emigration from Canada is also one of the demographic components of a city's ethnic change and character.

Each of these factors is in turn related to other noncultural aspects of Canadian cities. Internal migrants and immigrants are attracted to cities whose economies provide the types of jobs in which their cultural group specializes. Admittedly, Canadians of all ethnic origins are found in all occupations, but there is a degree of occupational specialization of ethnic groups which is translated into a weak relationship between ethnicity and the occupation structure of Canadian cities. Mining towns and cities with a lot of heavy manufacturing industry, for example, usually have high percentages of

Italians. Jews are found in cities where their managerial, professional, and salesmanship skills are most in demand.

The most basic component of cultural differentiation of Canadian cities, however, is the regional dimension. Regional differences greatly exceed urban size differences in all the broad categories of cultural measures used – birthplace, mother tongue, ethnic origin, and religion. These differences penetrate nearly every aspect of Canadian life. People and policies in Canada are, and need to be, sensitive to the cultural diversity of urban Canada.

NOTES AND REFERENCES

[1] Porter, John, *The Vertical Mosaic* (Toronto: University of Toronto Press, 1965); Wallace Clement, *The Canadian Corporate Elite,* (Toronto: McClelland and Stewart, 1975).

[2] Strictly speaking, the immigrant and foreign-born population are not identical since the small number of Canadian-born who leave Canada and subsequently return as immigrants are not included in the foreign-born total. Unless specifically stated otherwise, however, the terms "immigrant" and "foreign-born" are used interchangeably and exclude Canadian-born returning immigrants.

[3] The term "urban Canada" refers to 137 census metropolitan areas (CMA's), census agglomerations (CA's), and other urban areas over 10 000 population in 1971, and differs from the official Statistics Canada definition of urban.

[4] Quoted in Canada, Department of Manpower and Immigration, *The Immigration Program,* Volume 2 of the "Green Paper" on Immigration (Ottawa: Information Canada, 1974), p. 10.

[5] Most of the Asian-born in Leamington are from the Middle East, especially Lebanon, whereas the Chinese, Indians and Pakistanis are the major Asian groups in most Canadian cities.

[6] The identification of ethnic origin poses many problems. For example, some people do not know their ethnic origin whereas others have a mixed ancestry. In addition, problems of changing census definitions and procedures sometimes make it difficult to make accurate historical comparisons.

[7] Prior to 1951, racial origin (the term previously used for ethnic origin), was determined by colour for Indians and Eskimos. Persons of mixed white and nonwhite parentage were considered to belong to the nonwhite group until the 1951 census when the usual role of tracing ethnic origin through the male was adopted for persons of mixed colour. In the 1941 census, persons of mixed white and North American Indian ancestry were recorded as "Half-breed". In 1951, persons of mixed Indian and white parentage were enumerated as Native Indian if they lived on a reserve and according to their male ancestry if they lived elsewhere.

[8] The index of ethnic diversity for each urban area was calculated according to the following formula:

Index of Ethnic Diversity $= 1 - \sum P_i^2$

where P_i equals the proportion of an urban area's population in the ith ethnic group. In other words, one sums the squares of the proportion of the population in each ethnic group and subtracts this total from one. The index assumes a minimum value of zero when everyone in the urban area has the same ethnic origin. The maximum value of the index depends on the number of ethnic groups used in the calculation. Twelve groups were used here (the first nine shown in Table 7, plus Jewish, Russian and "other"), yielding a maximum value of 0.913 if each ethnic origin accounted for one-twelfth of the population.

[9] The figure for Newfoundland in 1961 is much higher than expected. The reason is that in Labrador City, which was little more than an encampment in 1961, many of the people were of "not stated" ethnic origin included in the "other" category, giving Labrador City an index of 0.637. Even in 1971, however, Labrador City was much more diverse than other Newfoundland cities.

[10] The index of ethnic diversity for Canada increased from 0.705 in 1961 to 0.708. The comparable figures for urban Canada are 0.703 in 1961 and 0.713 in 1971; and for nonurban Canada, 0.706 in 1961 and 0.693 in 1971. The values derived by calculating the index for urban Canada as a whole are not to be confused with the *average* indexes for the 137 urban areas, shown in Table 7.

[11] In 1968 the United Church also absorbed the much smaller Evangelical United Brethren Church, which accounted for 0.1 percent of the population of Canada in 1961.

[12] The addition of Newfoundland in 1949 increased the Salvation Army percentage to 0.5 percent in 1951, compared to 0.3 percent for Canada excluding Newfoundland in 1951, and also increased the percentage Pentecostal by 0.1 percent.

[13] The religious diversity index for Canada as a whole also increased almost imperceptibility from 0.727 in 1961 to 0.731 in 1971. The index for urban Canada as a whole was 0.727 in both years, while nonurban Canada's index increased from 0.723 to 0.734.

Some Myths of Canadian Urbanization: Reflections on the 1976 Census and Beyond

L. S. BOURNE

One of the occupational hazards of empirical research is that it depends so heavily on statistics and images from past periods. One result of this dependence is that emerging trends tend to be overlooked until they are obvious, and thus firmly established. Other trends pass us by without detection. Inevitably we find ourselves analyzing problems and processes of the past rather than those of the present or future; processes which may have changed their form or even their direction. When attention is turned to projecting the future on the basis of these analyses, the resulting errors in interpretation can become seriously magnified.

This paper examines some of these misconceptions as they relate to trends and future prospects in the study of urbanization in Canada. The examples selected for discussion derive from both research findings on urban growth as well as from positions evident in public policy statements. Admittedly, in the research context one of our difficulties is that there have been few systematic analyses of growth in the Canadian urban system.[1] Much of our literature, our generalizations, even our political platforms are built on a relatively thin inheritance of research.

This paper is a critique of current thinking, based on preliminary 1976 Census data[2] and on parallel studies underway on migration, economic growth, and spatial population forecasting. The purpose here is to debate, and thereby to assist in setting aside, some well-established misconceptions (perhaps they are myths) regarding urbanization in Canada, so that we can get on with the detection and analysis of emerging trends in urban growth, unencumbered by erroneous conceptual baggage from the past. The paper concludes with comments on some of the policy implications of these emerging trends, and a plea for more emphasis on futures-oriented and politically-sensitive research.

Background: The Conventional Wisdom

The literature on Canadian urbanization is now sufficiently diverse to provide any reviewer with ample opportunity for debate and criticism. The following discussion selects from this literature a few widely-quoted perspectives on present and future urban growth prospects in Canada. These serve as examples of our preoccupation with the immediate past, the misdirectedness of much of our policy thinking, and the relative impoverishment of our urban research. Many of the examples, based as they are on the conventional wisdom and empirical data of the 1950's and 1960's, now seem to be incorrect or at best simply misleading. Although it may not be possible at present to replace this collective wisdom with a new one, it is possible at least to raise questions about that wisdom as a beginning to defining alternative urban scenarios for Canada.[3] The positions outlined below then become hypotheses for which this paper argues the case for rejection or, in some cases, reserving judgement.

Among the most common preoccupations one finds in the recent urban literature, either explicitly or implicitly, and to which many of us have personally contributed, the following are particularly prominent.

1. *A growth fixation.* This position is most widely expressed in the persistent assumption that national population growth in general, and urban growth in particular, will continue, although perhaps not as rapid as

● LARRY BOURNE is Director of the Centre for Urban and Community Studies and is also in the Department of Geography, University of Toronto. This paper was presented at the Annual Meetings of the Canadian Association of Geographers, Regina, 1977, and is reprinted with permission.

previously, at a high rate in forthcoming decades. In other words, high rates of growth are still a problem in and of themselves. Interestingly, the same assumption seems to underlie the arguments of those who view growth of any kind as some sort of social disease.

2. *Metropolitan concentration.* This assertion is that the massive growth in population assumed above will continue to be concentrated, if not at an increasing rate, in the largest metropolitan areas. It is frequently asserted that over a third of all growth in Canada between now and the end of the century will be located in Monteal, Toronto, and Vancouver, and it is implied that this trend is both unusual and unfortunate.

3. *Regional depopulation and metropolitan growth.* The argument here is that the growth of these metropolitan centres is based on the depletion of rural areas and small-town Canada of their resident population, particularly in the Atlantic region, Quebec, and the Prairies.

4. *Net migration balance.* It is widely argued that the balance of interregional and interurban migration flows in Canada is, and will continue to be, highly skewed in favour of the major metropolitan areas.

5. *Heartland-periphery dichotomy.* A typical forecast of our urban future is that the established industrial and financial heartland of the nation will continue to increase its dominance of the national space economy by capturing the structural benefits, as well as income, deriving from agglomeration economies and technological innovations.

6. *Growth, stability, and size.* A more specific assertion is that the stability in the variance of urban growth rates over time is strongly and positively related to city size and to position in the urban hierarchy.

7. *Institutional reorganization.* The position voiced more and more frequently is that an effective national urban strategy (or strategies) is possible only by reorganizing the activities and coordinating the policy objectives of different levels and agencies of government.

These generalizations are of more than strictly academic interest, since each in turn leads logically to, or at least encourages, a certain defined range of political and policy responses. The first – the image of continued growth – led in the early 1970's to an almost panic-type response by both federal and provincial governments in Canada. The crisis of urban growth became a household phrase and an excuse for political action (or inaction). Hastily-conceived policies and band-aid programs were proposed to meet or avert the impending tide of population growth and urban expansion. The Canadian urban literature is replete with proposals for new towns, satellite cities, and growth poles. All require a continued input of new growth into the overall urban system.

Moreover, most policies and programs designed to achieve population redistribution and more balanced regional development are framed within a context which assumes that rapid growth will continue. This, in turn, leads governments to assume that simply influencing the marginal rate and distribution of new increments to the urban system will suffice to alleviate problems of interregional inequalities and growth instability. Growth, it seems, creates the rationale for implementation of macro-level urban policies (growth management), while at the same time it provides the necessary instruments (growth redistribution) for the success of those policies. Neither of these arguments now holds true, if they ever did; and both tend to divert attention from those more serious issues which are entrenched in our social system.

The second and third statements, on the relationships between metropolitan concentration and rural depopulation, focus the growth debate on specific locations within (and outside) the urban system. The obvious fact that an increasing proportion of our national population has tended to reside in the largest cities has led directly to proposals for limiting the growth of these cities and for the decentralization of jobs, industries, and population from these cities to new towns or development regions. The diffusion of planning ideas and policies from Europe to Canada has clearly had a lot to do with these emphases. It is also implied here that the successful application of growth-restrictive measures to the larger cities will at the same time solve, or at least stabilize, the second problem: that is, the depopulation of rural and peripheral regions. Unfortunately, the relationship between metropolitan containment on the one hand and accelerated growth in disadvantaged regions on the other is neither direct nor strong. In a relatively open spatial system such as Canada, restrictions on a growing metropolitan region may simply slow growth in the system overall, with little or even a negative impact on the declining regions.

The fourth position frequently leads policy-makers to recommend, in direct contradicition to their desire to restrict metropolitan growth, the improvement and expansion of urban infrastructure in the larger centres, such as in social services and housing, to accommodate the expected flood of new migrants. This is indeed a real dilemma. An improved urban infrastructure will *ceterius paribus* encourage more immigrants to those centres. To withhold the required investments in social infrastructure is to invite a deterioration in working and living conditions which is politically unacceptable. The above emphasis then is not wrong but oversimplified. It tends to ignore the complex interdependencies inherent in the growth process itself.

The fifth and sixth positions have lead to less direct and easily identifiable policy responses. However, enveloped in the heartland focus of most Canadian policy thinking is a clear emphasis on the production process and on manufacturing specifically, as the dominant component in generating urban growth. The traditional argument is that if manufacturing were spread more evenly across the country, particularly by shifting industry into slow-growth areas, then balanced urban development would follow. But manufacturing is no longer a dominant growth sector in the national economy.[4] In fact manufacturing employment is now almost at zero growth. What then do the policy-makers have to redistribute? Would such policies simply encourage more competition between individual cities and regions? In a no-growth or slow growth situation, does such competition become essentially a "no win" situation when viewed at the national level?

The sixth position listed above, on growth stability and city size, has also lead to an interest in establishing growth centres in lagging regions. Generally, such policies attempt to stimulate the development of existing urban centres or regional urban clusters, or to create entirely new centres, which are of sufficient size to ensure self-sustaining and consequently more stable growth. Clearly more stable growth is a valid and reasonable objective for public policy. The problem with induced growth centres, however, is that they seldom work, at least usually not in the way intended.[5] Most have been too narrowly conceived, often with an overly dominant focus on heavy industry and a limited range of incentives for relocation. Few have given adequate consideration to the need to relate the incentives to those economic sectors most vulnerable to external market instability. In any case a new question emerges: in a system which is not growing overall, do policies designed to encourage new growth centres become irrelevant?

If, as argued above, many of our conceptualizations of the trends and structural relationships on which these responses are based are highly oversimplified, increasingly out-dated, or simply incorrect, then the resulting political actions are likely to be equally misdirected and possibly counterproductive. The next section of this paper attempts to demonstrate that these assertions are, in large part, myths, deriving from inadequate information or prefunctory analysis, and that new and substantially different policy initiatives are required for managing future trends in urbanization.

Retiring the Myths?

With the mixed benefits of hindsight and more recent data, let us examine more fully each of these arguments in turn. The first point, our apparent fixation with the inevitability of urban growth, is now surely outmoded. Almost everyone is now aware that rates of overall population growth in Canada have declined steadily since the mid-1960's, and in a dramatic fashion in the last few years (Table1). Although most observers know this is happening, neither our research community nor our policy architects have done much to explain it or to assess its social and spatial consequences.[6]

The same trend of a declining population growth rate is prevalent throughout the developed world, and there are small yet significant signs of its emergence in the Third World. Zero population growth is now a reality in several European countries and in many other parts of the world.[7] Moreover, this decline has been most apparent in the metropolitan areas. Almost all major capitals

Table 1

HISTORICAL COMPONENTS OF POPULATION GROWTH, CANADA, 1851–1976

Decade	National				Urban System[a]			
	Initial Population (10³)	Average Annual Growth %	Ratio of Natural Increase to Total Growth %	Ratio of Net Migration to Total Growth %	No. of Urban Centres	Total Population (10³)	% of Total Population Urban	% Growth of Urban Population
1851–61	2 436	2.9	77.0	23.0	8	259	10.0	38.3
1861–71	3 230	1.3	132.6	-32.6	10	358	10.6	27.6
1871–81	3 689	1.6	108.5	-8.5	12	457	11.9	37.2
1881–91	4 325	1.1	128.7	-28.7	13	630	14.0	49.0
1891–01	4 833	1.1	124.2	-24.2	22	939	18.6	59.7
1901–11	5 371	3.0	55.9	44.1	28	1 500	26.8	71.8
1911–21	7 206	2.0	80.3	19.7	42	2 577	34.6	40.6
1921–31	8 787	1.7	85.5	14.5	50	3 619	40.0	34.2
1931–41	10 376	1.0	108.1	-8.1	59	4 855	45.6	15.8
1941–51b	11 507	1.7	92.3	7.7	65	5 621	47.6	36.7
1951–61	14 009	2.7	74.5	25.5	78	7 404	52.8	47.6
1961–71	18 238	1.7	78.3	21.7	105	10 926	60.0	27.3
1971–76	21 568	0.9	—	—	124	14 022	65.0	—

b Includes Newfoundland in 1951 (and thereafter), but not 1941.

a Cities over 10 000 population.

and urban centres in Europe now have stable or declining populations.[8] There are no doubt measurement problems here, but the direction of change is not in question.

In Canada, national population growth has dropped from 7.4 percent in the 1966-71 period to 4.7 percent in 1971-76. This represents the second lowest annual rate of growth since the decade beginning in 1851, the lowest being during the depression of the 1930's. The principal component in this recent decline appears to be the drop in the net fertility rate, which is now at or near the replacement level.[9] Corresponding declines in average household size have been even more dramatic, however, producing a net increase in households. Both trends have substantial implications for urban and regional planning, as well as housing and welfare policies. More revealing, when the recent declines recorded in the early age cohorts (0-4, 5-9 years) are extrapolated ahead, through the high fertility, young-adult population of the 1980's and 1990's, it is clear that the aggregate population growth rate will slide further toward zero growth.[10] What also now seems clear is that the advocates of zero population growth (ZPG) in the early 1970's had not anticipated that their goal was almost in sight. Nor had the ZPG supporters assessed the potential impacts of their recommendations on different groups and locations.

Immigration and Population Growth

From almost anyone's viewpoint, the rate of net foreign immigration into Canada will likely decline in the near future (as it has done this past year) from the high rates of the late 1960's. It will do so either by government fiat, or by in-migrant choice, or a combination of both. The recently tabled federal legislation on immigration tends to confirm this anticipation, although this issue remains highly controversial. Even the commonly-quoted target of a net balance of 100 000 persons annually, now seems to be a high estimate.[11] It could in fact be much lower, especially if emigration and re-emigration increase, if unemployment levels remain relatively high, and if ethnic and regional polarization increases. The conclusion: aggregate population growth in Canada is not going to be massive, and will almost certainly be much slower than projected by even the most conservative of current demographic scenarios.

The relatively small national population increase which might be projected into the 1980's can then be attributed almost entirely to two factors: first, to the movement of the large postwar age cohort through the high-fertility young-adult period; and second, to net foreign immigration. As the former component is to a considerable extent predictable or at least not susceptible to government policies,[12] the latter then becomes the key unknown variable in anticipating future rates of national population growth. From an overall political viewpoint this suggests that population growth, in theory, will become more rather than less of a policy variable in that it is subject to some overt control.

There are, it should be noted, impressive counter-pressures against further restrictions on immigration, many of which have potentially interesting political, as well as spatial, ramifications.[13] These pressures include the well-known arguments of manpower authorities who still see a labour shortage emerging in the early 1980's; or those from Quebec for more French-speaking immigrants; or of educators seeking to maintain student numbers; or of business looking for cheap labour. There are also pressures from diverse cultural cities who wish to see their populations maintained if not increased. To take a rather extreme example, the City of Toronto's new central area housing program depends in part on a steady flow of immigrant households to ensure a sufficient level of demand for housing in inner-city residential areas. That city will likely resist a further tightening of immigration restrictions. Other cities with visions of grandeur will also voice objections. Nevertheless, on balance, the odds do seem to indicate a much lower immigration rate in the future.

Size, Variability, and Zero Growth

What of the urban component of this reduced population growth rate? What will be its spatial distribution? No doubt the proportion of Canada's population which is urbanized will continue to increase in the foreseeable future (assuming continually revised census boundaries), but at a rapidly diminishing rate. That proportion (76.1 percent) is still somewhat below the standard of many Western industrialized countries, and is relatively lower in the Atlantic region (56.5 percent) and Prairie Provinces (i.e., Saskatchewan 53.0 percent; Manitoba 69.5 percent). But what do these figures mean, at least at the national level? As the urbanized component of the population increases (to say 85 percent in twenty years), the meaning of the term "urbanized" becomes of decreasing utility. The country has become, for most purposes, a spatial system dominated by urban-centred regions. It is here, in the urban system, that the most serious impacts of this declining growth rate will be felt. Although it is by no means clear what these impacts are, there is a growing feeling that they will be substantially different from those of past periods.

One immediate implication of this slower growth rate for the Canadian urban system is obvious, however. The system-wide mean growth rate for all cities in Canada has declined, and will likely continue to decline, shifting the entire frequency distribution of growth rates downward.[14] Assuming a constant variance of growth rates irrespective of the mean (another assumption we will challenge later), more and more cities will then decline absolutely in population, just as the Windsor and Sudbury CMA's have done between 1971 and 1976.

As numerous authors,[15] have shown for the U.S., this process of declining population growth, combined with shifts in migration flows, can produce dramatic changes in an urban system within a relatively short period of time. The map of metropolitan America is changing, through what Berry[16] called a process of "counter-urbanization". Nearly one-quarter of all Standard Metropolitan Statistical Areas (SMSA's) in the U.S. showed an absolute decline in population between 1970 and 1975. Population growth rates were highest for SMSA's in the West and South, for medium size SMSA's, and surprisingly, for rural areas well removed from the commuting fields of metropolitan areas.[17] This rejuvenation of small town and rural areas is perhaps the most striking element in the emerging urban pattern of the U.S.

What we must question at this point is whether the variance of overall growth rates remains stable in a slow or zero growth situation.[18] Does the variance increase (the divergence hypothesis), leading to even greater inequalities between regions and cities in opportunities and income, or does it decrease (the convergence hypothesis), thereby reducing such inequalities? Does it become more positively skewed (leading to greater spatial concentration) or more normally distributed? The answer is that we do not know; in fact the prevailing responses to such questions are based more on ideology than on any analysis, although there are exceptions.[19]

What we do know is that governments must start to plan for urban decline in some regions and at specific locations,[20] combined with near-zero growth in the urban system overall. More concern must be given to problems of redistribution rather than to growth itself, in terms of both population and income. These are not simple problems by any means, but they are (and will become more) critical issues.

Metropolitan Concentration

Now to the point of increasing metropolitan concentration, another enduring myth in the Canadian collectivity. For the moment we might sidestep the legitimate historical debate on the extent of economic expolitation of one region by another, and industrial concentration in the Great Lakes basin, and direct our attention here to population changes in the very recent past. In fact, to avoid misinterpretation, the concluding section of this paper re-emphasizes the importance of a long-term historical perspective on urbanization.

Throughout the 1960's and the early 1970's, academic researchers, politicians, and federal government departments, such as the Ministry of State for Urban Affairs, took the public position that Canada's major urban problem was the rapid growth of the three largest metropolitan areas – Toronto, Montreal, and Vancouver. The rapid growth of these cities was, in the words of the Minister responsible for urban affairs at the time, "bleeding the rest of the nation of its productive population" and, more to the heart of the question, the resulting population concentration was "threatening the very political fabric of the country". Part of this rhetoric by the Ministry of State, of course,

Table 2

POPULATION GROWTH, CENSUS METROPOLITAN AREAS (CMA's), CANADA, 1961–76

Rank (1976)	CMA	Population (10³)					% Change		
		1961*	1966**	(1971)**	1971***	1976***	1961-66*	1966-71**	1971-76***
1	Montreal	2110	2571	(2731)	2731	2760	15.5	6.7	1.1
2	Toronto	1825	2290	(2628)	2602	2753	18.3	14.8	5.8
3	Vancouver	790	933	(1082)	1082	1136	12.9	16.0	4.9
4	Ottawa-Hull	430	529	(603)	620	669	15.1	13.9	7.9
5	Winnipeg	476	509	(540)	535	553	6.8	6.2	3.4
6	Edmonton	337	425	(496)	496	542	18.9	16.4	9.9
7	Quebec	358	437	(481)	501	534	15.6	10.0	6.5
8	Hamilton	395	457	(499)	503	525	13.6	9.0	4.3
9	Calgary	279	331	(403)	403	458	18.5	21.6	13.5
10	St. Catharines-Niagara	217	285	(302)	285	298	5.8	6.3	4.3
11	Kitchener-Waterloo	155	192	(227)	239	270	24.3	18.0	13.1
12	London	181	254	(286)	252	264	14.4	12.9	4.6
13	Halifax	184	210	(223)	250	261	7.7	6.2	4.4
14	Windsor	193	238	(259)	249	243	9.5	8.5	-2.2
15	Victoria	154	175	(196)	196	212	12.5	11.7	8.5
16	Sudbury	111	137	(155)	158	155	5.7	16.0	-1.7
17	Regina	112	132	(141)	141	149	16.9	6.3	5.8
18	St. John's	91	118	(132)	132	141	11.0	12.1	6.9
19	aOshawa	81	106	(120)	120	134	30.8	13.0	11.3
20	Saskatoon	96	116	(126)	126	132	21.3	9.1	4.6
21	Chicoutimi-Jonquière	105	133	(134)	126	127	5.9	0.9	0.6
22	Thunder Bay	92	108	(112)	115	118	17.6	3.2	2.8
23	Saint John	96	104	(107)	107	110	5.2	2.4	2.8

*Based on 1966 Census area 1971 definitions.
**Based on 1971 Census area definitions.
***Based on 1976 Census area definitions.

aCMA created for 1976 Census; figures for preceding years are estimates.

Source: Statistics Canada, Preliminary Population Counts, 1976.

reflects other problems, and the fact that older and well-established government agencies already held responsibility for other related policy areas: such as those of peripheral regions and small towns, social welfare, pollution, transportation, and industry and trade. Initially at least, the Ministry needed to attract public attention to its existence and intended role through rhetoric. But it also indicates a failure to appreciate the full range and complexity of problems emanating from and through the urban system, and the emergence of new trends.

The view that these cities were growing too fast, and at the expense of the nation, was (and is) magnified by the prevailing Canadian mystique that big cities are somehow inherently bad. Although similar views are also held by a large proportion of Americans and Europeans, there has not been the same degree of outright rejection of the benefits of large cities. Most European researchers and governments seem to be more aware of the important role played by large cities in an increasingly competitive international economic system.[21] The debate continues on economic returns to scale in large European cities, but it is now focussed on the nature of these returns in light of stable or declining populations.[22]

Returning to the Canadian concern for spatial concentration, statistics again contradict the above assertions (Table 2). Preliminary population figures for individual urban areas in Canada for the period 1971-76 indicate that growth rates were highest among the medium-size metropolitan areas (Victoria, Edmonton, Calgary, Kitchener-Waterloo, and Oshawa) rather than among the national metropolises. The growth rates of Toronto and Vancouver declined substantially during this period, and Montreal did not grow at all. In fact, nearly one-half of Canada's census metropolitan areas (CMA's)

grew more slowly than the national average between 1971 and 1976 and, as noted above, two CMA's registered absolute population declines. Recall here that we are referring to the populations of spatially extensive metropolitan areas rather than municipal cities, and that these CMA's have frequently been extended with each census to include a larger geographic area and thus more of an existing resident population. At the same time, as a recent report by Gerald Hodge to the Ministry of State for Urban Affairs has shown, growth in small cities and towns has strengthened, despite the tendency for these towns to be neglected by those who are preoccupied by the "big-city" problem. This criticism is not intended to suggest that big cities do not have problems, nor that the national interest is best served by a concentrated urban or population pattern.

In any case, on almost any scale one cares to use, Canada is not a highly concentrated national territory either in population or economic terms, at least in comparison to most Western European nations, and certainly in comparison to the Third World. There may, of course, be other measures of concentration which would lead to a different conclusion. In fact the Canadian space economy and urban system are highly regionalized in response to local specialization in economic production. The extent to which the Canadian urban system is or has been spatially concentrated, instead reflects an overall lack of "maturity" in economic and regional integration, and the limited ecumene which is capable of sustaining a relatively dense population.

An examination of the aggregate growth of Canadian metropolitan areas also reveals that metropolitan population concentration has not increased over the last quinquennial census period (Table 3). The three national metropolises held roughly 29.7 percent of

Table 3

METROPOLITAN POPULATION CONCENTRATION CANADA, 1941–1976

	1941	1951	1961	1971	1976
Population of National Metropolises	2 449	3 244	4 725	6 415	6 649
Total National Population	11 507	14 009	18 238	21 568	22 598
% in National Metropolises	21.3	23.2	25.9	29.7	29.4
$ in 23 CMA's	40.2	44.9	48.3	55.4	55.5

Table 4

POPULATION GROWTH BY LEVEL IN THE URBAN SYSTEM, 1971–76

Level in Hierarchy	Number of Cities	Population 1976	Growth Rate 1971-76
National Metropolises	3	6 649	3.6
Major Regional Centres	8	4 745	6.3
Regional Centres	14	3 393	2.8
Small Regional Centres	36	4 233	6.1
Local Centres	64	3 571	3.8
Totals	125	22 591	4.7

the Canadian population in 1971 and 29.4 in 1976, even when allowing for boundary extensions. Granted, this relative decline is largely attributable to Montreal's recent stagnation, but it is also due to adjusting for boundary changes, symptomatic of a trend toward decentralization which has been emerging for some time. By the time that politicians discovered the big-city problem, the pendulum of population growth had already swung in the opposite direction.

An alternative view of the distribution of urban population growth is provided through the use of Simmons'[23] hierarchical classification of 125 urban regions in the Canadian urban system with populations over 10 000 (Table 4). These extended regions combine city and hinterland, and are essentially exhaustive of the national ecumene. This table shows that there is no systematic relationship between growth rate and city size at any given point in time. The highest rates of growth in the 1971-76 period were found among the major regional centres (Edmonton, Calgary, Quebec) and the small regional centres (Victoria, Oshawa, Prince George). The national metropolises are now next to the lowest category in terms of proportional growth rate.

Although the massive decentralization of population outward from large urban areas, already dominant in the U.S. and Western Europe, is not yet evident in Canada, the direction of change appears to be the same. The policy problem, if any, in this context seems to be in the lack of growth in the middle-size regional centres, with populations from 45 000 to 100 000, which could act as alternative growth poles, but have not done so to date. Among the twenty centres in this size range in Canada, only six (Guelph, Barrie, Prince George, Kamloops, Kelowna, and Lethbridge) showed substantial growth in the 1971-76 period.

Migration Flows

In part this shift in growth rates reflects changes in the volume and direction of interregional migration. These changes would appear to negate, if not contradict, the trends and relationships assumed under points 3) and 4) above. Although migration statistics are notoriously deficient and unstable over time, the most recent census data (1966-1971) on intermetropolitan population flows indicate that several of the large urban areas, including Toronto and notably Montreal, had migration losses along with other smaller metropolitan areas and cities, particularly with cities in Ontario and the West (Table 5). Toronto's population, for example, continued to grow, but primarily through natural increase and foreign in-migration (and a very substantial net migration inflow from Montreal), not through rural depopulation or skewed interregional flows.

More recent but less comprehensive data from labour force surveys and income tax files suggest that this trend has been maintained, if not increased, in the mid-1970's.[24] Rural areas in Canada may still be losing population, unlike many of their U.S. counterparts, but their population is no longer flooding into the large metropolitan areas. The conclusion: a decentralization of urban population is and has been taking place for some time in Canada through migration, at least at the national level, without (or more appropriately, in spite of) government policy. This does not mean that intraregional population concentration at a smaller scale is not continuing apace, as will be noted later.

There are many and varied reasons for these changes, far too many to review let alone debate here, and some excellent analyses are finally beginning to appear in the literature.[25] Yet it appears reasonable to argue in summary that pressures, from both

Table 5

INTERMETROPOLITAN MIGRATION FLOWS, SELECTED METROPOLITAN AREAS, CANADA, 1966–1971

Place of Residence 1966	Place of Residence, 1971											Totals (All 22 CMA's)[a]
	1	2	3	4	5	6	7	8	9	10	11	
1. Calgary	0	7 575	255	320	850	890	85	680	2 300	7 565	1 620	26 030
2. Edmonton	10 605	10 325	350	405	895	1 270	50	820	2 365	9 390	1 780	31 905
3. Halifax	650	485	9 240	500	1 815	2 640	175	65	3 510	1 055	595	15 775
4. London	455	330	255	6 610	790	1 105	40	65	6 725	1 025	460	17 050
5. Montreal	2 400	1 790	2 240	1 700	269 990	12 935	9 475	255	25 315	8 970	2 010	78 885
6. Quebec City	1 100	1 300	1 100	1 305	6 005	34 740	1 125	165	8 440	3 025	1 015	30 060
7. Quebec	130	155	260	85	11 765	2 025	33 915	0	740	230	75	17 150
8. Saskatoon	2 910	2 630	45	125	350	500	20	0	920	2 590	1 345	15 010
9. Toronto	3 050	2 465	2 295	8 560	9 435	8 870	380	485	153 135	11 635	3 345	84 795
10. Vancouver	3 610	3 830	370	385	2 245	1 665	70	550	6 030	99 130	2 170	28 625
11. Winnipeg	4 590	3 445	480	655	2 440	2 515	335	995	5 670	10 435	16 325	38 075
Totals (All 22 CMA's) a	36 105	28 430	10 750	21 560	44 920	41 480	15 260	6655	95 335	69 230	19 835	501 115
b	36 105	38 755	19 995	28 180	314 920	76 225	49 180	6650	248 465	168 350	36 155	

a excluding main diagonal (i.e. intermunicipal moves within a metropolitan area)

b including main diagonal

Note: Minor differences due to random rounding error.

Source: Statistics Canada

133

the push of labour surplus and the pull of labour demand and amenities, continue to play a role. To these must be added the spatial imprint of inflation and fluctuating commodity markets, not to mention political uncertainity in Quebec and overall economic stagnation. The industrial heartland, the so-called urban "axis", in particular has suffered a sustained employment slump during the current recession, along with other manufacturing areas in the developed world. When there are fewer jobs in Ontario and Quebec, Maritimers in turn stay home. The West, that is Alberta, has benefitted from oil and the inflation in resource prices generally. Some companies have shifted their headquarters from Montreal to Ottawa or Toronto, others have been attracted to Calgary or Edmonton, as locational considerations and markets change. Smaller cities have also become more attractive as the costs of housing and social services and the uncertainties of living in the metropolis increase. Perhaps social values and locational preferences have changed as well.

In any case migration flows have shifted, and will almost certainly continue to do so in the future. The dominance of the larger metropolitan areas relative to smaller centres, in terms of internal migration, has been shown to be largely a myth. Thus, along with many of our analytical models, our established images of population growth and distribution may be slowly coming apart at the seams.

Variability by City Size and Region

Finally, as noted above, our data show only a weak relationship between urban growth and city size (Table 4). In fact, Table 2 and Table 4 indicate that wide variability exists between rates of growth and position in the urban size hierarchy at any point in time, as well as over time. There appears to be little predictability in the growth performance of any single member of the Canadian urban system from one decade to another. This localized short-term variability, couched in the apparently conflicting context of long-term stability in the aggregate urban system, has long been typical of Canadian urbanization.[26] Again this reflects the high degree of specialization among regional (and urban)

economies in Canada and the extreme sensitivity of those economies to fluctuations in the external demand for primary staple products.[27]

It is also not a recent phenomenon that this variability in growth has a marked regional differentiation. Here the question of population concentration does reappear as a critical and persistent concern. The growth of rural regions and smaller towns in the 1971-76 period was highest in B.C., south-central Ontario, Alberta, and the Maritimes. The latter two represent somewhat of a reversal of trends from the previous census period. Alberta, for obvious reasons, has witnessed widespread urban growth throughout the province, through higher in-migration rates as well as the retention of potential out-migrants who would otherwise have gone to Calgary or Edmonton, or to B.C., or perhaps Arizona. More generally, a national map of population change now shows a sharp boundary between B.C. and Alberta and the rest of the Prairies, and between south-central Ontario and all other eastern regions. The Maritimes, however, have grown at a somewhat higher rate than previously, primarily because of lower out-migration rates. In fact all three Maritime provinces have recorded net migration inflows since approximately 1972. Whether this is a short- or long-term trend is uncertain.

Implicit in this entire discussion on population growth and concentration is the notion that one pattern of population distribution is somehow superior to another, in either social or economic terms, or both. Perhaps there is an optimal pattern, but there is little concrete evidence favouring one distribution over another. On specific criteria, such as minimizing transport costs or maximizing social accessibility to a finite range of public services, or the preservation of diversified labour markets, an optimal pattern could be defined. But not over several and unequal criteria. There is certainly no evidence that a different or less concentrated population distribution than the present one would stimulate national economic growth as has been suggested.[28]

Institutional Reorganization

The seventh of our predispositions, which necessitates only brief elaboration here, is the prevailing fixation with institutional reorganization.[29] Often these reorganizations are seen in themselves as solutions to problems of national urban growth and population distribution. Although likely to be a prerequisite for the implementation of almost any appropriate solution to such problems, as numerous authors have noted,[30] institutional change in no way assures that new policies will be carried through. In fact a primary focus on institutional forms *per se* can lead to a benign neglect of other more fundamental problems, for example, by diverting attention away from such questions as the appropriate distribution of economic and political power, rising aspirations, or income equality.[31]

One interesting aspect of the experience of other countries in developing policies to manage urban growth at the national level is that the role of institutions varies so widely. Remarkably different institutional forms can be used to achieve the same ends, and conversely, similar forms can produce markedly different results depending on timing and the particular socio-political circumstances involved.[32] The point here is not that institutional organizations have been unimportant in the Canadian urbanization process. Clearly they have been. The price system, for example, as Innis[33] pointed out, is an organizational structure for allocating resources, and no one would deny that that system has been a fundamental element in shaping the geography of economic activity and population in Canada. Here we must differentiate between the way society and its institutions are organized and the organization of institutions concerned with delivering urban goods and services. Institutional change must be seen as an instrument, a facilitating mechanism for change, not as an urban or population policy in itself.

Conclusions and A Look Ahead

The above discussion poses the question of whether we are entering a new period of urbanization in Canada. Is this period markedly different from those which preceded it? To conclude in the affirmative would perhaps represent an over-reaction to recent and essentially short-term trends, the same criticism that has been directed at the conventional wisdom. To ignore these trends would,

however, be missing an important indicator on our urban future.

A decline in overall population growth rates in Canada seems almost certain. Zero population growth is now a real possibility.[34] Shifts in regional population growth and metropolitan dominance, as well as in the symmetry of migration flows and in the spatial diffusion of economic growth, are already apparent. Add to these trends the serious, but as yet largely unanalyzed, spatial impacts of continued monetary inflation, wage and price controls, heightened regionalism and separatist feelings, and even greater economic uncertainty, and the climate for urban growth in the next decade does indeed become very different. What, for instance, are the implications of a greater decentralization of federal powers? Do the economic forces shaping the urban system become more prominent; or do the political forces? How does a slow-growth spatial system alter the balance between the conflicting forces of concentration and deconcentration inherent in a market economy? How does one allocate benefits in a social system which is no longer growing rapidly?

The above trends have not, to date, drastically altered the spatial structure of the Canadian urban system. Nonetheless they represent substantial changes, in both geographical and political dimensions, which are worthy of careful analysis and continual monitoring. Signs of the massive "counterurbanization" phenomenon now apparently prevalent in the U.S. and in most parts of Europe, are in fact present in the Canadian urban fabric, but do not yet appear to be firmly entrenched. Whether or not these trends become dominant in the next decade is open to debate. My purpose in this paper has been to encourage and contribute to that kind of debate.

What does seem to be an obvious and defensible conclusion, however, is that those conceptions of urbanization, which are so firmly rooted in the research literature, data, and growth-oriented mentality of the 1950's and 1960's, are now outdated. How does one plan for population decline and economic instability, when so many of our policy instruments assume continued growth?[35]

What we need is a new framework within which to appreciate the emerging geography of urban Canada in the 1980's. We also require a new set of analytical tools and a revised set of benchmarks, on which to build an understanding of the processes which are now shaping that geography. These new approaches must include a strong geopolitical interpretation, an explicit recognition of uncertainty and instability in spatial systems, as well as an improved capability for spatial monitoring and forecasting. In a recent and highly critical review of the literature on urbanization in the Third World, Friedmann and Wulff[36] emphasize the decreasing marginal utility of research based on traditionally narrow disciplinary perspectives and on "laissez-faire" social theories of the development process. These criticisms are increasingly true of studies of urbanization in Canada, and of policies based on those studies. We lack not only appropriate theories, but also the strong analytical background on which interdisciplinary and policy-oriented research might build.

One additional requirement for this revised research strategy is the removal of our preoccupation with large metropolitan growth and rejection of our simplistic assertions on metropolitan population concentration and the causes of differential urban growth. Instead research should focus on the dynamic, often turbulent, behaviour of the entire Canadian urban system, and its strengthening regional subsystems, and the full range of social and political problems embedded in those systems. Specifically it must look more to the nature of redistribution mechanisms operating in an urbanized society, rather the mechanism of simple population growth, in an environment of increasing uncertainty.

I personally do not have an agenda for the appropriate research or policy focus, but it is clear that we should be seriously looking for one (or more than one). Far too often our conceptions of the processes of urban change are enveloped in and by short-term time series data and images which, even if they capture the essence of geographies of the immediate past, will miss the emergence of future geographies in the present.

NOTES AND REFERENCES

[1] Simmons, J.W., "Migration and the Canadian Urban System: Part I, Spatial Patterns", *Research Paper 85* (Centre for Urban and Community Studies, University of Toronto, 1977).

[2] Subsequent releases by Statistics Canada will no doubt require revision of the following population aggregates and consequent growth rate calculations. There is no evidence, however, that any bias in the preliminary figures is not relatively uniform across cities and regions.

[3] Bourne, L.S., *et al.*, eds. *Urban Futures for Central Canada. Perspectives on Forecasting Urban Growth and Form* (Toronto: University of Toronto Press, 1974); H. Lithwick, *Urban Canada: Problems and Prospects* (Ottawa: C.M.H.C., 1970); Ministry of State for Urban Affairs, *Canadian Settlements. Perspectives* (Ottawa: Information Canada, 1976); M. Yeates, *Main Street* (Toronto: MacMillan, 1975); and D. M. Ray, *et al.*, *Canadian Urban Trends* (Toronto: Copp Clark, 1976).

[4] Some observers in fact refer to this trend as a "de-industrialization" process. Current research underway at the Science Council, directed at the question of an appropriate industrial strategy for Canada in a no-growth context, may also contribute to our debate on the further options for urban development.

[5] Hansen, N., "An Evaluation of Growth Centre Theory and Practice", *Environment and Planning A 7*, 1975, pp. 821-832.

[6] Barber, C.L., "Some Implications of the Move Towards Zero Population Growth in Developed Countries Upon the Level of Capital Expenditures", *Discussion Paper No. 19*. (Ottawa: Economic Council of Canada, 1975).

[7] Alonso, W., "Urban Zero Population Growth", *Daedalus, The No-Growth Society*, Special Issue, 102; 1973, pp. 191-206; and B.J.L. Berry, ed., *Urbanization and Counterurbanization*. Urban Affairs Annual Review No. 11 (Beverly Hills: Sage Publications, 1976).

[8] Vining, D.R. Jr. and T. Kontuly, "Increasing Returns to City Size in the Face of Impending Decline in the Sizes of Large Cities", *Environment and Planning A*, forthcoming.

[9] The gross birth-rate actually increased slightly in 1975 and 1976 to 16.0 births per thousand population, up from a low of 15.4 in 1974. The total natural increase in population has been below 200 000 since 1972, but has stabilized in the last two years. Both trends in part reflect age-cohort growth rates rather than changes in the fertility rate

[10] Science Council of Canada, *Perceptions 1: Population* (Ottawa: The Council, 1975).

[11] Department of Manpower and Immigration, *Immigration Policy Perspectives*. Volume 1 (Ottawa: Information Canada, 1974).

[12] Stone, L.O. and C. Marceau, *Canadian Population Trends and Public Policy Through the 1980's* (Montreal: Institute for Research on Public Policy, 1977).

[13] Aside from the debate on the number of immigrants to be admitted each year, there have been numerous suggestions for directing those immigrants to specific regions, or at least away from certain locations (cities) within the country.

[14] Simmons, J.W., "Migration and the Canadian Urban System: Spatial Patterns".

[15] Gans, H., "Planning for Declining Cities", *Journal of the American Institute of Planners*, 41; 1975, pp. 305-307; P. Morrison, "Migration and Rights of Access: New Public Concerns of the 1970's," *Rand Paper Series*, P-5785 (Santa Monica: Rand Corporation, 1977), and B.J.L. Berry and Q. Gillard, *The Changing Shape of Metropolitan America* (Cambridge, Mass.: Ballinger, 1977).

[16] Berry, B.J.L., *Urbanization and Counterurbanization*.

[17] Morrison, P., "The Current Demographic Context of National Growth and Development", *Rand Paper Series*, P-5514 (Santa Monica: Rand Corporation, 1975).

[18] Robson, B., *Urban Growth* (London: Methven, 1973).

[19] Massey, D., "Restructuring and Regionalism: Some Spatial Implications of the Crisis in the U.K.", paper presented to the North American meetings of the Regional Science Association, Toronto, November, 1976.

[20] Lithwick, N.H., *Urban Canada: Problems and Prospects*.

[21] Bourne, L.S., *Urban Systems: Strategies for Regulation. A Comparative Analysis of Urban Policy in Australia, Britain, Canada and Sweden* (Oxford: Oxford University Press, 1975).

[22] Vining, D.R. and T. Kontuly, "Increasing Returns to City Size".

[23] Simmons, J.W., "The Growth of the Canadian Urban System", *Research Paper 69*, Centre for Urban and Community Studies, University of Toronto, 1975.

[24] Department of Manpower and Immigration, *Internal Migration and Immigrant Settlement* (Ottawa: Information Canada, 1975).

[25] Department of Manpower and Immigration, *Internal Migration and Immigrant Settlement*; J. W. Simmons "The Growth of the Canadian Urban System"; Stone, L.O. and C. Marceau, *Canadian Population Trends and Public Policy Through the 1980's*, and Economic Council of Canada, *Living Together: A Study of Regional Disparities* (Ottawa: Ministry of Supplies and Services, 1977).

[26] Simmons, J.W., "Migration and the Canadian Urban System: Part I, Spatial Patterns".

[27] Neill, R., *A New Theory of Value. The Canadian Economics of H. A. Innis* (Toronto: University of Toronto Press, 1972), and L. Gertler and R. Crowley, *Canada's Changing Cities* (Toronto: McClelland & Stewart, 1977).

[28] Joint Committee of the House and Senate on Immigration Policy, *Proceedings* (Ottawa: Information Canada, 1975).

[29] Lithwick, N.H., *Urban Canada: Problems and Prospects*, and D. Cameron, "Urban Policy", Chap 9 in G.B. Doern and V.S. Wilson, eds. *Issues in Canadian Public Policy* (Toronto: MacMillan, 1974), pp. 228-252.

[30] Lithwick, N.H., "Urban Policy Making: Shortcomings in Political Technology", *Canadian Public Administration*, 15, 1972, pp. 571-584, and L.O. Gertler and R. Crowley, *Canada's Changing Cities*.

[31] Ministry of National Health and Welfare, *The Distribution of Income in Canada: Concepts, Measures and Issues*, Research Report 04. Ottawa: Long Range Welfare Planning Directorate, Policy Research and Long Range Planning Branch, 1977.

[32] Bourne, L.S., *Urban Systems: Strategies for Regulation*.

[33] Neill, R., *A New Theory of Value*.

[34] This does not mean that zero growth is an appropriate objective. In fact zero growth could not be achieved in any complex system, nor maintained even if achieved. The term is used here to denote relatively slow growth, a steady state situation, in which growth rates would fluctuate around and converge on zero.

[35] For example, in terms of spatial and regional development policies can we turn our attention from job generation to job-upgrading; and in social development from income growth to income redistribution?

[36] Friedmann, J. and R. Wulff. *The Urban Transition,* (London: E. Arnold, 1976).

RURAL CANADA

Introduction

Agriculture is still the prevailing land use in terms of hectarage in the settled parts of Canada. However, the countryside is changing, notably in the vicinity of the expanding urban centres. Farms, on average, are increasing in size; as a consequence there are fewer farms and bona fide farmers each decade. Not all rural areas, however, are being depleted of their populations; in fact, some are growing due to the migration of city dwellers to rural hamlets and villages and to rural nonfarm residences, hobby farms, and vacation homes. In some quarters alarm has been expressed about the loss of agricultural land and the potential for increasing agricultural production from a decreasing resource base, particularly in the densely urbanized parts of the country. In the first paper in this section, Travis Manning assesses the liklihood of future productivity gains in Canadian agriculture by evaluating the adequacy of the resource base including land, water, energy, and labour. He is not optimistic about the continuing potential of technological innovation, as most innovations will have to be energy- and labour-saving. Any major developments will therefore require long-term research. He concludes that given a "combination of *most favourable conditions* it would be possible to increase agricultural production by 50 percent by the end of the 1980's".

Rural settlement patterns are changing, and the direction of change is different from place to place. In the Quebec City-Windsor axis the spread of urbanites into the countryside is changing the character of many rural areas. On the other hand, there are areas where the rural population is decreasing as farms become fewer and larger, as transportation services are improved and rationalized (i.e. rail line abandonment). These changes have had a profound impact on many rural service centres whose economic base has depended upon the local agricultural community. The analysis of the decline of rural service centres in the Canadian Prairie region is the subject of Jack Stabler's contribution. On the basis of his analysis he concludes that service centres with fewer than thirty distinct functions and less than 2000 population will continue to decline, barring widespread government intervention.

The loss of agricultural land through abandonment, speculation, outright urban development, hydro corridors, etc. has been debated at the federal, provincial, and local levels for sometime. Much of the debate has been based more on emotion than fact. However, even facts can be arranged and manipulated and interpreted to show loss or gain of agricultural land. It was not until the completion of the Canada Land Inventory, in which mineral soils were is classified into seven capability classes on the basis of their capability for growing general field crops, that a nationwide inventory existed.

Three papers have been selected to reveal different facets of the agricultural land problem in Canada. In the first paper, Williams, Pocock, and Russwurm analyze the spatial association, on a nationwide basis, of the agroclimatic resource base, the distribution of urban population, and farmland areas based on the Canada Land Inventory. It is clear from this study that areas exist in Canada with a very high agroclimatic resource base with limited farmland resources that are under severe pressure from urban development including Vancouver, Toronto, and Montreal. They conclude that this exploratory study "is but a first step in the development of effective planning for the protection and preservation of our limited agroclimatic resource for present and future generations".

In the second paper Stephen Rodd poses the question: "Is there an agricultural land crisis in Ontario?". He concludes, on the basis of his analysis, that there is indeed a crisis. However, he points out that the crisis and problems are not uniform across the province but differ regionally. He concludes by describing briefly the basis for developing an agricultural planning strategy for the province.

The competition for land between agriculture and urban uses is probably more intense in British Columbia than in any other part of Canada. This is so because of the concentration of population in areas of very limited high capability agricultural land. In an attempt to protect its limited and economically important resource base from total urban erosion, the province introduced the B.C. Land Commission Act in 1973 – the only legislation of its type in Canada. The concluding paper in this section by G. G. Pearson discusses the rationale, objectives, and impact of this legislation.

The Agricultural Potentials of Canada's Resources and Technology

TRAVIS W. MANNING

Canada's agricultural production capabilities may be constrained by the increasing scarcity of resources in the near future. While aggregate output may not decline, its composition may change, and future increases may be more difficult to achieve than were past increases. Much of the past productivity gains in agriculture were achieved by the adoption of technology permitting the substitution of less scarce resources for more scarce resources. With increasing scarcity of once plentiful resources, different kinds of technology will be required to achieve future increases in productivity.

In the period following the Second World War, the notion that we had at last achieved the power, the knowledge, and the resources to control our social and economic destinies became popular. Many people accepted as articles of faith that economic growth would cure most social ills, and that scientific and technological capabilities and the supply of resources were adequate for indefinitely sustained economic growth. The first article of faith was shaken by the persistence of poverty amidst plenty at home, and the Malthusian spectre of exponential population growth and recurrent famine abroad. The second article of faith was weakened by the increasing awareness of the wastes of war, industrial pollution, and conspicuous consumption. It was further weakened by the growing realization that resources were finite and serious external costs were inherent in the basic institutional structure of our economic system. Even the more complacent and optimistic members of society became aware of the "limits to growth" when the Arabs acted to remind them that petroleum resources, at least, are limited and growing more scarce.[1] The Organization of Petroleum Exporting Countries (OPEC) has formalized the constraints on energy supplies for the foreseeable future. However, the ultimate reserves of fossil fuel resources are not affected by these constraints, since the reserves are finite and not increasable by human action. The actions of OPEC will have the long-run effect of somewhat extending the depletion of petroleum reserves. Although we are not facing the complete physical depletion of fossil fuel reserves, we are likely to experience continually rising fuel costs. Fuel prices will rise with or without monopolistic pricing because extraction and transportation costs will increase as reserves are depleted. As fuel costs rise, economic pressure will increase on other resources, most of which are also limited in quantity. All natural resources are likely to become scarcer, either absolutely or, at least, in relation to human needs. While new technology may slow the increase of pressures on resources, it cannot be expected to stop or reverse them.

Resource Scarcity

Economists have employed a number of resource scarcity models — varying from very pessimistic to very optimistic. Barnett and Morse described three basic scarcity models which they labeled as *Utopian*, *Malthusian*, and *Ricardian*.[2] The basic assumptions of these models are unlimited resources for the Utopian, fixed resource quantities for the Malthusian, and large quantities of varying qualities for the Ricardian. While it is difficult to believe that anyone would seriously entertain the notion of unlimited availability of resources, economic behaviour for the past three decades seems largely consistent with that assumption. The Malthusian concept has had few proponents among economists, mainly because Malthus did not foresee the effects of technology on productivity. This rejection of the Malthusian thesis appears to have been accompanied by a faith that technology can overcome

● TRAVIS MANNING is Professor, Department of Rural Economy, The University of Alberta. This paper is reprinted, with minor revisions, from the *Canadian Journal of Agricultural Economics*, Vol. 23, No. 1, 1975, pp. 17-29, with permission of the author and the publisher.

resource scarcities indefinitely — the *new Utopianism.*

The Ricardian scarcity model presupposes an order of resource use, beginning with the highest qualities and proceeding to successively lower qualities as higher qualities are used up. Increasing quantities of labour and capital will be required per unit of standardized product as the quality of natural resources diminishes. It should be noted that this model leads progressively toward the results of the Malthusian model, as the better resources are used up and lower and lower qualities have to be employed. However, the mitigating effects of technological innovation need to be considered.

The basic Ricardian scarcity model needs considerable elaboration to make it more applicable to the analysis of actual resource scarcities. Provision must be made for a large variety of resources with varying degrees of substitutability, for many different kinds of capital and labour with varying degrees of substitutability, and for substitution among natural resources, capital, and labour. In addition, allowance must be made for changes and substitutions in final products. In using this model, it must be recognized that knowledge about the quantities and qualities of various natural resources is still quite limited in many important cases. Even in the cases of such highly visible resources as land and water, Canada has lacked a complete and comprehensive national inventory. A model useful for long-range projection needs to take account of potential changes in technology and institutions that may mitigate resource scarcity.

The role of energy in relation to other resource scarcities deserves special attention. The use of lower quality material resources (of which recycling materials are a special case) requires more energy per unit of output than the use of higher quality material resources. This relation is particularly important for agriculture. During the last half century, Canada's agricultural output has increased considerably, with little change in land input, a substantial decrease in labour input, and a large increase in capital inputs. The production and employment of these capital inputs have required large amounts of energy. Most of the technological changes that have occurred in agriculture have required increases in applied (mainly fossil) energy inputs. Past increases in agricultural productivity can be attributed largely to the substitution of fossil energy for animate energy. Energy requirements for continued high productivity in agriculture are a major concern of the future.

Resource Constraints

Scarcity of land, water, energy, and labour resources are likely to constrain Canada's agricultural production in the 1980's and probably also in the 1990's. Since these resources are partly complementary and partly substitutable, it is necessary to consider them separately *and* jointly. A further complication is that most resources used in agriculture have alternative uses, so that the supplies available to agriculture are affected by demands from competing uses. In addition, the availability and productivity of resources may be influenced by institutional and technological changes, many of which are difficult to foresee.

Land

The popular notion of fixed land supply (zero elasticity of supply) is based on several misconceptions about the economic nature of land. Land is usually measured in terms of its horizontal surface, rather than its productive capabilites. The heterogeneity of land makes it difficult to describe a useful supply function in terms of areal measurement. The various characteristics of land (such as topography, location, structure, and fertility) have differing relative importances for different uses. Consequently, a quantity-quality (productivity) index for one economic purpose (such as wheat growing) may be inappropriate for competing uses (such as recreation or forestry). Surface area seems to be the only characteristic of land that is important to all competing uses. Consequently, it is the sole quantity measure presently available for an aggregate land supply function.

The aggregate land supply function, defined in terms of areal measurement, is very inelastic because the physical quantity of land can be increased only at very great cost. Land supply functions for specific purposes, such as crop production, tend to be considerably more elastic, even in terms of areal measurement, because of the possibility of shifts between competing uses. Land supply, in terms of productive potential, may be even more elastic because of the possibility of changing productivity through capital investment. Agricultural examples of capital investment to improve productivity include clearing, leveling, drainage, and the like.

The total land area of Canada is about 922 × 10^6 ha,[3] of which only 7.4 percent was in farms in 1971. About 64 percent of the land in farms (44 × 10^6 ha) was classified as improved land in the 1971 census report.[4] Some preliminary figures have been compiled from various provincial sources that indicate a total of about 65 × 10^6 ha of "potentially arable" land in Canada. (Potentially arable land was defined to include Canada Land Inventory capability classes 1 through 4.) This potentially arable land appears to be about 50 percent greater in area than the improved land in farms in 1971. Forty-three percent of the additional potentially arable land is located in Alberta, and most of it is marginal climatically as well as in fertility. Much of it will require timber clearing, rock clearing, or drainage. If the Alberta land is fairly typical of the rest of Canada, the remaining potentially arable land will be brought under cultivation only at relatively high cost, and it will be less productive than average land already cultivated. The potentially arable land has a production capability considerably less than 50 percent of that of the present improved land. Thus, its potential contribution to Canada's agricultural output is seriously limited by costs and technical possibilities.

It should be noted that "potentially arable" is an elastic term which depends upon product prices relative to improvement costs and production costs. At higher relative product prices, more land will become potentially arable – in effect, a supply response to increases in the derived demand for arable land. Technological improvements (such as better adapted crops, more efficient land clearing methods, and improved production techniques) could also increase the amount of potentially arable land – in effect, a shift in the supply. It should also be noted that not all productive land is, or needs to be, arable. Vast areas of nonarable land provide grazing

for livestock, and the potentials for expansion have not been exhausted.

Although the supply of land for agricultural use is not absolutely inelastic, the quantity of improved land is not likely to increase under present price-cost relationships. Indeed, further shifts of land from agriculture to urban, industrial, and recreational uses seem likely to occur, unless political means are used to prevent them. Even if price-cost relationships do improve, it may be more advantageous to use present agricultural land more intensively than to expand the area of land in farms. Some increases in crop production could be achieved by full adoption of present technology.[5] Similarly, the carrying capacity of range land could be increased substantially by the use of better grass and legume mixtures, and further increases could be obtained through greater use of fertilizer.

Water

Most of Canada's improved agricultural land area is adequately supplied with precipitation. However, there are some notable exceptions, such as the Palliser Triangle, the Peace River country, and some interior valleys in British Columbia. Nevertheless, many areas would respond positively to applied water. Some 300 000 ha in southern Alberta are accessible to irrigation water, although only about 250 000 ha are irrigated in an average year. Estimates of potentially irrigable land in southern Alberta range up to 800 000 ha. Substantial areas in Saskatchewan, Manitoba, and British Columbia are potentially irrigable. Irrigability, like arability, is an elastic concept depending on economic, as well as physical, conditions. Given past and present price-cost relationships, limited crop varieties adapted to Canadian climatic conditions, intensive labour requirements, and high capital costs for irrigation installations, it is not surprising that irrigation has been restricted in area.

The supply of irrigation water is influenced by physical, political, and economic factors. The fresh water resources of Canada are among the largest in the world. The annual runoff in Alberta alone has been estimated at about 120 × 10^9 m^3 (120 km^3).[6] Water availability in many of the drier agricultural areas is sufficient for a substantial level of irrigation, and much more could be made available at a high cost by inter-basin diversion. The supply of irrigation water is largely a function of the costs of surface water impoundment and diversion (or groundwater drilling and pumping) and on-site costs, such as ditching, land leveling, drainage, sprinkler systems, and other specialized irrigation facilities and equipment. The supply function for water delivered to the farm gate would likely be relatively elastic for quantities that are not greatly different than present usage. It should be noted that irrigation requires substantial energy inputs to impound (or pump) and distribute water. Higher energy costs may inhibit further expansion of irrigation unless farm commodity prices increase sufficiently to make expansion profitable.

Large scale water developments are usually financed by governments because the public interest is involved, large investments are required, and private financing for such high-risk projects is not readily available. Political decisions on water resource developments are influenced by various considerations, including net social benefits, local needs for water, and competing financial needs for other public services.

Energy

Energy is one of the most critical factors in modern agriculture, and it has become increasingly critical in recent years. Downing and Feldman estimated that the total applied energy input (excluding solar energy) in Canadian crop production in 1971 was 358.2 $\times$ 10^{12} kJ.[7] Direct use of fossil fuels by farmers accounted for 30.5 percent and fertilizer production accounted for 48.9 percent of the total. Total crop energy output was estimated to be 1.15 $\times$ 10^{15} kJ for 1971, giving an energy output-input ratio of 3.21. Energy output-input ratios were estimated at 7.47 for Saskatchewan and 2.05 for Ontario, reflecting largely lower relative fertilizer inputs in Saskatchewan. However, the net energy gain per hectare was more than twice as high in Ontario as in Saskatchewan, due to higher crop energy output per hectare in Ontario.

Although they cannot fully support their conclusions with data, some researchers believe that further increases in food production from increasing energy inputs will be more and more difficult to achieve.[8] Several writers have also questioned the efficiency of modern high energy-using agriculture, and by implication, the ethics of high energy use have been raised as well. Perelman reported that Chinese rice farmers produce more than 50 units of food energy for each unit of human energy expended, while U.S. agriculture produces only one-fifth as much energy in food as it uses from fossil sources.[9] Any conclusion that Chinese rice farming is far more efficient than North American agriculture must be tempered by the realization that human labour energy cannot be equated with fossil energy on a cost basis. The cost of human labour was about 200 times as high as fossil energy on an energy unit basis at 1970 prices in the U.S. Furthermore, each agricultural worker in North America produces enough food for fifty people, while workers in primitive agriculture produce little more than enough to feed themselves. A shift toward substitution of labour for other energy forms would require either a larger proportion of the labour force in agriculture or a reduction in agricultural output – both alternatives would likely reduce standards of living.

World petroleum prices have increased sharply during the past years, largely as a result of OPEC policies. However, OPEC can be blamed only for having speeded up an inevitable increase in energy costs. A conservation case can be made for raising present energy prices in order to spread fossil resources further into the future, but there is also an equity question of who should benefit from such increases. Surveys of world energy resources indicate that most of the petroleum reserves could be used up within fifty years and most the gas reserves could be depleted even sooner.[10] Coal reserves are relatively much larger, but their extraction and use pose some difficult environmental problems. The cost of producing "clean" fuel from much of our coal reserves will be higher than present energy costs. Energy from nuclear fission likewise poses dramatic environmental problems. Uranium reserves are quite limited, although they could be extended if "safe" breeder reactors are developed. The technological problems of controlled fusion energy have yet to be solved. While world reserves of thermonuclear fuel are enormous, their conversion to useful energy may be neither cheap nor pollution-free. The nonpolluting sources of energy (such as solar radiation, wind, tides, thermal currents, and possibly geothermal) pose many technological problems of conversion to usable forms and transportation to use locations. It seems very likely that energy costs will continue to increase regardless of new energy sources, and that the era of abundant cheap energy has nearly ended.

Energy prices, especially for liquid and gaseous fuels, seem likely to continue increasing for the remainder of this century, although new discoveries of major fossil fuel deposits or new methods of energy conversion could restrain (but not likely reverse) the price increases. Increasing costs of liquid fuels pose special problems for agriculture. Every effort should be made to use energy more efficiently in agriculture, and there are many ways to increase energy efficiency in both the design and use of power units and equipment. The real question, however, is not how to reduce energy use in agriculture but how to provide the energy necessary to produce the required amount of food.

Labour

A strong case can be made for the contention that too much labour is engaged in farming. Census figures reveal a large number of low-income farms on which labour is excessive in proportion to land and capital. Consequently, there is a popular notion that there is excess labour in all agriculture. However, many commercial farms are faced with a tight labour situation, and much of the mechanization of agriculture has been a response to the unavailability of labour for hire at prevailing farm wage rates and amenities. While it is likely that many low-income farm operators could earn more by working for commercial farmers, strong cultural factors inhibit such changes. There is considerable prestige and pride in owning even a small low-income farm. European culture, from which North American culture is derived, places high social value on land ownership and entrepreneurship. Hired farm labour occupies the "lowest rung" of the prestige ladder in agriculture. Traditionally, a farm labour job has been regarded as a stepping-stone or an apprenticeship leading first to the operation of a rented farm and

finally to an owned farm. Those who remained on the lowest rung were often considered unambitious or incapable of more responsible roles in the farm community.

The low prestige of farm labour, together with long and irregular hours, strenuous physical work under adverse conditions of temperature extremes and other unpleasant physical situations, low wages, inadequate amenities, lack of disability compensation, inadequate or nonexistent retirement pensions, and generally less satisfactory conditions than those enjoyed by workers in industrial and service occupations, have caused most people to consider farm labour undesirable as a satisfactory or attractive career. Although the traditional farm "ladder" has largely disappeared and a few farm workers can look forward to becoming owner-operators, farm labourers generally remain temporary and mobile. The better workers seem to move out of agriculture and into urban occupations, leaving behind those who lack the skills necessary for outward mobility plus a smaller number for whom rural life has a sufficient attraction to more than compensate for the disadvantages of farm labour.

The supply of farm labour does not seem likely to increase under present conditions. It seems reasonable to assume that the supply curve is very inelastic in the vicinity of current wage rates. The average age of low-income farmers is quite high, and most of the exodus of this group is through retirement and death. Young people leaving low-income farms seem to aspire primarily to urban jobs. Higher wage rates might induce more of them to seek jobs as farm workers, but many of them need additional skills for work on modern commercial farms. A shift in the supply of farm labour is highly unlikely unless farm work becomes more attractive relative to industrial and service occupations.

A number of things could be done to increase the farm labour supply. Some of them require political action and some require changes by farm employers to make working conditions more attractive. Political actions that seem worthy of consideration include an expansion of farm training programs in agricultural schools, regulated apprenticeship programs possibly coupled with examinations for advancement to journeyman and master worker status, extension of wage and hour laws to farm labour, disability compensation, a retirement pension program with portability provisions, health and medical insurance, a system for adjudicating grievances, a farm labour housing program, and other programs designed to improve farm labour conditions to levels existing for urban labour. Farm employers could take a number of actions that would improve the farm labour supply. These include cooperating to make the government labour programs more effective, providing more amenities of various kinds, instituting systems of bonuses and profit sharing, and responding positively to settle labour grievances. These programs would not likely have a dramatic effect immediately because it would take many years to raise the prestige of farm labour to the level of craft and industrial labour. However, farm employers should be able to take advantage of the renewed popularity of rural life and living close to nature, the interest in the natural environment, and the growing desire of many people to escape the congestion, crime, pollution, and other disagreeable aspects of urban living.

The farm labour situation is not likely to improve much in the next decade, even if positive labour improvement programs are initiated immediately. If such programs are not started soon, it will become increasingly difficult to halt or reverse the decline in farm labour, and it will become harder and harder to increase farm output.

Technological Possibilities

Most modern agricultural technology was designed to economize on scarce labour by substituting cheaper and more abundant forms of energy. Much of it also involves extensive land use. Whether efforts to expand agricultural output involve improving new land or intensifying use of present farmlands, large additional quantities of applied energy and possibly more labour will be required if present technology is used. Most of the additional farm energy demand will be for high density liquid fuels (i.e. petroleum products). If petroleum prices rise to levels two to four times as high as 1973 prices (as some have predicted), it will be difficult even to maintain present production levels without

substantially higher farm commodity prices. Higher farm commodity prices in conjunction with higher energy prices could lead to such effects as:
(a) A return to techniques requiring less energy.
(b) A higher demand for farm labour.
(c) The development of new energy-conserving technology.
(d) New crop varieties and management practices to make more efficient use of water, nutrients, and solar energy.
(e) New livestock strains and management practices to make more efficient use of forage, byproducts, and other feed sources that do not compete for resources that can be used for direct production of human food.

Among known techniques that could be used to conserve energy and material resources are better management, use of manure and sewage sludge instead of chemical fertilizer, use of legumes and inoculants for greater biological nitrogen fixation, recycling of organic garbage and processing byproducts, and use of machinery with lower energy requirements. These things have been characterized as reducing waste, but "waste" is very difficult to define operationally. For example, it could be argued that it is more wasteful to use manure than to use chemical fertilizer *if* more resources are required to distribute manure than to produce and distribute chemical fertilizer. A common characteristic of most energy- and materials-saving techniques is greater labour requirements. The cost of labour is far greater than the cost of an equivalent amount of energy from other sources. To increase the quantity of labour would necessitate substantially higher costs for wages, supplementary benefits, and amenities of various kinds.

The choice between higher energy costs and higher labour costs places agriculture on the horns of a dilemma. If higher energy costs lead to higher farm commodity prices and both lead to higher living costs, which (in turn) stimulate higher wage rates, the substitution of energy-saving, labour-intensive technology for labour-saving, energy-intensive technology may not be economically feasible. Such a situation demands new technology that is both energy-saving and labour-saving.

Considerable energy could be saved by making greater and more effective use of present scientific, engineering, and economic knowledge. Farm power units could be designed to achieve higher energy conversion to useful work. Better matching of power units and equipment and better matching of machinery size to the job to be done would reduce both the investment in machinery and operating costs. Combined operations (such as discing, seeding, and fertilizing) could lower the costs of labour and fuel. Minimum tillage techniques could reduce labour and fuel costs considerably, although these savings might be partly offset by herbicide costs and investments in new minimum tillage equipment. There have been few incentives to develop machinery and techniques to minimize fuel costs because fuel has been cheap and labour has been costly. With high and rising fuel costs, there will be more incentive to develop energy-efficient machinery and operating techniques.

New energy-saving techniques are possible in farm machinery manufacturing, fertilizer production, transportation, and food processing. Improvements in food distribution, more efficient methods of food preservation and preparation, and minimal packaging also would help conserve energy. Suggestions to minimize the output of convenience foods are not necessarily meritorious because home preparation may consume more energy than factory preparation. Much home cooking, refrigeration, and food preparation equipment was not designed for energy efficiency, and many users have not been adequately instructed in the efficient operation of home food preservation and preparation equipment.

Higher energy costs are likely to stimulate some shifts in the product mix of agriculture, and higher food costs may stimulate shifts in the food consumption mix of Canadian and export buyers. Lower energy-demanding crops may be substituted for some higher energy-demanding crops. Additional plant breeding work is needed to develop new varieties having lower energy requirements. For example, faster maturing grains could reduce the need for grain drying in areas with shorter growing seasons. Research in biological nitrogen fixation could have an immense payoff. The development of strains of free-living, nitrogen-fixing bacteria adapted to Canadian conditions, or strains that live in symbiotic relation with cereal crops, could reduce dependence on high energy cost manufactured nitrogen fertilizers. More emphasis on breeding and adapting high protein food crops to Canadian conditions would lessen the need for energy-expensive animal products.

Improvements in plant geometry to maximize leaf exposure to sunlight would be especially beneficial under Canadian climatic conditions. Improvements in photosynthetic efficiency could revolutionize crop production potentials. An average of only 1 percent of the solar energy striking leaf surfaces is converted to plant tissues — including roots, stems, leaves, fruit. Corn, which is one of the more efficient grain crops, converts to grain only 0.4 percent of the solar energy striking its leaves. The practical limits for improving photosynthetic efficiency are unknown, but some scientists are quite optimistic about the theoretical potentials. A number of possibilities for crop improvements are potentially achievable, including: more efficient utilization of plant nutrients; greater tolerance to drought, temperature extremes, and diseases; better germination, growth characteristics, and fruit setting; and adaptation of crops to minimum tillage and more efficient harvesting. It seems conclusive that greatly expanded research is needed in crop breeding and plant biochemistry.

Animal product production efficiencies vary tremendously among various species and husbandry methods. Grain fattening of beef cattle is one of the least energy-efficient methods of food production. Range-fed cattle, in contrast, compare favourably with many field crops, and additionally, much range land is not suitable for crop production. Grain-fed hogs are relatively inefficient energy converters; China (with one of the largest hog populations in the world) feeds hogs primarily on garbage and similar materials that otherwise would be wasted. Some potential exists for adapting wild animal species (e.g., antelope, bison, caribou, deer, elk, goat, moose, musk ox, sheep, etc.) for human food production. Most of the indigenous herbivores are more efficient foragers and better adapted to their physical environments than the domesticated species that have been imported into Canada. Indigenous species do pose some difficult management and harvesting problems as well as some legal and cultural taboos, but there may be considerable possibilities for managed production in the boreal forest and mountain areas. Feeding livestock grains and other materials that could be used directly for feeding humans has been challenged on ethical grounds and may become uneconomical as well.

Numerous possibilities exist for improving feed-energy conversion ratios, production rates, disease control, stress tolerance, husbandry, and other efficiencies of animal product production. Greater efforts could be made to use resources that are not suitable for producing plant food for human consumption. Animal research, together with plant research, needs to take new dimensions with a greater emphasis on energy use efficiency. This research should focus on both more efficient animal converters and more efficient feed production.

In summary, better use of present technology could increase food production considerably, but greater emphasis must be placed on the development of new technology that is both energy saving and labour saving. Considering the long lead time needed for research conception, scientific experiments, technological development, and adaptation, new research emphases are needed immediately to meet the increased food production requirements of the late 1980's and thereafter. Without these efforts, food costs are likely to rise to levels that seem incredible by present standards. The proportion of personal income that Canadians spend for food is likely to rise, despite the best efforts of politicians, administrators, scientists, technologists, agriculturists, and others. Anything less than "best efforts" could result in Canada becoming a food deficit country before the end of the century. Given Canada's agricultural potential, such a result would be unconscionable.

Conclusion

Canada may have the potential for increasing agricultural output by as much as 50 percent by the end of the 1980's, but this potential is likely to be realized only under a combination of most favourable conditions, including:

1. No further loss of prime agricultural land to alternative uses, such as urban sprawl, industrial plants, roads and highways, strip mining, parks and recreational areas, airport, wildlife preserves, reforestation, and military reserves.

2. Development of remaining unimproved lands that have high agricultural capabilities, at least through CLI land capability class 3.

3. Further irrigation development of lands having high irrigability potentials, including some further water impoundment, rehabilitation and improvement of diversion systems, and additions and improvements to farm irrigation facilities.

4. Allocation of adequate fuel supplies, fertilizers, and agricultural chemicals to farmers at "reasonable" prices — achieved, if necessary, by means of mandatory allocations at controlled prices. It is not suggested that farmers should pay artificially low prices, but that they be protected from monopolistic price increases.

5. Improved farm labour supply to be achieved through such means as extending to farm workers training programs, wage and hour regulations, workman's compensation, mandatory pension plans, health insurance, an adequate housing program, and other modern amenities.

6. Design and production of farm machinery and equipment to achieve greater energy and labour efficiency, together with a price monitoring system.

7. An improved financing system to permit farmers to modernize faster and make better adjustments to higher energy and labour costs.

8. A "stop-loss" system of farm product price guarantees that would allow farmers to make output expanding decisions without the fear of market price uncertainties. Such a program should be carefully designed and regulated to avoid encouraging overproduction of some commodities and underproduction of others.

9. Improved and well regulated systems of agricultural marketing, processing, and transportation, together with the elimination of restrictive controls on production, marketing, movement, and prices of both raw and processed farm products.

10. A greatly expanded extension program with more emphasis on well trained extension specialists and more effective transmission of new research results, technological information, and management assistance to producers.

11. A substantial program of fundamental and applied research designed to improve the overall efficiency of the food systems and integrated as closely as possible with the extension program.

Canada's agricultural potentials for the 1990's will depend upon the success of programs designed and implemented for the 1980's as well as on those programs whose desirability becomes apparent during the next fifteen years. If the full potentials of the 1980's were realized, further production increases in the 1990's could result only from new scientific knowledge and technological adaptations. It is difficult to over-emphasize the long and unavoidable lag between the initiation of research and the practical application of its results to the food system. Political and administrative decision-makers need to realize that Canada cannot continue to rely heavily on research and development efforts of other countries. There are many problems peculiar to the Canadian food system that will be solved only through Canadian research and development.

NOTES AND REFERENCES

[1] Recognition that resources are finite and cannot sustain growth indefinitely need not imply acceptance of the insistently gloomy conclusions of D. H. Meadows et al., The Limits to Growth (New York: Universe Books, 1972).

[2] J. Harold Branett and Chandler Morse, Scarcity and Growth: The Economics of Natural Resource Availability (Baltimore: John Hopkins Press, 1963).

[3] Note: 1 ha equals approximately 2.471 acres.

[4] 1971 Census of Agriculture, Statistics Canada, Ottawa.

[5] J. C. Woodward, "Canadian Production Potentials", Canadian Farm Economics, Vol. 9, No. 1, Feb. 1974, pp. 18-23; Hannah, A. E., "Production Possibilities for Grain in Western Canada", Canadian Journal of Agricultural Economics, Vol. 16, No. 1, Feb., 1968, pp. 71-76.

[6] Note: 1 m³ equals approximately 35.3 cubic feet or 220 imperial gallons, and 1 km³ equals approximately 810 713 acre-feet.

[7] C. G. E. Downing and M. Feldman, "Energy and Agriculture", Canadian Farm Economics, Vol. 9, No. 1, Feb. 1974, pp. 24-31. Note: 1 kJ equals approximately 0.948 British thermal units or 0.239 kilocalories.

[8] John S. Steinhart and Carol S. Steinhart, "Energy Use in the U. S. Food System", Science, Vol. 184, No. 4134, Apr., 1974, pp. 307-316.

[9] Michael J. Perelman, "Farming with Petroleum", Environment, Vol. 14, No. 8, Oct., 1972, pp. 8-13.

[10] There is very large range in the estimates of remaining reserves of petroleum and natural gas. Despite the possibility of large undiscovered deposits beneath the oceans and unexplored land areas, caution suggests the use of more conservative estimates. Even if such large deposits are discovered, extraction and transportation costs are likely to be high and possible uneconomical.

The Future of Small Communities in the Canadian Prairie Region

JACK C. STABLER

The viewpoint taken in developing this topic is that the cities of a region constitute a system in which several classes may be distinguished on the basis of functional specialization. In equilibrium, the role or function of each class of city within the system is fairly well defined. Through time, however, changes in the technical, social, or

● JACK STABLER is Professor of Economics, Department of Economics and Political Science, University of Saskatchewan. The article is reprinted from Contact, Vol. 9, No. 1, Spring 1977, pp. 145-173, Journal of Urban and Environmental Affairs, Faculty of Environmental Studies, University of Waterloo, with permission.

economic environment may alter the way in which goods are produced or transported, or the way in which people conduct their affairs. This, in turn, may bring about changes in the role or function performed by some or all classes of centres in the system.

The forces that bring about these changes may be broadly divided into two categories. The first category consists of a group of forces that arise from technical, economic, or behavioural change and which potentially affect all cities within the region's and even the nation's system of cities. For example, a substantial improvement in transport technology, which results in a lowering of the cost of moving goods or people, will almost certainly have an influence on the functions performed by communities of each class in the system. The second category consists of factors which are peculiar to a particular locale. An example of a factor of this sort is the discovery and exploitation of a mineral resource. This would have important consequences for all cities in the immediate vicinity, but it would not have a particularly strong influence on a given class of city or on the entire system of cities in a region.

The influence of the changes in the first category is pervasive and is therefore suitable for systematic analysis. The influences of the second variety of changes are much more particular and are consequently much less subject to general systematic analysis. To determine their importance, an examination of the local situation, on a case by case basis, is required. Fortunately for analytic purposes, empirical research suggests that the first set of changes has been more important than the second in shaping the environment in which small communities, as a class, exist and function.

It is the influence of the first category of changes that is discussed in this paper. The paper will argue that the forces arising from these changes, over the past couple of decades up to the present, have led to the consolidation of people and economic activity in a few large centres in the Prairies, at the expense of the region's smaller communities. The paper will argue that administrative and political considerations, as well as economic forces, have contributed to this consolidation and that they continue to do so. Finally, it will be argued that, while provincial governments have for many years expressed concern over the decline of small communities, they are unlikely to take the steps which would be necessary to halt or reverse the continuing process of consolidation.

The Setting

Most measures of economic well-being identify the Prairie region as one of Canada's more prosperous areas. Large and efficient agricultural and natural-resource-based industries provide the foundation for the already well-developed service sectors and for a variety of developing manufacturing industries. The wages and salaries paid to Prairie residents compare favourably with the national figures; unemployment is persistently below the national average; and the region's major cities have been among the fastest growing communities in Canada for several decades.

In view of these generally accepted measures of prosperity, the following observations may be somewhat surprising. While several of the larger Prairie cities have been expanding, over half of the region's more than 800 incorporated communities experienced a loss of population between the 1961 census and the census taken in 1971. In more than 600 of these incorporated communities, the number of business establishments in operation also declined between these two dates. In Saskatchewan, the loss of businesses in small communities was great enough to reduce the total number of business establishments in operation in the province between 1961 and 1971.

Factors Influencing the Economic and Spatial Structure of the Prairie Economy

During the past thirty years the Prairie economy has undergone a major structural transformation. Virtually every industrial sector and every locality within the Prairie region has been affected, as can be illustrated by a few comparisons. Thirty years ago, approximately 50 percent of the labour force in the three provinces was engaged in agriculture, while less than 40 percent found employment in the service industries. Today, more than 60 percent of the Prairie labour force is employed in service industries, while agriculture accounts for less than 20 percent. The structural reorganization of employment has been accompanied by a geographic shift of population from a primarily rural to a predominantly urban setting. Less than one-third urban in 1941, the Prairie population was approximately two-thirds urban in 1971. Improvements in production and in transport technology and shifting patterns of final demand have been important determinants of this transformation.

An important characteristic of all advanced industrial economies is the rapid adoption of labour-saving production processes. During the past three decades, such improvements have been more important in goods-producing industries, where standard products are produced through continuous, routine operations, than in service industries, where the product is more often tailored to individual tastes, and its delivery is less subject to continuous or repetitive procedures. At the same time, the effective demand for many goods, especially agricultural products, has not kept pace with the growth in the demand for services. The changes in the structure of the Prairie economy during the past three decades are thus part of a national process in which labour is shifting away from industries which are experiencing rapid gains in productivity, but only modest growth in demand, towards those in which the growth of productivity is low, but the growth in demand is rapid.

The observed impact of this shift on Prairie agriculture has been an increase in average farm size, from 160 ha in 1941 to over 300 ha in 1971, and a reduction in the agricultural labour force, from 462 000 to 226 000 workers between the same dates. This reduction in the agricultural labour force has meant, in the absence of offsetting rural development, a substantial absolute decline in the region's rural population. In absolute numbers, the rural population of the three Prairie Provinces declined by 435 386 people, from 1.6 million in 1941 to less than 1.2 million in 1971. This decline has obviously meant a loss of support for the small trade centres which serve the rural population.[1]

Of equal or perhaps even greater importance, however, has been the substantial improvement in the personal mobility of rural dwellers which was occasioned by the upgrading of the intercity road network, the increase in automobile ownership, and the improvement of the automobile itself. During the past twenty-five years, the number of kilometres of paved roads increased by a factor of ten. Gravel roads were increased by almost as much. Over one-half of the existing, hard-surfaced intercity roads and approximately one-third of the high quality grid roads were raised to their present standard during the 1960's. All-weather roads now reach into every corner of the region. At the same time, the number of persons per motor vehicle registration declined from approximately seven in 1941 to about two in 1971 for the three provinces combined. These improvements have greatly increased the options available to the rural dweller as a consumer, and have substantially altered his traditional spatial shopping patterns. The classic trade centre system, in which consumers shopped for the desired item in the closest centre at which it was available, is in the process of adjustment. In the pattern which is emerging, rural consumers still purchase convenience goods at the closest available point, but when forced to go beyond the immediate centre for goods which are not available there, they tend to drive to the nearest place where a complete range of consumer goods is available. Often intermediate size communities, which provide some of the items actually purchased in the larger, more distant centre, are bypassed along the way. These changes have enhanced the viability of regional shopping centres which are larger and favourably situated, but have been generally detrimental to smaller centres. Most smaller Prairie centres and many poorly situated communities of intermediate size have consequently lost a substantial portion of their commercial outlets during the past ten to fifteen years.

In the public sector, a process of consolidation has also been in effect in response to forces similar to those which have led to a reorganization of consumer-oriented commercial activities. In the interest of improvement in quality and, usually, reduction in cost (at least in relative terms), most of the high schools in smaller communities have been replaced by consolidated schools in larger, centrally-located communities. The conversion to a dial telephone system has eliminated the need for local switchboards and operators. Hospitals, highway maintenance facilities, liquor outlets, RCMP detachments, and post offices have all been consolidated to some extent in fewer, larger places. Finally, and to a large measure in response to the changes noted above, rail services to most Prairie communities have altered during the period in question, and additional modifications are currently being considered. In general terms, rail service to smaller communities now consist almost exclusively of the pickup of raw materials. Less-than-carload movement of nearly all freight into and out of communities of fewer than 5000 persons is handled by truck. Rail passenger service, too, has long since been discontinued, as the private automobile and bus lines have proven to be more attractive alternatives. The smaller centres are still served, perhaps even better than formerly, since greater convenience and flexibility are characteristic of the newer modes of transportation. The substitution of highway transport for rail transport has meant the closing of the rail station, however, and a further, visible decline in the status of the communities affected. Railroad employment in many communities has been further reduced by the conversion from stream to diesel locomotives. Whereas steam locomotives were usually run from one division point to the next, which were perhaps 95 km apart, diesels are run much greater distances, for example, from Winnipeg to Edmonton. Therefore, engines are no longer serviced or turned around at division points. Section crews at numerous locations have also been replaced by "floating" gangs that work out of larger centres and cover a much greater area.

The background to the loss of population from rural areas and the decline in the number of functions performed by smaller communities is thus defined in terms of the changes in production and transport technology and in terms of the shifts in preference patterns that have occurred during the past three decades.

Changes in the Prairie Trade Centre System, 1961-1971

The persistent decline in the number of farms between 1961 and 1971 has been accompanied by continuous migration from rural areas. Of the region's fifty-three census

divisions, thirty-six experienced absolute population losses between these dates. Further, the only census divisions which experienced *relative* gains were those in which the region's five major cities are located (and three divisions in the north, where high birth rates and the lack of migration out of the area by native people contributed to small relative gains).

The decline of the small communities, described above in general terms, was more pronounced during the 1961-1971 decade than in any preceding period. This can probably be attributed to the fact that a number of the changes discussed above had their greatest impact on these communities during the years following the middle 1950's. The decline in farm population has, of course, been more or less regular since the middle 1940's. Reorganization of the educational, communications, health, and postal systems, however, is a more recent development, as is the improvement in the intercity highway network. Further, it seems likely that the adjustment of consumer shopping habits to changed circumstances is something that does not occur immediately. Rather, it proceeds gradually and with a lag, as the consumer develops an appreciation of all the alternatives available to him. Thus, it seems reasonable to suggest that the changes described above had their greatest impact on the smaller Prairie communities during the 1960's.

In the following tables, some of the changes occurring in the region's system of cities between 1961 and 1971 are selectively quantified.[2] In the first of these, the size distribution of incorporated centres is shown. For the region as a whole the total number of communities was slightly larger in 1971 than in 1961. This increase is primarily due to the fact that, between these dates, the number of communities which were incorporating was greater than the number disincorporating. The table also reflects an increase in the number of communities in both the smallest and the largest categories. This increase is due to the middle being pushed toward both ends; that is, a majority of communities in the bottom two categories declined in population, while most of those above this level increased in size. Thus, the number of communities in categories toward either end of the distribution grew at the expense of those in the middle.

SIZE DISTRIBUTION OF INCORPORATED PRAIRIE COMMUNITIES 1961 AND 1971

| Pop. Class | Number | | 1961 Percent | | 1971 | |
	1961	1971	%	Cum.	%	Cum.
< 200	226	285	27.6	27.6	33.2	33.2
200 — 499	293	258	35.7	63.3	30.0	63.2
500 — 999	146	141	17.8	81.1	16.4	79.6
1 000 — 2 499	91	93	11.1	92.2	10.8	90.4
2 500 — 4 999	35	48	4.3	96.5	5.6	96.0
5 000 — 14 999	17	22	2.1	98.6	2.6	98.6
15 000 — 49 999	7	7	0.9	99.5	0.8	99.4
50 000+	5	5	0.6	100.1	0.6	100.0
Totals	820	859	100.1		100.0	

Source: Census of Canada, 1961, 1971.

Table 1

It should be pointed out that there are a number of unincorporated centres in the Prairie region whose experience is not explicitly identified here. The vast majority of these communities are very small, the mean population probably being under 100 persons. There are a few, however, that would otherwise fit into the bottom two categories of the tables. The majority of these larger unincorporated centres are located in Manitoba, where incorporated status is a less common characteristic of smaller communities than in the other two provinces. To the extent that the unincorporated centres were of the same size as their incorporated counterparts, their experience over the past decade was similar. For the majority of the much smaller unincorporated centres, a very rapid rate of decline was common.

In Table 2, a comparison of the changes in population of the *centres* themselves is made by holding each centre in the category that it occupied in 1961. Only those centres that were individually incorporated in both 1961 and 1971 were used in this comparison. Thus, the total number used is slightly smaller than the number of incorporated centres identified as existing in 1961 in Table 1.

For the region, only the smallest category registered an absolute loss of population. However, in Manitoba and Saskatchewan, an absolute population decline between 1961 and 1971 was the common experience for centres which, in 1961, occupied the two smallest categories. The third category from the bottom, (population class 500-999), remained approximately stable in Manitoba but realized a small gain in Saskatchewan. In Alberta, the bottom class was relatively stable, while the second and third categories realized small gains. Above the third category, population growth was the common experience in all provinces.

A further comparison of population change is provided in Table 3 where the *relative* gain or loss of communities in each category is identified. The comparison in these tables is again of communities that were individually incorporated in both 1961 and 1971.

For the region as a whole (and for each province), there was a shift of population to the larger centres. The relative decline is concentrated in the three smallest categories, while the region's five largest centres accounted for the majority of the upward shifts.

A better indication of the changing role as trade centres of communities in the various size categories is gained by comparing the average number of commercial functions present in 1961 with 1971. This done in Table 4. For the region, a substantial decline in the number of business outlets was experienced by communities in the two

Table 2

POPULATION DISTRIBUTION, PRAIRIE CENTRES,[1] GROUPED BY 1961 SIZE CLASSIFICATION 1961 AND 1971

Pop. Class	No. of Centres	Average Population		Population Distribution					
		1961	1971	1961			1971		
				Total	%	Cum.	Total	%	Cum.
< 200	215	120	100	25 794	1.3	1.3	21 532	0.9	0.9
200 – 499	292	331	334	96 596	4.9	6.2	97 463	4.0	4.9
500 – 999	145	694	754	100 661	5.1	11.3	109 360	4.5	9.4
1 000 – 2 499	91	1 569	1 892	142 789	7.3	18.6	172 149	7.1	16.5
2 500 – 4 999	35	3 189	4 145	111 600	5.7	24.3	145 086	6.0	22.5
5 000 – 14 999	13	8 839	10 227	114 906	5.8	30.1	132 947	5.5	28.0
15 000 – 49 999	6	27 515	31 146	165 090	8.4	38.5	186 877	7.7	35.7
50 000+	5	241 812	310 328	1 209 062	61.5	100.0	1 551 629	64.2	99.9
Totals				1 966 498	100.0		2 417 053	99.9	

[1] Centres incorporated in both 1961 and 1971, for which data were available for both dates.

Source: Census of Canada.

Table 3

POPULATION SHIFTS BETWEEN CLASSES OF INCORPORATED CENTRES,[1] PRAIRIE REGION 1961–1971

Pop. Class	Population		Change		Expected	Shift[2]		Percent Total Shift	
	1961	1971	Absolute	%		Up	Down	+	−
TOTAL	1 966 498	2 417 053	450 555	22.91					
< 200	25 794	21 532	-4 262		5 910		-10 172		13.84
200 – 499	96 596	97 463	867		22 132		-21 265		28.94
500 – 999	100 661	109 360	8 699		23 063		-14 364		19.55
1 000 – 2 499	142 789	172 149	29 360		32 715		- 3 355		4.57
2 500 – 4 999	111 600	145 086	33 486		25 569	7 917		10.77	
5 000 – 14 999	114 906	132 947	18 041		26 327		- 8 286		11.28
15 000 – 49 999	165 090	186 877	21 877		37 825		-16 038		21.83
50 000+	1 209 062	1 551 639	342 577		277 015	65 562		89.23	
TOTALS						73 479	73 480	100.0	100.01

[1] Centres incorporated in both 1961 and 1971 for which data was available for both dates.

[2] Population shift for a category is calculated as follows:

$$S_C = PC_t - \left(\frac{P_t}{P_o}\right) PC_o$$

where PC is population of the category; P is the total population for all categories combined; the subscript "o" refers to the original observation, i.e., 1961, and "t" the terminal observation, i.e., 1971.

155

Table 4

DISTRIBUTION OF BUSINESS OUTLETS,[1] PRAIRIE CENTRES,[2] GROUPED BY 1961 SIZE CLASSIFICATION 1961 AND 1972

| Pop. Class | No. of Centres | Average Number Business Outlets | | Distribution of Business Outlets | | | | | |
		1961	1972	1961 Total	%	Cum.	1972 Total	%	Cum.
< 200	215	6	3	1 229	3.0	3.0	726	1.6	1.6
200 — 499	292	15	11	4 357	10.5	13.5	3 169	7.2	8.8
500 — 999	144	27	26	3 948	9.5	23.0	3 730	8.4	17.2
1 000 — 2 499	90	51	54	4 627	11.1	34.1	4 862	11.0	28.2
2 500 — 4 999	34	94	110	3 206	7.7	41.8	3 731	8.4	36.6
5 000 — 14 999	13	195	211	2 535	6.1	47.9	2 745	6.2	42.8
15 000 — 49 999	6	490	518	2 941	7.1	55.0	3 106	7.0	49.8
50 000+	5	3741	4440	18 705	45.0	100.0	22 201	50.1	100.0
TOTALS				41 548	100.0		44 270	100.0	

[1] Business outlets, as the term is used here, refers primarily to commercial outlets. Thus public and quasi-public organizations are excluded.

[2] Centres incorporated in both 1961 and 1971 for which data were available for both dates.

Source: Dun and Bradstreet, Reference Book, March, 1961 and March, 1972.

Table 5

SHIFT IN BUSINESS OUTLETS[1] BETWEEN CLASSES OF INCORPORATED CENTRES, PRAIRIE REGION 1961–1972

Pop. Class	Outlets		Change		Expected	Shift		Percent of Total Shift	
	1961	1972	Absolute	%		Up	Down	+	−
TOTAL	41 548	44 270	2722	6.551					
< 200	1 229	726	-503		81		-584		22.21
200 – 499	4 357	3 169	-1188		285		-1473		56.01
500 – 999	3 949	3 730	-218		259		-477		18.14
1 000 – 2 499	4 627	4 862	235		303		-68		2.59
2 500 – 4 999	3 206	3 731	525		210	315		11.98	
5 000 – 14 999	2 535	2 745	210		166	44		1.67	
15 000 – 49 999	2 941	3 106	165		193		-28		1.06
50 000+	18 705	22 201	3406			2271		86.35	
TOTALS					1225	2630	2630	100.00	100.01

[1] Does not include public and quasi-public organizations.

Source: Dun and Bradstreet, *Reference Book*, March 1961 and March 1972.

smallest categories. Even the third smallest category (population size 500-999) realized a small absolute decline. Above this level, relative stability or a very modest growth in the number of business outlets was the tendency in communities up to those in the 50 000+ category, which experienced very substantial gains.

Relative changes in the number of business outlets is provided in Table 5. The concentration of the downward shift in the three smaller categories is again emphasized, as is the concentration of the upward shift in the 50 000+ category. The change in the total number of business outlets, shown in the first row of these tables, adds another dimension to the discussion. In total, those centres making up the three lowest categories in the trade centre network lost 1909 business outlets between 1961 and 1971. For the region as a whole, this represents a 41 percent loss for the smallest category of centres, and losses of 27 and 6 percent for the second and third smallest categories respectively. It is noteworthy, too, that the loss experienced by centres with a population of less than 1000 was nearly double the gains realized by those with a population of more than 1000 but less than 50 000. It was only the gains by cities with a population of 50 000+ that led to the 6.5 percent increase in business outlets which was realized by the Prairie region in total.[3]

In summary, both the population and business outlet data reflect a serious decline in the status of communities in the two smallest categories. For the third category (size 500-999) the population data indicate greater stability than do the business data. The business data are probably the better indicator in this case for several reasons. To begin with, the population increase, in most cases, reflects only a small net movement *into* these centres by people who already lived within the immediate area. In fact, as indicated above, there has generally been a net decline in the rural population served by most communities. The loss of business outlets reflects both the effect of population loss from the market area and the effect of consumers bypassing smaller communities in travelling to regional shopping centres. Not all communities within each category are identical, of course, and the experience of many individual centres differed from that described in the preceding aggregate comparisons.

For all classes of smaller centres, the chances for stability or growth were enhanced by certain combinations of economic and spatial characteristics. Spatially, a location close enough to a major city to serve as a dormitory for the larger community contributed to the prospect of stability or growth. At the other extreme, an isolated location, outside the market area of a larger community, enhanced the prospect of viability. Economically, greater diversification contributed to stability. For example, communities which housed manufacturing firms had a greater than average complement of public facilities (particularly high schools, hospitals, and retirement homes), or those that served as dormitories were more stable than communities whose sole or primary function was that of an agricultural service centre. Narrow specialization, particularly as an agricultural service centre, or a location within the market area of a larger community, but too far away to serve as a dormitory, were associated with a greater than average rate of decline.

Further, the gain or loss of activities, both commercial and public, does not typically occur on an unrelated, one activity at a time, basis. Rather activities are often interrelated, either by function or by the size of market required to support them. Growth or decline is thus often characterized by the gain or loss of a cluster of related activities which occurs in a rather short period and is followed by an interval of stability until the next advance or decline is experienced. The decision to close a hospital in a small community, for example, would probably cause the local medical doctor(s) to move to a centre which had a hospital. The drug store would not likely survive long in that event. The loss of the drug store would lessen the attraction of the community as a place to obtain goods and services of an intermediate order, and would affect the viability of other activities of an intermediate order.

Referring back to the tables once more, communities with a population between 2500 and 50 000 have generally demonstrated greater stability than other communities, although some individual centres have experienced serious losses (e.g., Moose Jaw), and others experienced important gains (e.g., Lethbridge). The largest cities and/or their metropolitan areas experienced impressive gains between 1961 and 1971.

Implications of Present Trends for the Prairie Trade Centre System

The changes in the trade centre hierarchy discussed above describe a system which is in disequilibrium. A consideration of the factors responsible for this disequilibrium implies that the adjustment now in process still has some way to go.

The combination of agricultural incomes, which for many smaller farmers are below those earned by nonagricultural workers, with the general possibility of a significant per unit cost reduction through increased farm size, particularly in grain-producing areas, provides a powerful incentive for further expansion for those farmers who are able to do so. This remains true even in the context of recent high prices for Prairie agricultural products. Consequently, additional consolidation is to be expected.

On the other hand, many of the other changes affecting the decline of smaller communities are nearer to completion. In particular, the potential for increased personal mobility must be nearly exhausted, given present technology. Another doubling of the region's paved roadways, which is an unlikely prospect, would not have anything near the impact that paving the basic inter-city network did in the 1960's. Consolidation of the region's high schools is near completion as well. In smaller communities, the loss of employment and removal of facilities associated with railroad reorganization have also pretty well run their course. The questions of rail line abandonment and the reorganization of the grain delivery system, however, have yet to be decided. In both cases, recent analyses have indicated that their potential for causing a change in the trade centre system are minimal. This is because the loss of employment for any community would be very small (one to three jobs in most cases) and because these attributes are not determinants of rural shopping patterns.

Nevertheless, the influence of the last mentioned set of factors will continue to be felt for several additional years, even though the changes are themselves nearly complete. For example, the loss of market potential does not immediately cause all the business firms affected to close down. Rather, it is frequently the case that the owner and his family will continue to operate the business as long as returns are sufficient to cover out-of-pocket costs and also to make some profit. This is particularly likely if the owner has some additional source of income (e.g., as school bus driver, postmaster, farmer, etc.) or is near retirement age. As these owners quit or retire, however, the firms will generally go out of existence. Discussions with numerous businessmen in smaller communities suggest that this is not an uncommon state of affairs. Closure of a large number of business outlets in the smaller communities can, therefore, be anticipated over the next several years.

Continuation of present trends would likely mean that nearly all centres presently in the under 200 population category will be on the verge of nonexistence by 1985, and that, by the same date, most centres presently in the 200-499 category will perform only a "corner grocery store" function, similar to that presently performed by centres with a population under 200. For communities in the next category (population 500-999), location and local leadership will be of critical importance. Those in situations which have good access to their local market area, but are well away from a higher order centre (50 km, for example), will probably remain relatively stable, given imaginative local leadership. Lacking any of these characteristics, decline seems more likely than stability. The prospects for the stability or modest growth of communities presently in the 1000-2499 category seem better, although here, too, poor local access, a location too near a larger centre, or ineffective local leadership could mean decline. For larger centres (2500+), at least stability seems likely and modest growth is a reasonable possibility.

The pattern actually realized will depend, to an important extent, on both provincial and federal government policies. This topic is considered in the following sections of the paper.

Provincial and Federal Government Policies Affecting the Status of Small Communities

It is probably fair to say that up to this point in time, none of the three provincial governments has had either an explicit subregional development program or a trade centre system program. Government departments are, for the most part, organized along functional lines, such as education, highways, or agriculture. Even departments of municipal affairs have historically been inactive in analyzing changes in the region's trade centre system or in formulating plans for their orderly development.

In each of the provinces, there have, of course, been numerous programs designed to improve the well-being of rural dwellers, and these programs have necessarily had a subregional impact. In most cases, however, the objectives of the programs were not explicitly designed to strengthen or to facilitate adjustment in the economy of an area of the province. More often, the objective was the satisfaction of selected functional criteria. The influence on the subregional economy was secondary and does not appear to have been a criterion that received systematic consideration. Most provincial government expenditures in the Prairie region are probably still made according to functional criteria, in the absence of specific consideration of their impact on the relevant trade centre system or subregional economy.[4] Consequently, while most provincial government programs may have been and usually were considered successful according to the criteria that they were specifically designed to satisfy, their impact on the region's trade centre systems and local economies has been equivocal.[5]

Since efficiency has, correctly, been emphasized in the pursuit of functional goals, the factors which led toward consolidation in the private sector have encouraged a similar pattern in government programs. On balance, it would appear that most provincial programs, insofar as they have a spatial influence, have been conducive to the consolidation of economic activity into fewer and larger centres. In fact, this consolidation may have been accelerated by the tendency to "spread around" provincial public activities. If a hospital is built in community A, for example, the expectation is often created that community B will get the regional high school, community C the regional retirement home, etc. By spreading these facilities around instead of concentrating them in one centrally-located regional centre, the potential attraction of this cluster of activities is dissipated, and the centralizing pull of the larger cities is not offset to the extent that it might otherwise have been.

As in the provincial governments, most federal government departments are organized on a functional basis. To a probably greater extent than provincial governments, however, the federal government is sensitive to the regional allocation of its expenditures. The nature of the federal system demands that this be the case. The federal government's perception of regions, however, is that, for the most part, they consist of provinces or groups of provinces rather than subregions within the provinces or rural areas versus urban. The programs of the various federal government departments may, of course, have an influence on the subregions of the various provincial economies. It would not be possible to assess the impact of the numerous federal government programs in this respect for this paper. Nevertheless, it seems reasonable to suggest that the impact of most federal government activity in the provinces has had an internal centralizing influence, for the same set of reasons advanced earlier in discussing the causes of centralization of private and provincial government activity.

One federal government agency, the Department of Regional Economic Expansion (DREE), does warrant specific consideration in the present discussion. Created in 1969, the primary objective of the department has been to initiate self-sustaining economic expansion in areas which have experienced prolonged periods of stagnation, evidenced by either persistently high rates of unemployment or slow growth resulting in people leaving the region. Under the DREE program, all of southern Manitoba, selected areas in Saskatchewan, including Saskatoon and Regina, and parts of Alberta, excluding Calgary and Edmonton, were initially designated for special assistance. Between the beginning of the incentive program and June, 1972, approximately 250 grants, totalling more than $40 million, had been made to business firms, primarily in the

PERCENTAGE DISTRIBUTION OF MANUFACTURING EMPLOYMENT WITHIN THE
PRAIRIE REGION AND ALLOCATION OF DREE INDUSTRIAL GRANTS TO JUNE 1972
BY PROVINCE

| | Manufacturing-Employment | | | Allocation of Grants | |
	1961	1966	1971	Jobs	Funds
Manitoba	43.9	43.7	42.4	58.0	38.0
Saskatchewan	14.2	14.1	13.2	22.0	17.0
Alberta	41.8	42.2	44.4	20.0	45.0
	99.9	100.0	100.0	100.0	100.0

Source: Statistics Canada, *The Labour Force*, various issues, and *Hansard*, various numbers.

Table 6

manufacturing sector, which were locating or expanding in the Prairie region.

A comparison of the distribution of employment in manufacturing within the region with the allocation of industrial incentive grants by province indicates that the influence of the grants on the distribution of activity *within* the Prairie region has been very small. These figures are shown in Table 6. Rather, the influence of the grants has been to improve the position of the *Prairie* economy relative to other major regions.

Further, within each province, the allocation of the DREE grants has not been one which would substantially alter the existing distribution of manufacturing activity vis-à-vis smaller communities. Statistics showing the concentration of employment in manufacturing in major centres and the allocation

of grants by size of community are shown in Table 7.

A program directly oriented towards the strengthening of service centres was worked out during 1972 by the Department of Regional Economic Expansion, in consultation with the provincial governments. Some sixty-six centres in the Prairie region were identified as "agricultural service centres". This program was based on the "need to afford greater economic security to people in the agricultural areas served by these centres . . . to contribute to economic expansion and social adjustment".[6] The criteria for the selection of centres were that the centres must:
(a) be outside metropolitan areas;
(b) be located in settled agricultural regions;
(c) be outside special areas;

Table 7

CONCENTRATION OF MANUFACTURING EMPLOYMENT IN MAJOR CENTRES AND
ALLOCATION OF DREE GRANTS TO JUNE 1972, BY SIZE OF COMMUNITY,
PRAIRIE REGION

| | Manufacturing Employment Centres of 50 000+ | | Allocation of DREE Grants by Community Size-jobs | | |
	1966	1971	50 000+	2500-49 999	2500
Manitoba	80.3	80.9	76.0	16.0	8.0
Saskatchewan[1]	53.4	52.0	72.0	23.0	1.0
Alberta[2]	69.8	69.1	—	45.0	13.0

[1] The Meadow Lake Special Area accounted for 4 percent of the jobs created.

[2] The Slave Lake Special Area accounted for 42 percent of the jobs created.

Source: Statistics Canada: *Manufacturing Industries of Canada*, various issues, and *Hansard*, various numbers.

(d) be capable of providing an adequate level and range of services for a growing agricultural service area;

(e) have population characteristics with demonstrated growth potential;

(f) have significant potential for nonagricultural employment;

(g) have prime access to principal transportation facilities;

(h) have an infrastructure base which can be readily expanded to accommodate future growth.[7]

For the most part, the centres named were the largest sixty-six in the region, excluding the five major centres. Most exceed 2000 in population. The communities selected were to be eligible for aid in constructing or improving water supply and waste disposal facilities. Beyond this, of course, the preparation of "a list" becomes a potentially powerful suggestion to other government departments or private investors.

While providing a decisive lead in establishing criteria for selection of communities according to demonstrated viability and potential for future development, the specific selection of communities may be criticized in that spatial distribution appears to have a lower priority than the size of centre. If circles of 40 km radius are drawn around each of the centres chosen, to approximate to their local market areas, a large number of blank spaces remain, especially in Alberta and Saskatchewan. Within these blank spaces, it is well over 80 km in any direction to an agricultural service centre. However, communities with a population of more than 1000 providing thirty or more distinct functions, and having records of stability or modest growth, are to be found in these areas.

The Future Trade Centre System in the Prairie Region

Economic Forces

In the preceding sections of this paper, some of the reasons for the decline of the trade centre role of smaller communities were identified, and the magnitude of the decline during the 1960's was selectively quantified. It was evident that centres with a population under 500 had, as a group, experienced substantial absolute losses during this period, while those in the 500-999 category experienced modest losses. In terms of relative change within the trade centre hierarchy, all classes of centres below 1000 in population declined markedly during the 1960's.

In the case of centres in categories above 2500 population, relative stability was evident up to the largest class (50 000+), which accounted for nearly all of the relative gain.

If the future of the trade centre system were to be determined primarily by economic forces, as has been the case in the past, it is almost certain that the trends identified above would continue. Centres which do not currently provide a combined total of at least thirty distinct retail, government, and professional functions would continue to decline in trade centre status until they provided only a "corner grocery store" function to the population resident in or near them. A continued, upward shift in the age-structure and a gradual decline in population could be expected to follow the decline in trade centre status.

This decline would likely be more rapid in areas where grain production provides the major source of income than in areas where the economy is more diversified. Larger centres, i.e., those currently providing thirty or more distinct functions and generally having more than 2000 residents, would strengthen – relative to smaller communities – their position as regional centres for the provision of common trade, government, and professional services. These centres, well-located with respect to local markets and spaced at 50-80 km intervals, would meet the everyday needs of rural residents for most items except the more specialized varieties which would be obtained in the region's major cities. Except for the largest fifteen to twenty centres in this category, the attraction of completely new activity would not be a significant factor in their development. The five major cities would continue to gain an increased share of the region's existing population and economic activity, and also to attract the majority of new residents and new activity.

Policy Options

Each of the three provincial governments has recently expressed concern over the level of agricultural incomes, the possible emergence of corporate farm ownership, and the decline of "smaller communities".

With respect to agriculture, they have initiated programs which would facilitate diversification, ease the tax burden of farmers, strengthen their position regarding the marketing of products, and ease the transfer of ownership from one generation to the next. It seems likely that the corporate ownership of agricultural land will also be restricted.[8] The successful execution of these programs could be expected to raise agricultural incomes, perhaps reduce the disparity between agricultural and nonagricultural incomes, and strongly contribute to the preservation of family ownership of agricultural land. It is not evident that these programs would alter the relationship between farm size and production costs, however, and this is one of the factors responsible for the continuing trend to larger farms, fewer farmers, and declining smaller communities.

With respect to the communities themselves, provincial governments will likely pursue programs which will extend to a large number of additional communities the type of assistance that is being provided to the farm service centres. In this manner, the number of communities in the Prairie region which receive direct government assistance, specifically designed to enhance their ability to cope with changing circumstances, may increase to 150-175. In addition, it is likely that the provincial governments will attempt to persuade some business firms that would otherwise locate or expand in the major cities to locate in a secondary centre. It would seem that this is about as far as the provincial governments are prepared to go at the present time.

Improvements of local water and sewerage facilities and the sporadic location of a business firm in a secondary centre, however, will have little influence on the evolution of the trade centre system. If this is all that the governments do, the system will continue to evolve primarily in response to economic forces which are producing the maximum degree of centralization. It is a problem of doing too little to alter the process, although it is not necessarily too late for bold measures to have an effect.

In order to be realistic, programs designed to alter the evolution of the trade centre system in a meaningful way would have to be developed in a manner which recognizes the following conditions:

1. the strong likelihood of further consolidation in the agricultural sector;
2. the certainty that past change will have a lingering influence; and
3. the very great attraction of the region's major cities.

Such programs would recognize the inevitability of the further decline of the majority of the region's incorporated centres, particularly those at the lower end of the scale, and the need to *greatly* strengthen a small number of the more favourably situated, more viable, intermediate-size centres in order to provide a measure of counter-attraction to the pull of the region's major cities. The steps in such a program would include measures like those envisioned in the farm service centre program, but they would also involve the deliberate, selective concentration in intermediate-size centres of that portion of government and government-supported activity which is conducted outside the major cities. Voluntary relocation of personal residences at government expense, from communities that are clearly no longer viable into those that are still viable, is a further possibility.

If the provincial governments were to concentrate their efforts in this manner, the trade centre pattern that evolved would allow more activity to be conducted in a select number of viable, intermediate-size, rural communities than would occur if market forces alone dictated the direction of change. In addition, these communities would be considerably more attractive places to live and work than they would otherwise be, and they would consequently provide a greater counter-attraction to the major cities than is possible under existing circumstances.

While such an approach would not involve major conceptual difficulties, from a planning perspective it obviously raises that complex set of political considerations with which Western governments have proven incapable of dealing effectively in the management of the space-economy. For this reason, it seems unlikely that anything resembling this degree of deliberate involvement by provincial governments in the future of small Prairie communities will be seriously considered.

Instead, in a manner which is more or less undirected, economic forces will dictate the direction in which the trade centre system in the Prairie region evolves, and centralization of the type that has occurred elsewhere will run its course.

NOTES

[1] Although per capita real income more than doubled in the Prairie region between these dates, this has done little to offset the loss of numbers. The common, everyday goods and services provided by the smaller centres are those for which per capita consumption is relatively constant above a very low minimum income.

[2] In Tables 1 through 5, the population and business outlets of Bowness, Forest Lawn and Montgomery are included with those of Calgary, and those of Beverly and Jasper Place with those of Edmonton. The population and business outlets of Brooklands, St. Boniface, St. James, Transcona, Tuxedo, East Kildonan, and West Kildonan are included with those of Winnipeg.

[3] This experience, while generally similar for each province, differed in detail. One notable exception to the general pattern was the absolute decline in the number of business outlets for the entire province of Saskatchewan during this period.

[4] Such generalizations obviously need to be made with considerable caution. The ARDA (federal provincial) programs in the Manitoba Interlake area, and the DREE special area programs, for example, do not come within this generalization.

[5] This should not be taken as a criticism of the functional approach as such. It only points to one of its characteristics.

[6] Department of Regional Economic Expansion, News Release, July 31, 1972. Reprint.

[7] *Ibid.*

[8] With the exception of family and cooperatively owned corporations.

SELECTED REFERENCES

John R. Borchert, *The Urbanization of the Upper Midwest, 1930-1960*, Urban Report No. 2 (Minneapolis: Upper Midwest Economic Study, 1963).

_____ and Russell B. Adams, *Trade Centers and Trade Areas of the Upper Midwest*, Urban Report No. 3 (Minneapolis: Upper Midwest Economic Study, 1963).

Canada Department of Agriculture, *Prairie Regional Studies in Economic Geography*, (Regina: Economics Branch, Canada Department of Agriculture, various issues, 1970-1975).

Gerald Hodge, "The Prediction of Trade Center Viability in the Great Plains", *Papers of the Regional Science Association*, 15, 1965, pp. 87-115.

_____, "Branch Line Abandonment: Death Knell for Prairie Towns?", *Canadian Journal of Agricultural Economics,* 16, 1968, pp. 54-70.

H. G. Kariel, "Analysis of the Alberta Settlement Pattern for 1961 and 1966 by Nearest Neighbor Analysis", *Geografiska Annaler,* 52B, 1970, pp. 124-130.

Saskatchewan, Royal Commission on Agriculture and Rural Life, *Movement of Farm People*, Report No. 7 (Regina: Queen's Printer, 1956).

_____ Service Centers, Report No. 12 (Regina: Queen's Printer, 1956).

Jack C. Stabler, *et al*., *Economic Effect of Rationalization of the Grain Handling and Transportation System on Prairie Communities* (Saskatoon: Underwood-McLellan and Associates, 1972).

Jack C. Stabler, "Transportation and Prairie Communities", *Canadian Transportation Commission Research Publication* Series, 1973.

_____ and Peter R. Williams, "The Dynamics of a System of Central Places", *Geographical Papers,* 22, University of Reading, 1973.

S. W. Voelker and T. K. Ostenso, *North Dakota's Human Resources: A Study of Population Change in a Great Plains Environment* (Fargo, North Dakota: Agricultural Experiment Station, 1968).

The Spatial Association of Agroclimatic Resources and Urban Population in Canada

G. D. V. WILLIAMS, N. J. POCOCK, AND L. H. RUSSWURM

Canada, with a 10×10^6 km^2 surface, is the world's second largest country. With a population of 23 million and a population density of less than three persons per square kilometre compared with a world average of about 27 persons per square kilometre, it might be assumed that Canada has no problems involving land shortages. But at best only about 13 percent of Canada's land is suitable for agriculture. Moreover, only about half of this small percentage can be used for field crops, the other half being suited only for pasture[1]. Since population in Canada can be expected to increase at least to the year 2025, the urban expansion pressures on our agricultural land are likely to continue. Our goal in this paper is to assess quantitatively the spatial association of Canada's agroclimatic resource base in relation to such urban expansion pressures on agricultural land.

Most food presently comes from agricultural production. It is unlikely that other sources will contribute very substantially to food supplies in the foreseeable future. Economic and energy limitations[2] and ecological hazards[3] will make it more difficult to greatly increase agricultural yields per hectare in the future.

Most of the limited remaining land in Canada that has potential for new agricultural development occurs in areas having inferior climates and/or soils[4]. Given these limitations of our land resource, we need to maintain our best agricultural land as the renewable resource that it is. Until recently, very little effort was made to do so, with urban uses being given absolute dominance. Increasing emphasis is now being given, in several parts of Canada, to the maintenance of the better agricultural lands. Census figures for 1976[5] show that from 1971 to 1976 overall farmland loss was much less than from 1966 to 1971. Better economic returns for farmers, combined with increased planning and political concern for maintaining farmland, are likely reasons for the reduction in the rate of loss of the farmland.

Unfortunately, however, Canada's growing cities are mostly located in areas having the best climates and soils for agriculture. The Great Lakes-St. Lawrence Lowlands and the Lower Fraser Valley have the most climatically favoured farmlands. In these areas are located the three large metropolitan centres with populations exceeding one million, Toronto, Montreal, and Vancouver (Table 1), that together account for nearly 30 percent of Canadian population. From 1961-71 these three accounted for 1 531 043, or 46 percent of Canada's population growth of 3 330 068. Census results for 1976 show that the populations of Toronto and Vancouver are still growing at a faster rate than Canada as a whole, though Montreal's population growth rate is well below the national rate.

The proportion of Canada's population that is urban is expected to increase from the current 76 percent to 90 percent by the year 2000[6]. Urban population in Canada includes all people living in places of 1000 or more and in municipalities surrounding such places having a density of 386 per square kilometre. A national population increase from today's 23 million to the expected 30 million by the year 2000 would mean, by the

● Contribution No. 906 of the Chemistry and Biology Research Institute. G.D.V. WILLIAMS is in the Chemistry and Biology Research Institute, Research Branch, Agriculture Canada, Ottawa; N. J. POCOCK is with the Geographical Services Directorate, Department of Energy, Mines, and Resources, Ottawa, formerly with the Lands Directorate, Environment Canada, Ottawa; and L. H. RUSSWURM is in the Department of Geography, University of Waterloo, Waterloo, Ontario. The analysis of the data began as a course project at Carleton University. A number of individuals in several organizations contributed assistance and advice.

Table 1

CHANGES IN URBAN POPULATION AND FARMLAND HECTARAGE, 1961 TO 1971, FOR CANADIAN CENSUS DIVISIONS THAT CONTAIN OR ARE NEAR METROPOLITAN CENTRES

Census Divisions,[a] 1971 basis, used for each centre	Farmland 10^3 ha				Urban Population thousands				Ratio (farmland decrease / urban pop. increase)
	1961	1971	Change	% Change	1961	1971	Change	% Change	
St. John's: Nfld. 1	8.4	14.9	+6.5	+77.4	118	136	18	15.4	-0.358
Halifax: N.S. 8	37.7	26.9	-10.8	-28.6	171	205	34	19.8	0.318
Saint John: N.B. 6,11	126.6	84.8	-41.8	-33.0	84	103	19	22.9	2.177
Quebec: Que. 37,47,53,54	217.3	143.9	-73.4	-33.8	374	503	128	34.3	0.325
Montreal: Que. 11,15,18, 28,35,36,70,71,72	345.3	262.0	-83.4	-24.1	2198	2764	566	25.7	0.147
Chicoutimi: Que. 16	109.2	79.4	-29.7	-27.3	123	128	5	3.9	6.208
Ottawa: Ont. 33,43 Que. 24,25	387.4	308.3	-79.1	-20.4	431	571	140	32.4	0.566
Toronto: Ont. 32,36,49,54	372.3	303.9	-68.4	-18.4	1884	2619	736	39.1	0.093
Hamilton: Ont. 16,53	146.3	120.6	-25.7	-17.6	419	539	120	28.6	0.214
St. Catharines: Ont. 28[c]	118.0	102.2	-15.8	-13.4	229	335	106	46.5	0.148
Kitchener: Ont. 51	112.0	100.8	-11.2	-10.0	148	223	75	50.6	0.149
London: Ont. 26	281.7	267.9	-13.7	-4.9	182	240	59	32.4	0.234
Windsor: Ont. 9	153.8	142.9	-10.8	-7.0	210	247	36	17.3	0.298
Sudbury: Ont. 46	74.2	44.5	-29.6	-40.0	118	146	28	23.4	1.070
Thunder Bay: Ont. 47	65.9	43.2	-22.6	-34.3	111	126	15	13.7	1.484
Winnipeg: Man. 1,5,6,9,20[b]	1329.2	1295.9	-33.3	-2.5	494	563	69	14.1	0.479
Regina: Sask. 6[b]	1629.6	1626.1	-3.4	-0.2	115	143	27	23.7	0.126
Saskatoon: Sask. 11[b]	1449.6	1435.4	-14.3	-1.0	98	132	34	34.5	0.421
Calgary: Alta. 6[b]	1236.9	1160.8	-76.1	-6.2	284	415	130	45.9	0.584
Edmonton: Alta. 11[b]	1059.3	1121.2	+61.9	+5.8	349	494	145	41.5	-0.427
Vancouver: B.C. 5,11,15[c]	111.1	76.6	-34.5	-31.1	773	1002	229	29.7	0.151
Victoria: B.C. 3,10[c]	34.0	27.9	-6.1	-18.0	160	187	26	16.6	0.230

[a] The divisions to be included were selected on the basis of studies by Bryant (1976), Martin (1975b) and Russwurm.

[b] Large areas for centres in the Prairie Provinces occur because of large census divisions; and area of 1.6×10^6 ha represents a radius of approximately 71 km or a square having sides of approximately 126 km.

[c] Ont. Div. 28 (Niagara) was Lincoln and Welland in 1961 census. In B.C. among the 1961 census divisions, Div. 4 was used for Vancouver and part of Div. 5 for Victoria.

above definition, an additional 11 million urbanites. A moderate urban density is 2500 people per square kilometre, which is about the density for the city of Hamilton. At such a density, if half of the expansion continued to be as it has been in the areas with the climatically-best 5 percent of the farmland[7], more than 200 000 ha of the land with the best agroclimate would be lost in accommodating the additional 11 million urbanites. This amount of land could feed half a million to a million people. A considerable portion of this loss could be prevented by shifting urban development to poorer land. Such shifts are practicable around most cities.

Although useful studies have been made of the loss of agricultural land in the growth of individual cities[8], in rural-urban fringe areas[9], and in specific regions[10], not enough attention has been given to the association between the spatial distribution of urban expansion and the agroclimatic resource on a national scale.

Such an analysis at a national scale presents data problems. Air photo interpretation is quite useful in detailed local analyses of farmland loss due to urban expansion[11], but time series of air photos for a national scale analysis are incomplete. In any event, air photo interpretation, if applied to obtain a national overview, would be a large and expensive undertaking. An air photo interpretation study of rural to urban land conversion recently completed in the Lands Directorate, Environment Canada[12], helps to show the farmland loss problem around cities of 25 000 or more people.

Another approach is to examine the change over time in the amount of improved farmland[13]. This census data does not reveal whether or not the land lost was actually used for urban development, which is often not the case[14]. Improved farmland, as reported by Statistics Canada, is land that is regularly cropped. Total farmland is all the land occupied by farmers and therefore includes both the improved and unimproved farmland. Improved farmland actually increased in Canada 1961-1971; most of this increase resulted from bringing into cultivation land of somewhat inferior soils and climate in western Canada[15]. Elsewhere some of the agricultural resource base was irreversibly lost, as farmland — both improved and unimproved — was used for urban development. Exactly how much good agricultural land was taken for urban development is not known. Probably it amounts to less that 10 percent of the actual farmland lost[16]. While most of the farmland lost 1961 to 1971 was a result of marginal farmland retirement, the land used for urban expansion is usually the best farmland around any city because developers view such land as ideal for urban development.

A population density of 2500 per square kilometre is equivalent to 0.04 ha per person. Around Toronto the farmland reduction from 1961 to 1971 was 0.09 ha per capita increase in the urban population (Table 1). It can be assumed that about half of this land was actually built on, and the remaining 0.04 or 0.05 ha per person was removed from agriculture but not actually built on. It would be possible to return much of this latter to farming if the need arose, but it would be almost impossible to return to farming the land that had actually been built on. Around cities in areas with rather poor physical environments, for example Chicoutimi, Thunder Bay, and Sudbury, a large amount of land has been going out of agriculture in relation to the urban population increase, but probably only a very small part of this land is being used for urban development. In areas such as around St. John's and Edmonton, farmland losses due to urban expansion from 1961 to 1971 were masked by the general increases in the extent of agriculturally developed land in these advancing agricultural frontier districts.

Some local or regional studies have analyzed agricultural land losses in terms of the Canada Land Inventory (CLI) soil capability ratings[17]. The CLI classifies land into seven classes of soil capability for agriculture, with Class 1 being the best and Class 7 having no agricultural capability[18]. For a national study, however, much of the required data would not be available in suitable form and furthermore, as Shields and Ferguson[19] have remarked, the individual soil capability classes are not comparable from one region to another. This lack of comparability is partly due to climate. For example, largely because of climatic limitations, Class 1 land in the Prairies cannot grow the same diversity of crops as Class 1 land in southern Ontario or the Vancouver-Victoria areas of British Columbia. For national overview purposes the CLI ratings by themselves do not give sufficient weight to climate, but the fact that they reflect climate to some extent makes it difficult to use them in conjunction with some climatic indicator.

We have chosen to use farmland area as our indicator of agricultural land on the grounds that, at a national scale, it best represents the area of land occupied by farmers given the data currently available. It also implies some information on the suitability for agriculture since, for example, most of the land on the Canadian Shield and in other areas unsuitable for farming is not included in the farmland totals.

In providing a quantitative assessment of the spatial distribution of Canada's agroclimatic resource in relation to urban expansion pressures, we used the following data and measures. First, urban population and farmland areas for 1971 were compared at the census division level. (In eastern Canada census divisions are equivalent to counties.) This comparison of urban population and farmland provides a measure of the spatial association of the direct urban threat to farmland resources. The farmland-urban population distribution and that of climatic resources for agriculture are then examined. The number of days from barley ripening to first fall freeze is used as an indicator of the potential of the agroclimatic resource. The development of an agroclimatic resource index to further help quantify the resource loss problem at a national and regional scale is also described.

Methodology

Census Urban Population, 1971

Urban population as defined for the 1971 Census[20] includes all persons living in: (a) incorporated cities, towns, and villages with a population of 1000 or more; (b) unincorporated places of 1000 or more having a population density of at least 386 per square kilometre; and (c) the built-up fringes of (a) and (b) having a minimum population of 1000 and a density of at least 386 per square kilometre. The same definitions apply for the 1976 Census. The rest of the population is classed rural by Statistics Canada. Rural population consists of rural farm and rural nonfarm. The latter consists of the people living in villages and hamlets of less than 1000 people and the scattered exurbanite population living in the countryside.

Urban population is used because its spatial distribution serves as a sound surrogate for the spread of urban development onto agricultural land[21]. Once land is converted to built-up urban uses, it is almost surely removed from agricultural use forever, although it is still possible that some urban open space (lawns, playgrounds, parks, etc.) could be used to produce food if the necessity should arise. Total population was not used for several reasons. Urban population is growing whereas rural population is stable or decreasing. Most urban population growth diverts farmland to urban use; changes in land use patterns by the rural population are less likely to reduce the potential agricultural land resource base, although scattered rural nonfarm, i.e., country residential development[22], may reduce this resource base.

To eliminate problems of lack of comparability resulting from changes in census boundaries, the 1971 urban population was used rather than the change in urban population from 1961 to 1971. The urban population increase tends to be highest in areas of greatest population concentration (Table 1). Because of this relationship, the spatial distribution of the increase in urban population will be quite similar to the distribution of the urban population itself at any given time. This fact provides further justification for the use of the 1971 census data instead of data on the 1961 to 1971 urban population increment for this study. This ongoing concentration of population in the larger cities has already been noted by others[23]. The 1976 census figures[24] show that this concentration is stabilizing with the CMA's growing 6.8 percent 1971-1976, compared with a national growth rate of 6.6 percent. In absolute numbers, however, the twenty-three CMA's accounted for 57 percent of the population added 1971-1976 (814 320 of 1 424 293).

Census Farm Area, 1971

The term "census farm", as used in the 1971 Census, refers to an agricultural holding of 0.4 ha or more, having sales of agricultural products during the twelve-month period prior to the census of $50 or more. (For the 1976 Census the definition is $1200 and 0.4 ha or more.) Census farm hectarage for 1971 consequently refers to the number of hectares of land that were in census farms at that time according to the 1971 definition. This measure reflects a current use. It does not reflect land capability, as rated by the CLI, nor economic viability as defined by the hectarage of census farms having agricultural product sales of $2500 or more. While data on improved land hectarage better reflects the intensity and commercial viability of present use, total farmland area consisting of all land occupied by farmers better reflects the amount of farmland readily available to meet future needs. For this reason, the total hectarage of census farms rather than only that of improved farmland was used in this study.

The census farm category as defined in the 1976 Census seems much less useful in land resource analyses than that based on the 1971 definition. Much land that is being used in agricultural operations that are not at present part of economically viable farming enterprises would be excluded under the new definition, but this excluded land will often have just as much potential for agriculture in future as has a lot of the 1976 census farm hectarage. Also, in making comparisons from one census to another, inflation may move some land from the less than $1200 category to the $1200 or more annual sales category. This probably explains why census farm hectarage in Canada increased from 64.9 to 67.2 × 10^6 ha from 1971 to 1976 according to the 1976 definition, but decreased from 68.7 to 68.4 × 10^6 ha during that period according to the 1971 definition[25].

Climatic Resources for Maturing Barley

Dansereau[26] defined resources as "conditions or elements of the environment exploitable by living beings", and cited the example of the use by plants of soil water and nutrients, oxygen and carbon dioxide, solar heat, winds, and the vegetation mat. Temperature and photoperiod (day length) are elements of particular importance in maturing a crop, and one effective way of quantifying their combined effect, when used as a resource by a crop, is to compute how long before first damaging fall freeze the crop will normally reach maturity. Given weather variability from year to year, this indicator of average conditions expresses the relative margin of safety for the crops in question.

In this study, the indicator crop used was barley; the values given are the estimated number of days from barley ripening to first fall freeze. The interpretation of the values is that the shorter this duration is, the poorer is the climatic resource.

The derivation and mapping of these data for the Canadian Great Plains and subsequently, for all of Canada, are discussed in previous work[27]. A map of climatic resources for maturing barley was superimposed on a census division map, and the representative value was determined by experienced judgement for each census division. It was difficult to be objective in this selection in British Columbia or the north shore of the St. Lawrence River in Quebec, where most of the urban population and the majority of the farmland are concentrated in one part of a census division. In such cases, the climatic resource value was selected to attempt to represent the populated part rather than the division as a whole.

In an earlier study by Williams[28], corn was used as an indicator because its distribution corresponds to the areas with the most favourable climates for agriculture in Canada. However, it was found that approximately half of Canada's farmland was in areas for which the climate had not even been mapped for corn because those areas were too cool for this crop. Since barley can usually be matured in practically all occupied census divisions in Canada, it provides a more suitable indicator for the country as a whole.

Barley has the widest distribution of all our cereals. The Olli variety, on which the computations are based, is an early maturing variety. The fact that in areas such as southwestern Ontario, crops with higher heat requirements would generally be grown, does not detract from the usefulness of the data employed here. In further work, the analysis should be expanded to include a number of additional crops.

Climate is a "flow" resource, like the water in a reasonably reliable river system, as opposed to a "fund" resource like coal in a

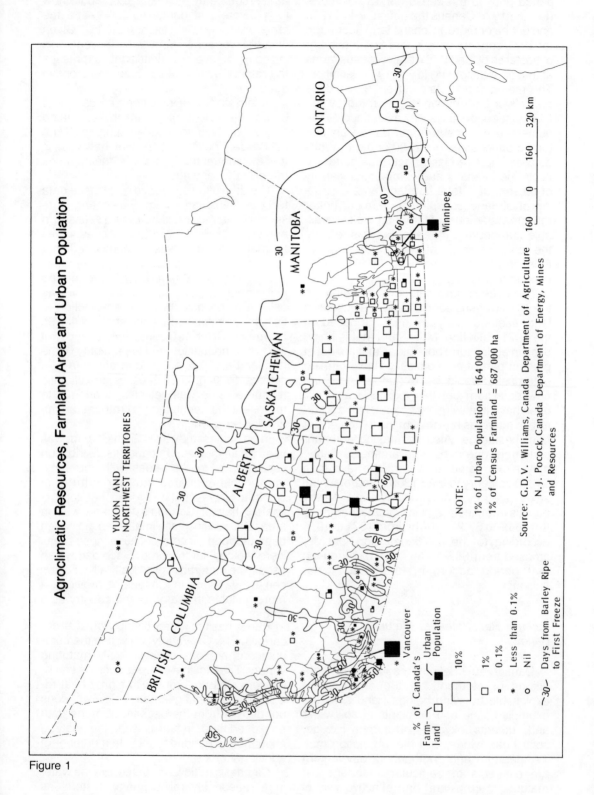

Agroclimatic Resources, Farmland Area and Urban Population

YUKON AND
NORTHWEST TERRITORIES

BRITISH COLUMBIA

ALBERTA

SASKATCHEWAN

MANITOBA

ONTARIO

Vancouver

Winnipeg

NOTE:

1% of Urban Population = 164 000

1% of Census Farmland = 687 000 ha

Source: G.D.V. Williams, Canada Department of Agriculture
N.J. Pocock, Canada Department of Energy, Mines
and Resources

% of Canada's
Farm-
land

Urban
Population

10%

1%

0.1%

Less than 0.1%

Nil

-30- Days from Barley Ripe
to First Freeze

160   0   160   320 km

Figure 1

170

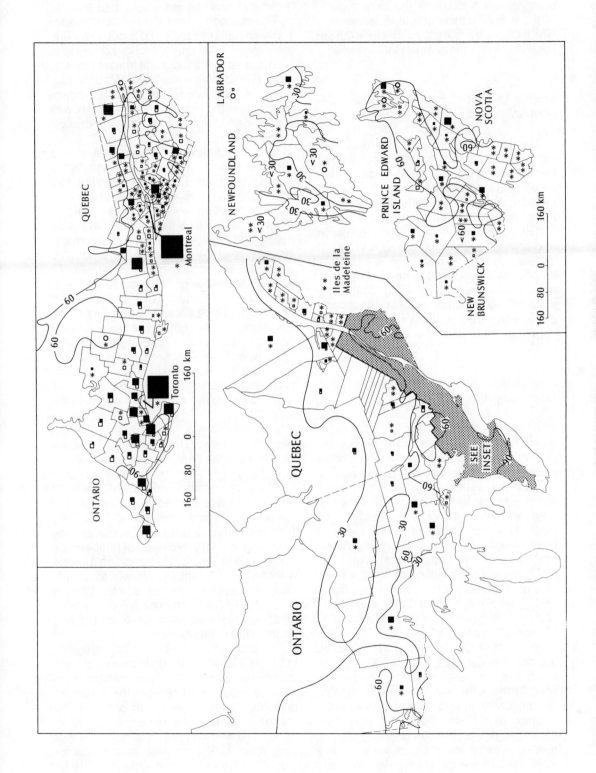

QUEBEC

ONTARIO

60

60

90

Toronto

Montreal

160 km

160   80   0

LABRADOR

NEWFOUNDLAND

30

30

<30

<30

<30

30

30

PRINCE EDWARD ISLAND

NOVA SCOTIA

60

60

<60

NEW BRUNSWICK

160 km

160   80   0

QUEBEC

ONTARIO

Iles de la Madeleine

SEE INSET

60

60

30

30

30

60

60

30

60

mine that will ultimately be used up. From the standpoint of agriculture, the climate is part of the overall agricultural land resource, of which other physical aspects include the soil, landforms, and the areal extent involved.

Spatial Analysis of Farmland, Urban Population, and Climatic Resources

The basic methodology involves comparing numbers of days from barley ripening to first fall freeze with farmland hectarage and urban population, using census divisions as the spatial unit. The spatial distribution of the urban population, the farmland hectarage, and the climatic resources, are examined and compared cartographically and by percentage classes. Comparison is thus being made between, on the one hand, climate and area, two aspects of the fixed location agricultural land resource complex, and, on the other hand, urban population, representing a conflicting high value site use.

Development of an Agroclimatic Resource Index (ACRI)

A further stage in the national scale methodology was to develop an index which permitted more specific quantitative land resource value comparisons than were possible with the number of days from barley ripening to first freeze, and which combined both the heat and moisture aspects of climate. Use was made of a map showing freeze-free season durations in days, and a map showing a climatic moisture index which ranged from values of 30 or 40 for a very dry climate to 80 or 90 for a very humid one. These maps have been published[29] and their derivation has been described by Sly and Coligado[30].

In much of Canada moisture limitations are not a major factor for agriculture. Too much moisture is about as frequent as too little. In the drier parts of the Prairie Provinces, however, and in some valleys in the interior of British Columbia, moisture shortages are a chronic condition, and the relative agroclimatic resource value is lower than the freeze-free season length would indicate. From examination of the moisture index map it appeared that the zone of

chronic moisture shortage would be that with a climatic moisture index less than 65.

The durations from the freeze-free season map were divided by 60 to obtain a value that ranged from one (60 d freeze-free season) in northern agricultural frontier areas to three (180 d freeze-free) near Windsor, Ontario. This value was then adjusted downward where the climatic moisture index was less than 65 to obtain ACRI, the agroclimatic resource index.

The adjustment factor was based on a linear regression analysis of average district wheat yield against the moisture index. In a case where, for example, the freeze-free season duration was 120 d (the middle range of freeze-free days), and it was estimated that lack of moisture would normally limit yields to 70 percent of the yields that would be obtained where the moisture index was 65, the value of ACRI would be 0.7 × (120/60) = 1.4. Where ACRI is as low as 1.0 in northern fringe areas it is because of the cold temperature, but where such a value occurs in a small part of southwestern Saskatchewan it is mainly due to the dry climate. Both sufficient heat and moisture are essential for agriculture; the ACRI index combines these two dynamic and critical climatic variables.

In coastal areas, where the freeze-free season may be very long but where there is a lack of summer heat, an alternative procedure for determining the temperature contribution was deemed necessary. For such areas ACRI was derived by dividing annual normals of degree days above $5.6°C$[31] by 833. Growing degree days, derived by summing the daily excesses of temperature above a specified base temperature, provide a measure of the heat available to a crop. Different bases are appropriate for different crops, but in North America $5.6°C$ has been most widely used as a base for growing degree day computations.

In summary, the agroclimatic resource index, ACRI, is based on freeze-free season duration except in coastal areas where growing degree days are used, and the values are adjusted downward in dry regions to reflect the lower value of the climatic resource for agriculture where moisture is lacking. It was developed at the request of the Science Council of Canada[32], and a map of ACRI for Canada has been published by the Science Council[33].

Results

In Figure 1 (pages 170-171), the spatial distributions of urban population, farmland hectarage, and the agroclimatic measure based on barley are visually represented. The concentration of divisions with the largest farmland hectarages, as indicated by large left-hand squares in Figure 1, is in the Prairie Provinces, where the number of days from barley ripening to first fall freeze is generally between 30 and 60. In the Great Lakes-St. Lawrence Lowlands (see upper inset, Figure 1), where this duration exceeds 60, farmland amounts are small but significant. In this region some divisions have larger percentages of Canada's farmland than of its urban population, but the striking feature is the number of highly urbanized divisions, as indicated by the large right-hand squares. Canada's most urbanized divisions – Ile de Montréal-Ile Jésus and Toronto, each with 13 percent of our urban population, and Greater Vancouver with 6 percent – all have particularly favourable agroclimates, with 80 d or more from barley ripening until first fall freeze.

A histogram (Figure 2) shows the percentage of Canada's farmland and urban population by 10 d climatic resource value intervals. The divergent nature of the two distributions is very noticeable. For instance, approximately three-quarters of the farmland occurs in the climatic range of 40 to 60 d, while approximately one-half of the urban population occurs within the climatic range of more than 80 d.

Clearly the poorer areas of Canada, from an agroclimatic standpoint, contain the

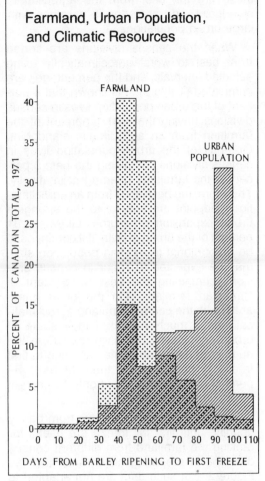

Farmland, Urban Population, and Climatic Resources

Figure 2

largest amounts of farmland. Those areas with more favourable climates encompass relatively small amounts of farmland, but

Table 2

CUMULATIVE PERCENTAGE DISTRIBUTIONS OF URBAN POPULATION AND FARMLAND FOR DECREASING CLIMATIC RESOURCE LEVELS

Climatic Resource Value at least:	100	95	82	80	70	65	60	52	50	48	45	35
Percent of Canada's Urban Population	4	31	43	50	62	65	74	77	81	81	87	98
Census Farmland	1	1.5	4	5	11	12	20	40	52	66	82	97

these are the best from the agroclimatic standpoint, and are often in divisions with large urban populations.

When the census divisions are sorted from best to worst agroclimatically using selected intervals, and the percentages are cumulated (Table 2), it is shown that 4 percent of the urban population lives in census divisions having the best 1 percent of the farmland from an agroclimatic standpoint. One-half of the urban population lives in census divisions containing the best 5 percent of the farmland, rated agroclimatically. Therefore, the best lands from an agroclimatic standpoint are subject to the strongest urban expansion pressures. Likewise, 81 percent of the urban growth, if it continues to be proportional to existing population size, may be expected to occur in census divisions containing the best 52 percent of Canadian farmland. On the other hand, areas with the poorest farmland agroclimatically have much less than their share of urban population. The divisions where the number of days from barley ripening to first freeze is normally less than 48 have 34 percent of the farmland but only 19 percent of the urban population.

Data on the number of days from barley ripening until first fall freeze are useful for ranking the farmland from different census divisions in preparing histograms such as Figure 2. But such data are not adapted to making comparisons of the magnitude of agroclimatic resource values at different locations. One can make such comparisons, however, using the agroclimatic resource index, ACRI. This index has the further advantage of including moisture as well as temperature. From preliminary study of field crop statistics and ACRI, it appears that agricultural productivity does tend to be in proportion to the ACRI.

The use of ACRI facilitates comparison of the agricultural land resource loss around urban centres with increases elsewhere. As an example, it might be thought that the gain (76 704 ha) from 1961 to 1971 in Alberta census division No. 13 north of Edmonton, was greater than the loss (41 079 ha) in the counties of Peel and York (includes Toronto) in Ontario. However, if the farmland change amounts are converted to the equivalent numbers of hectares of land where ACRI=3, they become $(2.5/3) \times 41\ 079 = 34\ 233$ ha

lost in York and Peel, and $(1.2/3) \times 76\ 704 = 30\ 682$ ha gained in Alberta Division 13. It is thus concluded that when climatic differences are taken into account, the real agricultural resource loss in York and Peel was greater than the gain in Alberta, Division 13.

By far the largest part of our farmland has ACRI less than 2.0 (Figure 3). The major part of the farmland in Saskatchewan and Alberta has ACRI less than 1.5, which means that it is likely to have less than half the productive potential of land in southwestern Ontario where ACRI is 3.0.

The agricultural land resource chart (Figure 3) shows the distributions in relation not only to ACRI but also to the CLI soil capability ratings. For common field crops, CLI Class 1 soil performs about twice as well as Class 4, and Classes 2 and 3 are intermediate in performance between 1 and 4[34]. Comparisons made among regions using Figure 3 may somewhat over-emphasize the climatic factor because much of the poorer class land in the Prairies has been given a low rating, partly because of climatic limitations in relation to the most favoured agroclimates of that region. Nevertheless this chart can help to illustrate some important aspects of the conflict between urban expansion and the agricultural land resource.

In the Prairie and Atlantic Provinces, where the conflict between urban and agricultural uses is the least serious, relatively large proportions of the farmland have Class 5 or 6 soils, which are not considered suitable for cultivated crops and in general are the poorest of the agricultural soils. In Ontario, where urban pressures are great, much of the farmland not only has some of Canada's best agroclimates (A and B in Figure 3), but also has Class 1, 2, or 3 soils.

Discussion

The census divisions containing Canada's three largest metropolitan areas — Montreal, Toronto, and Vancouver — all have high agroclimatic resource values and small farmland hectarages. Particularly well illustrated (Figure 1) is the fact that farmland areas with the highest agroclimatic resources are under strong urban pressures from already large populations within those census divisions. More than a dozen census

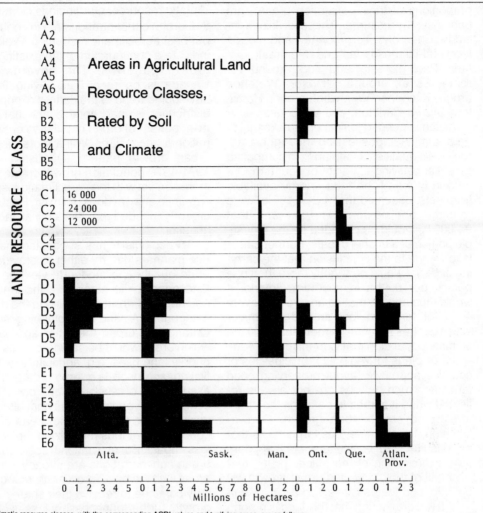

Figure 3

The agroclimatic resource classes, with the corresponding ACRI values and typifying areas, are as follows:

A (ACRI ⩾3) Essex and Kent Counties, Ont.
B (ACRI 2.5 to 2.9) SW. Ont., areas near Lake Ontario and Georgian Bay W. part of Montreal Plain.
C (ACRI 2.0 to 2.4) Lower Fraser Valley and Victoria area. Manitoba Plain N. of Winnipeg and Portage, Central Ont., Ottawa Valley, Manitoulin Is. Central St. Lawrence Lowlands of Que.
D (ACRI 1.5 to 1.9) S. interior valleys of B.C. and Vancouver Is. Parts of Upper Peace R. Valley, about half of the part of S. Alta. lying along the E. of an Edmonton-Calgary-Lethbridge line. Most of Sask. S. of Pr. Albert and E. of a North Battleford-Regina-Weyburn line. Most of Man. Plain and Interlake area. Thunder Bay, Algonquin Park and Clay Belt areas of Ont. Lac St. Jean and Riv. du Loup areas and area E. of Sherbrooke, Que., P.E.I. N.S. and S. part of N.B.
E (ACRI 1.0 to 1.4) N. interior valleys of B.C. Most of Peace and Athabasca River Valleys, parts of SE Alta. Most of Sask. N. of Pr. Albert and Sask. W. of a North Battleford-Regina-Weyburn line. Riding and Duck Mts. in Man. Abitibi and Laurentides area of Que. NW., NB. Most of Is. of Nfld.

The soil capability classes rate soils as follows:

1 No limitations for general field crops.
2 Moderate limitations that restrict range of crops or require moderate conservation practices.
3 Moderately severe limitations.
4 Severe limitations; suitable for only a few crops, or else the yield for a range of crops is low, or high risk of crop failure.
5 Perennial forage at best, but improvements feasible.
6 Perennial forage, improvements not feasible.

Data is not yet available for British Columbia, estimates for B.C. are 70 000 ha in Class 1, 398 000 ha in Class 2, 1×10^6 ha in Class 3, 2.1×10^6 ha in Class 4, 12×10^6 ha in Class 5, 11×10^6 ha in Class 6.

divisions, parts of census metropolitan areas in British Columbia, Ontario, and Quebec (the provinces of Canada having the largest proportions of their population urban), are under such pressures. These divisions include Chambly, Terrebonne, and Ile de Montréal-Ile Jésus around Montreal; Ontario, Peel, Toronto, and York around Toronto; Essex around Windsor; Waterloo around Kitchener; Wentworth around Hamilton; Ottawa-Carleton; and Vancouver.

Action is needed to limit urban expansion onto farmlands possessing the highest agroclimatic qualities. Hoffman[35], in computing potential carrying capacity of southern Ontario in terms of the food energy available from oats, assumed that 4.2×10^6 kJ were required per person per year and that 94 000 kJ would be provided by the digestible portion of a bushel of oats. Further calculation leads to the conclusion that the carrying capacity would be slightly more than one person per 0.4 ha in southern Ontario. It would thus appear that as an average every 40 ha of farmland lost around our major urban centres reduces by more than 100 the ultimate population that Canada could feed.

While plans for regional municipalities or counties appear to favour preserving good farmland, such policies are not always reflected when it comes to detailing the structure of projected urban growth. Ottawa-Carleton, for example, has large areas that are unsuitable for agriculture, but it appears from examination of the official plan[36] that over half of their projected urban growth will be onto land that is Class 1 to 3 for agriculture. It is worth noting that this land is among the best 8 percent of Canada's farmland from the agroclimatic standpoint. Gierman's study[37] in the Ottawa region confirmed that urban development from 1964 to 1973 took place primarily on Class 1 to 4 land.

Perhaps action at a provincial level, such as the establishment of the Agricultural Land Reserves in British Columbia[38], will be more effective in preserving farmland than will action at a subprovincial or regional municipality level, the approach taken elsewhere in Canada. In Ontario, recent official statements indicate provincial concern over agricultural land, if not necessarily strong action as in British Columbia[39]. Quebec, which next to British Columbia faces the most severe pressure on a limited agricultural land resource, is also considering legislation[40].

In this study we have quantified on a national scale the spatial association of urban population distribution and the agroclimatic resource. Our findings clearly show that urban concentrations are in conflict with our best agroclimatic resources. Projections indicate a continuation of the existing urban growth pattern[41]. The government of Canada is on record as saying that the developing trend of the pattern of population distribution is unacceptable, and that a more even pattern of urban population growth is a national necessity[42]. Will a more dispersed urban growth pattern suggested by the 1971-1976 population growth create less or more conflict with the agroclimatic resource? A definite answer does not exist. Research is urgently needed on this critical land use question.

An ecologist[43] has suggested that landuse planners can try either to provide moderate quality and moderate yield on all the landscape, or to categorize the landscape and apply different strategies to the different categories. A land use planning study in Ontario has urged the latter approach, by which areas would be categorized as having priority land uses, e.g. agriculture, with conflicting land uses, e.g. urban, denied[44]. The assumption of such an approach is that conflicts between urban and agricultural land use could be limited to the areas of urban concentration, thus minimizing the farmland loss impact. But at a national scale our major urban concentrations are already located in areas with the best agroclimatic resources. If they continue to receive their share of urban growth, further unnecessary losses of some of the best land for agriculture will be inevitable. If, instead, future urban development could be channelled onto land that is least suitable for farming, it would entail a minimal reduction of the land resource base for agriculture. In order for this desirable objective to occur, provincial, regional, and local jurisdictions must all accept and then implement land use plans which protect good agricultural land.

The use of land for transportation and other urban network facilities must also be taken into account. The construction of a large modern airport on land that is suitable for agriculture may use so much land that it reduces the number of people that Canada could ultimately support by many thousands. In Alberta, where population is relatively evenly dispersed over a large area and therefore requires extensive surface transportation networks, three times as much land is used for roads, highways, and railroads as has been used for urban development[45].

Specific research findings such as those of this study, together with findings relating to soil, socio-economic and other aspects of the problem[46], are but the first step in the development of effective planning for the protection and preservation of our limited agroclimatic resource for present and future generations. The Canadian Federation of Agriculture favours the preservation of agricultural land and has called for national land use planning[47], as has the Agricultural Institute of Canada[48]. A strategy to protect prime agricultural land should be developed as part of an overall national land use policy[49]. The first initial probings for such a policy are being made by the Interdepartmental Task Force on Land Use, and in the Lands Directorate of Environment Canada.

Such planning is particularly important in view of the expanding world population, and the accompanying increasing need for agricultural production, and also in view of the possibility that our climate may deteriorate[50]. If the climate cools, causing the agricultural frontier to contract, farmland under the greatest urban expansion pressure in the Great Lakes-St. Lawrence Lowlands and the Lower Fraser Valley will still be the best in Canada, agroclimatically. Consequently, this prime agricultural land resource would become of even greater national importance than at present.

REFERENCES

[1] McKeague, J.A., "Canada Land Inventory: How Much Land Do We Have?" *Agrologist*, 4 (4), 1975, pp. 10-12.

[2] MacEachern, G.A., "Biomass Energy in the Canadian Agricultural Economy", *Proceedings International Biomass Energy Conference* (Winnipeg: Biomass Energy Institute), 1973, pp. xv:1-11.

[3] Odum, E.P., "Ecosystem Theory in Relation to Man", in *Ecosystem Structure and Function*, J.A. Wiens, ed. (Corvallis, Oregon: Oregon State University Press), 1972, pp. 11-23.

[4] McKeague, J.A., pp. 10-12

[5] Statistics Canada, *1976 Census of Canada: Agriculture, Number and Area of Census Farms by Census Divisions*, Catalogue 96-857 SA-7 (Ottawa, 1977).

[6] Science Council of Canada, *Population, Food and Resources*, Report No. 25 (Ottawa: Supply and Services Canada, 1976).

[7] Williams, G.D.V., "Urban Expansion and the Canadian Agroclimatic Resource Problem", *Greenhouse-Garden-Grass*, 12 (Spring, 1973), pp. 15-26.

[8] Crerar, A.D., "The Loss of Farmland in the Growth of the Metropolitan Regions of Canada", in *Resources for Tomorrow*, Supplementary Volume (Ottawa: Queen's Printer, 1962), pp. 181-195.

[9] Russwurm, L.H., *Development of an Urban Corridor System Toronto to Stratford Area 1941-1966*, Research Paper No. 3, Regional Development Branch (Toronto: Queen's Printer, 1970).

[10] Howard, J.F., *The Impact of Urbanization on the Prime Agricultural Lands of Southern Ontario*, M.A. Thesis (Waterloo: Department of Geography, University of Waterloo, 1972); Krueger, R.R., "Changing Land-Use Patterns in the Niagara Fruit Belt", *Transactions of the Royal Canadian Institute*, Vol. 32, part 2, 1959, (Toronto), 140 pp.; Krueger, R.R., "Recent Land Use Changes in the Niagara Fruit Belt", in *Applied Geography and the Human Environment*, R.E. Preston, ed., Publication Series No. 2 (Waterloo: Department of Geography, University of Waterloo, 1973), pp. 164-182; and Lewis, W.S., *The Windsor-Quebec Axis*, Map and Text (Ottawa: Urban Affairs Canada and Environment Canada, 1974).

[11] Gertler, L.O. and Hind-Smith, J., "The Impact of Urban Growth on Agricultural Land: A Pilot Study", in *Resources for Tomorrow*, Supplementary Volume (Ottawa: Queen's Printer, 1962), pp. 155-179; Martin, L.R.G., *Land Use Dynamics on the Toronto Urban Fringe*, Map Folio No. 3 (Ottawa: Lands Directorate, Environment Canada, 1975); and Gierman, D.M., *Rural Land Use Changes in the Ottawa-Hull Urban Region*, Occasional Paper No. 9 (Ottawa: Lands Directorate, Environment Canada, 1976), 85 pp.

[12] Gierman, D.M., *Rural to Urban Land Conversion*, Occasional Paper No. 16 (Ottawa: Lands Directorate, Environment Canada, 1977).

[13] Lewis, W.S., *The Windsor-Quebec Axis;* Found, W.C., *The Disappearance of Ontario Farmland* (Toronto: Department of Geography, York University, 1976, mimeo), 10 pp.; and Yeates, M., *Main Street, Windsor to Quebec City* (Toronto: Macmillan, 1975).

[14] Guelph, University of, Centre for Resources Development, *Planning for Agriculture in Southern Ontario*, ARDA Report No. 7 (Toronto: Ontario Ministry of Agriculture and Food, 1972), 331 pp.; Russwurm, L.H., "The Countryside in Ontario: An Overall Policy and Planning Viewpoint", in *The Countryside in Ontario*, M.J. Troughton, J.G. Nelson and S. Brown, eds. (London: Department of Geography, University of Western Ontario, 1975), pp. 161-177; and Urban Development Institute, *Planning Goals: Food, Employment, Housing* (Toronto: Urban Development Institute, 1977).

[15] Dorling, M.J. and Barichello, R.R., "Trends in Rural and Urban Land Uses in Canada", *Canadian Journal of Agricultural Economics*. 1975 C.A.E.S. Workshop Proceedings on Agricultural Land Use in Canada (Ottawa: Canadian Agricultural Economics Society, 1975), pp. 33-63.

[16] Gray, E.C., "Direct Urban Land Needs in the Decades Ahead", *Notes on Agriculture*, 10, University of Guelph (April, 1974), pp. 20-21; and Beaubien, C and Tabacnik, R., *People and Agricultural Land*, Perceptions 4, Study on Population, Technology and Resources (Ottawa: Science Council of Canada, 1977), 137 pp.

[17] Gierman, D.M., *Rural Land Use Changes in the Ottawa-Hull Urban Region* and Guelph, University of, Centre for Resources Development, *Planning for Agriculture in Southern Ontario*.

[18] Canada, Department of the Environment, *The Canada Land Inventory Soil Capability Classification for Agriculture*, CLI Report No. 2 (Ottawa: Environment Canada, reprinted 1972), 16 pp.

[19] Shields, J.A. and Ferguson, W.S., "Land Resources, Production Possibilities, and Limitations for Crop Production in the Prairie Provinces", *Symposium on Oilseeds and Pulse Crops* (Calgary: Western Cooperative Fertilizers, Ltd., 1975), pp. 115-156.

[20] Statistics Canada, *1971 Census of Canada: Population, Urban and Rural Distributions*, Catalogue 92-709, Vol. 1, 1973, Part 1 (Bulletin, 1:1-9) (Ottawa).

[21] Bryant, C.R. *Farm-Generated Determinants of Land Use Changes in the Rural-Urban Fringe in Canada, 1961-1975* (Ottawa: Lands Directorate, Environment Canada, 1976), 172 pp.

[22] Russwurm, L.H., "Country Residential Development and the Regional City Form in Canada", *Ontario Geography*, 10, 1976, pp. 79-96.

[23] See, for example, Stone, L.O., *Urban Development in Canada* (Ottawa, Dominion Bureau of Statistics, 1967); Lithwick, N.H., *Urban Canada: Problems and Prospects*, (Ottawa: Central Mortgage and Housing Corporation, 1970); Bourne, L.S., MacKinnon, R.D., Siegel, J., Simmons, J.W., eds., *Urban Futures for Central Canada: Perspectives on Forecasting Urban Growth and Form* (Toronto: University of Toronto Press, 1974); and Simmons, J. and Simmons, R., *Urban Canada*, 2nd ed. (Toronto: Copp Clark, 1974).

[24] R.E. Preston and L.H. Russwurm, "The Developing Canadian Urban Pattern: An Analysis of Population Change 1971-1976", in L.H. Russwurm, R.E. Preston, and L.R.G. Martin, *Essays on Canadian Urban Process and Form*, Geography Publication Series No. 10, (Waterloo: Department of Geography, University of Waterloo, 1978).

[25] Statistics Canada, *1976 Census of Canada: Agriculture, Number and Area of Census Farms by Census Divisions*, and *1971 Census of Canada Advance Bulletin: Agriculture, Number and Area of Census Farms*, Catalogue 96-727 AA-10 (Ottawa, 1972).

[26] Dansereau, P., *Biogeography: An Ecological Perspective* (New York: Ronald Press, 1957).

[27] Baier, W., Davidson, H., Desjardins, R.L., Ouellet, C.E. and Williams, G.D.V., "Recent Biometeorological Applications to Crops", *International Journal of Biometeorology*, 20, 1976, pp. 108-127; and Williams, G.D.V., "Physical Frontiers of Crops: The Example for Growing Barley to Maturity in Canada", in *Frontier Settlement*, R.G. Ironside, V.B. Proudfoot, E.N. Shannon and C.J. Tracie, eds. (Edmonton: Department of Geography, University of Alberta, 1974).

[28] Williams, G.D.V., "Urban Expansion and the Canadian Agroclimatic Resource Problem".

[29] Agrometeorology Research and Service, *Agroclimatic Atlas of Canada* (Ottawa: Chemistry and Biology Research Institute, Research Branch, Agriculture Canada, 1977).

[30] Sly, W.K. and Coligado, M.C., *Agroclimatic Maps for Canada, Derived Data:Moisture and Critical Temperatures near Freezing* (Ottawa: Agrometeorology Research and Service, Chemistry and Biology Research Institute, Research Branch, Agriculture Canada, 1974), 31pp. plus maps showing part of western Canada.

[31] Chapman, L.J. and Brown, D.M., *The Climates of Canada for Agriculture*, Canada Land Inventory, ARDA Report No. 3 (Ottawa: Queen's Printer, 1966), 24 pp.

[32] Science Council of Canada, *Population, Food and Resources*.

[33] Geno, B.J. and Geno, L.M., *Food Production in the Canadian Environment*, Perceptions 3, Science Council of Canada (Ottawa: Supply and Services Canada, 1976), 71 pp.

[34] Hoffman, D.W., "Soil Capability Analysis and Land Resource Development in Canada", in *Canada's Natural Environment: Essays in Applied Geography*, G.R. McBoyle and E. Sommerville, eds. (Toronto: Methuen, 1976), pp. 140-167.

[35] Hoffman, D.W., *Crop Yields of Soil Capability Classes and Their Uses in Planning for Agriculture*, Ph.D. Thesis (Waterloo: School of Urban and Regional Planning, University of Waterloo, 1972).

[36] Ottawa-Carleton, Regional Municipality of, *Official Plan, Ottawa-Carleton Planning Area* (Ottawa: Planning Department, Regional Municipality of Ottawa-Carleton, 1974).

[37] Gierman, D.M., *Rural Land Use Changes in the Ottawa-Hull Urban Region*.

[38] Pearson, G.G., "Preservation of Agricultural Land: Rationale and Legislation – the B.C. Experience", *Canadian Journal of Agricultural Economics*. 1975 C.A.E.S. Workshop on Agricultural Land Use in Canada (Ottawa: Canadian Agricultural Economics Society, 1975), pp. 64-73; and Rawson, M., *Ill Fares the Land, Land-Use Management at the Urban/Rural/Resource Edges: The British Columbia Land Commission* (Ottawa: The Ministry of State for Urban Affairs, 1976), 45 pp.; and British Columbia Land Commission, *Annual Report* (Burnaby: British Columbia Land Commission, 1976).

[39] Ontario Ministry of Agriculture and Food, *A Strategy for Ontario Farmland* (Toronto: Ontario Ministry of Agriculture and Food, 1976), 16 pp.; and *Green Paper on Planning for Agriculture: Food Land Guidelines* (Toronto: Government of Ontario and Ontario Ministry of Agriculture and Food, 1977), 8 pp.

[40] *Financial Post,* Special Report on Quebec (June 11, 1977), p. 39.

[41] Ray, D. Michael *et al.,* (eds.) *Canadian Urban Trends,* Vol. 1 (Toronto: Copp Clark, 1976).

[42] Danson, B., Minister of State for Urban Affairs, *An Urban Strategy for Canada*, Address to the Conference Board in Canada, Winnipeg, April 2, 1975 (Ottawa: Ministry of State for Urban Affairs, 1975, mimeo), 16 pp.

[43] Odum, E.P., "Ecosystem Theory in Relation to Man".

[44] Ontario Ministry of Treasury, Economics and Intergovernmental Affairs and County of Huron, J.F. MacLaren Consultants, *Countryside Planning, a Methodology and Policies for Huron County and the Province of Ontario* (Toronto: Local Planning, Policy Branch, Ministry of Treasury, Economics and Intergovernmental Affairs, 1976), 232 pp.

[45] Alberta Land Use Forum, *Land Use Policy = Population, Growth*, Technical Report No. 8 (Edmonton: Alberta Land Use Forum, 1974), 51 pp.

[46] See, for example, Gertler, L.O. and Crowley, R.W., *Changing Canadian Cities: The Next Twenty-Five Years* (Toronto: McClelland and Stewart, 1977); Russwurm, L.H., *The Surroundings of our Cities* (Ottawa: Community Planning Press, 1977) and "The Urban Fringe as a Regional Environment", in L.H. Russwurm, R.E. Preston and L.R.G. Martin, *Essays on Canadian Urban Process and Form*, Publication Series No. 10 (Waterloo: Department of Geography, University of Waterloo, 1977); Troughton, M.J., *Landholding in a Rural-Urban Fringe Environment: The Case of London, Ontario*, Occasional Paper No. 11 (Ottawa: Lands Directorate, Environment Canada, 1976), 162 pp.; Martin, L.R.G., *A Comparative Urban Fringe Study Methodology,* Report (Ottawa: Lands Directorate, Environment Canada, 1975); Nowland, J.L. *The Agricultural Productivity of the Soils of the Atlantic Provinces,* Monograph No. 12 (Ottawa: Research Branch, Agriculture Canada, 1975), 19 pp.; and *The Agricultural Productivity of the Soils of Ontario and Quebec,* Monograph No. 13 (Ottawa: Research Branch, Agriculture Canada, 1975), 19 pp.; Beaubien, C. Rivers, P. and Lash, T. *Population, Technology and Land Use,* Sourcebook for a Working Party (Ottawa: Science Council of Canada, 1975); Québec, Université du, Institut National de la Recherche Scientifique, Centre de Recherches Urbaines et Régionales, *Region Sud: Agriculture* (Québec: Office de Planification et de Développement du Québec, 1973); *Land Seminar Proceedings* (Montreal: Canadian Council of Resource and Environment Ministers, 1973); Archer, P., *Urbanization on Agricultural Land: Trends and Implications for National Housing Policies,* Background paper (Ottawa: Central Mortgage and Housing Corporation, Policy Resource Group, 1976); Spurr, P., *Land and Urban Development* (Toronto: James Lorimer, 1976).

[47] Bursa, M., "The Politics of Lands Use", *Agrologist*, 4 (4), 1975, pp. 27-28.

[48] Agricultural Institute of Canada, "A Land Use Policy for Canada", *Agrologist,* 4 (4), 1975, pp. 22-25.

[49] Russwurm, L.H., "Land Policies Across Canada: Thoughts and Viewpoints", in *Battle for Land, Conference Report* (Ottawa: Community Planning Association of Canada, 1975), pp. 25-29.

[50] Williams, G.D.V., "An Assessment of the Impact of Some Hypothetical Climatic Changes on Cereal Production in Western Canada", in *World Food Supply in Changing Climate*, Proceedings Sterling Forest Conference, Dec. 2-5, 1974 (Sterling Forest, New York, 1975), pp. 88-102.

The Crisis of Agricultural Land in the Ontario Countryside

The purpose of this article is to examine trends and conditions in Ontario agricultural land use, in order to identify the nature of the changes taking place and to decide whether there is some kind of "crisis" which will require planning action.

The assumption is that by understanding the factors influencing farmland uses, it will be possible to define the nature of the planning issues and hence to outline future steps to improve efficiency in land-use patterns. As a final step, the article sketches one proposal for a planning solution, but a separate article would be needed to discuss some current thoughts in those directions.[1]

The agricultural land situation is very complex, with facets not generally familiar to those outside agriculture. Even in the field of agriculture, there is a lack of understanding of the relationships in the agricultural/rural systems among the socio-economic and biophysical elements. Most important, there are varying interpretations of the "facts" about farmland. It is hoped that this article provides a comprehensive, if brief, perspective.

The question of "crisis" is whether we are at a "turning point" from which events could become better or worse; in other words, we must analyze whether we are at a "moment of danger or suspense" with respect to agriculture in the Ontario countryside. The author agrees that there is a crisis, but feels that the real crisis has not been properly identified hitherto. Many of the existing arguments have missed the point, having been limited to symptoms rather than causes of social issues.

The unrecognized crisis is with the future costs and productivity of Ontario farming and our ability to compete against other regions, to attract new farmers and new capital, and to keep land resources in the industry. In the countryside certain conditions have been created during the past decade which, unless relieved, will inevitably weaken the future size and vitality of the Ontario agricultural economy, with cumulative social loss in a variety of forms. These consequences lie in the future, arising from causes which exist today, and can be averted by the right kind of planning action.

An immediate caveat is needed, however, to stress that changes in rural land use cannot be viewed simply as a case of economic choice where we examine individual reactions to individual opportunities for gain and loss. Agricultural land must be examined from the social point of view, and from the planning point of view, because the issues go beyond the dollars and cents of markets and the decisions by producers and investors relating to financial incentives and rewards.

With apologies to most readers, it is acknowledged that this article has put to one side a number of important rural matters of concern, including preservation of scenic beauty, protection of hazard lands, enhancement of wildlife, hazards of monoculture, and rural poverty. By and large, these are micro-level concerns which can be (more) readily treated in a macro-framework of agriculture and rural strategy.

The Role and Economic Importance of Ontario Agriculture

It is essential to consider agriculture in the national context because all Ontario farming is affected by Canada's net export position in international trade and by national policies.[2] Trade in and out of Ontario is an essential mechanism for greater efficiency in resource use, creating higher incomes for society. A loss of efficiency by Ontario farmers will

● STEPHEN RODD is in the School of Agricultural Economics and Extension Education, University of Guelph. The article is reprinted from *Plan Canada*, Vol. 16, No. 3/4, Sept./Dec., 1976, pp. 160-170, with permission.

reduce output here and raise output by farmers in other provinces or countries eager to expand their incomes.

Ontario has about 10 percent of the land in Canadian farms, accounts for 33 percent ($1.4 billion) of national gross sales, and receives about 25 percent of the national total net income from farming. It is relatively specialized in heat-related crops such as corn, soy beans, fruit, and vegetables, and in livestock. Favorable economic access to markets and services, combined with above average soil and climate, enabled production in 1971 of $314 of farm products per hectare of improved land, which was three times the national average.

Within the whole Ontario economy, agriculture is a major generator of personal incomes, both directly and indirectly in other industries, which can be estimated by input-output analysis from the recent provincial study of the 1965 interindustry relationships.[3]

(a) The income realized by farm operators in 1965 was $350 million.

(b) In eight major industries, it is estimated that their sales to farmers generated $260 million of "value added" in the form of wages, salaries, dividends, interest, rent, and undistributed producer earnings.

(c) In five major industries, the required inputs of raw materials had to come from Ontario farms; their 1965 value added $400 million.

(d) The preceding three items amounted to just over 5 percent of the total provincial value-added, but this kind of estimate should be raised by perhaps one-third or one-half by taking account of income created in other linked industries.

(e) If we consider the further income created by the subsequent circulation of the household income in items a to d, the total income related to agriculture might be estimated as at least 8 percent, and possibly as high as 15 percent, of the provincial economy. There were at least three dollars and possibly five dollars of personal income created in other industries for each dollar of net income generated within farming.

It is hard to say how much income "depends on" farming, because everything is interdependent. If farming suddenly ceased to exist in Ontario, the short-run impacts would be of the size estimated above.

Table 1

CHANGES IN FARMLAND AREA IN ONTARIO 1951 TO 1975

	Improved Land in Census Farms	Seven Grains, Hay and Improved Pasture
	percent per year	
1951—1966	-0.4	N.A.*
1966—1971	-2.0	-1.9
1971—1975	N.A.*	-0.4
	hectares	
1951—1966	- 279 023	N.A.*
1966—1971	-461 580	- 383 466
1971—1975	N.A.*	- 54 961

Note: The seven grains are winter wheat, oats, barley, mixed grain, grain corn, fodder corn, and soybeans. In 1975 these grains plus hay accounted for 3 001 050 ha out of the 3 127 815 ha of "principal field crops" sometimes mentioned in Ontario Ministry Agriculture and Food (OMAF) statements on land changes, excluding only spring wheat, dry peas, rye, buckwheat, dry white beans, and tobacco.

*N.A. Not available.

Source: Based on *Census of Agriculture* and *Agricultural Statistics for Ontario*, various issues.

Farmland Area: Variable Trends and Regional Patterns

Agriculture displays substantial changes from one period to another, and in its spatial dimension reveals very large differences in the kinds of adjustments taking place. Three main recent periods are distinguishable: (i) the 1950's and early 1960's, (ii) the late 1960's, and (iii) the early 1970's.

The first period reflects a long-run trend of gradual erosion in the remote and marginal areas, such as the Shield and the North, with output shifting to the areas with better markets, soils, and climates.[4] During the fifteen years, 1951 to 1966, the general pattern across the province was for most decrease in the land in farms to be in the unimproved areas (0.97×10^6 ha), either by being improved for crops or by being sold to nonfarm uses, with rather minor decreases in the amount of improved land in crops and improved pasture (0.28×10^6 ha).

Figure 1 delineates agricultural regions which are generally different in terms of quality of resources and historical adjustments to changing conditions. The Southwest and Central regions have been distinguished from each other because of their different rates of change in farming from 1966 to 1975 and because of the relationships associated with the Toronto-Centred Region. The map portrays with symbols the latest data available on the changes from 1966 to 1975 in farmland area (major grains, hay, and improved pasture) so as to indicate the individual counties in which major reductions or increases took place. These two time periods are combined on the map, despite their dissimilarity, so as not to focus unduly on only the earlier five years and to give a fuller perspective.

Figure 1

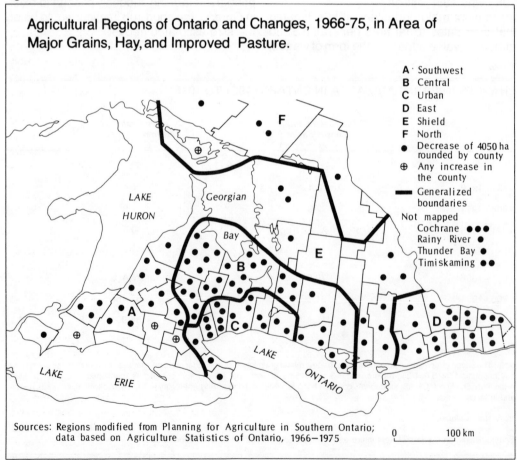

Agricultural Regions of Ontario and Changes, 1966-75, in Area of Major Grains, Hay, and Improved Pasture.

A · Southwest
B Central
C Urban
D East
E Shield
F North
● Decrease of 4050 ha rounded by county
⊕ Any increase in the county
▬ Generalized boundaries

Not mapped
Cochrane ●●●
Rainy River ●
Thunder Bay ●
Timiskaming ●●

Sources: Regions modified from Planning for Agriculture in Southern Ontario; data based on Agriculture Statistics of Ontario, 1966–1975

0   100 km

AREA (IN HECTARES) OF MAJOR FARM CROPS, PROVINCE OF ONTARIO,
1966 TO 1975

	1966	1971	1975
Grain Crops	1 696 319	1 843 662	1 907 550
Hay	1 384 453	1 096 339	1 093 500
Improved Pasture	1 188 956	946 261	830 250
Total	4 269 728	3 886 262	3 831 300

Note: The crops listed in this Table accounted for about 90 percent of total improved land in Census farms in 1966 and 1971, excluding only fallow, barnyards, lanes, idle farm fields, and selected crops equal to 2 to 3 percent of total improved land.

Source: *Agricultural Statistics for Ontario*, various issues.

Table 2

The second recent stage started around 1966 but was not documented until the 1971 Census results became available in 1973. A marked acceleration occurred in the loss of improved land in farms, with a fivefold jump in the rate of loss over the previous periods as shown in Table 1.[5] There was little recognition of the scope and regional distribution of the loss because the annual estimates by the Ministry of Agriculture and Food in the late 1960's completely missed the enormous decline in improved pasture and underestimated the decline in hayfields by about 243 000 ha and 60 750 ha respectively.

The continued cost-price squeeze of the later 1960's[6] induced farmers to shift land from pasture and hay to more intensive and valuable grain crops, or to sell out their entire farm for nonfarm residential and recreational uses (Table 3). The reductions were smallest in the extremely versatile soils and climate of the Southwestern Region (Figure 1). There were heavy losses of farmland throughout the entire area of the Central Region and the Urban Region (B and C in Figure 1), i.e., within a 130 to 160 km radius of Toronto, despite their superior soils and climate, good access to markets, complete agribusiness service infrastructure, and progressive farmers.

The cost-price squeeze ended abruptly in 1972, as the prices on world grain markets jumped upward when poor harvests in some nations were coupled with the sudden exhaustion of the reserve grain stocks accummulated during the 1960's. The 1973-1975 average farm prices in Ontario for principal field crops, including grain corn, and hay, were twice as high as the levels for 1971 and the previous ten years.[7]

Table 2 indicates that since 1971 the provincial grain area has risen, the drop in hay has been stopped, and the decline in improved pasture has slowed.

The dramatic but only partial cessation of the loss of farmland does not mark a return to the conditions before 1966. First, the causes of the world boom were clearly a fortunate coincidence (for grain and fodder growers) of temporary factors and the prices have started to decline in 1976. Second, many of the other causes of the loss of farmland still persist in Ontario and will be discussed in later sections.

It can be argued on the evidence of Table 3 that the rise in grain and hay prices did not have nearly as strong an effect as one would expect, especially when it was combined with the shift to low "farm" assessment of land for property tax if some crop such as hay

ANNUAL PERCENTAGE RATES OF CHANGE IN FARM AREA BY TYPE OF USE AND BY REGION, 1966-1971-1975.

Ontario Regions	Seven Grains		Hay		Improved Pasture		Total*	
	66-71	71-75	66-71	71-75	66-71	71-75	66-71	71-75
Southwest	3.0	1.0	-4.6	-0.7	-5.3	-5.7	-0.3	-0.5
Central	-0.4	1.4	-3.8	-0.3	-4.1	-2.4	-1.9	-0.2
Urban	0.9	1.3	-5.5	-1.4	-6.6	-3.2	-3.1	-0.5
East	0.4	-0.7	-4.8	0.9	-2.5	-1.7	-2.7	-0.4
Shield	-3.5	-0.8	-3.8	0.5	-4.0	-0.2	-3.8	—
North	-0.7	-1.6	-5.4	1.5	-3.4	-0.8	-4.1	0.3
Province	1.7	0.9	-4.6	-0.1	-4.5	-3.2	-1.9	-0.4

*Sum of seven grains, hay and pasture. See note to Table 1.

— less than 0.05

Source: Calculated from *Agricultural Statistics for Ontario,* various issues.

Table 3

or corn were on the land. Instead of accelerating upwards, grain crops, other than corn, continued to fall, and even the growth of corn area slowed from the 1966-1971 rate of 8.7 percent per year to only 3.4 percent per year from 1971 to 1975.[8]

In Table 3, we can see the different responses and patterns of adjustment in the different regions in the two recent periods. There is greater general stability in the total land in these crops in the 1970's (columns 7 and 8), especially in the Central and Urban regions in which there had been unusually large land losses, possibly related to the "urban field" of influence of the Toronto-Hamilton metropolitan areas, within a radius of almost 160 km.

In that area it seems certain that the losses were slowed when planning policies were changed to end the upheaval of the "4 ha minimum lots".[9] A notable figure in column 2 is the retarded rate of grain expansion in the Southwest Region (even if the recent increase in dry white beans were included the Southwest figures for columns 2 and 8 would be 1.4 and −0.2).

The last two columns of Table 3 reveal that the balance among all the major categories of farmland continued the decreases of the 1960's but much more slowly. Only in the North was there a net increase, due to a recent increase in hay. The net decrease in the Southwest was due to a large continued cut in pasture.

The prices of farm products move continuously in response to many market factors, and producers' short-run reactions are reflected in land-use changes together with the long-run trends in land allocation arising from technology, social trends, economic growth, and a host of other factors. Each region reacts differently according to its local stimuli, its resources, and its relative opportunities, shifting land among different farm enterprises or into other uses. However, the picture is not fully random, and generalized relationships can be protrayed in the next section.

Causes of Loss of Agricultural Land

Not all of the causes of loss of agricultural land emerge from outside agriculture. The causes of reduction of farmland are listed here, not in order of their importance or the size of their impact. Some of the major issues are discussed in detail later.

1. The most obvious cause is the enlargement of our cities in the form of subdivisions and suburbs. This explains a relatively small part of the total reduction in farmland, judging from Figure 1 where there is no correlation between the losses and the location of cities.

2. A fundamental and different factor is the growth of the total economy, with more production of all kinds, which must be serviced

by larger capacity in airports, highways, electrical systems, waste disposal sites, sand and gravel pits: all of these located in the countryside.

3. The rising income of the average Canadian means more spending on leisure and recreation, second homes, travel, and prestigious homes with more privacy.

4. In prolonged inflation, people protect the value of their savings by buying physical assets. Rural land is one of the favorite hedges against inflation.

5. A number of weaknesses in the quality of urban management have had serious side-effects for rural areas, by making it increasingly attractive to city workers to build their homes far out in the countryside. Inadequate renewal of the fabric of existing towns and cities has deteriorated the urban environment for many people. There have been delays, costs, and failures in planning city growth. The costs of land and servicing for new residences are excessive inside the existing built-up areas. The cost of local government in the older central areas is relatively high. All of these circumstances can be escaped simply by building a country home outside the urban jurisdiction.

6. All of agriculture has been subject to internal adjustments to changing market prices for inputs and outputs and to changes in technology, reducing the amount of land needed in farming. Marginal areas such as the Pre-Cambrian Shield become submarginal and fall out of production. Because of different rates of change in productivity relative to other parts of the world, in some lines of product Ontario farmers have lost sales to competitors in other regions.

7. Another group of factors involves reaction by agriculture to its operating environment. Uncertainty has pervaded many farming areas; farmers have become unsure about what the next five or ten years might hold for farming in their area. Uncertainty will always reduce investment, which reduces productivity, which reduces incomes, and makes farming less attractive.

8. Farming must react when forces outside agriculture push the price of land far above the value that can be justified by any kind of agricultural production feasible on that land. Thus, in many personal circumstances, a farmer will consider it unwise to continue farming and preferable to sell for some other use.

9. An increasing proportion of land is held by absentee landlords and by investors from foreign countries. These owners of rural land have quite different motivations and will tend not to continue farming to the same degree that the original Ontario farm operator would do.

10. The Agricultural Code of Practice of Ontario[10] has definitely had an effect in reducing the amount of land in agriculture, as more and more rural nonfarm residences infiltrate the countryside, conflicting with farming and, in some instances, making it impossible to expand and receive acceptable rates of return.

The reasons for the loss of farm land in southern Ontario vary from area to area. This is a very fundamental observation which must be recognized in attempting to make a diagnosis of the problems, the causes, and the solutions.

Productivity and Yields

It is generally believed that there has been sufficient rise in farming productivity to offset declining area. Although true, such a relationship cannot be extrapolated indefinitely into the future.

The index of physical production in Ontario[11] has been rising in three successive plateaus from 100 in 1961, to 128 in 1971, and 133 in 1975. However, this index of physical production includes livestock production as well as crops. It therefore includes not only the effects of greater crop yields per hectare of land but also greater livestock numbers, plus the greater efficiency, in kilograms, of meat (or milk or eggs) per kilogram of feed. In the issue of farmland, the rise of this combined index is partly an illusion because it introduces a nonland dimension.

The index of Ontario physical production contains two other illusions. As the land base has decreased, the proportion of the production from the warmer regions with good soils becomes more important. This shift away from hay and pasture further exaggerates the rise in productivity as the average mix shifts toward crops, even as the total area shrinks.

It is crucial to consider the "agronomic" productivity of the land being used by farmers, in the way they are actually operating,

185

by looking at the yields of crops, excluding livestock, by crop and by region, each separately. The highest increase in yield has been grain corn; its average yield rose by 39.5 percent between the 1956-60 period and the 1966-70 period. For hay, fodder corn, barley, and winter wheat the increases were 22 to 25 percent. For other major grains the gains were even slower. The achieved crop productivity has risen more slowly than the index of physical production.

The scientific and technical means exist to produce any given crop at much higher physical outputs per hectare than the actual average achieved by farmers (perhaps at higher costs). However, those higher potential levels could be achieved only under quite different management and socio-economic conditions than actually exist today. Our agriculture achieves what current producers, on average, can achieve with the physical and human resources available within the existing social and economic conditions and incentives.

Ownership and Tenure of Farm Land

A change in farmland ownership took place in the 1966-1971 period which was much more dramatic than the decrease in land used by Census farms, and it appears to have continued since 1971. Comparing the censuses of 1966 and 1971, there was an increase, by almost 24 percent, in the amount of land which was rented by farm operators, at the same time as the amount of land owned by the operators of farms fell by over 15 percent. Nonfarmers increased their holdings of the land in Census farms by 221 940 ha at the same time as farm operators decreased their land ownership by 2 976 455 ha within five years.

The rate of increase in rented farmland was fastest in the Urban Region and in the Southwest. Also, the relative concentration of rented area is highest in the Urban and Southwest regions, which is perhaps a paradox because these two regions have different kinds of real estate market pressure. No data exist to explain the concentration of rented land in the counties bordering Lake Erie, but one guess is that foreign buyers are drawn to the warmest areas.

There are many potential implications from the change in land ownership and tenure which need research to identify the effects on:

1. agricultural efficiency and costs of production;
2. fragmentation and size of land parcels;
3. levels of investment in land improvements, buildings, and equipment;
4. adoption of new technology and new enterprises;
5. long-run security of the individual operator;
6. elasticity of supply of farm products;
7. supply price of land;
8. commitment of land to farming;
9. management posture with respect to environmental and biological issues;
10. posture and political action with respect to changes in the social community and to local planning objectives.

Rental of land by farm operators can be very desirable but it requires suitable traditions, laws, leases, and owners to make rental an aid to efficiency and soil conservation. The circumstances in Ontario make rental a very mixed blessing, with some serious shortcomings.

Price of Farm Land

The most neglected and the most dangerous issue related to farmland is the question of the price of land. It has the potential to destroy farming in most of southern Ontario and has already created serious harm.

There has not been recognition of the threat posed by the price of land for several reasons. Data are extraordinarily lacking. There has been little recognition, measurement, analysis, or discussion of the fact that the rural land market is no longer dominated by the land demand of farmers and that land prices are determined by nonfarm forces. The study of land economics is a neglected area in both agricultural economics and economics, and there is a tendency to consider the land market as no different from, say, the soybean market. There is among both farmers and economists a basic ambivalence about whether a high price for land is good or bad.

There has been some grudging admission among economists that the general level of rural land prices in much of southern Ontario has gone well above the long-run equilibrium level, pushed by disequilibrating speculation. The eventual correction to a long-run equilibrium will be painful and destructive in agriculture.

It is not true in much of Ontario that the value of farmland is the present value of the future stream of earnings generated by farmers competing in the market place. It is not even true, in many areas, that the value is related to any stream of productive income. The values are very much influenced by (a) demands to use land for personal enjoyment and consumption and (b) demands which reflect pure speculation or a search for an inflation-proof savings account. The outward manifestation of much of this nonfarm demand is the proliferation of nonfarm rural houses, estates, hobby farms, and recreational properties. Less visible is the purchase of land by foreign and other absentee owners.

The sale of the single piece of land for, say, a new rural nonfarm house will immediately create a shadow of higher values on a surrounding area much larger than the fragment actually used for the house. The opportunity costs for existing farmers will rise, and the real cash costs will be higher for new or expanding farmers. Farmers will behave as if that real estate value is a true value and it will eventually become built-in to the cost structure of agriculture. The current farm operators will continue to farm while watching the real estate market rise, since they are protected to some extent from the effect of that high value by the current rules of the assessors.

The higher value of land will have a number of harmful effects, potentially avoidable, for agriculture. The high purchase price for land will inhibit the entry of new farmers. It will inhibit the purchase of land to enlarge current farms. It will accelerate the exit of farmers who are less efficient, or older, or less emotionally committed to farming. The high price and the resultant uncertainty will inhibit investment in modernization by farm operators, gradually eroding competitive efficiency and productivity over wide areas.

It appears impossible that agricultural product prices, productivity, and agricultural incomes could be raised so far as to enable farming to pay prices for land equal to those which nonfarm people are willing to pay in southern Ontario. Certainly that could not be achieved by raising Ontario farm product prices, because of the ever-ready competition from other regions which will come in to satisfy the Ontario demands for farm products.

As all of the preceding effects take place and farms begin gradually to sell out to other uses, we find an erosion of the market for agricultural services and an erosion of the infrastructure of services and institutions to supply the farming industry, both of these hindering the remaining farms.

Current policies in rural areas and in agriculture put pressure on farmers to sell and to quit farming. There is a self-fulfilling and self-destructive fallacy in current thinking. There is an opinion that government cannot restrict the market for farmland unless the farmer can "make money" from farming; the paradox is that the lack of a land policy reduces farm efficiency and incomes, making it even more attractive to "cash in the chips". Farmers sell out and the state of agriculture in an area becomes worse and worse, pushing the area down the spiral.

Another common fallacy regarding land price is that farmers must be free to sell to anyone in the market because the appreciation in land value provides the only source of retirement income for the farmer. We must distinguish between the sale by the retiring farmer to another (younger) farmer and the sale to a nonfarm purchaser. To secure a retirement income, it is not necessary to have the freedom to sell to a nonfarm purchaser; farmers have always been able to secure a retirement income by selling to other farmers, with capital value rising during their lifetime as a result of the upward trend in farming productivity. Only in the submarginal and most remote regions has the retrenchment of farming reduced that retirement nest egg in recent years, and farmers need social assistance such as the ARDA purchase program in order to escape.

Infiltration by Low-density Nonfarm Housing

The health of agriculture in large areas of southern Ontario is affected by the increasing impact from low-density rural nonfarm housing, both the so-called "estate" housing as well as scattered homes, whether used on a year-round basis or only seasonally.[12] One township 50 km from central Toronto recently had 320 identifiable properties which were in use for farming, but it had approximately 1400 nonfarm properties, almost all of which were low-density housing; these properties were all outside the built-up areas of hamlets and villages. This ratio of nonfarm properties to farm properties is rather extreme, but a recent study by the author has shown that the open countryside for 160 km from Toronto is dominated numerically by nonfarm properties.[13] The farming areas have been infiltrated with low-density nonfarm houses.

A particularly insidious and debilitating effect is the increased value of all land in the area. This is a planning problem which could be ameliorated by providing a clear spatial separation of the two markets.

Farm fields become smaller in size and irregular in shape as pieces are cut out by severances. It becomes slower and more dangerous to move farm machinery across or along roads from one part of the farm to another. Labor and machine costs are increased for all field operation.

The traditional farm-based community characteristics are diluted and eventually dislocated, with many farm families feeling alienated from their traditional social framework. The political scene changes as councils and committees become more influenced by the interests, needs, and opinions of the new urban-oriented inhabitants. Local tax rates rise as the new nonfarm residents demand more and better services of all kinds, paid for by tax revenue secured from the farm properties as well as from the new nonfarm houses. There are conflicts and complaints between the farms and new neighboring uses, particularly with respect to the question of smells and sounds of farm operations. It has been demonstrated many times that the new rural urban-oriented residents, before very many months, find that the idyllic rural surrounding has unanticipated aspects which they find unpleasant and which they will do their best to get changed, even if it means restricting the farmers' operations. There are some areas in which farm operators may not operate their machinery after 22:00 in the evening, regardless of whether it is seeding time or harvest time.

One of the greatest effects is the feeling of instability and uncertainty which begins to pervade the area formerly characterized by stability conducive to the operations of farming.

Recent studies have indicated that the rural nonfarm housing phenomenon extends as far as 145 to 160 km from Toronto, and somewhat smaller distances (80 km?) around each of the major cities of southern Ontario, such as Kitchener-Waterloo, London, or Windsor. Over very wide regions the loss of a small part of agricultural land is creating harmful effects which might, in the long run, endanger the viability of virtually all Ontario farming.

The amount of land used for houses by urban-oriented people is actually a very small percentage of the land in most cases, but the effects of those scattered houses are pervasive over very wide areas. Demands by urban-oriented people for low-density housing could be satisfied in small parts of the countryside in other ways which would reduce the impacts on farming. Ideally, we could satisfy both the production and the consumption demands to use the countryside, with little or no reduction in production potential.

Policy Issues: Implications For Planning

The causes of the loss of farmland reveal both the existence of a problem and the nature of the solution. Major stress has been placed on scattered rural nonfarm housing, escalated land prices, and changing land ownership because their complex impacts have not yet been fully felt, and the impacts will be cumulative and virtually irreversible. The nature of the factors and their consequences indicates that the solution requires planning action, and dictates its form because there is no foreseeable end to the processes at work, and the harmful effects are not transitory or incidental.

Planning is a multi-faceted process of guidance with a three-fold set of functions: (a) to give us analytical predictions or projections of the future possibilities which might be realistically expected, (b) to provide information for all of the participants in the situation so that all can agree on facts and act with common information, and (c) to provide some regulations and controls to limit behavior or actions of individuals. All three are clearly needed for rural areas.

The situation with rural land requires planning because it involves avoidable inefficiencies and harmful effects which are not only unacceptable but preventable. The sacrifices and costs incurred in meeting our needs for housing, recreation, privacy, and physical goods can be reduced by planning. If we provide guidance to those making decisions on the countryside, whether they are individual households, farms, municipalities, or ministries, we can create a larger total set of all kinds of satisfactions than if those individual decisions were unguided. The increased satisfactions include both private and public benefits, monetary as well as nonmonetary.

What is "crisis"? Its Greek root "krino" means "to decide". Thus we must conclude that we do have a crisis in the countryside because we are in a position to make decisions about the future of Ontario farming and about rural housing, scenic amenities, wildlife, environmental stability, and all of the conservation issues of mankind. More than that, our decisions will affect whether we get a larger or smaller total of benefits from all the resources and efforts we will use.

My belief is that we have a crisis, but it is not a black-and-white issue. It is based on assessing a variety of considerations, which may be assessed differently by others. Moreover, there are different crises in different regions of Ontario.

Conclusion

The nature of the problem dictates the form of the solution. There must be a strategy for development of the countryside, at a macro-scale, which provides the priorities and policies needed before detailed work is done on the official plan, zoning by-laws, and secondary plans. The time horizon must be very long-term, preferably forty years. The political framework in the process must be appropriate, with a partnership between the province and the county/region on a joint and more equal basis, balancing both regional and provincial interests together with the local interests. (The *BNA Act* limits the national role in planning.) The strategy must be based upon the characteristics of the resources and the resource-based activity systems. Above all, the strategy must set a macro-level priority for each area, deliberately making a selection from the alternative systems to determine the dominant viewpoint or perspective for the area's planning decisions.

In a nutshell, that is the only way to ensure a bright future for farming in southern Ontario as an important and efficient economic activity, providing food for everyone and attractive incomes to the industry. This approach is also the only way to provide an opportunity to satisfy the other diverse and legitimate demands for land in the rural areas.

The means exist to accomplish these two objectives, but they will require strong action.[14]

A macro-strategy is needed to underpin the justification and mechanisms at the micro-level for the imperative spatial separation between farming and rural nonfarm housing. In an area where agriculture is to have a long-term future, farms and scattered houses are incompatible and the latter must be concentrated in large blocks of low quality farmland where there are suitable building and community conditions.

Basis for an Ontario Agricultural Strategy

The following objectives are proposed at this time for purposes of discussion as a basis for Ontario's agricultural strategy:

Item 1: Identify areas and regions where farming has demonstrated an ability to provide acceptable levels of living to those engaged in it on a long-run basis, or where there is the potential to do so if provided with reasonable access to efficient management, labor, capital, and supporting services, both public and private.

Item 2: Identify those areas where agriculture will be accorded by governments a dominant priority in the planning system over the competing systems of recreation, urbanization, and forestry. In those areas encourage and improve the availability to agriculture of the land resources which will be demanded, in the long run, by the industry.

Item 3: Initiate planning policy which will make land available to the agricultural industry at costs which are more related to the values of the products, to the returns to farmers of average efficiency, and to the returns needed to attract efficient labor, management, and capital.

Item 4: Provide an operating environment and community in which farming will be able to adjust as needed in response to future technologies and economic opportunities without externally imposed costs or restraints on farm operations.

It is very important to recognize that in our rural areas we have some resources which are of purely local or regional significance, while we have other resources which are of provincial and national significance. We must recognize different degrees of significance of the resources, and we must reflect the level of significance in the kind of strategy we accord the different countryside regions of Ontario.

NOTES AND REFERENCES

[1] Rodd, R.S. and W. van Vuuren, "A New Methodology for Countryside Planning", *Canadian Journal of Agricultural Economics*, 1975 Workshop Proceedings, pp. 109-140.
See also James F. MacLaren Ltd., *Countryside Planning: A Pilot Study of Huron County.* Ontario, Ministry of Housing (Toronto: The Queen's Printer, 1976).

[2] Gray, E.C., *A Preliminary Paper on Canadian Agricultural Land Use Policy*, Reference Paper No. 3, Food Prices Review Board, Ottawa, February 1976.

[3] Ontario, Department of Treasury, Economics and Intergovernmental Affairs, *Ontario Economic Review*, (Toronto: The Queen's Printer, 1972-1974 issues).

[4] Centre for Resources Development, *Planning for Agriculture in Southern Ontario*, University of Guelph, A.R.D.A. (Toronto: The Queen's Printer, December 1972).

[5] Rodd, R. S. "A Remarkable Change in the Rural Land Market", *Notes on Agriculture,* Guelph, Vol. 10, no. 2, April 1974, p. 21.

[6] During the ten years before 1971 only in three years did the index of prices received by farmers in Ontario for agricultural products rise faster than the index of prices they had to pay for farm inputs. This prolonged and cumulative economic pressure on farmers' returns from their labor, capital, and land predictably led to a flight of these assets in many areas, despite improvements in productivity. Ontario, Ministry of Agriculture and Food, *Agricultural Statistics for Ontario 1974* (Toronto: The Queen's Printer) pp. 8-9.

[7] Ontario, Ministry of Agriculture and Food, *Agricultural Statistics for Ontario 1974*, pp. 17, 22, 24, and *Monthly Crop and Livestock Report* (Toronto: The Queen's Printer, January 1976).

[8] Based on *Census of Agriculture and Agricultural Statistics for Ontario*, various issues.

[9] In a counter-productive attempt to reduce rural ribbon development during the 1960's, it was provided in the Ontario Planning Act that new parcels of land being created in rural areas must be over 4 ha in size unless they had been approved under subdivision control exercised by a planning body or by consent by a Committee of Adjustment.

[10] To reduce the potential for pollution, the Code of Practice provides guidelines for livestock producers and local governments concerning the location and expansion of livestock buildings and the methods of handling manure and dead animals. A 1976 edition has been issued by the Ontario Ministries of Agriculture and Food, Environment, and Housing to replace the 1970 and 1973 Codes.

[11] Statistics Canada, *Index of Farm Production*, Cat. No. 21-203 (Annual), (Ottawa: Information Canada, 1974).

[12] Although some localities have reduced severances, no real solution has been set in place and much of the past harm cannot be undone.

[13] Rodd, R.S., "The Use and Abuse of Rural Land", *Urban Forum/Colloque Urbain*, Canadian Council on Urban and Regional Research, Fall 1976, it summarizes and gives conclusions from a University of Guelph report forthcoming Fall 1976.

[14] For models of the needed approach, see the following reports: *Northumberland Area Development Strategy*, Northumberland Task Force, Ontario, Department of Treasury, Economics and Intergovernmental Affairs, (Toronto: The Queen's Printer, May 1976); *Countryside Planning: A Pilot Study of Huron County*, James F. MacLaren Ltd., for Ontario, Ministry of Housing (Toronto: The Queen's Printer, 1976), and R. S. Rodd and W. van Vuuren, *op. cit.*

Preservation of Agricultural Land: Rationale and Legislation – The British Columbia Experience

G. G. PEARSON

Shortly after World War II, concerns began to be expressed about the future of farmland lying in the path of predicted urban development in the Lower Fraser Valley. A number of steps were taken at the municipal and regional levels to cushion the impact of urban expansion on agricultural land. However, by the early 1970's it had become evident that "many local jurisdictions have not been able to withstand pressure to change zoning and it is at this point that all known land preservation schemes had failed".[1]

On December 21, 1972, the Provincial Government implemented a farmland freeze under the Environment and Land Use Act as an interim measure. On April 18, 1973, the Land Commission Act became law. The Act provided for the zoning of an Agricultural Land Reserve (ALR) in each of the twenty-eight regional districts in the province. Upon establishment of each reserve, the land freeze would be lifted. All of the ALR's have now be designated.

Provisions of the Land Commission Act

The Land Commission Act ranks second only to the Environment and Land Use Act in terms of authority. The Act has four major objectives:
1. to preserve agricultural land for farm use;
2. to preserve greenbelt lands in and around urban areas;
3. to preserve certain land bank lands having desirable qualities for urban or industrial development;
4. to preserve parkland for recreational use.

Only in the preservation of agricultural land does the Land Commission have zoning powers, and in no instance does the Act provide for expropriation.

The Act provides for the designation of Agricultural Land Reserves after a process involving public hearings held in various regional districts, a review by the Land Commission, and finally after consideration, an ultimate approval by the Cabinet. Only after designation are the uses of agricultural land limited to bona fide farming and certain other uses compatible with the preservation of land for farm use.

Regulations have been adopted under the Act outlining procedures and criteria for excluding land from an Agricultural Land Reserve – appeals, exemptions, inclusions, subdivision, and compatible uses.

Agricultural Land Reserves have been established on the basis of Canada Land Inventory Classifications of agricultural capability based on soil quality and climatic conditions. Agricultural Land Reserves include CLI Classifications 1 to 4 and contiguous or adjacent 5's. Parcels of less than 0.8 ha are excluded. Land designated in these categories comprises approximately 4.5×10^6 ha or 4.9 percent of the provincial land area.

The Land Commission is a statutory body established by the Legislation and functions independently of the Department of Agriculture. The Commission consists of a chairperson and four members. The Commission has statutory funding of $25 million to purchase land freely offered to it to assist in fulfilling it's mandate.

Rationale for Agricultural Land Preservation in British Columbia

The main reasons for preserving agricultural land are:
1. The supply of prime agricultural land in British Columbia is extremely limited.

● G. G. PEARSON was the Director of Policy Development and Planning, British Columbia Department of Agriculture, when this article was prepared. He is now a land resource specialist with the Department of Agriculture, Province of Saskatchewan, Regina. The article is reprinted from the *Proceedings*, Canadian Agricultural Economics Workshop, Banff, Alberta, 1975, pp. 64-73, with permission.

2. The loss of prime agricultural land prior to the land freeze had reached a rate of 4000 ha per year.

3. Increasing uncertainty as to the future reliability of external food supplies necessitates "keeping the options open" for British Columbia farmland.

4. The world demand for food continues to grow at a steady pace reflecting increasing population and income.

5. Maintenance of food production capability in British Columbia helps to maintain the province's bargaining position for food supplies.

6. Agriculture is an integral part of the British Columbia economic system. Farming has an employment-creating and income-generating effect up and down the food chain equivalent to three or four times the farm value of agricultural production.

7. A minimum level of farm production is necessary to sustain a viable infrastructure and farm product processing sector.

In addition, preservation of farmland has the beneficial effects of:

1. preserving the integrity of public investments in land improvement plans such as irrigation and flood control;

2. minimizing inconvenience for farmers associated with urban sprawl;

3. encouraging in-filling and development of land within urban areas;

4. improving the efficiency of supplying municipal services in urban areas;

5. facilitating comprehensive and integrated land use planning.

British Columbia Agricultural Policy

The Land Commission Act as of itself does not constitute a farm policy, but does provide mechanisms for ensuring the integrity of the farmland base.

Three other legislative measures were introduced within months of the Land Commission Act. These legislative measures are designed to facilitate development of a viable agricultural industry on the land so preserved. The additional legislative measures include the Farm Income Assurance Act, providing for the establishment of government- and producer-financed income assurance plans linked to basic costs of production; the Agricultural Credit Act, providing for

loan guarantees and interest reimbursement; and the Farm Products Industry Improvement Act, providing for loans, loan guarantees, and direct government participation in agricultural enterprises which process agricultural products and supply inputs to farming.

Taken together, these legislative measures constitute important cornerstones of a comprehensive agricultural policy aimed at the longer term development of British Columbia agriculture.

Some Economic Issues

Farm Land Prices

Economic theory predicts that the price of land reflects the economic rent that can be generated from the use of the land in various ways. Barlow[2] outlines a generalized profile of land use related to land values. Generally, lower land values are associated with use of land for agriculture, and higher values are associated with land capable of residential and industrial use.

Where land is capable of several alternative uses, its value might be expected to lie on Barlow's continuum somewhere between agricultural and urban use.

If, through zoning, the potential use of land is restricted to agriculture, then one would expect the value of that land to fall or at least stabilize in relation to the economic rent that can be generated from agricultural use. This, in fact, was one of the major concerns of agricultural land owners at the time of the land freeze.

While limited data exist, there are indications that farm land values in ALR's have not decreased. Because of uncertainty associated with the land freeze, the rate of increase in land values appears to have slowed in 1972 and 1973 and then increased in 1974.

While it is still too early to predict the ultimate effect of zoning on agricultural land values, a continued increase may be explained by renewed confidence in the farming community, a greater ability of farmers to pay for land associated with measures to stabilize farm incomes and improved access to farm credit, the nation wide upward trend in agricultural commodity prices, and continued spill-over of urban

192

demand for land, particularly for smaller parcels existing prior to the land freeze.

Urban Land Prices

Economic theory predicts that elimination of the supply of agricultural land for urban development would put greater pressure on nonagricultural land, thereby increasing the price of such land. In fact, the price of nonagricultural land for urban development has increased substantially in British Columbia in the past few years. However, it is a trend which is not out of line with a trend experienced around urban centres in eastern Canada, where no such limitations on the use of agricultural land presently exist.

In a recent analysis[3] of the factors affecting house prices in the Greater Vancouver area, it was concluded that demand and not supply factors has been driving up market prices for housing. While noting that the national trend for house prices was similar to the Vancouver region, the study concluded that house prices were increasing due to the decreasing availability of rental stock forcing people into the purchase market, greater availability of mortgage money, increased buyer confidence in the general condition of the economy and the housing market, with rapid increases in housing prices themselves increasing the attractiveness of the investment. The effect of the land freeze on nonagricultural land prices was marginal, limited mainly to stimulating a psychological fear that there was a limited supply of developable land.

In fact, recent studies[4] show that, on the basis of past population trends, there is enough developable land in and around the lower mainland for sixty years without utilizing any farm land.

In-filling of vacant land within Vancouver alone could supply sufficient housing for eighteen years.

Compensation

At the time of the land freeze, many proposals were put forward to compensate land owners for the loss of value anticipated by the establishment of Agricultural Land Reserves. In fact, there is no basis in law for compensating individuals for perceived losses due to zoning. To recognize the principal of compensation in zoning matters would create an impossible financial burden for tax payers.

In any event, the fears that land values would decline as a result of agricultural zoning have not materialized.

Private Versus Public Decisions

In 1973, in considering Senate Bill S-268 entitled "Land Use Policy and Planning Assistance Act of 1973", the United States Congress is stated to have found: "the increased size, scale, and impact of private actions have created a situation in which land use management decisions of wide public concern are often being made on the basis of expediency, tradition, short term economic consideration, and other factors which too frequently are unrelated or contradictory to sound environmental, economic, and social land use considerations".[5]

Bosselman and Callies, in a report prepared for the United States Federal Council on Environmental Quality, attribute one reason for the trend towards shared decision-making to the fact that "the entire pattern of land development has been controlled by thousands of individual local governments, each seeking to maximize its tax base and minimize its social problems, and caring less what happens to all the others".[6]

Economists have long supported the concept of resource and output pricing on the basis of freely operating demand and supply factors in a market economy. This concept finds its place in land economics in the principle of "highest and best use". Barlow states that "land resources are at their highest and best use when they are used in such a manner as to provide an optimum return to their operators or to society ... Land is ordinarily considered at its highest and best use when it is used for that purpose or that combination of purposes for which it has the highest comparative advantage or the least comparative disadvantage relative to other uses."[7]

This concept is argued from the point of view of efficiency in resource allocation; however, when decisions are left totally in the hands of private land owners, the universal trend is for agricultural land to be gradually converted to urban uses. Where there is an abundant supply of agricultural land, such private decision-making can be tolerated. However, where the supply of agricultural land is extremely limited, then such decisions have important implications for other members of that society. And increasingly, it becomes necessary for other citizens, via their governments, to guide and share in land use decisions.

The subject of how agricultural land will be used in the future is one in which economists will be increasingly required to wrestle with the reconciliation of micro-economic theory and welfare economics. In the future it is likely to become mandatory for economists to pay attention to social benefits and costs associated with private decision-making on land use matters.

The Future[8]

The Mandate of the British Columbia Land Commission* is both constructive and extensive. The Commission has been described as the coordinator of activity at the boundary where town and country meet. Now that the major thrust of the Commission in setting up ALR's has been completed, much of its work will deal with those land use conflicts that arise where growing communities infringe upon neighbouring farmland.

The encouragement of family farming is one of the main objectives of the Land Commission Act. In the administration of ALR's, the Commission will give increasing attention to the impact of regulations on the operations of family farms. The Commission expects to assist farmers to adjust property lines, and in some cases, assemble land in order to achieve more viable production units.

The general policy regarding land owned by the Commission will be to place property in the hands of farm families and new farmers who do not possess the capital resources required to purchase a farm. A leasing policy is being developed and will be administered in conjunction with the Department of Agriculture. In this respect, the Commission will be concerned with new dimensions in farm land tenure.

Part of the task of the Commission in preserving British Columbia agricultural land will necessitate continued cooperation with towns and cities in planning for urban and industrial development on nonagricultural land. This will involve increasing attention to the objectives of the Land Commission Act pertaining to green belt land bank and parklands.

The Agricultural Land Reserve may be viewed in the long haul as a fail-safe device to conserve land for food production. However, much of the land in the ALR's is suitable for integrated use without compromising the land's food production capability. The identification of such uses will be an important aspect of the work to be done in the future in cooperation with local governments and provincial government departments.

*Ed. Note: See also Mary Rawson, *Ill Fares the Land* (Ottawa: Ministry of State for Urban Affairs, 1976), for a brief over-view of the Commission, problems, and prospects.

REFERENCES

[1] British Columbia Department of Agriculture, Background paper. "Bill 42: The Land Commission Act", 1973.

[2] Barlow, R., *Land Resource Economics* (Englewood Cliffs, N.J.: Prentice-Hall, Inc., 1965), p. 14.

[3] Bysse, J. L., "Why Did Prices of Homes Go Up So Much?", *Real Estate Trends in Metropolitan Vancouver, 1973-1974* (Vancouver: Real Estate Board of Greater Vancouver, 1972).

[4] Pearson, N., "The Fraser Valley", *Land Uses in the Fraser Valley — Whose Concern?* (Vancouver, U.B.C. Centre for Continuing Education; 1972); and Baxter, D., *British Columbia Land Commission Act — A Review* (Ottawa: Canadian Council on Urban and Regional Research, 1974).

[5] B.C. Land Commission, *First Annual Report, Vancouver, 1973.*

[6] *Ibid.*

[7] Barlow, *Op. Cit.*

[8] The B.C. Land Commission, *Keeping the Options Open*, Vancouver, 1975.

RESOURCES CANADA: SOME ISSUES, DEVELOPMENTS, AND QUESTIONS

Introduction

The Canadian natural resource base is rich and diverse and its exploitation has been and continues to be of major significance to Canadian economic growth and development. In this section a broad range of resource developments, problems, and questions are explored.

In the introductory paper, John Chapman discusses major developments that have occurred in the forestry, water, and mineral (including energy) sectors of the economy during the period 1970-75. Perhaps the overriding conclusion to emerge from this review is the recognition that the resource base *is* limited and that a very different management strategy has been evolving in terms of public policy and resources development in Canada.

Since 1975, or thereabouts, there has been a marked decline in the outcry against multinational corporation ownership or control of Canadian resources and manufacturing activities. No doubt one of the factors to dull the opposition to foreign ownership has been the recent downswing in the national economy accompanied by unemployment rates exceeding or approaching those in the Great Depression. A job is a job whether corporate ownership resides in the United States, England, Japan, or Canada. In his paper, Charles Barrett discusses recent trends in direct foreign investment by foreign investors. While still substantial, this is decreasing, and Canadian investment abroad is increasing. Some of the possible implications of a continuing decrease in direct foreign investment are suggested.

The Great Lakes basin, particularly the lower lakes area, is under enormous population pressure from the United States and Canada, and growth projections indicate that this pressure, accompanied by increased industrial growth, is likely to continue. For decades both the United States and Canada have been grappling with the question of how best to manage the lakes. In his paper, George Francis tackles some of the major questions regarding the future management options for the Great Lakes. He concludes that the key to the future is "constraint planning – the recognition for development planning and urban design of physical, biological, and cultural landscape features which need to be integrated into design or accepted as constraints on development".

One of the major water development schemes that excited the minds of Canadians, particularly those living in the drought-prone portions of Saskatchewan, was the completion of the South Saskatchewan River Project in 1967, and the creation of the 225 km long reservoir, Lake Diefenbaker. The project was promoted as a major impetus to the development of irrigation agriculture, hydroelectric power for urban and industrial use, and, latterly, recreational potential. In his paper, Howard Richards examines the background leading to the development of the project and its goals, and analyzes its performance against objectives. He concludes that, to this point in time, the project can only be described as a modest success.

Recent economic development in northern Canada has been spear-headed by the gas, oil, and mineral industries. These developments, real or proposed, have not been implemented without relatively widespread vocal concern. Among the questions raised are native land rights and the economic, social, and environmental impact of proposed transportation and pipeline corridors on native and white life styles. The Berger Commission report, *Northern Frontier, Northern Homeland (1977),* dealt with many of these social and cultural issues, in addition to recommending a ten year moratorium on pipeline construction in the Mackenzie Valley. Shortly after the release of the Report, the federal government approved the proposal of Foothills Pipe Lines (Yukon) Ltd. to construct a pipeline along the Alaska Highway route; however, enabling legislation has not yet passed through the House of Commons. Doubtless other pipeline proposals will be made and other Berger-type commissions will be struck before the complex issues of northern economic development, land rights, life styles, and environmental issues are resolved – if indeed that is a possibility.

Three selections are included here to illustrate some of the problems associated with northern development. One of the major issues underlying all development in the North is the question of: "Development for whom?". In the first paper, Keith and Fischer identify five major areas of concern in northern development: technological, environmental, economic, social, and political. They provide a tidy frame of reference around which discussion can take place and policy issues can be identified.

In the second paper, Hugh Brody focuses on the past, present, and potential social-economic impacts of northern development on native peoples. His paper is based upon a presentation commissioned by the Berger Commission.

Native land claims have emerged as one, if not the most, significant issue in northern development and elsewhere in Canada. It comes as a shock to most Canadians to appreciate the full impact of the fact that the white man did not conquer what we know as Canada and that others had prior claim to the land. In his paper, Harold Cardinal explores the issues of land claims from the Indian point of view in Alberta and suggests how the white man will have to adjust to these claims.

Many of the papers in this section and in the others that precede it have dealt with environmental concerns in one form or another — resource development, water, urban growth, loss of agricultural land, etc. In the concluding paper in this section, Pierre Dansereau outlines briefly the major environmental problems and crises in Canada. The range of issues is broad and he concludes that Canada, while still relatively healthy compared with many countries and regions of the world, is in need of a comprehensive environmental management program based on austerity, scientific research, planning, information, and consultation.

Natural Resource Developments in Canada, 1970-75

J. D. CHAPMAN

Resource development in Canada frequently serves as a focal point for many of the larger concerns of Canadian society. For example, current resource trends involve such long-standing issues as foreign versus domestic control, "continentalism" versus "Canada first", and federal versus provincial conflicts. Issues, such as the concern for environmental quality, native rights, multipurpose management as a policy objective, and the inclusion of the public in decision-making, have sharpened around recent resource exploitation proposals.

A review of the current dynamics of the Canadian resource scene is aided by first identifying the functional, spatial, and institutional dimensions which are common to all resource sectors. The functional dimension of each resource system includes the facts of production, processing, and consumption, which in turn relate to such elements as productive capacity, efficiencies of conversion, price fluctuations, market demands, and social perceptions and preferences. Secondly, the geographically familiar elements of spatial distribution, association, and interaction, taken together, constitute the spatial dimension of resource systems. Thirdly, there is the institutional and organizational dimension which includes government policy and administrative agencies, company objectives and strategies, and most recently, the aims and aspirations of citizen groups.

Three resource systems – forest, water, and mineral – have been selected to illustrate some of the trends in Canadian natural resource developments in the past five years. Each of these resources is unique in its physical and spatial characteristics, and thus generates different management and policy issues. Forest resources are areally very extensive and constitute a dominantly visible element of much of the Canadian landscape. Forests are seen to have many values beyond harvesting for wood and fibre, and may be managed so as to achieve policy objectives within a time span commensurate with human affairs. Being a biotic resource, their character varies with age and they are subject to natural hazards such as disease, thus requiring long-term management just for preservation. Assessment of the nature and extent of the forest resource involves large-scale survey and inventory by age, size, and species.

Of all resources, water has the greatest variety of uses, and provides society with the most fundamental benefits. It can also constitute a devastating hazard. Spatially, water resources occur in areal, point, and linear form; they are mobile, and temporally they vary greatly in form and magnitude. Assessment of the size and character of water resources requires monitoring and surveillance networks that must be both extensive and long-lived.

Mineral resources are quite different. At the scale of this paper the great majority occur at point locations, do not constitute a part of the visible landscape, and result from natural processes operating over a time scale which is orders of magnitude greater than those relevant to man. Mineral resources do not generally constitute "environment" for plant, animal, or human life in the direct way that forests and water do, and they have little real value to society unless exploited.[1] Furthermore, the essentially "hidden" character of mineral resources means that assessment of their nature and extent is complex, costly, and uncertain.

Forest Resources

Approximately one-third of Canadian land space is classified as "forest land". Of this some 60 percent is considered to be "primary forest land" and 5 percent is reserved for

● JOHN CHAPMAN is Professor, Department of Geography, University of British Columbia. Reprinted from *The Canadian Geographer,* Vol. XX, 1976 (Spring), with permission. The article has been updated by the author and editor.

PROVINCIAL DISTRIBUTION OF THE FOREST RESOURCE*

	Area of Nonreserved Forest Land (10^6 ha)	(%)	Volume of Mature Timber (10^6 m^3)	(%)	% of Total Soft	Hard
Newfoundland	12.2	(3.8)	380.3	(2.3)†	86	14
P.E.I.	0.4	(0.1)	4.1	(0.3)	68	32
Nova Scotia	4.1	(1.3)	257.6	(1.5)	70	30
New Brunswick	6.5	(2.0)	574.0	(3.4)	60	40
Quebec	69.7	(21.6)	2 525.6	(16.0)	72	28
Ontario	49.0	(15.2)	4 051.6	(25.0)	62	38
Manitoba	15.4	(4.8)	173.6	(1.1)	77	23
Saskatchewan	10.5	(3.3)	288.4	(1.7)	59	41
Alberta	27.5	(8.5)	271.6	(1.6)	63	77
B.C.	55.9	(17.3)	7 414.4	(45.5)	97	3
Yukon ⎫	71.7	(22.2)	165.2	(1.0)	n.a.‡	n.a.
N.W.T. ⎭			98.0	(0.6)	n.a.	n.a.
Canada	322.4	(100.0)	16 204.4	(100.0)	79	21

*The percentage of the national total is given in parentheses.
†Includes Labrador.
‡Not applicable.
Source: *Canada's Reserve Timber Supply* (Ottawa, 1974).

Table 1

such purposes as parks, game refuges, and watershed protection.[2]

Based on area, the Northwest Territories and Quebec each have more than one-fifth of the national forest land, followed by Ontario and British Columbia, but in terms of *volume* British Columbia has nearly double that of the next province, Ontario (Table 1).

Forest Resource Depletion

By 1974 the total depletion of the Canadian forest resource had reached nearly 1.6 × 10^6 ha per year made up of over 0.8 × 10^6 ha harvested, 0.6 × 10^6 ha burned, and the remainder lost to insects, disease, flooding, and other hazards. By volume, the total annual depletion is now of the order of 140 × 10^6 m^3, of which approximately 126 × 10^6 m^3 are harvested and more than 8 × 10^6 m^3 burned.

Commercial roundwood harvesting has been increasing by approximately 3.5 × 10^6 m^3 per annum in recent years, a large proportion of which has been supplied by the major producing province, British Columbia (Table 2). The introduction and diffusion of new technologies have permitted increased efficiency in both the harvesting and initial processing phases of the industry, and are

Table 2

PRODUCTION OF ROUNDWOOD AND WOOD PRODUCTS, 1964-68 AND 1969-71

	Nfld.	P.E.I.	N.S.	N.B.	Que.	Ont.	Man.	Sask.	Alta.	B.C.	Yukon and N.W.T.	Canada
Roundwood (10^6 m^3)												
Average 1964-68	2.7	0.2	3.1	5.9	27.2	16.4	1.1	1.5	3.5	44.4	0.1	105.9
Average 1969-71	2.5	0.2	3.2	6.7	29.9	16.5	1.4	2.3	4.1	54.5	0.1	132.2
Lumber (10^3 m^3)												
Average 1964-68	59.0	14.2	512.1	677.3	3478.6	2006.0	99.1	193.5	759.9	17 362.5	16.5	25 228.4
Average 1969-72	73.2	30.7	247.8	712.7	4321.2	2237.3	132.2	278.5	1147.0	19 934.9	16.5	29 162.5
Wood pulp (10^3 t)												
Average 1964-68	n.a.*	n.a.	n.a.	n.a.	5161.8	3186.4	n.a.	n.a.	n.a.	3 276.4	n.a.	13 979.1
Average 1969-71	n.a.	n.a.	n.a.	n.a.	5825.5	3536.4	n.a.	n.a.	n.a.	4 268.2	n.a.	16 706.4

*Not applicable.
Source: *Forest Resources and Utilization in Canada to the year 2000* (Ottawa, 1971).

WOOD PROCESSING PLANTS IN BRITISH COLUMBIA, 1969 AND 1972

		Sawmills				Veneer and
Year	Number	Daily Capacity $(10^3 m^3)$	Chippers	Attached Barkers	Pulpmills, Number	Plywood, Number
1969	974	112.3	227	194	19	22
1972	603	128.7	305	272	23	30
1976	659	139.7	365	326	25	30

Source: Annual Report, 1976 (B.C. Forest Service), p. R41.

Table 3

having a notable effect upon the spatial organization of the forest product industry in two ways.[3] First, the amount of nonmarketable material from each tree which enters the sawmill has been greatly reduced, thus reducing the need to carry out the initial sawmill operation as close to the logging operation as possible. The result is to reduce the number and dispersion of sawmills (Table 3). A second result is to bring about a closer spatial and functional association of saw and pulp mills. Indeed in British Columbia in 1974, 60 percent of fibre-feed to the pulp mills was provided by chips produced as a by-product of medium- to large-scale sawmill operations. Although most notably developed in British Columbia, this trend is rapidly spreading to other parts of Canada.

Depletion of the forest resource by losses from fire has remained fairly constant in recent years, despite increased investment in fire detection and fighting equipment and large-scale public information programs. Although major infestations of insects occur from time to time and require concerted action to prevent large-scale depletion of the resource, there is no evidence to suggest that there has been any significant change in the nation-wide incidence of such outbreaks in recent years. Current concern over the dangers of extensive aerial spraying, however, has resulted in increasingly active research programs aimed at replacing insecticides with carefully directed pathogenic and biological forms of control.

The demand for forested land for such purposes as parks, wildlife habitat reserves, watershed protection areas, and grazing areas has accelerated in recent years. This demand is being met by the establishment of reserves in which no commercial harvesting is permitted or in which controlled logging may be carried on only in conjunction with other uses. To a greater or lesser extent, these "social" uses of forest land are now recognized and respected all over Canada, and a growing proportion of the forested domain is either being reserved or placed under multiple-use management. Conflicts arise because many social uses require the same degree of accessibility as the tree-harvesting use, and in some local areas strong competition between the two has arisen (e.g., in Ontario and Nova Scotia). On the national and regional scale, however, the withdrawal of forest land for social uses has not yet been sufficient to reduce significantly the stock of merchantable timber available for commercial harvesting.

Forest Resource Adequacy

Many Canadians are concerned as to whether the forest resource can support prevailing depletion rates and whether it is capable of supporting even greater harvesting on at least a sustained-yield basis.

Two recently published reports are addressed to these questions.[4] One contains the following statement:

> The forest services in several provinces do not have the kind of inventory data which is really necessary for systematic planning, especially on timber land which is privately owned or under long term lease. Even the better inventory systems fail to include in their surveys sufficient economic criteria on diameter class, log quality, topography, and so on. Furthermore, re-inventory cycles are quite long and in some cases exceed fifteen years. These circumstances seriously handicap forest management.[5]

In the face of this statement, it is encouraging to learn that large-scale re-inventories are under way in all the western

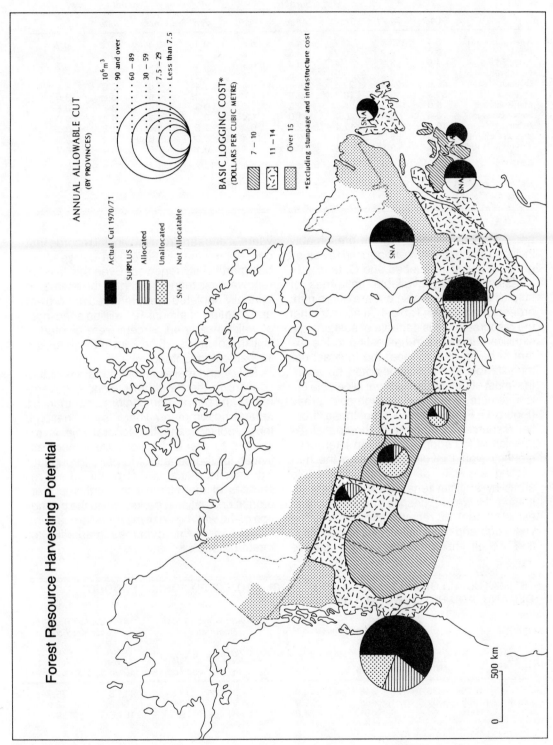

Forest Resource Harvesting Potential

ANNUAL ALLOWABLE CUT
(BY PROVINCES)

10^6m^3
- 90 and over
- 60 – 89
- 30 – 59
- 7.5 – 29
- Less than 7.5

- Actual Cut 1970/71

SURPLUS
- Allocated
- Unallocated
- SNA Not Allocatable

BASIC LOGGING COST*
(DOLLARS PER CUBIC METRE)
- 7 – 10
- 11 – 14
- Over 15

*Excluding stumpage and infrastructure cost

0 500 km

Figure 1

ANNUAL ALLOWABLE CUT (AAC) AND SURPLUS (10^6 m^3)

	AAC	Cut 70-71	Surplus Total	Surplus Alloc.	Surplus Unall.	Product Ven.	Product Lumber	Product Pulp	Product Chip
Newfoundland	6.4	3.8	2.7	n.a.	n.a.	0.1	0.4	2.2	0.2
P.E.I.	0.4	0.2	0.2	n.a.	n.a.	—	0.03	0.2	—
Nova Scotia	4.6	3.2	1.4	n.a.	n.a.	0.1	0.4	0.9	0.2
New Brunswick	12.4	6.2	6.2	n.a.	n.a.	0.3	1.5	4.4	0.5
Quebec	46.6	23.9	22.8	n.a.	n.a.	2.4	7.6	12.8	3.1
Ontario	32.9	16.2	16.7	8.9	7.8	2.3	7.6	6.8	3.0
Manitoba	6.0	1.3	4.7	2.2	2.5	0.3	1.4	3.1	0.5
Saskatchewan	8.4	2.0	6.4	1.0	5.4	0.3	1.7	4.4	0.6
Alberta	15.8	3.4	12.4	2.8	9.7	2.5	8.1	1.8	3.2
British Columbia	91.0	55.9	35.1	16.7	18.4	4.9	25.0	5.2	9.0
Yukon & N.W.T.	1.4	0.1	1.3	0.3	1.1	0.1	1.5	0.4	0.2
Canada	225.9	116.0	109.9	n.a.	n.a.	13.2	55.1	41.6	20.3

n.a. – Not applicable.
Source: See Table 1.

Table 4

and Atlantic provinces and are presumably included in the recently announced revisions of forest policy in Quebec and Ontario.

Assuming that available estimates of surplus annual allowable cut are of the right order of magnitude (Table 4), it appears that the resource base is capable of sustaining a significant increase in harvesting to the extent of being able to meet the forecast demand (Table 5). It should be noted, however, that most of the surplus annual allowable cut is in remote, and therefore high-cost, areas (Figure 1). Future harvesting of these "frontier resources" would mean the spatial expansion of log supply areas. An alternative strategy would involve increasing the harvesting capacity of currently accessible areas by artificial renewal of the forest resource. However, despite the widespread establishment of renewal programs some years ago and the relatively rapid acceleration of their implementation in the last five years, it appears from provincial records that only approximately 60 750 ha per year are being artificially renewed. Even this level of renewal (less than 5 percent of the estimated annually depleted area) required the growth and planting of almost 125 million seedlings, of which nearly 40 percent were planted in British Columbia, 25 percent in Quebec, and 12 percent in Ontario.

Policy and management decisions of this sort are in the hands of provincial governments and private companies rather than the federal government. There is no "national forest policy", and the federal role is restricted to research and development activities for the benefit of forestry throughout Canada, to management of the forest resources of the Yukon and Northwest Territories and national parks, and to the promotion of the well-being of the forest processing industry and the overseas marketing of forest products.[6]

Table 5

ESTIMATED ALLOWABLE ANNUAL CUT (AAC), PROJECTED FOREST PRODUCT DEMAND, AND ROUNDWOOD REQUIREMENT

AAC ($10^6 m^3$)		Forest Product Demand Lumber ($10^6 m^3$)	Forest Product Demand Plywood ($10^6 m^2$)	Forest Product Demand Pulp (10^6 t)	Roundwood Volume Required For Wood Products ($10^6 m^3$)	Roundwood Volume Required For Pulp (r'wood equiv.)	Roundwood Requirements Met From Forest Harvest ($10^6 m^3$)	Roundwood Requirements Met From Sawmill Residue (r'wood equiv.)
	1975	30.4	315	16.9	67.2	72.8	114.8	25.2
226.8	1985	37.8	468	22.9	84.0	100.8	156.8	28.0
	1995	44.4	684	28.8	100.8	126.0	187.6	39.2

Sources: Annual allowable cut from Canada's Reserve Timber Supply (Ottawa, 1974). All others from Forest Resources and Utilization in Canada to the Year 2000 (Ottawa, 1971).

TENURE OF NONRESERVED FOREST LAND (%)

	Provincial Crown	Federal Crown	Private
Newfoundland	93.0	—	7.0
P.E.I.	5.0	—	95.0
Nova Scotia	24.5	—	75.5
New Brunswick	45.0	1.0	54.0
Quebec	89.0	1.0	10.0
Ontario	89.0	0.5	9.5
Manitoba	97.0	1.0	2.0
Saskatchewan	95.0	1.0	4.0
Alberta	96.0	1.0	3.0
British Columbia	93.0	3.0	4.0
Yukon	—	100.0	—
N.W.T.	—	100.0	—
Canada	70.0	23.0	7.0

Source: Mimeo, Canada Forest Service, 1973.

Table 6

Provincial forest policies are formulated and administered by forest services which are usually part of a department concerned with a broader range of natural resources. In recent years almost all of these agencies have come to administer policies which are dedicated to sustained-yield management, stabilization of the wood supply, regeneration, multipurpose use, and reservation of forest land for other than commercial harvesting. In the last few years some have embarked on programs aimed at drastically reducing wood waste and fostering the development of integrated processing operations (e.g., British Columbia), at provision of access (e.g., Quebec), or at large-scale increase of harvesting (e.g., Ontario). In all provinces the role of government regulation is expanding, although the over-all effectiveness of such measures is influenced by the proportion of the forest resource which remains under public jurisdiction (Table 6). In this respect the three Maritime Provinces have a large proportion of their forested land in private ownership, and direct government control is thus more restricted than elsewhere in Canada. Furthermore, it should be noted that area data, such as those in Table 6, do not give any information about tenure in relation to volume of timber.

On balance, the spatial aspects of recent forest resource developments may be summarized as involving a continued extension toward "frontier reserves" and, at the same time, a tendency for the concentration of sawmilling and an agglomeration between saw and pulp mills. The large-scale plans of

Ontario and Quebec may represent the beginning of an infilling process of forest resource exploitation based upon intensive management. Functionally, multipurpose use of the forest resource is established as a policy objective, and nonharvesting uses now compete effectively (but not without conflict) with commercial harvesting. The development of highly integrated operations, extending from the woods to the market, is expanding from western Canada into the central and Atlantic provinces, as are increasingly efficient harvesting and processing techniques. Institutionally, many provinces seem to have been content to leave the development of the forest resource to the private sector, and the federal government has had little influence other than in a protective and marketing role. The latter does not appear to be going to change, but there are clear indications that most provincial governments have now realized the importance of holding the initiative in setting policy in the "public interest" and of securing a high rate of return from forest depletion. In too many instances, though, the initiatives are in the study or planning stage rather than being in the implementation phase.

Water Resources

Although water supply problems have existed in some parts of Canada for a long time (e.g., the Prairies) and local supply difficulties are arising now in some other areas (e.g., the Lower Fraser Valley, B.C.), supply is not currently a major national or regional problem. More significant are the recently delineated issues related to water quality, the expansion of large-scale flow-uses and diversions of water, and jurisdiction over salt water fish stocks.

Water Quality

Data on withdrawal uses of water and on waste discharge are only now becoming available on a national basis. The first issue of the *Canada Water Year Book* (1975) contains some of the first nation-wide data on water withdrawal and municipal waste treatment (Tables 7 and 8).[7]

The largest-scale and most dramatic deterioration of water quality was formally recognized in 1969 in the lower Great Lakes. Eutrophication was identified as a major

WATER WITHDRAWAL, CANADA 1974 (10^6 L/d)

	Municipal and Rural	Manu- facturing	Mining	Agriculture	Thermo- electric	Total
Atlantic	773	3 100	459	55	546	4 810
Quebec	3241	5 151	236	345	64	9 037
Ontario	3555	12 029	423	550	21 753	38 309
Prairies	1227	1 241	1400	5000	6 574	15 443
British Columbia	1137	4 287	314	1318	786	7 842
Yukon and N.W.T.	27	—	59	—	—	86
Total	9960	25 808	2891	7269	29 722	75 650

Source: *Canada Water Yearbook* (Ottawa, 1976).

Table 7

MUNICIPAL WASTE TREATMENT AND DAILY WASTE LOADINGS BY REGION, 1972

	B.C.	Prairies	Ontario	Quebec	Atlantic
Percentage of municipal population served by					
Primary systems	32	15	17	3	2
Secondary systems	30	83	77	14	38
Tertiary systems	1	—	—	—	—
Total	63	98	94	17	40
Net suspended waste loading after treatment (10^3 kg/d)					
B.O.D.*	84.4	42.2	148.3	327.5	59.9
Solids	97.5	48.1	169.6	384.7	62.1
Phosphates	6.3	7.5	21.2	15.6	4.3

*Biochemical oxygen demand.
Source: *Canada Water Yearbook* (Ottawa, 1975), pp. 179-80.

Table 8

problem, and the high rate of addition of phosphorus from municipal sewage (particularly human waste and laundry detergents) was recognized as the principal cause.

In June, 1970, the International Joint Commission (IJC) recommended to the governments of Canada and the United States that a nutrient control program was required, involving both waste treatment and replacement of phosphorus in detergent.[8] Reaction was swift and positive. In Canada, nutrient control provisions were embodied in the Canada Water Act (1970). The Canada-Ontario Lower Great Lakes Water Quality Study Agreement was signed in August, 1971, and the Canada-U.S. Agreement

on Great Lakes Water Quality in April, 1972. Under the Canada Water Act the section regulating cleaning agents came into force in August, 1970, and those sections dealing specifically with phosphorus in January, 1971. The Canada-Ontario Agreement provided funds for joint research into sewage treatment, and authorized $250 million from both governments for the construction of sewage treatment facilities. As a result of this international program it is estimated that by 1975-76, 60 percent of the 1971 sewered population on the U.S. side and 98 percent of the 1975 sewered population on the Canadian side will have adequate sewage treatment.[9] Table 9 indicates, in gross terms, the

PHOSPHORUS LOADINGS IN THE GREAT LAKES, 1971 AND 1975 (kg/d)

	1971		1975	
	Reported	*Target*	*Reported*	*Target*
Lake Erie				
U.S.A.	63 800	63 800	33 570	28 600
Canada	7 900	7 900	4 650	5 500
Total	77 400	77 400	46 320	40 100
Lake Ontario				
U.S.A.	17 100	17 100	4 210	9 500
Canada	15 600	15 600	7 320	13 100
Total	44 600	44 600	22 830	34 500
Lake Superior				
U.S.A.	3 670*	—	3 140	3 500
Canada	2 240*	—	2 800	2 110
Total	5 910*	—	5 940	5 610
Lake Huron				
U.S.A.	3 810*	—	6 720	3 540
Canada	3 340	—	3 880	3 040
Total	7 150*	—	10 600	6 580

*Data for 1972.
Source: *Great Lakes Water Quality, 1974 Annual Report*, pp. 78-9.

Table 9

progress being made in controlling phosphorus inputs. Abatement of industrial waste disposal has been more difficult to achieve, but control now extends over all major discharges, and programs in both countries are shifting to monitoring and enforcement.

Over all, the scene, as reported in 1974, gives the impression that a significant proportion of the abatement elements of the total program are in place or under way, but what remains must be pursued with the utmost vigour. It is emphasized that the effectiveness of the program on a lake-wide basis is difficult to assess because of the long response times of such large bodies of water. To overcome this, as well as to assess the more responsive local problem areas, the report strongly advocates that monitoring and surveillance activities be stepped up.[10]

Protection of water quality from ship-generated wastes and accidents in such areas of shipping concentration as Puget Sound and the Maine-Atlantic Provinces coastal waters, in which both Canada and the United States have interests, has not progressed as formally or as far as the Great Lakes agreement. Joint planning programs are only in the preliminary stages, and joint emergency action programs remain essentially at a consultative and technical level.

Flow-Use of Water

Few countries in the world have the hydroelectric potential of Canada and still fewer have relied so heavily on the development of that potential to meet their electrical energy needs. This flow-use of water has increased considerably in absolute terms since 1970 in British Columbia, Manitoba, Quebec, New Brunswick, and, most dramatically, in Newfoundland (Table 10). Future hydro developments include over 10 000 MW in Quebec (10 000 MW in the La Grande first stage and the remainder on the Manicouagan-Outardes system), almost 4500 MW in British Columbia (Columbia and Peace rivers), and lesser additions to the Churchill River (Labrador) and the Nelson-Churchill river system in Saskatchewan and Manitoba (Figure 2). Despite the large scale of these developments, the contribution of hydroelectricity to the total Canadian electricity supply system is declining; indeed, by 1985 it is expected to constitute little more than half (55.8 percent) of the total national capacity (Table 10). Newfoundland, Quebec, Manitoba, and British Columbia will continue to depend heavily upon hydro up to 1985, but beyond that date the latter two provinces have declared their intention to turn to thermal sources (nuclear and lignite,

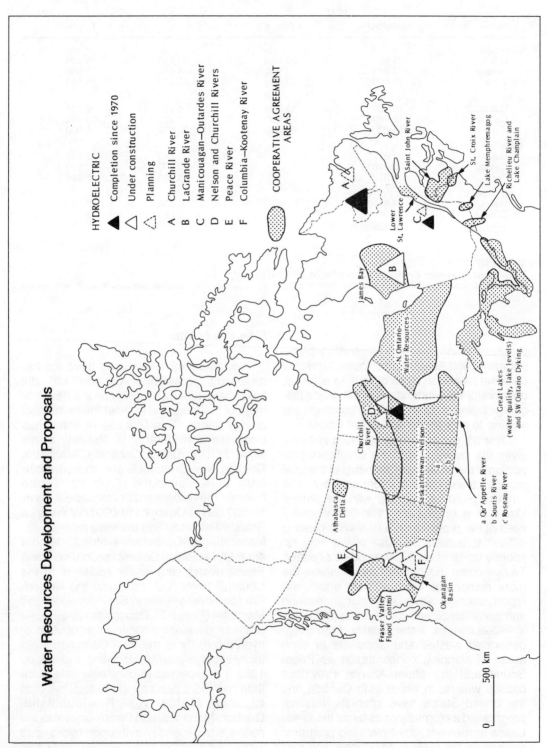

Water Resources Development and Proposals

HYDROELECTRIC

▲ Completion since 1970
△ Under construction
△ Planning

A   Churchill River
B   LaGrande River
C   Manicouagan–Outardes River
D   Nelson and Churchill Rivers
E   Peace River
F   Columbia–Kootenay River

COOPERATIVE AGREEMENT AREAS

James Bay

B

Lower St. Lawrence

Saint John River

St. Croix River

Lake Memphremagog

Richelieu River and Lake Champlain

A

C

N. Ontario-Water Resources

Great Lakes (water quality, lake levels) and SW Ontario Dyking

Churchill River

Saskatchewan–Nelson

D

a Qu'Appelle River
b Souris River
c Roseau River

a
b
c

Athabasca Delta

E

Okanagan Basin

F

Fraser Valley Flood Control

0   500 km

Figure 2

206

ELECTRIC GENERATING CAPACITY, CANADA AND PROVINCES, 1970 AND 1975

Area	Year	Grand total (1)	Hydro Total (2)	Hydro % of (1) (3)	Thermal Total (4)	Thermal % of (1) (5)
Canada	1965	29 348	21 771	74.0	7 577	26.0
	1970	42 816	28 293	66.1	14 523	33.9
	1975*	60 980	37 333	61.2	23 647	38.8
	1985†	94 900	52 800	55.8	42 100	44.2
B.C.	1970	5 471	3 949	72.2	1 522	27.8
	75	6 941	5 328	76.8	1 613	23.2
Yukon and N.W.T.	1970	143	58	40.6	85	59.4
	75	174	61	35.1	113	64.9
Alberta	1970	2 674	616	23.0	2 058	77.0
	75	3 405	718	21.1	2 687	78.9
Saskatchewan	1970	1 533	567	37.0	966	63.0
	75	1 875	567	30.2	1 308	69.8
Manitoba	1970	1 793	1 319	73.6	474	26.4
	75	3 033	2 559	84.4	474	15.6
Ontario	1970	13 699	6 976	50.9	6 903	49.1
	75	20 713	7 033	34.0	13 680	76.0
Quebec	1970	14 046	13 282	94.6	764	5.4
	75	15 109	13 997	92.6	1 112	7.4
New Brunswick	1970	1 201	570	47.5	631	52.5
	75	1 755	790	45.0	965	55.0
Nova Scotia	1970	930	162	17.4	768	82.6
	75	1 205	160	13.3	1 045	86.7
P.E.I.	1970	78	—	—	78	100.0
	75	118	—	—	118	100.0
Newfoundland	1970	1 248	974	78.0	274	22.0
	75	6 631	6 200	93.5	431	6.5

(megawatts)

*All 1975 figures estimated.   †1985 projected.

Source: Electric Power in Canada, 1973 (Department of Energy, Mines and Resources, Ottawa, 1974).

Table 10

respectively). In Manitoba there will be little potential left after the development of the Churchill-Nelson rivers, but in British Columbia, even after the full development of the Peace and Columbia systems, there is a very large potential on such rivers as the Fraser, Homathko-Chilco, Skeena, and Liard, as well as possibilities involving the southward diversion of the Yukon. Yet it is not proposed to develop this potential. For the first three river systems this decision has been made primarily to protect them as the breeding environment for Pacific salmon, even though a high dam on the Fraser would probably provide substantial flood control (and, possibly, irrigation) benefits to a large proportion of the population in the Lower Fraser Valley. The potential of the Yukon, Liard, and other northern rivers is not included in development proposals for the next decade or more, at least in part because of the high capital costs involved in construction at such remote sites.

The various grandiose schemes for continental-scale diversion of waters from Canada to the United Stated, although in some cases carried to quite advanced stages of conceptual design by their proponents, aroused much more public interest than they did official recognition. None was referred formally to the IJC and, although they remain a matter of record, few people give serious consideration to them in 1975.[11] A potentially contentious issue is the U.S. Bureau of Reclamation proposal known as the Garrison Division Unit. This involves the transfer of water from the upper Missouri River into three streams (including the Souris River) which form part of the Red River drainage basin and thus will impinge directly upon Canada in the province of Manitoba.[12] This matter has now been referred to the IJC.*

*Ed Note: On 19 September, 1977, the IJC recommended that all work affecting Canada be halted until appropriate changes and modifications in the project could be made. These included: modifications to prevent the introduction of unwanted biological materials; reductions in the amount of highly saline soils being irrigated (the salts in the drainage water would be carried into Manitoba); restoration of lost wetlands used by ducks; lining the Velva Canal to prevent seepage into Manitoba; solution of the problem of potential additional nitrogen entering the Souris River; payment by the United States to Canada for any work done in Canada to counteract the effects of the project; assurances that North Dakota farmers use proper irrigation management techniques; and that negotiations be undertaken to develop water quality agreements for the Red and Souris rivers.

CANADIAN SALT WATER FISHERIES, 1965-74

	1965	1970	1971	1972	1973	1974
Landings (10^6 kg)	1 042.4	1 188.2	1 111.7	1 007.3	938.2	790.9
Landed value (10^6 $)	145.1	191.7	191.9	220.4	320	285
Marketed value (10^6 $)	291.5	401.1	436.6	512.9	733	696
Fishermen	62 335	53 404	50 741	49 643	50 775	n.a.
No. of vessels	47 508	39 358	38 400	35 552	35 604	n.a.

n.a. – Not available.

Source: Summary of Canadian Fisheries Statistics, 1972, and Canada's Fisheries in 1974 (Mimeo, Fisheries and Marine Service, Department of the Environment).

Table 11

Salt Water Fisheries

A broadly-based international issue over water has arisen in the last five years as a result of the dramatic decline in Canada's share of the catch from the salt water fish stocks off the Pacific and Atlantic coasts (Table 11). This decline has occurred despite the existence of three international commissions,[13] and despite a variety of programs by the government of Canada to improve the economic and technical capability of Canadian fishermen to retain or augment the quantity of fish caught. In the international context Canada is attacking the problem from two directions by (a) working through the commissions to achieve catch allocations which are both more equitable to Canadian fishermen and more cognizant of the condition of the fish stocks, and (b) working through the Law of the Sea Conference to extend Canadian jurisdiction over offshore resources.

At the 1974 Law of the Sea Conference, Canada's over-all position was to argue that specific jurisdictions should be established for specific purposes, and that the rights and responsibilities of coastal states and states in general should be balanced. The application of this position involves the assumption of a management authority rather than the extension of Canadian sovereignty over coastal and adjacent areas. The exercise of this management authority requires the grouping of living marine resources into categories so that appropriate management regimes may be established for each.[14*]

In addition to taking a well-articulated and leading position in these international matters relating to fish, Canada has recently embarked upon a large-scale salmon enhancement program on the Pacific coast, which, although completely domestic in scope, will require the successful outcome of negotiations at the international level if its benefits are to be realized by Canadians.[15] The objective of the program is to double the salmon stock over a ten-year period by means of hatcheries, artificial spawning channels, and other facilities at shore and inland locations along the British Columbia coast. The program will be undertaken in two stages, commencing in the spring of 1975, with a two-year planning study which will prepare a detailed program of construction and development with implementation commencing in 1977.

Cooperative and Comprehensive Planning

Domestically both federal and provincial governments have taken initiatives toward developing stronger control via regulation and toward greater concentration and integration of administrative structures concerned with water. The most far-reaching piece of legislation, the Canada Water Act (1970), delineates the federal role in managing Canadian water resources.[16] Other new legislation, such as the Northern Inland Waters Act and the Arctic Waters Pollution Act, and amendments to the Canada Fisheries Act and Canada Shipping Act, further increase the control and regulatory powers.

In the provinces, recent legislative and administrative changes have moved toward more specific effluent regulations, financial incentives to control emissions, the establishment of environmental advisory councils,

*Ed Note: On 1 January, 1977, the Government of Canada declared a 360 km offshore management zone, and plans and programs are being actively pursued to rebuild the fish stocks and extend Canadian fisheries along both east and west coasts.

and the introduction of mandatory environmental impact assessments in advance of any water project. Most provinces have taken measures to afford some over-all control of semi-autonomous agencies, such as electric utility systems, while, at the same time, diffusing responsibility for some aspects of water management to local and regional bodies.

One major objective of the Canada Water Act was to encourage and facilitate the development of comprehensive plans for the management of water resources by means of joint federal-provincial arrangements. The list of joint projects is long (Table 12), and when broken down by primary objective and stage gives an indication of the breadth of the total program. The widespread diffusion of a comprehensive approach to water management represents a significant development in a country where the single-purpose approach to water resources has for so long been dominant.[17] The recently announced (1975) National Flood Hazard program includes some significant new policy directions, including the adoption of nonstructural flood control measures.[18] The clear emergence of a concern for water quality is another recent and far-reaching development across Canada based, on the one hand, on "cooperative federalism" and, on the other, on the implementation of a set of national standards for effluent discharges. The recent adoption of the Environmental Assessment and Review Process by the

Canadian government, via the Department of the Environment, represents another important step forward which, coupled with varied, but similar, moves in most provinces, has at least established the principle of concern for environmental impact.[19]

Progress appears to have been slowest in comprehensive studies related to the coastal zone. Coastal zones are by definition transitional areas with respect to biological and human processes; they are also transitional in the sense that they fall between several jurisdictions in government, which, in part, may account for the fact that comprehensive planning studies are still in the "under consideration" phase.

It is also notable that Canadians may be witnessing the completion of a seventy-five year construction phase of the use of water to generate electrical energy. At the same time we are watching the beginning of new programs involving the management of commercial fish populations. The enhanced-yield program for Pacific salmon and the Canadian initiatives toward the sustained-yield management of fish stocks along the Atlantic coast represent important new policy directions.

Mineral Resources

In the past six years, production of Canadian mineral resources has expanded considerably for all groups except precious metals. Since 1968, coal, potash, and sulfur

Table 12

FEDERAL-PROVINCIAL COOPERATIVE WATER RESOURCE PROJECTS

Stage	Comprehensive	Flood	Quality	Impact	Use
Implementation	Qu'Appelle River Okanagan Basin	Lower Fraser Valley Flood Control S.W. Ontario Dyking Flood Hazard Mapping Program	Lake Ontario — Lake Erie Canada - U.S. Great Lakes Water Quality*	Athabasca Delta	Prairie Provinces Water Apportionment
Under study Completed or nearing completion	Saskatchewan-Nelson (1973) Qu'Appelle Basin (1973) Okanagan Basin (1974) St. John River (1974) N. Ontario Water Resources (1975)	Richelieu R. - Lake Champlain (1974)* Great Lakes Water Levels* Fraser Valley Flood Control Upstream Storage (1975)		Peace-Athabasca Delta Project (1972) Roseau R. Basin (1975)* Lake Winnipeg Churchill & Nelson R.(1976)	
Continuing and in progress	Souris R. Basin (1975)	Montreal Region (1974) National Flood Hazard (1975)	Great Lakes Water Quality Board (1973)* Mining Pollution, New Brunswick Lower St. Lawrence R. (1974) L. Memphremagog*	James Bay (1973) Churchill R. (1973)	Prairie Provinces Water Apportionment Agreement St. Croix R.*
				St. Croix R.*	
Under consideration	Coastal zone - Atlantic Provinces Gulf of Georgia Athabasca R. Basin				Prairie Provinces Water Use

*International studies involving the UC.

production have more than doubled, and natural gas, oil, and gypsum have increased by over 50 percent. Prices have climbed steeply in the same period for almost all mineral commodities, thus contributing to a

Table 13

MAJOR MINERAL GROUPS, CANADA, 1968 AND 1974 (millions of dollars; percentage of total in parentheses)

	1968	1974
Fuels	1342 (28.5)	5 169 (44.5)
Base metals	1025 (22.0)	2 431 (21.0)
Iron and alloys	1098 (23.0)	1 755 (15.0)
Ind. & struct.	472 (10.0)	933 (8.0)
Precious metals	253 (5.3)	537 (4.5)
All other	535 (11.2)	793 (7.0)
Total	4725	11 618

Source: *Canada Mineral Yearbook*, 1968; *Canadian Mineral Survey*, 1974.

Table 14

VALUE OF MINERAL PRODUCTION BY PROVINCE, 1965 AND 1974 (percentage of national total)

	1968	1974
Alberta	23.1	38.1
Ontario	28.7	20.8
Quebec	15.4	9.9
British Columbia	8.2	10.2
Saskatchewan	7.6	7.1
Newfoundland	6.6	3.9
Manitoba	4.4	3.8
Northwest Territories	2.4	2.0
New Brunswick	1.9	1.9
Yukon	0.5	1.6
Nova Scotia	1.2	0.7
Prince Edward Island	0.02	0.01

Source: *Canada Mineral Yearbook*, 1968; *Canadian Mineral Survey*, 1974.

dramatic increase in the value of production. The product-mix, by value, has recently experienced an even greater dominance by the fuel group and the rise of base metals above iron and alloys (Table 13).

Provincially, Alberta has become the leading mineral producer by virtue of the rapid increase in production of hydrocarbons and associated sulfur. British Columbia has displaced Quebec for third position, largely as a result of rapidly rising coal and copper production (Table 14).

Mineral Resource Developments, 1968-74

Analysis of developments in the location of the mining industry between 1968 and 1974 by major geological regions and mineral-producing areas (Table 15 and Figure 3) indicates that the Shield has retained its position as the major mining region of the country, albeit with a reduced dominance. Within this great region the Ontario-Quebec metals belt has experienced a significant absolute decline in the number of precious metal mines, but over-all milling capacity has been sustained as the result of expansion of nickel and base metal capacity. In the northwestern Manitoba-northeastern Saskatchewan district of the Shield, both the number of mines and milling capacity more than doubled in the six-year period. In the near future two more mines are scheduled to come into production, several other prospects are in advanced stages of delineation, and production of uranium commenced in 1976 from the near-by Wollaston Lake area in Saskatchewan.

In recent years, mineral resource exploitation has increased rapidly in the Cordilleran

Table 15

MINING ACTIVITY BY MAJOR GEOLOGICAL REGION, 1968 AND 1974 (percentage of national total)

Major Geological Region	Milling Capacity		Number of Mines*		Hydrocarbon Production					
					Coal		Oil		N. Gas	
	1968	1974	1968	1974	1968	1974	1968	1974	1968	1974
Shield	61.0	50.5	56.0	50.0	—	—	—	—	—	—
Appalachian	20.0	16.6	17.0	17.0	29.0	8.0	—	—	<1	<1
St. Lawrence Lowlands	2.0	1.3	5.5	5.0	—	—	1	1	<1	<1
Interior Plains	2.0	3.9	8.5	11.0	52.0	46.5	99	99	99	99
Cordillera	15.0	27.7	13.0	17.0	19.0	45.5	—	—	—	—

*Excluding coal, sand and gravel, and clay.
Source: Operators lists, Mineral Resources Branch, Department of Energy, Mines and Resources.

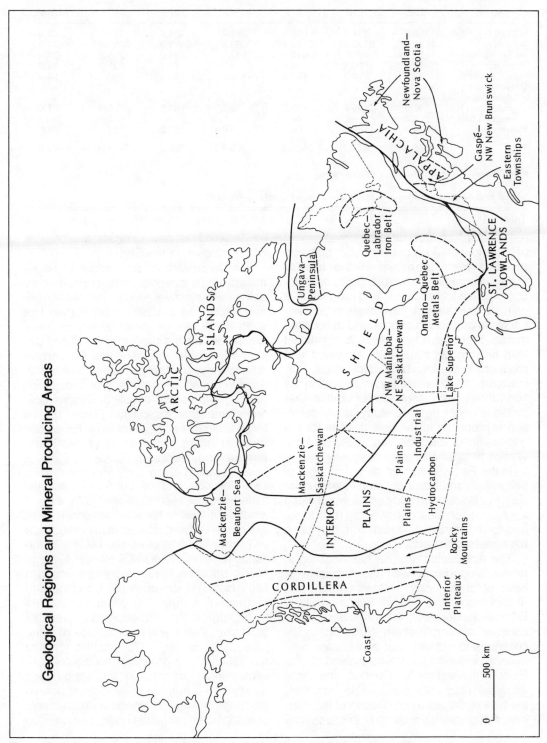

Geological Regions and Mineral Producing Areas

Figure 3

OIL AND GAS DRILLING, 1969-74 (10^3 m)

Area	1969	1970	1971	1972	1973	1974	1976(P)
B.C. (N.E.)	258.5	274.3	289.6	348.4	264.6	231.6	283.0
Alberta	2620.1	2370.4	2403.7	3063.5	3872.2	3657.6	4872.0
Saskatchewan	1113.1	755.9	612.0	553.2	534.0	215.2	230.0
Manitoba	44.5	16.8	9.8	5.2	13.7	21.9	14.0
N.W.T. & Arctic	83.5	110.3	142.6	175.0	225.9	153.3	84.0
Offshore							
West Coast	11.3	—	—	—	—	—	—
East Coast	4.0	45.4	62.2	58.2	100.9	56.4	23.0
Hudson Bay	1.5	—	—	—	—	2.9	—
Other	129.8	104.5	86.6	96.9	82.0	82.1	34.0
Total	2426.5	3677.7	3606.4	4300.4	5093.2	4419.9	5589.7

Note: Figures may not add up exactly due to rounding.
(P) Preliminary.
Source: 1969-74, *Canadian Mineral Yearbook;* and Canada Mineral Reprints, 1976.

Table 16

region, chiefly as a result of the development of large, low-grade copper-molybdenum and precious metal deposits in the Interior Plateau subregion. A number of deposits are under active investigation, and production is planned from several new mines in the near future. In addition to copper and associated metals, there have been several significant lead-zinc developments in recent years, and more are in the feasibility study stage. Furthermore, an extensive lignite deposit in the Hat Creek area of southwestern British Columbia is under detailed study, and production is proposed in the next decade for the generation of electricity and possible conversion to synthetic gas.

In the Rocky Mountain area a number of delineation and feasibility studies are under way. In the southern and central areas these are focused on the extensive coal resources, and to the north, in Yukon Territory, upon base metal prospects.

The Appalachian region has declined in relative importance with the reduction of base metal production in Newfoundland and of coal production in Nova Scotia and New Brunswick. However, the large base metal complex of northeastern New Brunswick continues to expand, and at least one new mine is projected for Newfoundland. In the Eastern Townships of Quebec, the large asbestos resources continue to represent the bulk of Canadian production of this mineral, though the dominance of this area may

be declining as new deposits are brought into production in the Shield.

The hydrocarbon and major industrial mineral resources of the Interior Plains were essentially discovered and delineated before 1968. But the period since then has seen the full-scale development of these resources with the production of oil and gas increasing by 60 percent and 80 percent respectively, and that of coal, elemental sulfur (produced at natural gas processing plants), and potash doubling. Despite continued intensive exploration (Table 16), no large-scale gas or oil discoveries have been made in this region for some years.* Furthermore, the reserve/production ratio has been declining since 1969 for oil and since 1972 for gas, and oil production actually decreased between 1973 and 1974, a trend which continued in 1975. In this context there has been renewed interest in the Athabasca oil sands. The first production facility in the sands (Great Canadian Oil Sands: average 1976 production 7520 m^3/d) was licensed in 1964, came on stream in 1968, and, despite setbacks, began to demonstrate its technological (if not economic) viability thereafter. With the dramatic increase in the price of oil in late 1973 and the growing realization of the potential shortage of conventional oil, interest in the sands rose rapidly and a number of projects appeared on the proposed list. However, to date only one additional project is under construction

*Ed Note: In 1977 potentially large reserves of oil (described as the largest find in ten years) were discovered in the West Pembina field, west of Edmonton, and a potentially major discovery of gas in the Elmworth region of Alberta (400 km northwest of Edmonton). In fact, in 1977, gas reserves increased marginally with respect to production.

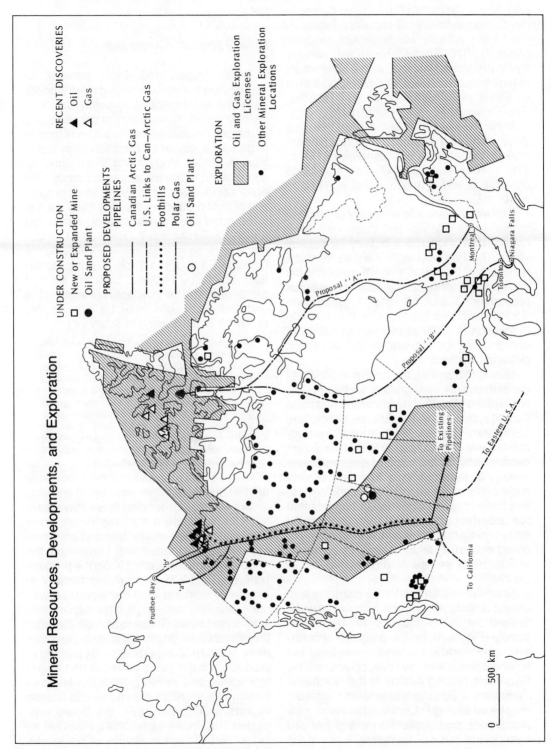

Mineral Resources, Developments, and Exploration

UNDER CONSTRUCTION
□ New or Expanded Mine
● Oil Sand Plant

PROPOSED DEVELOPMENTS
PIPELINES
——— Canadian Arctic Gas
–·–·– U.S. Links to Can—Arctic Gas
········· Foothills
——— Polar Gas
○ Oil Sand Plant

RECENT DISCOVERIES
▲ Oil
△ Gas

EXPLORATION
▨ Oil and Gas Exploration Licenses
● Other Mineral Exploration Locations

Prudhoe Bay

Proposal "A"

Proposal "B"

Montreal

Toronto

Niagara Falls

To Existing Pipelines

To Eastern U.S.A.

To California

0   500 km

Figure 4

(Syncrude) and two more have received Albertan government approval (Petrofina and Shell). Imperial Oil Ltd. is active in developing a heavy oil recovery facility at Cold Lake Alberta which should be completed in 1978. Shell Oil recently (February 1978) expressed interest, with a consortium of ten other companies, in developing a third oil sands plant northeast of Fort MacKay. (*The Financial Post*, 4 March, 1978).

In the Mackenzie Delta-Beaufort Sea area there is an active exploration program which to date has established reserves of 168 to 196 × 10^9 m³ of gas.* Oil has also been discovered, with a minimum reserve of 300 × 10^9 m³.

At present there is little mineral production in the Arctic Islands but there has been considerable exploration activity for several years. This has resulted in the discovery of a minimum of 336 to 364 × 10^9 m³ of natural gas and at least three significant base metal properties. One of the latter, at Strathcona Sound on Baffin Island, was brought into production in 1976 at the rate of 1364 t/d, and the two others are in the advanced delineation stage.

During this period, the centre of gravity of mineral activity has shifted westwards by virtue of the growth of hydrocarbon and base metal production in the Interior Plains and Cordilleran regions. Both these developments have been in response to large export demand and, in the case of coal and base metals, to the continued evolution of very large scale open-pit mining technology. In fact there is some evidence to suggest that the underground/open-pit ratio is changing more rapidly than ever in favour of open-pit, giving rise to a different and, in some respects, more severe environmental disturbance.[20]

Speculations on the future distribution of mining activity are fraught with uncertainty. Judged by the distribution of exploration activity (Figure 4), by the progress in northern exploration and exploitation technologies, and by the comparatively favourable tax regulations in the Northwest Territories, a significant extension northward may be emerging.[21] On the other hand, new discoveries and production plans are still being made in the older mining areas of the Shield, and the over-all trend cannot be solely represented as dispersion to the frontier.

Mineral Resource Adequacy

Unlike forests and water, mineral resources are not renewable and new deposits have first to be discovered before they can be inventoried. Assessment of the adequacy of mineral resources for the future is thus in large part a matter of establishing probabilities. A recent study along these lines for some of the major metal and industrial minerals emphasizes not only the small proportion of total production which is consumed in Canada but also the necessity of maintaining a high discovery rate if expected future demand is to be met twenty-five years hence (see rows 3 and 4 in Table 17). If expectations are to be realized, both the level of exploration and the availability of information and analytical skills must be high. In this context it is disquieting to learn that:

> Canada does not have an effective early warning mechanism to assess the adequacy of mineral exploration and discovery rates [and] the changing picture of national and regional resource sufficiency . . . Information available to governments on actual reserves is incomplete and is greatly inferior to the quality of information on, for example, oil and gas and forestry resources. Certainly data available to the governments on mineral resources are inadequate to meet the full range of mineral policy questions.[22]

Despite the reference to the favourable relative quality of information on oil and gas, recent events surrounding these mineral resources constitute a dramatic illustration of what may happen where data and analytical capabilities are inadequate. Throughout the 1960's and into the early 1970's the producibility of Canadian oil was substantially in excess of demand, and the emphasis was upon securing sufficiently large export markets in the United States to service the large investments in exploration and development.[23] By 1972 exports were 64 percent of production but reserve additions had begun to decline, and in early 1973 the National Energy Board (NEB) commenced to license exports. In October 1974, the Board concluded that "in the early 1980's there will no longer be sufficient availability of indigenous

*Ed Note: In 1977 the potential gas reserve situation was further enhanced by three major strikes in the Beaufort Sea by Dome Petroleum Ltd., lending further support to the prediction that large total reserves await discovery.

Table 17

MINERAL RESOURCES IN RELATION TO CURRENT AND POTENTIAL OUTPUT

	Copper	Lead	Zinc	Iron	Nickel	Potash	Asbestos
1. 1970 domestic consumption as percentage of 1970 production	35.4	15.4	8.5	23.8	3.9	6.0	5.0
2. Estimated domestic needs to year 2000 as percentage of currently established reserves	14.0	18.0	16.0	1.0	13.0	0.04	0.3
3. Estimated domestic and export needs to year 2000 as percentage of currently established reserves	148.0	110.0	162.0	42.0	96.0	1.0	7.0
4. Estimated domestic and export needs to year 2000 as percentage of deposits that can be expected to exist	34.0	47.0	60.0	n.a.	38.0	n.a.	n.a.

n.a. − Not available.

Source: *Towards a Mineral Policy for Canada: Opportunities for Choice* (Department of Energy, Mines and Resources, Ottawa, 1975), pp. 39-42.

oil to meet Canadian feedstock requirements"[24] and that "if there is no significant increase in the producibility levels now forecast there will be a rapid phase-out of all exports".[25] In the same report it is concluded that a return to Canadian self-sufficiency will occur only by developing oil in the frontiers such as the Mackenzie Delta and Athabasca oil sands.*

A similar situation has developed for natural gas. Hearings before the NEB were commenced in November 1974, and in April 1975 it was reported that "Canadian demand for natural gas and existing export commitments are virtually certain to exceed the supply available until the connection of natural gas from frontier areas or from other major new sources of supply".[26] In accounting for this situation, the Board expressed the feeling "that more weight should have been given to deliverability, as distinguished from reserves; and that judgments as to the likely ratio of new discoveries and of their deliverability characteristics both contributed to the present situation".[27] Such abrupt shifts in matters of fundamental national importance must surely have few precedents in Canadian resource history and, in the absence of any penetrating analysis to

date, can only be attributed to the lack of adequate data and analytical capability, independent of that available from the multinational companies involved.†

It appears clear that large-scale changes are impending in the spatial structure of both oil and gas resource systems. First, there is the decline of exports and, within five to ten years, the prospect that existing transborder pipelines may be operating well below capacity or even not at all. Secondly, there is the extension of the Interprovincial Pipeline from Toronto to Montreal, intended to extend the range of Canadian oil beyond the Ottawa River and into the large Montreal market previously supplied solely by imports. As events concerning crude oil supply have transpired, some observers are already speculating on the possibility of the flow through this line being westward rather than eastward. Thirdly, the possibilities for future Canadian supplies include (a) extension to "frontier area" (e.g., the Mackenzie Delta), (b) extension to "frontier-sources" (the Athabasca oil sands and Lloydminster /Cold Lake heavy oil), (c) extension to "foreign sources" (e.g., Alaska, as well as greater reliance upon existing sources such as Venezuela), and, in the shorter term, (d)

*Ed Note: It is now expected that Canada will not regain self-sufficiency in the foreseeable future. In fact, by 1985 it is thought that net imports of oil could constitute over 40 percent of national requirements. The federal government has set a target of holding this proportion to no more than one-third. See: *An Energy Strategy for Canadians: Policies for Self-reliance,* Energy Policy Sector, Dept. of Energy, Mines and Resources, Ottawa, 1977.

†Ed Note: Recognition of this shortcoming led the Dept. of Energy, Mines and Resources to initiate a wide-ranging program of mineral resource studies. Those relating to petroleum have confirmed both the pessimistic outlook for conventional oil resources and the more optimistic position for gas. See: *Oil and Natural Gas Resources of Canada, 1976,* Report EP 77-1, Dept. of Energy, Mines and Resources, Ottawa, 1977; and *Oil Sands and Heavy Oils: The Prospects,* Report EP 77-2, Dept. of Energy, Mines and Resources, Ottawa, 1977.

intensification of production from "conventional supply" areas. At present, oil imports from foreign sources are increasing to serve Canada east of the Ottawa River; an expensive, but in realistic supply terms small-scale, expansion is under way on the Athabasca oil sands; and for both oil and gas much faith is being placed on frontier areas.

Established gas reserves in the Mackenzie Delta-Beaufort Sea area are still small (approximately 196×10^9 m^3, but, if augmented by the large reserves in the Prudhoe Bay district of Alaska, would be well above the threshold necessary to warrant the construction of a large pipeline. Such a connection would not only transport Canadian gas to Canadian consumers but would involve carrying gas from one part of the United States to another (Alaska to the Midwest and California), which, to the Canadian public, raises the spectre of continued export of domestic gas. Some consider the proved reserves plus discoverable potential in the Mackenzie Delta alone to be sufficient to warrant and all-Canadian line (Foothills, Figure 4). Arctic Island gas reserves are already much larger than those in the Mackenzie area and closer to the threshold amounts needed to sustain Canadian markets (Proposals A and B, Figure 4). Two other solutions are also being advocated. One is to link Mackenzie gas with Arctic Island gas eastward across the Northwest Territories and south to Ontario-Quebec markets, and the other is considerably to augment production from conventional reserves in existing fields in Alberta.

This whole complex scene is now before the National Energy Board, which will eventually recommend one set of solutions to the federal government. At the same time the biophysical and social impacts of a Mackenzie Valley gas pipeline are being considered by the Berger Commission, before which the issue of native land claims is being vigorously pressed.*

In the past five years both federal and provincial governments have moved into increasingly direct roles in the mineral resource sector, particularly in relation to fuels. These initiatives have included increased taxation, price setting, export controls, and direct involvement in the exploration, production, transportation, and marketing phases of the mineral resource system. The speed with which these initiatives have been

*Ed Note: The Berger Commission reported its findings in mid-1977 and recommended a ten year moratorium on pipeline construction in the Mackenzie Corridor. Subsequently the National Energy Board approved an alternate route proposal (the Alcan Route) by Foothills Pipe Line (Yukon) Ltd., a consortium of Alberta Gas Trunk Line Ltd., Westcoast Transmission Co. and possibly Trans Canada Pipe Lines Ltd. (their participation is still under negotiation). Northwest Pipeline Co. is the chief American partner and is responsible for the construction of the line in Alaska and south of the 49th parallel. The route follows the Alaska Highway with a proposed spur line (the Dempster Spur) following the Dempster Corridor to the Mackenzie Delta.

Following the Berger Commission report and the approval of the Foothills proposal, a three member commission of enquiry was established (The Lysyk Commission – so named after it's chairman, Dean Kenneth Lysyk of the University of British Columbia) to examine the potential effects of the pipeline in the southern Yukon. The enquiry concluded that, among other things, pipeline construction would have a net negative effect on the Territory and the people. Among the recommendations were:
1. A tax-free cash payment of $50 million to be made to Yukon Indians before a land claims settlement. (The Council of Yukon Indians has stated that land claims must be in place before any major pipeline construction begins.
2. A simple regulatory agency be established immediately with power to regulate engineering social and economic aspects.
3. No actual pipeline construction will start before 1 August, 1981 to allow sufficient time for the implementation of Indian land claims.
4. A minimum of $200 million is to be paid by Foothills Pipe Line (Yukon) Ltd. to be administered by and for Yukoners to compensate for unquantifiable social and economic costs and the deterioration in the quality of life in the Yukon.

The Polar Gas Project route has been modified several times, but an application was filed with the National Energy Board in December, 1977, for authorization to build a 105 cm pipeline from Melville Island to Longlac, Ontario, with a capacity of 850×10^6 m^3/d. While this application is before the NEB, two other approaches to the marketing of frontier gas are being explored. The first is a spur line along the northern coast to tap the Mackenzie Delta – Beaufort Sea (thus removing the need for the Dempster Spur associated with the Foothills project).The second is an examination of the feasibility of transporting Arctic Island gas in liquified form via tanker to an east coast port. Yet another scheme involves an examination of the Trans-Canada gas pipeline to eastern Quebec and the Maritimes, which was proposed in 1977 by Q&M Pipe Lines Ltd. (owned by Alberta Gas Trunk Lines Ltd. and PetroCan). The proposal is at the evaluation stage.

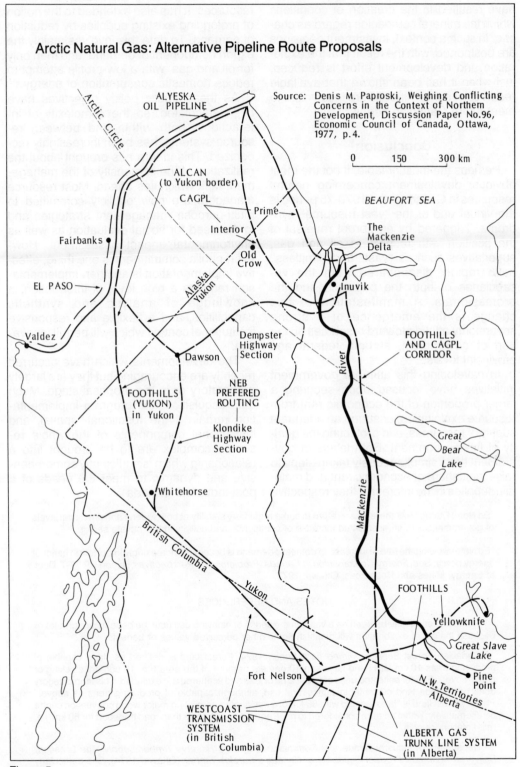

Arctic Natural Gas: Alternative Pipeline Route Proposals

ALYESKA OIL PIPELINE

Arctic Circle

Source: Denis M. Paproski, Weighing Conflicting Concerns in the Context of Northern Development, Discussion Paper No.96, Economic Council of Canada, Ottawa, 1977, p.4.

0   150   300 km

ALCAN (to Yukon border)

CAGPL

Prime

Interior

Fairbanks

Old Crow

EL PASO

Alaska Yukon

Valdez

Dempster Highway Section

Dawson

NEB PREFERED ROUTING

FOOTHILLS (YUKON) in Yukon

Klondike Highway Section

Whitehorse

British Columbia

Yukon

Fort Nelson

WESTCOAST TRANSMISSION SYSTEM (in British Columbia)

BEAUFORT SEA

The Mackenzie Delta

Inuvik

River

FOOTHILLS AND CAGPL CORRIDOR

Mackenzie

Great Bear Lake

FOOTHILLS

Yellowknife

Great Slave Lake

Pine Point

N.W. Territories Alberta

ALBERTA GAS TRUNK LINE SYSTEM (in Alberta)

Figure 5

undertaken and the complexity arising from conflicts over federal-provincial jurisdiction have resulted in the creation of conditions which the mineral companies regard as chaotic. In such a context, investment decisions are postponed, with the result that the exploration and development effort is reduced, just when it has been shown that available reserves of some minerals are insufficient to meet future Canadian needs.*

Conclusion

Perhaps the most notable, if not the most obvious, development concerning natural resources in Canada in the 1970-75 period is the virtual end of the "vast resource" syndrome. Triggered by an abrupt reversal of the position with respect to oil and gas, superlatives such as "vast" and "limitless" have rapidly disappeared from the vocabularies of both the populace and the professionals. A manifestation of this change is the emergence of a strong sovereignty attitude toward resources on the part of government, at both federal and provincial levels.

In developing this attitude, government initiatives have focused upon securing a larger proportion of the economic rent from resource exploitation, ensuring an adequate supply of resources, and protecting the quality of the environment. Both levels of government have simultaneously taken steps to raise their share of economic rent and maintain supplies in the interest of their respective jurisdictions. Maintenance of supply is seen to involve expanding exploitation to frontier resources. It has also extended to the notion of prolonging existing supplies by reduction of demand. To date this involves mainly the export component of demand, and then only for oil and gas, with a low-profile attempt to reduce domestic consumption of energy.†

As these public policy objectives have been articulated, so the complexity of interactions, both within and between resource systems, has been increasingly recognized. This in turn has brought about the realization of the immensity of the management task which lies ahead. Most resource managers are now publicly committed to multi-purpose management strategies and to the need for social evaluation as well as environmental impact assessment. However, public commitment is one thing; effective implementation is another. Implementation requires a high level of information, a knowledge of analytic and synthetic capabilities, and a flexible and responsive institutional context which will permit innovation.

The developments which have occurred recently are encouraging but they are largely preparatory or in the proposal stage. Much data acquisition and program implementation remain if the functional, spatial, and institutional components of the whole resource complex are to be brought into a relationship which is sufficiently comprehensive and dynamic to meet the needs of a post-industrial Canada.

*Ed Note: During 1977 there was a return to some regulatory stability at both federal and provincial levels of government, and an associated increase of confidence and activity in the private sector.

†Ed Note: Slowing the rate of increase of domestic demand is becoming increasingly important in national energy policy. See: *Energy Conservation in Canada: Problems and Perspectives,* Report EP 77-7, Dept. of Energy, Mines and Resources, Ottawa, 1977.

NOTES AND REFERENCES

[1] Nonexploitation may be seen as a value in the sense of delaying use until the benefits are greater or because the act of exploitation will diminish some other perceived values or benefits.

[2] Canadian Forest Service, *Mimeo Fact Sheet,* 1974. Forest land is defined as "land capable of producing trees 10 cm DBH and larger on 10 percent or more of the area (i.e. 10 percent stocking or crown cover). Shelter belts and units of forest 2 ha or less and scattered are excluded. This land category does not include land currently in agricultural use, although capable of producing trees as above." Primary forest land is "land located with 80 km of a designated or existing major wood conversion centre or, alternatively, within an area considered to have equal or lower costs than road transport for 80 km, i.e. rail, flotation, or ship."

[3] Department of Industry, Trade and Commerce, *Canada's Reserve Timber Supply: The Location, Delivered Cost, and Product Suitability of Canada's Surplus Timber* (Ottawa, 1974). See particularly chapter IX, "Product Suitability and Potential for Expansion", for comments on technological development.

[4] G. H. Manning and H. Rae Grinnell, *Forest Resources and Utilization in Canada to the Year 2000,* Publication no. 1304, Canadian Forest Service, Department of Environment (Ottawa, 1971), and Department of Trade, Industry and Commerce, *Canada's Reserve Timber Supply.*

[5] Note 3, pp. 193-4.

[6] The major federal role is exercised through the Canadian Forest Service, Department of Environment, and for the national parks and territories through the Parks Canada Directorate and the Northern Natural Resources and Environment Branch, respectively, both of the Department of Indian and Northern Affairs.

[7] Department of Environment, *Canada Water Year Book* (Ottawa, 1975).

[8] Boundary water issues between Canada and the United States are dealt with under the Boundary Water Treaty of 1909. Under this treaty the governments established the International Joint Commission to which they may refer matters for investigation, report, and recommendation. The role of the commission, which consists of three members from each country, is primarily investigative and judicial but may also be administrative.

[9] Great Lakes Water Quality Board, *Great Lakes Water Quality, 1974 Annual Report* (Windsor, Ontario, 1974), p. 123.

[10] See T. R. Lee, "The Decision to Control Eutrophication", and L. B. Dworsky, "Management of the International Great Lakes", chapters 5 and 11 in F.M. Leversedge (ed.), *Priorities in Water Management,* Western Geographical Series, 8 (Victoria, B.C., 1974), for a more complete review of the Great Lakes program.

[11] See, for example, R. Bryan, *Much is Taken, Much Remains* (Duxbury Press, North Scituate, Mass., 1973).

[12] Onno Kremers, "Prairie Madness: The Garrison Diversion", *Alternatives,* 4 (summer 1975), 28-31, 36-40.

[13] The three commissions are: International Commission for North Atlantic Fisheries (ICNAF), International Pacific Halibut Commission (IPHC), International Pacific Salmon Fisheries Commission (IPSFC).

[14] The four categories developed are: (1) sedentary species, or species which at the harvestable stage are either immobile on or under the seabed or unable to move except in constant physical contact with the seabed, e.g., crabs; (2) coastal species, or free-swimming species which inhabit nutrient-rich areas adjacent to the coast such as productive and relatively shallow areas of the continental shelf and slope, e.g., cod, herring; (3) anadromous species, or species which spawn in fresh water but attain most of their growth in the sea, e.g., salmon (for these species coastal states often undertake costly measures to maintain rivers in a condition suitable to support spawning runs and juvenile fish as well as facilities to produce them by artificial means); (4) wide-ranging species, or species which range widely over the world's oceans, e.g., tuna. Based upon *The Marine Environment and Renewable Resources,* Law of the Sea Discussion Paper, Fisheries and Marine Service, Department of Environment (Ottawa, 1973).

[15] Hon. Romeo Leblanc, *The Salmon Enhancement Program,* Press Release, Mimeo, Minister of State for Fisheries, Environment Canada (Ottawa, 24 March 1975).

[16] Part I of the Act provides for the establishment of formal federal-provincial consultative arrangements for water resource matters, for cooperative agreements with the provinces for the development and implementation of comprehensive plans for the management of water resources, and for the development of any aspect of research into water matters. Part II permits the establishment of a joint federal-provincial incorporated agency to plan and implement approved water quality management programs and to set national effluent standards for federally controlled water and, with provincial agreement, any designated water bodies or river basins. Part III provides for the banning of specific nutrients which contribute to the eutrophication of water bodies.

[17] See, for example, J. O'Riordan, "The Evaluation Process in Comprehensive River Basin Planning: An Analysis of Selected Water Quality Alternatives in the Mainstem Okanagan", Section II in H. D. Foster (ed.), *Okanagan Water Decisions,* Western Geographical Series, 4 (Victoria, B.C., 1972), for a brief review of water management strategies.

[18] Environment Canada, *Federal Flood Damage Reduction Program Outlined,* News Release, Mimeo (Ottawa, Sept. 1975). The approach is based on the principle that: flood-risk areas must be clearly defined and mapped; information on flood hazards must be communicated to the public, industry, and municipalities; construction of federal facilities, federal housing loans, and other grants and loans should not be allowed in flood-risk areas or should be made conditional upon adequate flood proofing or other damage-reducing measures; disaster assistance should be refused for further development in identified high flood-risk areas where the public has been made fully aware of the hazard; and provinces and municipalities should be encouraged to consider appropriate restrictions on land use in high flood-risk areas. *(Canada Water Year Book, 1976* Ottawa, 1976).

[19] P.J.B. Duffy, *The Development and Practice of Environmental Impact Assessment Concepts in Canada,* Occasional Paper no. 4, Planning and Finance Service, Environment Canada (Ottawa, 1975).

[20] For example, by 1974, 80 percent or over of the coal, iron, and asbestos mines of Canada were open pit.

[21] Department of Energy, Mines and Resources, *Canadian Mineral Survey, 1974,* Mineral Development Sector (Ottawa, 1974), pp. 19-23, and "Yukon High Grades, Low Taxes Draw Mine Firms", *Vancouver Sun,* 30 Oct. 1975, p.56.

[22] Department of Energy, Mines and Resources, *Towards a Mineral Policy for Canada: Opportunities for Choice* (Ottawa, 1975), p. 43.

[23] National Energy Board, *Report to the Honourable Minister of Energy, Mines and Resources in the Matter of the Exportation of Oil* (Ottawa, 1974). On page 2-1 producibility is defined as: "the estimated average annual ability to produce, unrestricted by demand but restricted by reservoir performance, well density and well capacity, oil sands mining capacity, field processing and pipeline capacity".

[24] *Ibid.,* p. 6-2.

[25] *Ibid.,* p. 6-8.

[26] National Energy Board, *Canadian Natural Gas: Supply and Requirements* (Ottawa, 1975), p.i.

[27] *Ibid.*

What is Happening To Foreign Investment In Canada? — Economic Trends and Public Expectations

CHARLES H. BARRETT

The implications for Canada's social and economic development of the very high level of foreign ownership and control of Canadian business has been a perennial topic of discussion and debate in this country. Interest in the subject shown by the media and the general public reached a peak during the early 1970's, and it was accompanied by an emergence of a renewed and strong spirit of nationalism. This, in turn, has had a profound impact on both federal and provincial government policy initiatives of such important fields as communications, natural resources, and direct foreign investment.

With the acceleration of inflation and the onset of recession in 1974, there has been more concern with these issues. Nevertheless, while the intensity of the debate on cultural and economic nationalism has subsided to some extent, public concern about foreign investment appears to be persisting. Concern has been expressed in various quarters about the effects of shifts in Canadian attitudes and policies on Canada's external relations, and particularly relations with the United States. Among the misgivings expressed in the United States about the apparent deterioration in the relationship between the two countries, American concern over Canadian attitudes to, and future treatment of, foreign investors is particularly significant.

Canadian businessmen and multinational companies are apparently increasingly considering the possibilities of shifting production from Canada to the United States in the light of Canada's recent unfavorable cost performance. The U.S. Ambassador has pointed out that the flow of Canadian direct investment into the United States is as high

● CHARLES BARRETT is Director of General Economic Analysis, The Conference Board in Canada, Ottawa. Reprinted from *Canadian Business Review,* Vol. 3, No. 5, Summer 1976, pp. 41-44, with permission.

or higher at this time than is the flow of new American direct investment into Canada. Businessmen in both countries have indicated some concern over the prospect of a decline in Canada's attractiveness to foreign investors at a time when it is expected that the capital requirements of the Canadian economy over the next several years will be massive.

The purpose of this article is *not* to provide definitive answers regarding the impact of recent events on the ability of this country to attract foreign capital. Rather, it represents an attempt, employing existing published data, to determine if any trend can be ascertained in the net flow of direct investment into Canada. A clear downward trend is evident in Canada's reliance on new net inflows of direct investment as a source of long-term capital when suitable allowance is made for changes in the level of aggregate economic activity in Canada. A number of factors which may have a bearing on this phenomenon are alluded to briefly below.

Ownership and Control of Canadian Industry

Although the broad facts of the extent of foreign ownership and control of Canadian industry are well known, it is worthwhile to re-examine them briefly here using the most recent published data on Canada's foreign investment position (which refer to the situation as it existed in 1972).[1] As is apparent from the accompanying table, both foreign ownership and control of Canadian nonfinancial corporations are substantial. The largest concentrations of foreign control are evident in the manufacturing sector, in petroleum and natural gas, and in mining. Much smaller shares of public utilities, distribution, and construction are foreign-owned or foreign-controlled. In the case of public utilities, this may be partly due to the long-established tradition of public ownership in those industries. Using data which are more highly disaggregated, further conclusions about the pattern of ownership and control of Canadian industry would emerge. In manufacturing, foreign ownership and control tend to be relatively highly concentrated in industries which produce differentiated or brand-name products, or which are technologically sophisticated, or both. Relatively high degrees of Canadian ownership are found in industries producing identical products or employing standard, well-established technologies. Thus, for example, foreign control predominates in the automotive industry, but the steel industry is largely Canadian-owned.

Data based on nonfinancial corporations alone tend to overstate the degree of foreign ownership of Canadian assets. First, Canadian ownership is more common in financial corporations, partially because there exist restrictions in foreign control in that sector of the economy. Second, the inclusion of personal assets such as real estate would raise the proportion of total Canadian assets owned by residents of Canada, probably to about 90 percent of the total. Nevertheless, despite these qualifications, it is true that an unusually high proportion of Canadian productive assets are controlled by nonresidents, and principally by individuals or corporations resident in the United States.

Both the level and the pattern of foreign ownership and control of Canadian business reflect the characteristics of direct foreign investment which set it apart from other forms of international capital flows. First, direct investment involves not only a transfer of capital but also the transfer of other specialized but intangible assets in the form of technology, marketing ability or access to markets, and management expertise. Second, direct foreign investment is only attractive if the advantages of the foreign firm over the indigenous firm, which stem from its possession of the type of specialized knowledge or abilities described above and which are unavailable to the local firm, are sufficiently great to overcome the risks and costs associated with operating in a foreign environment. Finally, in the Canadian case, additional factors, such as the presence of attractive natural resources and Canadian tariff laws, have been important determinants of the pattern of direct investment.

This suggests why Canada has, to date, been such an attractive site for American direct investment; given the broad cultural, social, and economic similarity between the two countries, the cost disadvantage to the U.S. firm of operating in Canada would tend to be small, as would be the perceived risks associated with investing in Canada. There are costs which, while difficult to identify specifically, must be borne by a firm when it begins to do business in an unfamiliar environment. Other things being equal, a local

Table 1

OWNERSHIP AND CONTROL OF CANADIAN NONFINANCIAL ENTERPRISES – 1972

	Canadian Residents	United States Residents	Residents of Other Countries
OWNERSHIP			
Manufacturing	48%	43%	9%
Petroleum & natural gas	43	45	12
Mining & smelting	44	46	10
Railways	85	7	8
Other utilities	81	16	3
Merchandising & construction[a]	91	7	2
Total	66%	27%	7%
CONTROL			
Manufacturing	42%	43%	15%
Petroleum & natural gas	26	57	17
Mining & smelting	42	46	12
Railways	98	2	0
Other utilities	92	5	3
Merchandising & construction[a]	89	8	3
Total	65%	26%	9%

[a] Ownership and control of merchandising and construction firms are subject to a wider margin of error than is the case of industries.
Source: Statistics Canada.

firm should have an advantage over the foreign firm because of its familiarity with local markets, customs, law, and other factors such as language and ease of access and communications. The industry pattern of foreign investment in Canada is also broadly consistent with what one might expect; it is concentrated in natural resource extraction, where access to world markets and sources of supply is the major determining factor, and in high technology manufacturing industries where access to markets, marketing ability, and the use of brand names is important.

Trends In Direct Investment In Canada

It is important to bear in mind that direct foreign investment represents much more than a transfer of long-term capital between countries. The value of the unique, intangible assets associated with direct foreign investment may be only partially reflected in the recorded flow of capital. This problem aside, data on long-term capital flows relating to direct foreign investment, which are found in the estimates of Canada's balance of international payments, provide only limited information on changes in the extent and pattern of foreign ownership and control of Canadian assets. This is true for several

reasons, the principal one being that the degree of control of Canadian assets may be increased by other means than by a new inflow of capital into Canada – such as, for example, by reinvesting previous earnings or by depending wholly or partly on Canadian sources of finance for capital expansion.[2] Nevertheless, one can argue that changes in the flow of direct investment, as estimated in the balance of payments, may reflect changes in Canada's attractiveness compared to other countries as a site for new investment prospects.

The net value of flows of direct investment into Canada, and of Canadian direct investment abroad, are shown in Figures 1 and 2 on an annual basis between 1950 and 1975. No discernible trend is apparent in either the total net capital inflow or the U.S. share of foreign investment in Canada over the past twenty-five years. Since 1970, however, direct foreign investment in Canada has fallen significantly, despite the fact that this period included a major capital investment boom in Canada between 1973 and 1975. Canadian foreign investment abroad, on the other hand, shows a clear upward trend after the mid-1960's, and especially since 1970.[3]

Direct investment, as reported in the balance of payments, is in current dollar terms and therefore takes no account of inflation.

Figure 1

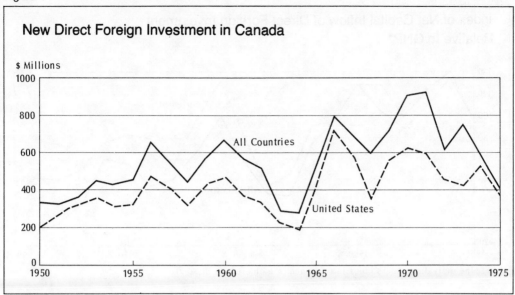

New Direct Foreign Investment in Canada

$ Millions

All Countries

United States

Figure 2

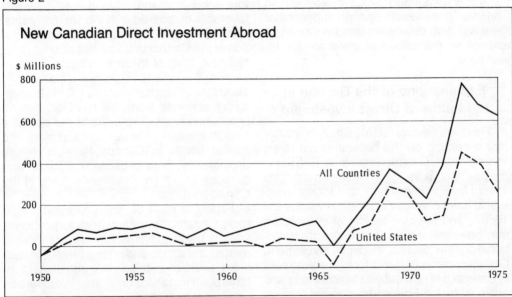

New Canadian Direct Investment Abroad

$ Millions

All Countries

United States

No suitable price index exists for deflating these capital flows to a constant dollar basis. An alternative approach is to examine changes in flows of direct investment relative to changes in aggregate economic activity in Canada. Figure 3 shows the net flow of direct investment into and out of Canada in index number form normalized using an index of current dollar GNP for the period 1955 to 1975. Four major cycles in direct investment are discernible. After the capital investment boom of the mid-1950's, the net flow of direct investment relative to GNP fell during the 1958 recession. It recovered considerably in 1960, but fell sharply over the next few years, perhaps because of the uncertainty associated with the devaluation of the dollar in 1962. A sharp increase occurred during the mid-1960's which was primarily associated with major investment prospects in the automotive industry following the signing of the Auto Pact in 1965. This was following by a more modest cycle in the 1970's.

Over the whole twenty-year period, and particularly after 1970, a very clear downward trend is evident in the net inflow of direct investment relative to GNP. During the last three years, the outflow of Canadian

Figure 3

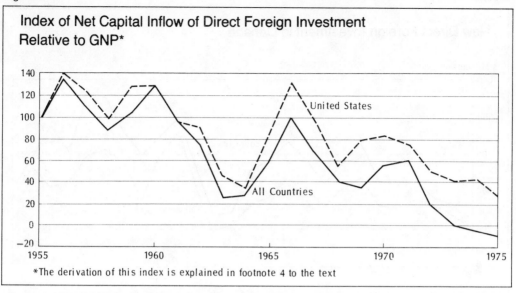

Index of Net Capital Inflow of Direct Foreign Investment Relative to GNP*

*The derivation of this index is explained in footnote 4 to the text

direct investment abroad actually exceeded the inflow of foreign direct investment in Canada. These results strongly suggest that there is a clear downward trend in Canada's reliance on net inflows of direct foreign investment.

Explanations of the Decline In Net Inflow of Direct Investment

There are several factors which may have had a bearing on the decline in net direct investment in Canada relative to GNP over the past few years. First, Canadian attitudes toward foreign investment and foreign ownership have changed dramatically since the 1960's, and the new feeling of nationalism may have either directly or indirectly contributed to the decline in direct investment. Certain aspects of direct foreign investment in Canada are now subject to screening and approval by the Foreign Investment Review Agency. Policies toward natural resource development have changed in Canada in a manner which has tended on balance to discourage new investment in the exploration for, and development of, natural resource assets, and thus encouraged a shift in capital and other resources from Canada for employment elsewhere in the world.

Other factors may have contributed as well. During the 1960's, when the Canadian dollar was pegged at a substantial discount to the U.S. dollar, investment in Canada was particularly attractive, because Canadian assets appeared relatively inexpensive to

foreigners, and because Canada's cost performance in manufacturing was relatively favorable compared with the United States. After 1970, partly because of the appreciation of the Canadian dollar, but also for other reasons, both of these conditions were reversed. Canada is no longer a particularly attractive investment location for manufacturing activities from the point of view of production costs. The costs to the potential foreign investor of acquiring fixed plant and capital goods in Canada have increased dramatically, both because of inflation and because of a higher exchange value of the Canadian dollar.

Finally, it must be borne in mind that in recent years the data reflect a number of special transactions which are not directly related to factors underlying private investment decision-making. In particular, there have been instances of reverse takeovers by Canadian institutions of assets in Canada which were formerly controlled outside the country. The most outstanding example of this to date was the acquisition by the Canada Development Corporation (CDC) of effective control of Texas Gulf Sulfur Inc. More recently, the CDC has been engaged in the repatriation of Canadian oil and gas assets which had been controlled in the United States.

Implications of These Developments

Many Canadians, no doubt, would welcome the decline in the inflow of direct

224

investment on the assumption that this may ultimately lead to a decline in the degree of foreign control of Canadian business. This may not necessarily be the case, however, at least in the short run, since the relationship between changes in the *stock* of foreign-owned and foreign-controlled assets in Canada and *net inflows* of foreign investment is complicated by a number of other factors, in particular the flexibility which the subsidiary of a foreign company has for financing capital needs in Canada. On the other hand, other Canadians may worry about the possible implications of the decline in direct investment for Canada's long-run ability to attract foreign capital. Given the recent high levels of Canadian long-term borrowing abroad, there is no evidence to suggest at the present time that Canada is experiencing any difficulty whatsoever in attracting foreign debt capital. If, however, one of the reasons for the decline in direct investment is uncertainty on the part of the foreign investor over the risks of investing in Canada, it can be argued that this may utimately have some effect on the ability of Canada to attract foreign equity capital.

Finally, it should be borne in mind that direct investment, especially in secondary industries, provides benefits in the form of such things as access to technology, to markets, and to management expertise. Some persons argue that it is not clear that direct investment has been the most efficient way for Canada to acquire specialized knowledge of this type, and this argument may be valid. Nevertheless, at a time when Canadian secondary industries appear to be suffering from serious structural problems, such as poor liquidity position, low productivity, and a lack of competitiveness in markets at home and abroad, a decline in direct investment may have the effect of cutting off Canadian industries from traditional sources of specialized knowledge, and this may well exacerbate some of these problems.

NOTES

[1] These data refer to the distribution of ownership and control of the stock of assets of Canadian nonfinancial corporations. More recent estimates are available only for the distribution of *control* of Canadian corporations. Since these data are essentially structural in nature, the results change very slowly. Data on *flows* of direct investment into and out of Canada are available up to and including the first quarter of 1976.

[2] Undistributed profits form a major part of Canada's international liabilities, as may be seen from data in *Canada's Foreign Investment Position, 1968-70.* (Ottawa: Information Canada, 1975). Indications are that Canada's international liabilities of undistributed profits have grown rapidly in recent years.

[3] A significant proportion of Canadian direct foreign investment abroad is made up of the investment in other countries by Canadian subsidiaries of multi-national firms.

[4] The index of net inflows of direct investment relative to GNP was computed in two stages. First, both the net inflow of direct investment and nominal GNP were converted to index number form with 1955 = 100. Second, the former index was divided by the latter to produce the index of net inflow of direct investment in relative terms.

On Looking Towards the Future Around the Great Lakes

GEORGE R. FRANCIS

"Futurism" is always a risky business to engage in, especially for situations which defy comprehension, let alone management or control. The developing situation in and around the Great Lakes of Canada and the United States can well serve as a case in point. However, unless one has an abiding faith in the virtues of self-equilibrating systems regardless of their outcomes, or a deep belief that whatever happens from being

● GEORGE FRANCIS is in the Department of Man-Environment Studies, University of Waterloo. This article is reprinted from the *Canadian Water Resources Journal,* Vol. 2, No. 1, 1977, pp. 18-32, with permission.

adrift in a sea of events is the best that could be hoped for, there is no option but to match wits with the unknown. It is the faith of all planners and managers, and the faith of "futurists" as well.

Some such confession of humility is called for when contemplating what is already known about the Great Lakes as natural ecosystems, about many interrelated problems arising from human activities in the surrounding Basin, and about the driving forces of change at work in the megalopolitan region which overlays this. There is already a rich body of writings which celebrate the geography of the lakes, their uniqueness among fresh water resources in the world, and their historical and contemporary importance for developments in the North American heartland. Interest in the future of this huge area stems from apprehensions about what is now becoming understood about the consequences arising out of use and abuse of the lakes, from the realization of the sheer magnitude and complexity of the human-societal-environmental interdependencies upon and around these lakes, and from a growing conviction that there are real opportunities to strive for a better future than whatever might arrive by drift and default. This will, however, require considerable improvement in our collective institutional capabilities to anticipate and intervene.

The Great Lakes as a Focal Point of Concern

Concerning the lakes themselves, it is now clear that the only feasible "management strategy" is one of adjustment to their thresholds through constraints on human activities. In terms of levels and flows, they are just not manageable in the way most river systems are, where dams and control works can be designed to manipulate levels and flows more or less at will. The outflows of Lake Superior and Lake Ontario are already regulated within a small range agreed to by Canada and the United States and managed under International Boards of Control, but as the recent International Joint Commission studies have shown, there is little that can be done feasibly to regulate large fluctuations over a period of years.[1] In addition, the attempt to manipulate levels gives rise to

difficult questions of benefits and costs accruing to different interests on different lakes,[2] and this has already become the subject of litigation in the United States. Attempts to provide protection from high water levels by engineering works often prove to be ineffective, prohibitively costly, and of questionable economic justification.[3]

The only other engineering solution to the problems of lake levels would be massive diversions of water in to and out of the Great Lakes system. Paper schemes featuring this were quite popular a decade ago, and the idea is still alive.[4] A completely open mind to future alternatives would not automatically rule these out. However, none seem to have gone beyond the proselytizing pamphleteer stage to acknowledge problems in the economic, social, ecological, and political feasibility of donating northern Canada water resources to the cause, not to mention those which would arise from flushing excess and polluted waters to other basins in the United States.

The only reasonable alternative to opening up something as horrendous and intractable as massive diversions is to adopt the strategy of adjustment to, not of, lake levels. This means that "coastal zone" management for human use and activities along the shorelines is required. The planning process to guide this is already well under way on the United States portion of the Basin, through the impetus given by the U.S. Coastal Zone Management Act, 1972. It has not yet started in a concerted manner in Ontario.

The long history of environmental abuse, to the lower lakes in particular, is leading to the progressive and, some would suggest, accelerating loss of their biological utility. This is reflected by the realization of the extent to which toxic and hazardous materials pervade the lake ecosystems, to the point that they are now understood and acknowledged to be the major threat to water quality and to the fisheries of the Great Lakes.[5] The polychlorinated biphenols (PCB's), for example, occur throughout the system. These compounds are acknowledged to pose a serious threat to human health;[6] a variety of fish, including sports species, have been routinely found to carry residues far in excess of maximum tolerance levels for consumption;[7] and PCB's have been implicated in the recently documented disasters among herring gulls in eastern

Lake Ontario. The eggs of these birds contained some of the highest organochlorine residue burdens yet reported for any biological system, and the tissues of adults were found to contain no less than twenty-nine identifiable contaminants.[8] At a time when considerable expense and effort is being placed on programs to rehabilitate and restructure fish communities in the Great Lakes,[9] health hazard warnings are being issued to caution against the consumption of catches.[10] Clearly, these biological indicators are showing red.

Suffice to say, when the strategy of "the solution to pollution is dilution" no longer works for bodies of water the size of Lake Erie or Lake Ontario, then the only feasible approach must be one of adjustment and constraint on activities. This is central to the underlying rationale for the Great Lakes Water Quality Agreement of 1972, which has been instrumental in bringing about the concerted binational efforts to control point sources of pollutants from municipalities and industries, especially the phosphorus loads which are accelerating the eutrophication processes in the lower lakes. There are limits to how far this can go, however, until the vexing problems of controlling or preventing nonpoint sources can be resolved. The work initiated under the "Pollution from Land Use Activities Reference Group" (PLUARG)[11] is of particular importance in terms of the implications for land use planning inherent in it. It is also encouraging to know that, at least in Canada, work under the Agreement will be extended and expanded to other water quality issues under the new federal-provincial environmental accord.[12]

All this stresses a point which is obvious but has interesting implications. That is, while the problems which are of such great concern show up in the water, they don't originate there, nor are they to be solved there. Yet the whole thrust of research on the lakes for the past twenty years is water-bound. While this is not to deny the importance of knowing the physical, chemical, and biological characteristics of the Great Lakes, should this be all anyone wants to know, it will be in vain. Future efforts will have to see how research on the dynamic changes in popluation, urbanization, and land use in the Great Lakes Basin can be linked to questions of their impacts on the lakes. This may require casting enquiries within a broader

policy-oriented and decision-making framework generally, so that results are usable and hopefully used. And this may warrant re-examining priorities for the water-based research in the context of what needs to be known for devising long term restoration and rehabilitation programs for whole lake ecosystems themselves.[13]

Looking to the Population Forecasts

One way to glimpse the magnitude of what we have to come to grips with in the future is to look at population forecasts and think of their implications. Forecasts have been done by several groups quite recently as the millenium year 2000 approaches. The current best guess is that by then Ontario will have some 11 million people,[14] over three million more than are already here, and the large majority of these will reside within the Great Lakes Basin. This will constitute about one-third of the total Canadian population. What has been recently called the Canadian "main street axis"[15] (because, for better or worse, much that happens economically and politically in Canada is determined there), runs from Windsor to Quebec City, i.e., along the lower Great Lakes-St. Lawrence system, and contains about 12 million people now, or 55 percent of the Canadian popluation. It is projected to have some 19 million of the 30-32 million Canadians by 2000; a proportionate as well as an absolute increase, and 6.6 million of these are expected to be living within 65 km of Toronto. On the United States side there are already some 30 million people within the Great Lakes Basin and this is projected to increase to over 42 million by 2000.[16]

This means that the Great Lakes Basin, including the upper lakes in terms of the influences such a population will have on them, is already and will remain of strategic importance to Canada. However absurd it may seem to have so many people crowding into such a small portion of such a large country, this is not a trend that can be easily reversed in an acceptable way by government policies. So the "main street axis" will endure, grow steadily, and consolidate during the foreseeable future, and in doing so will generate increasing and more conflicting demands for land, shoreline, and water resources as more people strive for access to social choice, material well-being, and environmental quality. And they will be doing this along with millions of their neighbours across the lakes.

A "Futures-Oriented" Response

A "futures-oriented" response to a situation of this magnitude would be to examine trends, try to visualize outcomes and consequences if they persist, and if this most likely situation is not desirable, try then to ascertain how a different course of action may be embarked upon and maintained to result in a preferred set of outcomes. This "future studies" approach has to make a number of assumptions about trend extrapolations, about the creative imagination needed to think of social and economic alternatives different from those we have become so accustomed to, and about the dynamics of social change which would indicate how we could steer a relatively painless passage from here to there. This is a heroic task. However, governments routinely, if only implicitly, make these kinds of assumptions whenever they devise and implement many of their policies. The problem is that all too often it is done entirely within the short range program interests of individual ministries, with the result that, like corporate planning, the future well-being they are pursuing is their own.

The "future studies" field is emerging with a wide variety of methods and techniques, ranging all the way from computer simulations of the quantifiable to analyses of science fiction. One approach, which when done well combines the virtues of meshing creative imagination with a sense of what is feasible, is the "scenario". The strength of scenarios is the plausibility with which a series of events can be traced through to different possible outcomes. In Canada, this is being explored under the rubric of the "conserver society" by the Science Council and, in considerably more detail, by the GAMMA group in Montreal.[17]

It would be a worthwhile exercise to strive over the next few years to devise some "broad-brush" scenarios for the Great Lakes Basin. These could draw in part on methods for devising "alternative whole futures projections"[18] to gain the wide-angled vision which is so necessary, along with some analysis of the on-going "technology assessment systems", to understand the institutional dynamics,[19] and a review of policy issues in terms of the "field concept of public management"[20] to point out feasible options and priorities. Such a suggestion may seem hopelessly academic, yet it may prove to be a most practical approach for ultimately coming to grips with a situation of such magnitude and complexity.[21]

A Spatial Perspective

The future, whatever it turns out to be, will exist in a distinct space. There are two spatial configurations of particular importance for the Great Lakes. One is the Great Lakes Megalopolis, and the other is the Great Lakes Basin. The Megalopolis is where the driving forces of change are at work, and the Basin embraces the main region where the resource consequences unfold and the management responses have to be implemented. Both are, of course, binational in extent, which makes the situation even more complicated in terms of formulating effective policies. It should also be noted that other definitions of relevant "regions" may well be necessary for certain purposes, especially for tracing economic interrelationships and also for air pollution issues.[22]

The Great Lakes Megalopolis embraces the lower Great Lakes, running from Chicago eastwards along both the north and south sides of lakes Erie and Ontario. On the Canadian side it embraces most of the "main street axis", and the portion from just east of Toronto through to Windsor-Detroit has evolved an urbanized structure and set of interconnections sufficient to be recognized by "ekisticians" as a genuine, all-Canadian megalopolitan structure on its own.[23] The Great Lakes Megalopolis in its entirety contains a population of about 75 million people, and in geographic extent is probably the largest of the seven great megalopolitan systems in the world.[24] For those of us residing around the Great Lakes, this will be our "city of the future". There is nobody in charge of it, and not even arrangements to monitor it.

Few have thought to look at this huge urbanizing system in its total extent. The question is, what might be gained from doing so. By way of an answer, the preliminary results from a recent overview of the megalopolis forming over five countries in northwest Europe, and containing some 80 million people, are of interest. The study sought in part to identify which factors of supranational dimensions needed consideration in formulating long-term spatial strategies for the continued development of this megalopolis. The main ones they came up with were: conservation of the megalopolitan and several intermetropolitan open spaces as defined in the study; the integration of transportation systems (i.e., integration of both the modes and the networks); the protection of sea coasts and estuaries from pollution caused by heavy industries and power stations; and concerted national efforts at water and air pollution control generally.[25] None of these concerns would be unfamiliar to residents around the Great Lakes.

While this type of overview for the Great Lakes Megalopolis has not yet been done, studies of the Detroit Urban Area (which embraces extreme southwestern Ontario) and the Northern Ohio Urban System pointed out the "chaotic networks" of transportation and utility corridors and the resulting waste of so much land. Creation of a coordinated transport and utility system has been recommended recently as a reasonable and necessary measure to guide the development of the Great Lakes Megalopolis as a whole, while retaining a considerable range of choice in the spatial development of the submegalopolitan parts.[26]

The Great Lakes Basin, on the other hand, can be looked at as the main resource and environmental region within which to respond to the multitude of demands and their consequences being generated by the dynamics of Megalopolitan growth. While the geographic area embraced by Megalopolis only directly overlaps the lower half of the total Great Lakes Basin, it would seem safe to assume that it already exerts considerable influence over much of the entire Basin, the more northern part of which can be seen as a kind of hinterland to Megalopolis itself.

A Planning Response for the Future

The key to the response strategy will have to be "constraint planning" defined in this context as the recognition for development planning and urban design of physical, biological, and cultural landscape features which need to be integrated into design or accepted as constraints on development; as the art of doing this improves, it may lead closer to an ideal of a much more systematic environmental planning and design process.[27]

In Ontario, the dilemmas of reconciling growth and constraints are quite evident in discussions over the Toronto-Centred Region, the government's recent trends and options papers, and the farmlands preservation debate.[28] For open space and recreation, they have been recognized in the creation of the Niagara Escarpment Commission, the Georgian Bay Recreation Reserve, and the joint federal-provincial Rideau-Trent-Severn recreation corridor study. They need now to be recognized for the Great Lakes shoreline and coastal zone as well.

The shoreline-coastal zone is clearly one of the most valuable resources of the Great Lakes, representing as it does the critical land-water interface. A glimpse into a very probable future, if matters continue to drift, can be seen in some problems which have already surfaced. In recent years around the west end of Lake Ontario, for example, there have been protracted struggles to save the last remnants of marsh and natural areas, such as Rattray's marsh (the tag-ends of which will cost $4938/ha to acquire); there are expensive programs to rebuild and redesign waterfront shorelines areas, such as is being done by the Metropolitan Toronto and Region Conservation Authority, and there are embittered conflicts over shoreline allocation processes, such as the one building up over the Oshawa second marsh.

To preserve options for the future and implement the necessary adjustments-to-the-lakes approach, the whole coastal zone of the Great Lakes in Ontario should be designated under the Ontario Planning and Development Act (S.O. 1973). No further allocations or alienation of this resource would then be allowed until a comprehensive assessment of the situation is made, including the drawing together of various *ad hoc,* single purpose studies reportedly going on, and the requisite zoning and development controls drawn up. This would, of necessity, be a joint federal-provincial undertaking which could be modelled on the Rideau-Trent-Severn study in terms of its organization. It would also be able to benefit considerably from the experience of work along the United States side of the lakes, notably in Wisconsin and Michigan.

To the extent that a creative form of constraint planning can become the main strategy for guiding the general direction and nature of urbanization over the Great Lakes Basin, hopefully with an increasingly clear consensus about the kind of future being sought through the process, there would then be a much clearer and firmer basis from which to respond to the continuous flow of independent initiatives and interventions being made from so many interdependent private and governmental decision centres. This is the very process by which the future gets created.

At the present time, all manner of concerns are handled as reactions rather than foresight. Key industrial location decisions along the north shore of Lake Erie by the steel and energy industries (Stelco, Dofasco, Texaco, Ontario Hydro) were followed by special studies on the social and environmental consequences which government has to provide for, and the costs of which it has to absorb; the multi-billion dollar expansion proposals of Ontario Hydro have to be sorted out and assessed by a special Royal Commission on Electric Power Planning (Porter Commission), and a standing reaction mechanism is about to be launched under the Ontario Environmental Assessment Act, passed in July 1975. The federal government has a comparable arrangement for environmental protection as a matter of policy.[29]

These are all mechanisms which at best can only raise issues of broad social purposes and objectives, after a commitment is made to do some one thing by government or the private sector, and usually only after the proposal is forced into the open through actual or anticipated political controversy. The Berger Commission hearings on the proposed Mackenzie Valley gas pipeline are the most dramatic example in Canada, where enquiry about a specific development scheme has been allowed to evolve into a futures-oriented enquiry and debate on the fundamental objectives and basic policies with regard to the well-being of people in a vast region of the country, issues which should have been aired long before billion dollar technical commitments were encouraged and embarked upon. Should the new provincial Environmental Assessment Act or procedures developed under comparable federal policy really not allow challenges to be made to assertions of the real need for specific proposals being assessed (as some critiques suggest), then the constant challenging of means and alternatives will become a proxy for this more basic question and will make satisfactory resolution of it, through the hearings process, difficult if not impossible.

The point is that while there is definitely a need for these reaction mechanisms and the provision of them is to be welcomed, they cannot and should not be relied upon alone to raise and resolve the more basic futures-oriented questions which need to be considered. It is here that there is need also to develop additional capability, in part through developing scenarios and specifying constraint planning guidelines, in order to air broad issues of purpose and objectives, and do so by raising them largely as the "front-end questions" in the policy formulation and development decisions process, rather than as by-products or afterthoughts. Planning to serve this should get beyond simple trend projections, although these are helpful in one context, to more detailed review of policies and decisions surrounding the "leading part" of systems change.[30] In the Great Lakes Basin this may be best represented by key industrial location decisions of the private sector, and public investment decisions in the energy and transportation sectors.

Towards Concerted Binational Planning and Management

Mention has been made several times of the binational character of the Great Lakes Megalopolis and the Great Lakes Basin. Clearly, there is no way to get a realistic overview of future prospects and possibilities if the international boundary running down the middle of the lakes continues to serve as one which also limits the perceptions, definitions, and concerns of either country about the shared resources used and abused by both. Yet both continue to do this, although informal communication channels lessen the mutual isolation somewhat. The U.S. Great Lakes Basin Commission, for example, has recently completed a twenty-seven volume set of framework studies to provide the basis for on-going "comprehensive, coordinated joint planning" of the Basin, but these of necessity stop in the middle of the Lakes. On the other hand, for the last ten years Ontario has discussed "designs for development" and policies to distribute several million people around the Great Lakes Basin, without any obvious urge to take note of fourteen million other people in the Lake Erie-Lake Ontario basins alone, the nearest of whom may be only about 2 km away and never more than 80 km. If there is the slightest concern about planning for development with an awareness that the lakes are already seriously stressed and they do have environmental limits and thresholds, then these two halves must be placed together as a matter of urgent binational awareness, and to set the only realistic framework within which major policies and decisions can be initially and properly assessed.

It is here that institutional innovations on a binational basis are called for. While there is certainly no lack of government around the Great Lakes — with two federal systems comprising three and often four layers of government, each with a number of agencies engaged in various functions and activities, this is assured — there is a lack of governance. No one is *in charge* of the Great Lakes Basin. The answer to this is not to create some supranational administrative and bureaucratic authority, but to work towards developing an anticipatory planning capability within the existing structures, on a binational basis, for the express purpose of tackling problems before they reach crisis proportions.

This would require provisions for a monitoring or surveillance function which would strive to give a wider measure of common awareness of what is going on in and around the lakes, and would be largely a matter of gathering certain information and data systematically on a regular and agreed-upon binational basis. It would be desirable also to provide for a mediation function, which could be served by open informal discussion to resolve difference of perception or interpretation of problems, and to try and agree on joint approaches for their

solution. In a very limited way, some functions are already being performed on a binational basis, but largely as an *ad hoc* response to particular questions, technically defined. There are as yet no formal provisions for maintaining some measure of a continuous watch over the whole Great Lakes Basin, with the broad futures-oriented perspective already referred to, in order to identify problems before they become unmanageable, look at future alternatives and opportunities, and exchange ideas and experiences for coping and mobilizing a concerted binational response. The response itself would have to be carried out through the existing institutional structures of both countries.

A proposal to build this capacity into the International Joint Commission's responsibilities has at best been greeted with only a very partial and cautious acceptance.[31] An alternative idea of devising a basin-wide policy planning body to assume some of these functions has been raised in recent years, but not followed up.[32] There is as yet little sense of urgency and little sense of the future among many of us who live in the midst of it all, and until there is a marked groundswell of concern and support for creating a better future, there will be a limit on how much official action and commitment can really be expected. The newly formed "Great Lakes Tomorrow" network among citizen groups on both sides of the lakes may provide a much needed impetus in this regard.[33]

Conclusion

It should be recognized that the past few years have been very productive in terms of consolidating the knowledge about the lakes themselves. Basic scientific data gathered during the International Field Year for the Great Lakes have been processed and made available;[34] the extensive studies on water quality problems embarked upon under the Great Lakes Water Quality Agreement are coming to fruition; case histories summarizing the available information on changes in fish populations in each of the Great Lakes have been prepared through the Great Lakes Fishery Commission;[35] and an impressive twenty-seven volume set of planning inventory materials for the United States portion of the Basin has been assembled and analyzed by the Great Lakes Basin Commission. Linkages between land use in the Basin and its impact on the water are being explored by IJC/PLUARG, and the phenomenon of the Great Lakes Megalopolis itself has come under discussion and scrutiny.

It would indeed be unfortunate if much of this has to sit on shelves, unused because of the lack of coherent and consistent interinstitutional arrangements to draw it into a policy and program formulation process. The process would be explicitly normative and "futures-creative" in its search and review of options and opportunities, and would build on the experience of binational cooperation in organizing a concerted response to the challenge.

REFERENCES

[1] International Joint Commission. *Regulation of Great Lakes Water Levels.* A Summary Report by the International Great Lakes Levels Board, 1974, and subsequent discussions.

[2] International Joint Commission. *Seminar on the IJC, its Achievements, Needs and Potential.* Montreal, June 1974.

[3] Nelson, J.C., *et al.* "The Fall 1972 Lake Erie floods and their significance to resources management". *Geographical Inter-University Resource Management Seminars*, 5:1-27, 1974-75, and *Canada — Ontario Great Lakes Shore Damage Survey*, June 1976.

[4] See: A *Continental Hydrology:* A continental design for flood and erosion control, and water for power, irrigation, transportation, recreation and climate modification (Technocracy Inc: Rushland, Pennsylvania, 1970), 12 pp. For summaries of earlier proposals see Alberta Department of Agriculture, Water Resources Division. *Water Diversion Proposals of North America.* Prepared for the Canadian Council of Resource Ministers. Edmonton. December 1968.

[5] Great Lakes Water Quality Board, Fourth Annual Report to the IJC, Windsor, July 1976.

[6] Ahmed, A.K., "PCB's in the environment", *Environment* 18(2):6-11, March 1976.

[7] Great Lakes Fishery Commission. *Lake Ontario Committee* Appendix XXII of Minutes. March 1976.

[8] Reference 4 and G. A. Fox, *et al., Herring Gull Productivity and Toxic Chemicals in the Great Lakes in 1975.* Toxic Chemicals Division, Canadian Wildlife Service, Manuscript Report No. 34, 1975.

[9] Great Lakes Fishery Commission. *A Management Policy for Great Lakes Fisheries.* Ann Arbor, August 1974. Ongoing work stresses continued control of sea lampreys and stocking the lakes with various species of salmonids.

[10] A recent example is one from the Ontario Ministry of the Environment. *Insecticide Mirex Found in Lake Ontario Fish.* July 16, 1976.

[11] The work is being done under a Reference to the International Join Commission made at the same time as the Water Quality Agreement was drawn up. This "PLUARG" group is to report back to IJC by 1978. See also: *International Reference Group on Great Lakes Pollution from Land Use Activities.* Annual Progress Report. July 1976.

[12] Canada-Ontario Accord for the Protection and Enhancement of Environmental Quality, signed in October 1975.

[13] For a discussion of this problem, see Henry A. Regier, "Environmental Biology of Fish: Emerging Science", in E. K. Balon (Ed.). *Environmental Biology of Fishes*, Vol. 1, March 1976.

[14] Government of Ontario. *Ontario's Changing Population, Volume 2, Directions and Impact of Future Change, 1971-2001*, Ministry of Treasury, Economics and Intergovernmental Affairs, Regional Planning Branch, March 1976.

[15] Yeates, Maurice, *Main Street, Windsor to Quebec City* (Macmillan Company of Canada: Toronto 1975).

[16] Great Lakes Basin Commission. *Great Lakes Framework Study*, Appendix 19, Economic and Demographic Study. Ann Arbor, 1975. These forecasts were based on data up to 1970, more recent data since then would require a slight revision downward from the "mid-range" forecast adopted by the Commission.

[17] Science Council of Canada, *Conserver Society Notes* 1(3), February-March, 1976, and "Toward a Conserver Society: A Statement of Concern", in *Science Forum*, 9(3), June 1976; and Montreal/McGill Interuniversity Futures Studies Group (GAMMA), *Tentative Blueprints for a Conserver Society in Canada.* Report on Stage 1. Montreal, July 1975.

[18] Drawing upon approaches suggested for example, by Gary Gappert, "The development of a pattern model for social forecasting", *Futures*, August 1973, pp. 367-382; and Russel Rhyne, "Technological Forecasting within Alternative Whole Futures Projections", *Technological Forecasting and Social Change* 6, 1974, 133-162.

[19] This approach was used by the Science Council of Canada in its study of policy and decision-making for northern development. See: Robert F. Keith, *et al. Northern Development and Technology Assessment Systems*, A study of petroleum development programs in the Mackenzie Delta-Beaufort Sea Region and the Arctic Islands. Science Council of Canada, Background Study No. 34, January 1976.

[20] See for example, Michael Chevalier and Thomas Burns, "A field concept in public management". Working Paper, March 1975 (York University).

[21] The recent formation of the Canadian Association for Future Studies represents a timely start in organizing a network of individuals across Canada interested in such questions. See the "charter issue" of *Futures Canada*, 1(1) 1976.

[22] Concerning air pollution questions, the relevant "regions" in terms of airsheds may need to be highlighted more in the future in view of the discovery of the relative importance of atmospheric loadings as a major source of contamination of water quality in the Lakes, and preliminary evidence to suggest that sources may be located as far away as St. Louis, Missouri. See: *Upper Lakes Reference Group, Reference Questions Summary and Recommendations*, and observations presented at the Fourth Annual Meeting of the IJC on the Great Lakes Water Quality Agreement, Windsor, July 1976.

[23] Leman, Alexander B. and Ingrid A. Leman (Eds.) *Great Lakes Megalopolis: From civilization to ecumenization*, Ministry of State for Urban Affairs, Ottawa, 1976, and also Alexander B. Leman, "Great Lakes Megalopolis – Canada", *Ekistics* 243: 114-119, February 1976.

[24] Gottman, Jean, "Megalopolitan systems around the world", *loc. cit.*

[25] Robert, Jacques, "Prospective study on physical planning and the environment in the megalopolis in formation in Northwest Europe", *Urban Ecology* 1:331-411, 1976.

[26] Doxiadis, C.A., *Organizing Efforts Towards a Desirable and Feasible Great Lakes Megalopolis.* Great Lakes Megalopolitan Symposium, Canada. World Society for Ekistics, Working Group – Canada, Toronto, March 1975. Summarized also in Leman and Leman (Eds.), *op. cit.*, pp. 77-79.

[27] Dorney, R.S., and S. G. Rich, "Urban design in the context of achieving environmental quality through ecosystem analysis". *Contact: Journal of Urban and Environmental Affairs*, 8(2): 28-48, May 1976.

[28] Government of Ontario, *Toronto-Centred Region Program Statement*, Design for Development, March 1976; *Central Ontario Lakeshore Urban Complex*, COLUC Task Force Report, December 1974; *Ontario's Future: Trends and Options*, March 1976; Ministry of Treasury, Economics and Inter-governmental Affairs. Also: *A Strategy for Ontario Farmland.* Statement by the Ministry of Agriculture and Food, March 1976.

[29] Slater, Robert W., "Development and Implementation of Federal Environmental Protection Requirements in Ontario", in: Ontario Ministry of the Environment. *22nd Ontario Industrial Wastes* Conference, Toronto, June 1975, pp. 198-214.

[30] Emery, F.E., and E. L. Trist, *Towards a Social Ecology: Contextual appreciations of the future in the present* (London: Plenum Publishing Co. Ltd., 1973).

[31] Senate of Canada. Proceedings of the Standing Senate Committee on Foreign Affairs. Proceedings respecting: Canadian Relations with the United States, Issue No. 10 (March 18, 1975) and Report of Committee. *Canada-United States Relations, Volume 1 – The Institutional Framework for the Relationship*, December 1975, pp. 51-55.

[32] Canada-United States University Seminar, 1971-1972. *A Proposal for Improving the Management of the Great Lakes of the United States and Canada.* Water Resources and Marine Sciences Center, Cornell University, January 1973, and also; Lyle E. Craine. *Final Report on Institutional Arrangements for the Great Lakes.* Report to the Great Lakes Basin Commission, Ann Arbor, March 1972.

[33] Great Lakes Tomorrow. *A Year-End Report – 1975.* Lake Michigan Federation, Chicago, January 1976.

[34] Ludwigson, John O., *Two Nations, One Lake – Science in Support of Great Lakes Management.* Objectives and activities of the International Field Year for the Great Lakes, 1965-1973. Canadian National Committee for the International Hydrologic Decade, Ottawa, May 1974, 145 pp.

[35] Loftus, K. H., and H. A. Regier (Eds.) *Great Lakes Fishery Commission*; Technical Report Series, Nos. 19-23, 1973.

Is Lake Diefenbaker Justifying Its Planners?

J. HOWARD RICHARDS

The first stage of the South Saskatchewan River Development Project was completed in 1967. Its large reservoir, 224 km long and named Lake Diefenbaker, then reached its full level and its water became available for irrigation, power development, and urban water supply.

For many individuals, municipalities, and government agencies, the completion of this state of the project brought to realization a dream which had extended over several decades. The reservoir, through the application of irrigation water, was expected to bring stability to the rural economy of this part of south-central Saskatchewan, where drought and other hazards confronted the dryland farmers. While according to a 1949 government report the lake was planned to provide irrigation eventually for some 200 000 ha, a 1962 estimate more than halved this.

However in 1973, when only 6500 ha were under irrigation, the provincial government informed householders in the area that, because of high costs, further construction for irrigation would be deferred. At the same time, it should be noted that by the middle 1960's many dryland farmers whose lands lay within the designated irrigation blocks did not welcome irrigation, and some militantly opposed it. Further, various post-reservoir studies cast doubts on future success for either cattle-fodder combinations or as a large specialty crop-producing area. The vision of the pro-irrigationists thus was dissipated in the harsh light of economic reality.

Both branches of the Saskatchewan River, the North and the South Saskatchewan, rise in the eastern ranges of the Rocky Mountains. Diverging until they are 480 km apart, they then unite east of Prince Albert, Saskatchewan, and finally empty in the north end of Lake Winnipeg, in Manitoba, whence the waters travel via the Nelson River to Hudson Bay. From the head of the Bow River in Alberta to the mouth of the Saskatchewan is 1940 km.

The only sizeable tributary entering the South Saskatchewan in Saskatchewan is the Red Deer River, which joins near the Alberta boundary.

The first major project to use the Saskatchewan River and its feeder streams for irrigation purposes was begun in southern Alberta in 1901; by the 1950's this had developed into the successful irrigation of 247 000 ha.

The South Saskatchewan Project, first recommended in a 1947 government study, was enormously expensive from the outset. It began in 1958 as a shared-cost program in which the province's share of reservoir construction costs was supposed to be 25 percent, but not to exceed $25 million. In fact, the federal share of all costs exceeded $100 million and Saskatchewan's share (including power station, irrigation works, irrigation development, recreation development, municipal water supply, and other benefits) rose to $75 million.

The Lake Diefenbaker reservoir is 440 km² in area and has a 800 km long shoreline. The main Gardiner Dam is of rolled earth, 64 m high at the river bed and 4.8 km long across the crest. Blocking the Qu'Appelle Arm is a smaller dam about 3.2 km long and 27 m high. The reservoir has a maximum capacity of nearly 9880×10^6 m³, of which about 3335×10^6 m³ constitute live storage, i.e., water available for release. Thus in April Lake Diefenbaker has a low level at 546 m above sea level but, after recharge from snowmelt and rainfall on the eastern slopes and foothills of the Rocky Mountains, achieves a "full supply" level 11 m higher in October. During the winter months water is drawn from the reservoir to drive the turbines in the associated hydroelectric power installation.

Near its "elbow", the South Saskatchewan River is only separated from the Qu'Appelle River system by a low divide. Thus, by

● HOWARD RICHARDS is Professor and Head, Department of Geography, University of Saskatchewan. This article is reprinted from the *Canadian Geographical Journal*, Vol. 91, No. 6, 1975, pp. 22-31 with permission.

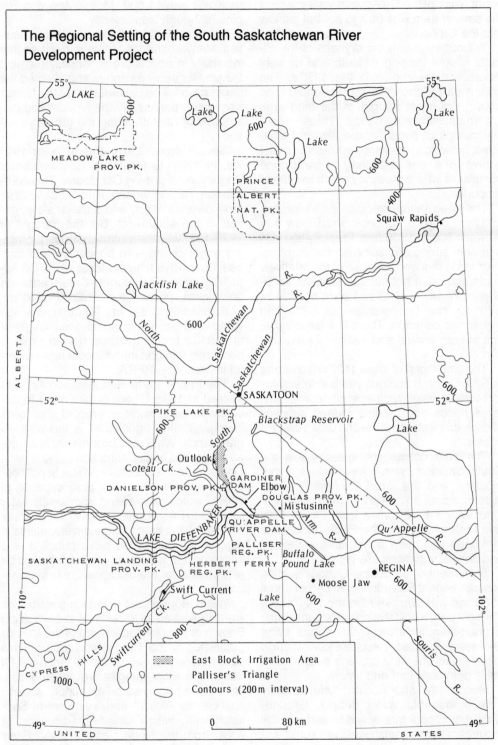

The Regional Setting of the South Saskatchewan River Development Project

Legend:
- East Block Irrigation Area
- Palliser's Triangle
- Contours (200m interval)

0   80 km

Figure 1

building a main dam across the South Saskatchewan it was possible also to flood the upper reach of the Qu'Appelle Valley, where the smaller dam was built to control outflow into the Qu'Appelle.

Without irrigation, the dryness of the climate makes farming difficult, and drought devastated the region in the 1930's. The area is part of the interior grasslands, lying toward the moist margin of the land area known as Palliser's Triangle. Trees – usually clumps of aspen – are few and scattered, growing in sheltered coulees, in moist depressions, and especially in sand dune complexes where they may reach woodland proportions.

The mean annual precipitation is less than 360 mm, while at Outlook and Elbow it is between 300 and 330 mm. Precipitation falls mainly in spring and summer, but its effectiveness is reduced because of the relatively warm and often hot summers. Mean temperature in July is 19°C and short hot spells, with daytime temperatures of 32°C and higher, are common. There is a large variation in both annual and seasonal precipitation.

The small total of snow (102 cm average at Swift Current) and rain yields little surplus for ground-water recharge or to maintain lakes, ponds, and streams. Indeed, supplies of drinkable water for rural and urban use are scarce.

The only permanent streams are the South Saskatchewan River, with a minor tributary, and the small Qu'Appelle River system. About 90 percent of the South Saskatchewan's flow comes from the foothills and eastern slopes of the Rocky Mountains in Alberta and, here again, great seasonal and annual variations are typical.

Settlement of the Elbow-Outlook section of south-central Saskatchewan began about 1905 and was completed in the 1920's. For the most part, a dry farming system stressing a grain-summerfallow rotation has been necessary. Even so, yields are low except in those years when the growing season rainfall is both sufficient and timely.

Moose Jaw and Regina suffered early from inadequate water supply. Groundwater was of poor quality and limited quantity "no matter to what depth wells are sunk" and local streams, even when dammed, were too small and undependable. Eventually, by 1955, water from the South Saskatchewan River was pumped to Buffalo Pound Lake in the Qu'Appelle Valley, and this enlarged reservoir served both Moose Jaw and Regina, although insufficiently.

Farm settlers also experienced difficulties. Many abandoned their holdings so that instability in agriculture continually plagued the smaller service centres and affected the larger towns. The big problem for the farmer was water, particularly the amount and distribution of rainfall during the growing season.

Severe drought in the years after 1917 led, in 1920, to the Royal Commission of Enquiry into Farming Conditions. This commission recommended land classification, farm diversification, and small-scale irrigation in the southwest. But the drought of 1929-1937 was the most disastrous, prompting major changes in population and land use. During this time stretches of the Qu'Appelle River and Swiftcurrent Creek ceased to flow, lakes, ponds, and swamps dried up, and soil drifting set in. Together with the world-wide economic slump, the drought of the 1930's brought Saskatchewan close to economic ruin and introduced a new powerful institution – PFRA.

The Prairie Farm Rehabilitation Act was passed in 1935. There is little need to describe the administration's record of success in rehabilitating agriculture in the western grasslands. What may be stressed is its role as a powerful engineering arm of the federal Department of Agriculture. Water supply became a proper and major preoccupation of PFRA, which progressed from small water projects to irrigation schemes in the Cypress Hills area, and then to large developments in Alberta. At that time a basic principle was accepted: the federal government should be responsible for major capital costs in irrigation.

Water storage and diversion possibilities were described in reports of the Irrigation Branch of the old Department of the Interior (abolished in the 1930's) and the branch's successor, the Reclamation Service. Of these, the most imaginative scheme conceived of a "stock watering" supply, serving east-central Alberta and west-central Saskatchewan, which called for diversion of water from the North Saskatchewan River through the Clearwater and Red Deer rivers in Alberta. This project was re-examined by PFRA but was discarded in favour of use of the South Saskatchewan River and of irrigation in Saskatchewan.

As a federal development agency employing special engineering skills and operating in a province which had experienced social upheaval, PFRA had the full support of Saskatchewan. Its 1947 interim report was well received. Indicating a cost exceeding $66 million for a main dam at Coteau Creek and for auxiliary structures, the report stressed irrigation of some 200 000 ha, hydroelectric power development, regulation of river flow, and diversion of water into the Qu'Appelle Valley.

By the end of World War II, Canada was in a position to consider major measures for regional development. On July 25, 1958, Prime Minister John Diefenbaker announced the South Saskatchewan River Project in the House of Commons. It was to be the third major project in a federal program directed toward satisfying regional needs and aspirations.

A royal commission, reporting on October 29, 1952, had calculated that the project would cost $250 million. The price was deemed too high and the project was put aside by the federal government. But in 1958, with a change of government led by a Prime Minister from Saskatchewan, a campaign promise was redeemed: the South Saskatchewan River Project would proceed.

The findings of the 1952 report were discounted collectively by the PFRA, the Saskatchewan government, and the new federal government of Progressive Conservatives (whose Minister of Agriculture, like Prime Minister Diefenbaker, was from Saskatchewan). Approval was based upon few new arguments, but corresponded with a period of optimism in the province. Saskatchewan, seemingly, was entering an era in which development of minerals and, possibly, secondary industry would balance the agricultural economy. It may be that such a change could have occurred, although against it were locational and institutional constraints (including transportation and banking), the same frustrations which currently give force to discussions of "western alienation".

It is hard to determine, let alone allocate, dollar values-to-costs-of or benefits-derived-from either the total project or its several parts, mainly because plans were modified and many agencies were involved. It is possible, however, to examine the gross claims of benefits made for the project and to assess broadly their success.

In 1949 PFRA indicated a potential of 194 000 ha of irrigable land (9700 ha in the Qu'Appelle Valley) but in 1962 reduced this estimate to 81 000 ha. The 1958 agreement required the province to have ready facilities for irrigating 20 250 ha one year after the reservoir was filled, i.e., by the spring of 1967 or 1968. Yet by 1973 only 6500 ha were irrigated, although construction work was completed to serve a total of 16 200 ha, all in the block of land located east of the South Saskatchewan River, surrounding Outlook and Broderick.

Eighteen million dollars has been spent by the province in developing this eastern block and another $150 to $300 per hectare has been invested by farmers. Costs of developing 6885 ha on the west side of the river were estimated at about $270 000 per 160 ha unit, the government paying $243 000 of this. These costs are thought excessive, and so development is to be confined to the east block where land now capable of being irrigated is more than adequate "to meet the limited market for special crops". (Note that $2.6 million had been spent on the west side by August, 1971.)

Most farms are 195 to 260 ha in size, of which 120 or 160 ha may be irrigable and the remainder used for dry farming. Government policy is to expropriate lands from those who choose not to irrigate and then make them available for settlement under generous lease-option agreements. Alfalfa is the most important crop, much of it grown for a cooperative cubing-plant which baled and sold 6820 t for animal feed in 1973. Another firm grew 200 ha of potatoes and has a cleaning and storage plant. Other crops include soft wheat, fababeans (horsebeans), silage corn, grass seed, and some root vegetables; in all cases average yields were excellent.

Irrigation has been unsuccessful, for some of the causes foreseen by the 1952 royal commission which expressed concern about adequate markets for specialty crops, high capital costs of irrigation, and possible difficulties in persuading dryland farmers to accept irrigation. In March, 1973, the Provincial Minister of Agriculture stated that the Saskatchewan government had large responsibilities in aspects of agriculture other than irrigation; in the eastern block, the primary obligations were to the irrigators developing it and to the taxpayers who bear much of the cost; and development of the west block must be "tied to future development of processing facilities" (no time scale was suggested for this).

It was generally believed by the federal, provincial, and local governments, together with most local organizations, that the reservoir would increase the number of farms and stabilize or increase the rural and village populations. These expectations have not been realized. From 1961 to 1971 the number of farms, together with farm, rural, and most urban populations, has declined (Figure 2).

A few small service communities grew promisingly during the main stages of reservoir construction, but declined as the work force left. Many now have smaller populations than they had in 1956; only Outlook has grown considerably, due to its role as service centre for a fairly large area and as headquarters for irrigation operation. The combined population of fifteen service centres (including Outlook) rose from 3617 in 1956 to 4759 in 1961, and escalated to 4916 in 1966. However, by 1971 the total had dropped to 4202, despite the fact that Outlook's population had risen from 885 in 1956 to 1790 in 1971.

Rural and village depopulation is typical of much of Saskatchewan; the reservoir plainly has not changed this characteristic in its vicinity, and irrigation has yet to prove its wonders.

Besides its irrigation function, the new reservoir was designed to provide hydroelectric power for pumping water for irrigation and to supply commercial power during the winter season. The Coteau Creek hydroelectric station, located below Gardiner Dam, was built at a cost of $40 million and started commercial production in December, 1968. Five penstocks were built but turbines were installed in only three of these, the two others were to be put in at a "later stage" (which now seems unlikely). The plant is used essentially as a winter peaking station, releasing water stored during spring and summer. Its capacity is 187 000 kW and in 1973 and 1974, respectively, it represented 11 percent and 16 percent of all electrical power generated by Saskatchewan Power Corporation (a Crown corporation) stations, both thermal and hydro.

Other power benefits were to be derived from the reservoir. Controlled releases of water during winter were expected to help increase generation at the Squaw Rapids hydro station (1963), which lies far downstream on the Saskatchewan River below the confluence of the North and South Saskatchewan rivers. This occurred, power production having been increased and seasonal differences in generating capacity reduced. Last, it was conceived that, between 1970 and 1990, the South Saskatchewan and Saskatchewan rivers would be developed into a ribbon of stepped lakes separated by six or seven dams between the Gardiner and Squaw Rapids dams; this was expected to provide ten times the power developed at the Gardiner Dam. Development has not occurred but the deferment of and controversy concerning hydroelectric development on the Churchill River has induced statements concerning a power dam at Nipawin, downstream on the Saskatchewan River. In this regard it is interesting to note the fight to preserve the physical and cultural environments on a major river on the Pre-Cambrian Shield and to contrast this with the apparent apathy to conserve one of the few beautiful river systems of the prairielands.

Recreational use of the reservoir was an afterthought. It caught attention after the 1958 agreement and was quickly assessed as the "big bonus of the project". A consulting geographer, appointed in 1959, identified the basic characteristics and problems inherent in the reservoir's use for recreation. Among the problems foreseen were that a new set of erosional processes would come into play as the lake was formed, and that not all sections would be suitable for development.

Further, because of the marked change in levels due to water management for power, it was realized that special care would be necessary in selecting development areas and in providing access, e.g., via boat ramps that could serve very different lake levels in spring and fall.

The benefits seemed to be obvious. The large reservoir was located where surface water was "absent", the shoreline and lake would provide for many recreation wants, and the location was excellent with respect to the larger centres (including Swift Current, Moose Jaw, Regina, and Saskatoon).

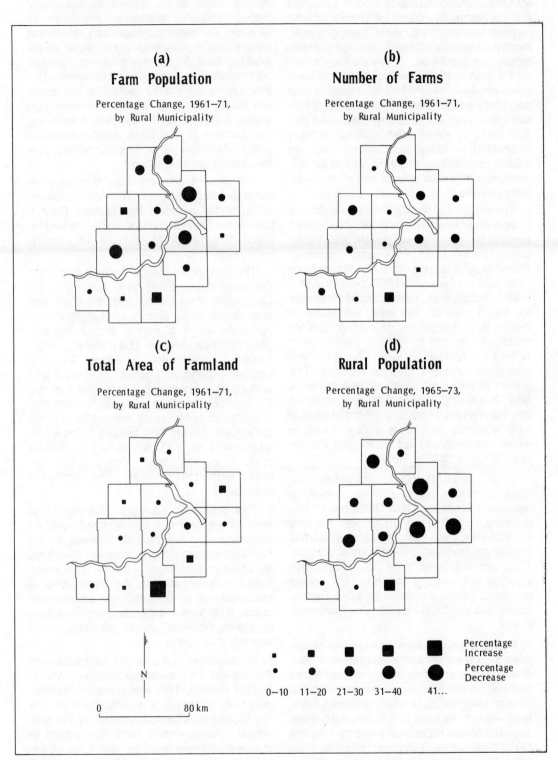

(a)

Farm Population

Percentage Change, 1961–71,
by Rural Municipality

(b)

Number of Farms

Percentage Change, 1961–71,
by Rural Municipality

(c)

Total Area of Farmland

Percentage Change, 1961–71,
by Rural Municipality

(d)

Rural Population

Percentage Change, 1965–73,
by Rural Municipality

N

0   80 km

Percentage
Increase

Percentage
Decrease

0–10   11–20   21–30   31–40   41...

Figure 2

Following other research, a master plan for Lake Diefenbaker, prepared in 1966, set out development plans for provincial and regional parks, institutional campgrounds, recreation sites and areas, cottage subdivisions, campgrounds, a wildlife sanctuary, and a road system. Two million trees have been planted (nearly half of these in one park) for esthetic and other reasons. Game fish have been successfully introduced into the lake; a small commercial fishery (whitefish) is likely to develop; and the wildlife populations of game and other animals, together with upland and water birds, have increased.

The success of recreational use is difficult to appraise. Relatively few lots have been occupied in the various private and public cottage subdivisions. For example, south of Elbow only a dozen or so cottages have been built on the plotted 250 lots of the Mistusinne subdivision; here, despite nearness to, and a view of, the lake, the essential feeling is of a stark prairie setting, and for most people, this is not the environment sought for recreation living. As trees are grown the rawness may be reduced. The impact of site, particularly of tree cover, is seen in the area adjacent to, and south of, this subdivision. Here is a delightful area of well wooded, extensive dunes, much of which has been allotted to Douglas Provincial Park and Wildlife Refuge.

In 1973 some 50 000 vehicles entered the three provincial parks and two provincial recreation areas on Lake Diefenbaker; this is nearly 18 000 less than entered the little 4.7 km² Pike Lake Park (focused on a small shallow oxbow lake), 30 km from Saskatoon. Total camping permits, at 3303, were also less than at Pike Lake. Indeed, of the sixteen provincial parks, the three on Lake Diefenbaker rank twelfth, thirteenth, and fourteenth in use.

Perhaps seven years are too few to assess the lake's value for recreation. It is a little too far and has too few significant facilities to attract day-users from Saskatoon and Regina, but a new road from Saskatoon will cut distance to 105 km. Also there are established recreation patterns − to the Qu'Appelle lakes, Cypress Hills, Jackfish Lake, Meadow Lake Provincial Park, Prince Albert National Park, and the Canadian Shield − that may be difficult to alter.

It is possible that too much planning and money have been applied to too many places, and that planners have failed to develop the facilities necessary to attract people and activities to two or three select areas. Certainly, tree-planting should accompany the building of subdivisions. The bald prairie cannot compete with the northern mixed woods or established recreation areas, although below the prairie level, on the surface of the clear deep recreation water, the view is of slopes, gullies, low headlands, and beaches.

The project also provides a system of water distribution, supply, and control. Since 1955, water pumped into Buffalo Pound Lake in the Qu'Appelle system has helped to supply Regina and Moose Jaw, as already mentioned.

The building of the Qu'Appelle Arm Dam permitted direct diversion of water into the Qu'Appelle, thus saving pumping costs and ensuring a large supply to Buffalo Pound Resevoir which is part of the Qu'Appelle glacial spillway system. This water is used by a solution potash mine near Moose Jaw, and use will expand with growth in Regina and with other demands. It is intended that the Qu'Appelle diversion will supplement the irregular, and often small, natural flows of the Qu'Appelle River, thus helping to maintain lake levels of long-established recreation areas in the Qu'Appelle Valley. It will also help to extend the small area now irrigated in this valley.

Problems have emerged. The canal and river diversion from the reservoir are inadequate to meet projected needs in the Qu'Appelle basin. There is need, therefore, for canal improvement. Further, if diversion from the Qu'Appelle into the Souris River is contemplated (one doubts this at present), competition for a scarce resource may cause problems between power generation and easterly diversions.

A latecomer to the South Saskatchewan Project was the Saskatoon-Southeast Water Supply Project. This was another multiple-purpose scheme, a counterpart of the Qu'Appelle diversion, approved by the provincial government in 1965. Legislation in that year affirmed need for, and rights of, the province to control the use of surface waters.

Towns were demanding adequate water, and several new potash mines needed large

Figure 3

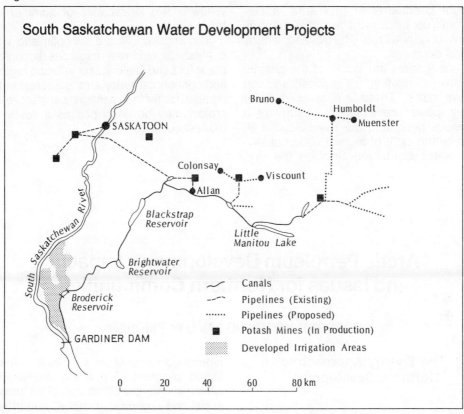

South Saskatchewan Water Development Projects

Legend:

~ Canals
--- Pipelines (Existing)
...... Pipelines (Proposed)
■ Potash Mines (In Production)
▨ Developed Irrigation Areas

0   20   40   60   80 km

amounts in their operations. A system of canals, reservoirs, and pipelines carries water northeast from Lake Diefenbaker and makes water available for urban, industrial, recreation, small irrigation, and wetland wildlife uses. Not all mines use the supply, while some expected new mines and demands did not materialize, so that the original scheme remains incomplete. There is little doubt that this has been a costly project, but at the time, both the plan and legislation were necessary to protect and rationalize the use of scarce water resources in southern Saskatchewan (Figure 3).

Much water is lost by evaporation from the open canals and reservoirs, but pipelines through sparsely populated areas are too expensive. Blackstrap Reservoir, designed to provide Saskatoon with water for recreation, fails to compete with the smaller Pike Lake because the warmth, shallowness, and high organic content of the water induce extensive algae bloom, thus making it less attractive. But, because of ease of access and the addition of winter facilities, its use is slowly increasing.

Regulation of river flow below the dam was another presumed benefit of the South Saskatchewan Project. This has been achieved in part. Downstream urban/industrial users, such as in Saskatoon, no longer experience huge variations in flood flow peaks (e.g., 3930 m³/s in 1953 compared with 283 m³/s in 1941).

The downstream size of the river is changed so that winter flows are slightly higher than those of summer, and seasonal differences are reduced; control has reduced downstream flood hazards but has not eliminated them.

Saskatoon derives another benefit in that the new river regime has induced channel changes which should encourage additional recreational use of the valley.

A stored water supply for this south-central region was necessary, but whether this location and project are best in terms of total regional water needs is difficult to establish. While it was developed separately rather than integrated into a Prairie water scheme, the lake obviously affects planning for the use of water in other parts of the west. The reservoir is there because of recurring drought conditions in this area, socio-economic upheaval in the province, the aspirations and power of PFRA, and federal political decisions.

If one ignores the failure of the primary objective — irrigation — it is possible to see positive gains. These are assured water supply, power development, control of a capricious river, and the psychological lift given by the sight of a great ribbon of sky-blue water stretching through the dry, yellow-brown landscapes of summer and fall.

The irrigationists are still optimistic, while manipulators of water resources recognize a place for Lake Diefenbaker in future regional and continental water arrangements. In this context, rather than a magnificent failure, the project may be described as a costly but modest success.

Arctic Petroleum Development: Impacts and Issues for Northern Communities

ROBERT F. KEITH AND DAVID W. FISCHER

The Energy Approach to Northern Development

Energy issues are among the most important faced by Canada today. International, national, and regional pressures have raised the problems of energy supply, demand, and deliverability to unprecedented levels of attention and action. The economic, social, political, and environmental issues which surround energy questions have made analysis and decision-making complex in the extreme.

In Canada, as elsewhere, the quest for secure supplies of petroleum has focussed attention upon the frontier regions. The Canadian Arctic has become one of the country's most active exploration regions, as the search for oil and gas in the Mackenzie-Beaufort Sea and Arctic islands regions progresses. While not the sole solution to both short and longer term energy needs, Arctic petroleum is seen as an essential component of Canadian petroleum policy and programs. Moreover, Arctic petroleum development is viewed, particularly by the federal government, as the basis of a sustained program of northern development which will lift the North out of its present social and economic "malaise" and provide the basis for a strong northern economy.[1]

In 1972, the federal government set forth its objectives for development of the Yukon and Northwest Territories.[2] While including a broad range of development objectives, the emphasis was upon social factors and the compatibility of northern aspirations and lifestyles with the overall development thrust. A number of critics[3] have taken issue with the federal government, inasmuch as northern development as it is practiced does not appear to reflect this priority. Rather, the emphasis is upon resource extraction on a very large scale, with social and environmental adjustments to be made in the wake of fundamental resource development decisions. The federal government's position can be understood in terms of a number of assumptions made by it about the North. They are:[4]

1. Large quantities of nonrenewable natural resources (oil, gas and minerals) exist in the Northern Territories.

● ROBERT KEITH and DAVID FISCHER are Associate Professors, Department of Man-Environment Studies, University of Waterloo. The article has been updated and is reprinted from a special issue (Arctic Land Use Issues) of *Contact*, Vol 8, No. 4 (Nov. 1976), pp. 118-127, Journal of Urban and Environmental Affairs, Faculty of Environmental Studies, University of Waterloo, with permissions.

2. Development of these resources must necessarily be large scale undertakings.

3. Development of these resources consists of a sequence of interrelated activities and spin-off developments.

4. Development will take place over a relatively long period of time.

5. Socio-economic benefits will flow to all those involved but especially to northerners.

6. These benefits will be the basis of a rising standard of living and better quality of life and opportunity for northern residents.

It has been upon the basis of these assumptions that the federal government has given strong encouragement to and has actively participated in northern petroleum development programs. From its 1959 decision to open the northern territories to petroleum exploration to the present, the Department of Indian and Northern Affairs, the Task Force on Northern Development, the Cabinet, and other government agencies have vigorously pursued oil and gas programs. What seems apparent now is that northern development, in the broadest sense, has been subordinated by the sense of urgency with which petroluem is being sought. Moreover, it can be argued that the present petroleum development programs may not be the basis of northern development. Rather, they represent only growth in a relatively narrow perspective rather than expansion through diversification of economic activity.[5] The apparent fixation of government and industry upon large scale energy-based developments appears to have distracted them from smaller scale, locally-based and renewable resource-oriented economic activities. [6]

Northern Development Issues

From discussions with many of the "actors"[7] in northern development, each with varying and sometimes conflicting objectives, information needs, and decision processes, five broad issue categories have been identified. These are:

1. Technological
2. Environmental
3. Economic
4. Social
5. Political

Table 1

NORTHERN DEVELOPMENT ISSUES

TECHNOLOGICAL:
- Drilling in the deep sea
- Transportation alternatives
- Oil clean-up under ice
- Alternative energy sources

- Pipelines and sea bottom scour
- Liquefaction in Arctic
- Laying pipeline through ice
- Icebreaking ship technology

ENVIRONMENTAL:
- Impacts on aquatic and terrestrial regimes
- Impacts on fish and wildlife
- Oil spills
- Lack of baseline information

- Waste disposal
- Impacts of infrastructures
- Pipelines in permafrost
- Artificial islands and marine ecosystems

ECONOMIC:
- Lack of local and regional benefits
- Impacts on prices, interest rates, exchange rates
- Impacts on financial capabilities
- Consumption pattern

- Incentive and controls
- Impacts on industry, labour, markets
- Industrial strategy
- Investment pattern (capital flows)

SOCIAL:
- Lifestyles
- International equity
- Aboriginal rights
- Native employment

- Transient populations
- Education priorities
- Health and welfare
- Community disruption

POLITICAL:
- Native peoples' institutions
- Foreign ownership and participation
- International realignment
- Industry-government relationships

- Petroleum exports
- Territorial sovereignty
- Regulatory uncertainty
- Responsible government in Territories

Table 1 lists selected examples of the issues surrounding petroleum development in the North. These issues have occasioned extensive research programs by governments, industry, and northern native groups. While some research on most of these issues has been undertaken, the feeling as late as 1974 was that social and community impacts had not received the attention that other issues have had. The inherent difficulties of social measurement, the absence of baseline social data in the North, and an early emphasis on environmental impact assessments have contributed to this deficiency. In recognition of local social concerns, the Mackenzie Valley Pipeline Inquiry (Berger Commission) heard considerable testimony from likely affected communities in the Yukon and Northwest Territories. In his now precedent-setting report, [8] Berger paid particular attention to social and cultural issues. His recommendations for a ten year moratorium on pipeline development in the Mackenzie Valley and the setting aside of environmentally sensitive areas for preservation in perpetuity, appear to reflect deep concerns for the future of northern natives' lifestyles, culture, institutions, and communities.

Impacts on Local Communities

Native Land Claims

Among northern native peoples, the single most important concern is the "land claim" issue, one that, in their view, must be settled prior to the granting of a right-of-way for a pipeline and subsequent pipeline construction. Pipelines and other major development projects appear to have hastened the day when the claims of Indian and Inuit groups must be settled. Reorganization and consensus-building among native groups have been among the first effects of petroleum development. As Usher notes:

> With oil exploration something new has happened. The outside world needs the North, or at least its oil and gas. . . . If native people have nothing to offer the oil companies, how can they bargain with them . . . . What they have is their land. That is what the native people in Alaska learned. They organized and they fought and they finally got a settlement for their land. Native people in Canada have been hearing that and wondering if they couldn't get something like it themselves.[9]

In taking the widest possible view of his terms of reference, Justice Berger indicated that the land claims issue had to be considered within the context of his hearings, inasmuch as the question of land claims represented a significant factor in consideration of the terms and conditions of pipeline development.

The land claims issue is of particular significance to northern communities. Though land settlements have yet to be reached, likelihood of both land and money (some perhaps in the form of royalties) seems great in view of precedents in the James Bay region of Quebec and the Alaskan agreement mentioned above. Whatever the settlement, significant capital (land and/or money) will accrue to native groups. Again if precedent is followed, the capital will be vested in native peoples' "corporations" designated perhaps by communities, regions, tribes, or some combination. In any event the acquisition of capital will permit a degree of local control in decision-making never before accorded native peoples.

Boom-Bust Cycles

Characteristic of much of the development process in the North has been periods of economic upswing followed by significant downturns in the economy. Technologically intensive programs are particularly susceptible to such patterns. Usher discusses the problem.

> Boom and bust is the white man's way of doing things. That's how it started with the whalers and that's how it is today with the oil companies. When the DEW line was built lots of people got jobs. Then the jobs were gone. Now there is oil exploration and maybe a pipeline. What happens after that?[10]

The Hon. Justice William G. Morrow cites the DEW line as one of the developments which "brought about the first movement of natives from their traditional homesites to the new developments".[11] Though one might argue that the fur trade had a similar effect many years before, military activity in the North has undeniably altered northern communities and native peoples' lifestyles. Wood cites unemployment, social dislocation, and increased welfare dependence as characteristics of the "bust" which followed the DEW line "boom".[12]

244

While it is probably fair to say that greater awareness of and concern for the social costs of development are now evident, uncertainty as to (1) petroleum reserves, (2) the availability of capital, (3) technological solutions to engineering problems, (4) the land claims issue, (5) potential environmental disruptions, and (6) corporate rivalries may produce an uneven pace of development, making further boom-bust cycles a real possibility.

Need for Local Economic Activity

The need for locally-based economic activity is seen by many, particularly northerners, to have received too little attention in comparison with large scale petroleum programs. From discussions with some northerners, both native and non-native, Keith, et al[13] identified a growing concern for smaller scale and renewable resource-based economic opportunities in the North. Logging and local saw mill operations in the Mackenzie Valley, the continuing harvest of furbearers, increased tourism, and local construction and house prefabrication are but some of the opportunities mentioned. Such opportunities are seen to capitalize on existing skills and knowledge, and to be reasonably compatible with the traditional native lifestyles and thus with their aspirations. Northern businessmen too have expressed concern over petroleum-related development. Though generally favourable to exploration and pipeline development, the pace and scale of development may preclude effective participation by some in various related economic opportunities.

Modernization of Communities

In an attempt to upgrade social services in the North, schools and educational programs, nursing stations and hospitals, electrification, housing, and welfare have been implanted. While these southern amenities have not been without their benefits, certain social costs have emerged. As more and more people congregate in and migrate to such communities, existing services become strained. Where native peoples constitute a significant number of the residents and where these people still pursue traditional

economic activities (hunting, trapping, fishing), the pressure on such resources is of concern. The educational system, which separates families and which consists of little native culture, has been identified by native groups as disfunctional.[14] As Justice Morrow points out, mobility among native peoples has traditionally been as groups not as individuals.[15] Social relationships and interdependencies have been transformed, though full participation in the wage economy has not been possible.

Company Towns

The problems of company towns, often based upon a nonrenewable resource, are not new to the North. As the fortunes of the company fluctuate so does the well-being of the community. Rankin Inlet, where a nickel mine operated for a brief period of time, is one such example. With the mine in operation the settlement grew, only to be faced with serious social and economic difficulties upon the closing of operations.[16] Reliance upon petroleum and mineral resources as the basis for an expanding northern economy may increase the incidence of such difficulties. Uncertain mineral policies and even world pricing systems make mining communities particularly vulnerable in this regard.

As the pace and scale of petroleum development increase, the shift of the economic base, both in the short and longer terms, for northern communities becomes more apparent and problematical. In existing communities more and more services and facilities are being utilized by petroleum development processes and personnel. Housing, power, water, schools, hotels, airway facilities, and local businesses feel the strain of growing demand, with little certainty as to what even the short term future will bring. In recognition of some of the social and economic disruptions brought on by the presence of large numbers of transient workers (especially during pipeline construction), present plans call for self-sufficient work camps, "isolated" from existing communities. While such a strategy may minimize various social impacts, economic "leakage" (wages, supplies, services) to the south may be increased.[17] Thus, while serving certain social objectives, the "company camp" strategy seems, in part, to run counter to the expectation that petroleum development will be the basis of an expanding northern economy.

Northern Development: A Lack of Overview

An assessment of the northern development goals and programs has been neither timely nor balanced in approach. Petroleum exploration and pipeline planning have proceeded since 1960, but only in the past few years has a concerted effort been made to assess the impacts. Following the introduction of the federal government's pipeline guidelines in 1970, extensive environmental, economic, social, and technical research programs were launched. The resulting information base, while apparently meeting the conditions of the guidelines, is incomplete. As one critic noted:

> The Mackenzie Valley Gas Pipeline has provoked a sheaf of engineering and environmental studies, economic analyses and public policy reviews on a scale that is claimed to surpass anything in our history . . . For the most part their purpose is not public education. Rather they are sponsored by investors, environmentalists and tax collectors attempting to overcome our ignorance of the complex technical, ecological, sociological, and economic consequences peculiar to northern resource development. As a result, they are fragmentary both in their coverage and in their point of view.[18]

The fragmentary and disjointed character of the assessments, the relative inattention to alternative energy sources and energy transportation technologies, and the lack of emphasis on smaller scale community and regional economic activities suggest basic problems. Dosman[19] charts the "drifting" course of northern policy and programs, and identifies the lack of a truly "national interest" in the North. Keith, et al[20] report how northerners, environmentalists, and industry all face major uncertainties. Several factors contribute to these conditions.

1. The lack of an effective overall policy mechanism for northern development is seen as an important limitation of the present institutional structure. No institution has a clear responsibility to develop, guide, and apply northern policies which take into account, in a comprehensive and balanced manner, the diverse and divided jurisdictions. The subordination of northern interests to federal perceptions of a national need attest to this fact.

2. Unresponsiveness to change, especially with respect to the pace and scale of northern development, at a time when flexibility to changing conditions is needed, mitigates against the assessment of a more varied set of program alternatives and policy options.

3. The lack of coordinated information systems has led to multiple and overlapping research programs, with timely access to information made difficult for those not directly involved in much of the research. The result is as suggested above – a fragmentary information base.

4. The lack of a mechanism to coordinate the interests of all legitimate actors in northern development has resulted in assessments which are incomplete and lacking balance and timeliness.

For northern communities the feeling is once again that a disproportionate level of "costs" will befall them as the rest of Canada benefits from its resources. Until further and significant steps occur in the political evolution of the North which give increased decision-making and policy formulation responsibilities to northerners, little change is expected. Until the native land claims have been settled, the original inhabitants of the North remain effectively without economic and political strength. As long as a relatively few interests dominate policy and program processes, the North will likely remain fundamentally a resource hinterland for the South, rather than a unique partner in the Canadian milieu.

REFERENCES

[1] For a discussion of the perceptions, beliefs and attitudes of the various interests ("actors") in the Mackenzie-Arctic Islands petroleum programs see Robert F. Keith, et al., Northern Development and Technology Assessment Systems: A Study of Petroleum Development Programs in the Mackenzie Delta-Beaufort Sea Region and the Arctic Islands, Science Council of Canada, Background Study No. 34, Ottawa, January 1976, Chap. V, pp. 57-104.

[2] "Canada's North: A Statement of the Government of Canada on Northern Development in the '70s", Presented to the Standing Committee on Indian Affairs and Northern Development, March 28, 1972.

[3] See M.M.R. Freeman and L.M. Hackman, "Bathurst Island, NWT: A Test Case of Canada's Northern Policy", *Canadian Public Policy*, Vol. 1, No. 3, pp. 402-414; P. J. Usher and G. Beakhurst, *Land Regulations in the Canadian North*, Canadian Arctic Resources Committee, Ottawa, 1973; D. H. Pimlott, *Oil Under Ice*, Canadian Arctic Resources Committee, Ottawa, 1976; and Robert F. Keith, *et al., op. cit.*

[4] Keith, *et al., op. cit.*, pp. 63-64.

[5] For a discussion of economic development and economic growth see K. J. Rea, *The Political Economy of Northern Development*, Science Council of Canada, Background Study No. 36, Ottawa, April 1976.

[6] See Keith, *et al., op. cit.,* pp. 82-91.

[7] Discussions and interviews held during the conduct of this study reported in Robert F. Keith, *et al., op. cit.*

[8] Berger, Thomas R., *Northern Frontier, Northern Homeland*, Vol. 1 (Ottawa: Supply and Services, 1977).

[9] Usher, Peter J., The Committee for Original Peoples' Entitlement, Ottawa, April 25, 1973, mimeograph, p. 95.

[10] Usher, *op. cit.*, pp. 13-14.

[11] Morrow, Hon. Justice William C., "Observations on Resource Issues in Canada's North", *Journal of Natural Resource Management and Interdisciplinary Studies*, Vol. 1, No. 1, Feb. 1976, p. 7.

[12] Wood, K. Scott, *Social Indicators and Social Reporting in the Canadian North*, Panel Presentation, Proceedings at a Seminar: Social Indicators, The Canadian Council on Social Development, Ottawa, January 1972, p. 51.

[13] Keith, *et al., op. cit.*

[14] Yukon Native Brotherhood, Together Today for Our Children Tomorrow, A Statement of Grievances and an Approach to Settlement by the Yukon Indian People, Whitehorse, January 1973. Also see the Hon. Justice William G. Morrow, *op. cit.*, p. 8.

[15] Morrow, Hon. Justice William G., *op. cit.*, p. 8.

[16] Morrow, Hon. Justice William G., *op. cit.*, p. 8.

[17] For a discussion of "leakage" see Stuart Jamieson, "Impact of an Arctic Pipeline on Northern Natives", in Peter H. Pearse (ed.), *The Mackenzie Pipeline: Arctic Gas and Canadian Energy Policy* (Toronto: McClelland and Stewart, 1974), p. 111.

[18] Pearse, *op. cit.*, p. xi.

[19] Dosman, Edgar J., *The National Interest* (Toronto: McClelland and Stewart, 1975).

[20] Keith, *et al., op. cit.*

Industrial Impact in the Canadian North

HUGH BRODY

A Northern Paradox

The vastness, coldness, and low biological productivity of the North have given rise to a very remarkable paradox. On the one hand, human population has always been small and widely scattered. The largest single aboriginal Arctic community was probably no more than 2000 strong, and entire culture areas in the Canadian North comprised population clusters of between ten and fifty persons. On the other hand, the qualities of

● HUGH BRODY is a Research Associate, Scott Polar Institute, Cambridge, England. This paper, which is based on material prepared for the Berger Inquiry, is reprinted from *Polar Record*, Vol. 18, No. 115, January, 1977, pp. 333-339, with permission.

the North have meant that for a long time it has been beyond the reach of agricultural interest or industrial possibility. Thus, these same fundamental qualities mean that once the industrial potential of the North is apparent, it can only be tapped economically by the application of huge amounts of capital and large-scale operations. Therefore, when industry does come to the North, we find the smallest, most isolated societies alongside some of the most costly and technically complex development projects in the world. Hence the paradox: the smallest alongside the largest, the most traditional alongside the most modern, and the most remote becoming involved with national or even international economic interests.

The interaction between industrial development and small, isolated communities has been the subject of social scientific concern since the 1920's. There is a vast literature dealing with problems of culture contact and colonization in all parts of the world. Yet the pattern in the Arctic does not fit easily into the best documented models. The reason for this is simple enough; industrial advance in the North does not represent a southern wish to make use of either native peoples or of vast new territories. It includes neither of the main ingredients of classic colonialism − the wish to profit by reserves of labour or by increased land. Instead Arctic "colonialism" is motivated by what lies under the ground, in comparatively restricted areas, and shows a preference for imported labour, which is often housed and supported on industrial sites. This means that the impact of industrial development on small northern communities does not necessarily have a great deal to do with the sudden penetration of a native community by overwhelming numbers of outsiders, nor does it mean the direct expropriation of land upon which native peoples have long depended.

But there are similarities in the pattern, nonetheless. For example, sociologists refer to an ideal-type of small community, which has a number of features that are of special relevance and importance. There are the economic factors: the small communities are poor − at least relatively so − and are economically dependent on the larger society for some essential goods. Local resources are not able, or are no longer able, to support the population's demand for goods (even though it is theoretically possible for

them to supply basic foodstuffs). Then there are the social factors: the small community is highly integrated, and local foodstuffs are shared in such a way as to maximize their use and minimize local inequalities. Family life is well regulated, and each generation grows into its expected roles without too much conflict. And, finally, there are the political factors: the small community is at least indirectly under the aegis of another, far more powerful social order, of which it is politically a part. These characteristics are typical of remote societies where an aboriginal culture is of not too distant historical importance. They do, therefore, apply to the settlements and camps of the Canadian North, and it is worth keeping this fact firmly in mind when looking at the effect of industrial development on such communities. It is not just smallness that is of relevance, but also the degree of remoteness, political subordination, economic dependence, and solidarity of community and family life.

It is also important to be as clear as possible about the kind of industry that is at issue. It is high wage, capital intensive, and dependent upon highly rationalized economies of scale. It is a frontier mode of economy, and accordingly has distinct ideological components, including individualism and encouragement to mobility of labour. As far as native peoples are concerned, it involves the view that it will provide things that native societies badly need − more money, opportunities for participation in the mainstream of Canadian life, and what is broadly thought of as "progress". Thus industry is often seen, and indeed is often justified, as a solution to the problems of small native communities. In this discussion, then, I shall look at some of the relations between these characteristics of northern industrial advance and the characteristics of small northern communities already discussed. I shall cover three broad areas of inquiry − the economic, the social, and the political − with specific emphasis on the questions of money and sharing, as well as on problems of individualism, identity and mobility, and the way in which industrial development tends to be totally intrusive. Throughout I shall be supposing a simple model of industrial development in which a large, elaborate, and costly program for a mine or an oil-and-gas site is either geographically or economically close to a small native community.

Prosperity and Poverty

It is often said that northern natives are poor, and that the obvious solution to the problem of poverty lies in providing more opportunities to earn higher wages. In fact, native income is not easy to calculate. Income distribution figures for the Mackenzie delta region, published in 1972 and cited by Professor Charles Hobart in his evidence to the Berger Inquiry, give annual per capita earnings for Dene[1] as $839.64, for Inuit as $666.89, and for Métis as $1146.52. These figures are compared with the $3554.61 per capita annual income for southerners. But the figures are puzzling. Other sources estimate that the annual income equivalent for country foods is in the region of $4000, which yields by itself a higher per capita figure than the total suggested by the 1972 study. Peter Usher's evidence to the Berger Inquiry indicates that the value of country foods in the Mackenzie delta is even higher and could, in fact, come close to $8000 per family per annum. A government-sponsored study of Indians living in the Great Bear Lake area shows that in the years 1970-75 nearly 60 percent of households depended on country foods and sales of fur, and that the dollar value of these gave a real income of at least $1500 per capita per annum. Even that study quotes an unrealistically low price for meat. If the figures are adjusted using more plausible price equivalents, the per capita income is increased by about 45 percent to $2175, which represents a household income of between $4350 and $10 000 per annum. Inuit and Dene peoples are not attached to their lands by sentiment alone.

These figures should not be interpreted as demonstrating that the inhabitants of small northern communities are wealthy, or that they lead economically secure lives. They do, however, raise some interesting questions about the impact of industrial development. Remembering that industry is capable of offering jobs at high wages and remembering, also, that members of the affected community are short of money they need to buy goods and services that they have come to regard as essential, one consequence of industrial development is likely to be a reduction of earnings or earning-equivalents from land-based and traditional activities. The amount involved could be as much as $5000 per family per annum.

Of course, industrial employment does not wipe out, at a stroke, all production of country foods; it does not even put an end to trapping and earnings from sales of furs. Indeed, because wage earners can afford to improve the technology they apply to the harvesting of renewable resources, high wages can actually be beneficial to traditional economic activity. But in the longer run, all the evidence suggests that whatever employment opportunities are created, whatever the levels of earning in the industrial sector, the use of land and production of country foods eventually declines. Frobisher Bay, the Hay River-Pine Point area, and Inuvik all exemplify this trend. In Frobisher Bay, for instance, there is now a persistent shortage of country foods, including seal meat and whale skin, whereas fifteen years ago the area was providing enough meat and fish for the subsistence of a community numbering 65 percent of the present Frobisher Inuit population.

Industrial employment also has an impact on the distribution of income within a small community. Like so many village dwellers who have lived, in the not too distant past, in relative economic isolation, Inuit and Dene peoples are proud of the ways in which they share the produce of the land. The activity of hunting may be comparatively individualistic, but is produce tends to be communal – at least in so far as those in want are able to approach successful hunters and ask for food. Also, the basic means of production – land – is regarded as communal. Money is not so readily shared. Wage earners tend to regard it as their own private property, and to spend it on their immediate families' personal needs. Consumer durable goods cannot easily be divided among neighbours. The shift towards a money economy thus creates a possibility for poverty that previously did not exist: those in want are more likely to stay in want, and substantial inequalities introduce themselves into native communities. It is possible, in an unequal society where the basis of wealth is not shared, for average per capita income to go up, while the number of households experiencing poverty is also increasing.

Identity and Mobility

Inequality is also evidenced by the way in which money affects the hunter's social position. High earnings can be used to maximize a hunter's mobility and reduce the time needed for making kills; a new snowmobile, supplies of fuel and spare parts, and rifles with accurate telescopic sights go a long way towards ensuring a successful hunt. But the families that are in a position to buy and maintain expensive equipment of this kind are the ones that have secured highly paid jobs. This means that those with most cash are also in the best position to be successful hunters. In fact, of course, their lifestyle and inclinations are frequently at odds with their realizing such potential, but the paradox nonetheless remains. Those who are most inclined to hunt, and who have carefully elected to maintain the hunter's lifestyle, are often least equipped to do so efficiently. Consequently, hunters feel at a technical disadvantage vis-à-vis wage labourers. And, if the families with the highest money incomes are not prepared to share their hunting produce, then the paradox becomes extreme: the hunters' families are the ones with least meat. Thus poverty begins to be associated with hunting, trapping, and the traditional options. Loss of prestige follows in the wake of economic disadvantage. Once hunting is associated with poverty, the hunter loses his status within the society, and the small community's sense of cultural distinctiveness is eroded. On a global scale, small agricultural villages have suffered as much as groups of seminomadic hunters and gatherers wherever industrial development has established a rival economic mode.

No amount of additional money income will resolve this social-psychological problem — indeed, further development of the wage labour mode of economy will aggravate many of the difficulties. Communities in which the process of cultural erosion has reached near extremes are now notorious. Frobisher Bay, in the southern Baffin region, is perhaps the best known example. As I have already mentioned, the 900 Inuit living in the town are always short of country foods, even seal meat, although earnings are high. If income and spending habits are taken as the standard, the level of material life in Frobisher Bay is higher than in any other settlement of that region. Yet despite their more or less complete incorporation into a money and labouring way of life, Inuit of Frobisher Bay express anxieties about their identity and say that they are hungry for "real food". The scale of the community's alcohol, prostitution, and family breakdown problems is all too well known. Less well known is the fact that a significant proportion of the community's Inuit wish to be able to spend more time hunting, and one group is trying to establish an Inuit community outside Frobisher as a permanent hunting camp. Over the past five years the problems in Frobisher have worsened and, in some sectors, are close to becoming critical. Demoralization and social disorganization are spreading gradually deeper and deeper into the Inuit community there. It is a process that will not be halted by more work or more money.

Moreover, the frontier ethic itself is especially disruptive to native people. Frontiers attract the "get-rich-quickers" at every level; they encourage — indeed, depend upon — a footloose work force, mobile capital, and all their idealogical concomitants. Individualism, uncertainty, and instability are part and parcel of the social and moral qualities of the frontier. Frontierism exaggerates and worsens the processes whereby traditional social controls are broken down, and aggressive, deviant, individuated, and more pathological kinds of behaviour become everyday features of life.

The frontier, both as a matter of economic fact and as part of a national spirit, has always been important in North America, and is especially important in Canada. It has its representatives throughout the country who, if they do not themselves go to the bush, at least encourage, or even idealize, those who do. Native peoples who live at the frontier are encouraged to participate in frontierism, and are urged to accept the attitudes and lifestyles that go along with frontier activities. These attitudes and lifestyles are radically different to those embedded in native tradition and a small, permanent community.

The effects of industrial development are not only to be found in its immediate path. The frontier encourages men to become mobile: work opportunities come and go, so the labourers must come and go as well. This is already happening in the eastern Arctic. During the later 1960's the wage labour sector was associated, in virtually every Eskimo settlement, with a number of part- and full-time jobs in and around the village itself.

However, the local community could provide only a limited number of jobs, so the new northern industries seemed to offer the opportunity to maintain the rising standard of earnings. But to take advantage of such opportunities, the workers have to go away from home. In the case of the oil and gas industries of the Arctic islands sites, they have to travel long distances and, because of the arrangements at these sites, are away for twenty days in each month. Approximately forty Inuit work each shift, which represents between 15 and 25 percent of the labour pool of the communities concerned. If nothing comes of the oil and gas search in the Arctic islands area, then the communities will either have to accept a sudden and drastic loss of income, or their workers will have to go to some other location. In fact, there have been rumours of a contingency plan which provides for the men to go to the lead and zinc mine now being established at Strathcona Sound. In order to work there, they will have to accept much longer periods away from home – perhaps as much as three months at a stretch. So the mobility of the work force becomes a condition of its ability to find work; those who do not want to be mobile must accept not earning; communities that want to avoid the effects of such mobility must accept that they cannot take advantage of the industrial frontier. But all the pressures, including those that stem more or less indirectly from governmental policies in the North, make it very difficult to decide against participation. As a result, industrial development tends to create an increasingly mobile work force. That mobility causes maximum disruption to the home community and disorientation to the native worker.

The Total Intrusion Effect

The debit side of economic impact is at its greatest when development is both large scale and, from the local employment point of view, short term. A boom in the labour and money economy very rapidly causes the sort of changes that this paper has been discussing. Evidence shows that in only a few years a community can become dependent on high earning levels and store-bought food and clothing. The speed of this transition can

itself be disruptive. It aggravates native feelings of nonparticipation and loss of control, for their life is suddenly altered with a minimum of consultation and agreement with those most directly affected.

The problem of consultation is of particular importance. There is an Eskimo word that characterizes the feeling that southerners inspire in Inuit. That word (or root) is *ilira*, and it is not easy to translate. It is a kind of fear, a blend of awe and intimidation. It is the feeling a strong and effective father inspires in his children; it is the feeling you have about a person whose behaviour you can neither control nor predict, but who is perhaps going to be dangerous; it is the feeling you have when you are in a room full of important strangers whose language you cannot understand; and it is the feeling inspired by the trader, the missionary, and the policeman, white strangers who were so obviously powerful, upon whom Inuit were so acutely dependent and who told people what to do and believe but were not often disposed to listen to what Inuit wanted to do and believe. Indeed, Inuit express their surprise and pleasure when they have dealings with a southerner who does not make them feel *ilira.*

In the course of two or more decades of dealings with southerners, Inuit came to have expectations and attitudes strongly influenced by the *ilira* they felt. They did not expect to be able to state their own opinions and criticisms of what southerners were doing; they tended to accept the decisions of traders and missionaries, and to avoid all possible confrontation. There took place what might be called political retreatism, as well as the careful preservation of a cheerful and obedient countenance. Native people came to present themselves as conciliatory and accepting. This meant that they were inclined to smile and look cheerful whenever they had dealings with southerners; it also meant that they did what they were asked to do, even when it was, in reality, something they thought wrong or foolish; and, in the end, it meant that they subordinated themselves to the changing whims of individuals no less than to shifts in prices or policies by which their lives were profoundly affected.

Retreatism of this kind is described by many Inuit who can recall the first introduction of schools. In some areas this was in the later 1950's, and in most areas followed directly on the trade and mission period, and represented one of the first major governmental initiatives in the North. In a series of forty-three discussions I had on this subject with parents who were asked to give permission for their children to go to a federal school in a settlement, but who were living at the time in camp away from the settlement, thirty-two said they wanted to say "no" to the officials by whom they were approached. All but three, however, said "yes", and, in many cases, against their own very strong feelings and quite different judgement, agreed to leave their children in school or moved to a settlement to avoid being separated from them. This represents approximately a 70 percent acquiescence rate. When describing their reasons for thus acquiescing, twenty-one of the twenty-nine parents used the work *ilira* to explain their behaviour, and fourteen used the work *kappia*, which means fear of danger. Eighteen talked of southerners as *angajuqaat,* bosses, and themselves as just not being able to do other than what they were told. Each person's description of the beginnings of the education program involved some more or less explicit reference to their subordination to, and dependence upon, southerners.

When Inuit talk about subsequent events, they indicate that these attitudes and their corresponding retreatism have tended to persist or, in some cases, even to be reinforced. Inuit have rarely felt able to oppose southerners' wishes. Since southerners represent their innovations as their wish, they thereby minimize the possibility of the kind of dialogue that genuine consultation must entail. What is more, the discussions that do take place are often bedevilled by misunderstandings: there are serious language problems, not to mention the vastly different traditions of dialogue and social exchange that govern the representatives of industry and the representatives of Inuit or Dene communities. Inuit are slow to decide, and prefer to wait for the gradual emergence of community opinion before expressing a definite point of view. They are also suspicious of mere opinion, and regard an error of judgement as a lie. But decisions about industry can rarely wait long – the harsh economic realities are always said to be pounding on the door. From the native point of view, their representatives seem to be stormed into making decisions, into giving agreements, and into expressing their wishes and conditions for the southerners' programs.

The case of the Nanisivik mine in north Baffin Island (Strathcona Sound) is illustrative. The mine is situated only 24 km from the community of Arctic Bay. In 1972-73 rumours reached the community that the mine was going to be opened. In 1973-74 the government considered an application for a subsidy to the company concerned. Although ores had been discovered as long ago as the 1950's, and in 1962 the development potential of the area had been examined, in all that time virtually no background work had been done on the availability of labour, the kinds of impact the mine would have, and the attitude of the community as a whole to its establishment. The chairman of the Arctic Bay community council had worked for some years with the mining company as a guide and assistant, and his views about a large-scale development were now canvassed. No in-depth or protracted investigations were carried out in either Arctic Bay or other settlements in the region from which labour was to be used. In 1974 a local Oblate missionary and a number of social scientists expressed concern about the impact, and reported that in some settlements there was growing unease about the possible effects of the mine.

Nevertheless, in June, 1974, an agreement was signed between the federal government, the mining company, and two representatives of the Inuit community. At about the same time the Arctic Bay community council wrote a formal letter protesting about several aspects of the mine as it was then planned, and also made clear its unhappiness at the way Inuit interests had not been given time to emerge. By the autumn of 1975, workers were going to the mine from Pond Inlet, Igloolik, and Arctic Bay. If the Arctic Bay community council has reported the consultation procedure accurately, then its members were pushed into agreeing to a project which they felt threatened the social well-being and possibly the very existence of their community. The councils of Pond Inlet and Igloolik were consulted only in so far as representatives of the federal government spent a few days in the area trying to point out all the advantages of the mine, and representatives of the mining company visited settlements to recruit workers.

It cannot be said too often that the relative scale of northern extractive industries on the one hand, and small native communities on the other, works very much to the detriment of the small community. As well as the damaging consequences of speed, there is the high level of labour recruitment. The impact of a large-scale development is felt by everyone in the community: there is not so much a selection as a total intrusion. In the case of small-scale developments, those who are particularly qualified for − or inclined towards − wage labour are selected or select themselves; in the case of large-scale programs, all available manpower is urged to move into the new job opportunities. Since those who live by hunting, trapping, and occasional labour are often regarded as partially or wholly unemployed, pressures are applied throughout the traditional sector of the economy. These direct pressures are intensified by the recurrent cash problem of those who have opted most firmly for a life based on renewable resource harvesting. So it is that persons − or even whole communities − most likely to have cultural and personal links with the land and its resources are most firmly pushed towards participation in industrial activities. Hence the total intrusion effect.

The paradoxical co-existence at the northern frontier of the grandest, capital-intensive development projects and the smallest, most isolated of native societies is likely to be short-lived: the native communities are in danger of being engulfed by the social and economic modes of the extractive industries. In some locations this process is already under way. But there is not necessarily a need for the enforced isolation of small communities. The pace of industrial development can be restricted, and time allowed for suitably protracted consultation with, and deliberation by, Inuit and Dene peoples. It may happen that an adjustment can be made to various forms of industrial enterprise. But it will be an adjustment rather than a response to pressure, or another acquiescence, only if time is allowed for present experience and local opinion to mature. Similarly, the nature of local participation in industry can be controlled. Those who want to take their place in the extractive industries should, of course, be able to do so. But those who do not want to, or those who are anxious about maintaining what they regard as the traditional basis of their community's social and economic practices, must also be able to realize their aims.

It is when people or even whole societies cannot do the things they regard as most important and most useful that human pathologies spring up − when retreatism, apathy, and futile violence become endemic. Industry does not exist alone, but has socially and politically determined forms. It is amenable to social controls, and can be put to the service of societies. It can, therefore, be either useful or destructive in the North. The larger the scale, the more urgent the impetus towards its development, the more scrupulously it must be monitored. And that requires time, patience, and intelligent regard to the experience at hand.

NOTE

[1] Dene is the word for "Indian" preferred by Indian peoples of the western Arctic. The Métis are descendants of Indian women who married 'whites', especially French Canadian *voyageurs*. They do not have the status of Treaty Indians, but have lived for many generations in northern communities, by hunting, fishing, and trapping.

Indian Land Claims in Canada

HAROLD CARDINAL

Amongst our people, the elders are the keepers of our knowledge; they are our source of knowledge; they are the transmitters of our knowledge. Not being an elder, and being a considerable time away from being an elder, I feel highly unqualified to talk on subjects as important as land use and Indian claims. At best, I hope to share some of the things that I am only now beginning to learn from our elders about us as people and our views of our relationship to this land and to our Creator. If I misinterpret any of this knowledge, it is not the fault of our elders, but mine. One of the things that has not been talked about in an interracial situation or within the Indian community is the concept of who we are as a people. This concept is not clearly understood by the majority of non-Indian people and, perhaps at this stage, by the majority of young native people.

One of our central concepts, as I understand it, is that our relationship as Indian people is closely tied to our concept of our Creator, the Great Spirit. In that sense, all the things that we relate to, or should relate to, are guided by our covenant with the Great Spirit as a people. Therefore, our elders' view of land use is probably much closer to that of the Hebrew than it is to the majority of Canadian people.

The presence of the white people on this continent was prophesied many centuries in advance of their arrival. The problems that would arise were also prophesied many years before the arrival of the white man. We are now just reaching the stage where some of our prophecies, and indeed, they are as valid as the prophecies of any other religious group, are beginning to unfold. One of the things that many of our people find hard to believe is that before the advent of roads, automobiles, airplanes, telephones, and railways, our elders predicted that the white people would be bringing with them these creations, and that they would in fact occupy this land. There would be as many on this land as there were trees or wildlife. I think they began to sense some of the problems that did eventually arise, and in this part of the country (Alberta), they agreed to fulfill one of their prophecies and that was to sign a compact or an agreement with the representatives of the white man's society. At that time, the agreement centred on how two societies, which were basically different, could coexist. On one hand, our people were largely in the hunting, fishing, and trapping era – they based their economy on wildlife, and based their way of living on the elements of their environment. On the other hand, the white people were bringing with them a different way of life, which at that time was a predominantly agricultural way of life.

A request was made of our elders that agricultural land be shared so that these new people would be able to grow their crops in Mother Earth. Also our elders were asked to make available productive land that would be needed for the cattle that the white people were bringing. Our "cattle" were the wildlife.

Part of that agreement was that in the cohabitation of the land, the Indian people would not interfere with the new lifestyle that was being brought by the foreign people who were coming to this land. But neither would those who came to this land interfere, or make laws that would interfere, with the way of life of the Indian people. For example, on the question of wildlife, our people agreed that the larger society would enact laws only insofar as those laws would restrict the access of members of the foreign society to our "cattle", but our people did not agree that the larger society was given a mandate to make laws which would in fact govern the way of life of our wildlife or those parts of the land that were not being used for agricultural purposes. Since that time, our elders have watched with increasing concern the development with which a lot of your people are now becoming concerned.

● HAROLD CARDINAL was President of the Indian Association of Alberta at the time this article was written. This paper is reprinted from *Proceedings,* Canadian Agricultural Economics Workshop, Banff, 1975, pp 4-14, with permission.

It was our elders' understanding that the people who came would bring with them a lifestyle from which they could earn a living and produce their food, and upon which their society would be based. But with the advent of the industrial age and the diversification of the nation's economy, we find now that the majority of people are gathered in relatively small areas of land, with the density of the population increasing in these areas of land. As the cities continue to grow, the functional productivity of the white society in terms of food is going down accordingly. Our people are beginning to wonder what is happening to the covenant that we had with the larger society. As the cities continue to grow and as resources which have not yet been surrendered are exploited, there is a chain reaction in terms of what happens to the land. What happens, for example, in terms of those people who came to exploit the resources from the mountains? How did they affect the climate which in turn has other chain effects? How does the industrial development of our mountains in this particular part of the country affect, or how is it going to affect in the future, our water supply and all of the basic elements that we need to survive as a people, whether we are Indian or White?

Our elders are beginning to view with extreme concern the fact that there is some indication that we are being threatened by a society that apparently is interested in reaping the greatest dollar return over the shortest possible time, without due regard for the long range consequences. We have not yet reached a stage as Indian people where we have begun to exercise our sovereignty and our covenant with our Great Spirit. We have not yet reached the time where we have begun to set up our own laws to govern land use in this country. I do not know when that time will come, but we have not at any point given up our sovereignty or our right to make those laws. And as such, our covenant with our Great Spirit will continue, and at some point we will reach the stage where we will have to begin looking at these things.

Planning must take place in order to insure that the development of our country will take place at a balanced rate and that it will be in harmony not only with the needs of our people, not only with Indian rights, but also with the needs of our environment – the needs which must be respected in our environment.

For those of you who are going to be pursuing this question in a serious way, I hope at some point in the future we have an opportunity to organize a session with our expertise (our elders), so that as Canadians we can begin not only to share the concerns, but also to tie in the expertise and knowledge that we possess and that presumably you possess as professional people in your own right. That type of development will necessarily have to occur. I do not think that Indians or Whites can begin looking at the question of land use, resource use, and other uses of our environment in isolation from one another. We are going to fulfill the covenant that was agreed upon between your sovereign and our people. We have to begin in a more active way, and perhaps in a more concrete and productive way, the exchanging of the knowledge and the expertise that we possess. No group of people in this country has a monopoly on knowledge. White academics do not have that monopoly, neither do the elders of the Indian people. For that knowledge to grow, for that knowledge to better this country, for that knowledge to be of use to our children and our future generations, we have a responsibility in our generation and in our time to begin putting together the expertise that we have on each side, within each cultural group, so that we can begin to help nurture the growth of the knowledge that we possess as Canadian people.

We talk of land claims amongst Indian people. There is a growing and serious discussion within the ranks of the Indian community regarding the following questions. What precisely is a land claim? What rights are we talking about? What do we want for our land claims? In fact, what do we expect, what obligation has the other side got, and what responsibilities have we got as people of this land?

There are approximately four categories of claims. One set of claims is related to land in the areas of British Columbia, the Yukon, the Northwest Territories, and Quebec, where no treaties have been entered into between the Indian people of those parts of the country and the Crown. There exists a legitimate claim, from our point of view, which calls for a settlement between the

Indian people of those areas and the Canadian government. Some negotiations are taking place and some are at quite an advanced stage, especially in the Yukon Territories and areas of the Northwest Territories. In British Columbia they are at a less advanced stage in their discussions. There is one set of claims under the title of "aboriginal rights". I think that this term is a misnomer, or at least a misunderstood term, both by Indian people and those on the government side of the negotiating team.

The second type of claim is found in the treaty areas where, prior to the last three years, it was assumed by government that the treaties extinguished all claims that the Indians had. The research that we have been doing over the last three or four years indicates that the only thing the treaties really did was to share agricultural land with the white society. The treaties did not surrender, from our point of view, our claims to our rights to mineral resources, to water resources, to the Rocky Mountains range, and to a number of other things as well. During the next few years we are going to be dealing in that area. For example, one area of examination is with regard to the Athabasca tar sands.

The third set of claims which we have with regard to land concerns "the lost reserve lands". There was a period after the reserves had been set up when the government coerced, bribed, lied, and cheated in terms of getting hoodwinked Indian people to surrender substantially large portions of land that had been created as reserves. We now feel that we have a legitimate claim to those lands, or their equivalent, and that they should be returned to the bands who lost their land base in the mid-twenties, mid-thirties, and mid-forties.

The fourth type of claim is in terms of individuals who were "given" land but who never received it, or who got land in a location where they did not want it. So we are now negotiating with the Canadian government to set up a process through which the various Indian claims can be dealt. We are reexamining, in a farily critical way, the treaties and their limitations, from our point of view, so that these areas will come up for negotiations with the Canadian government.

One of the things that we have to begin looking at, and that I think our people have to look at in an extremely serious way, is the fact that land claims should not be used by the Indian people of this country solely to acquire money, solely to stop development, or solely with the point of view that these are our lands and that these are our resources. Going back to our antiquity, going back to our covenant, we have to begin looking at the responsibilities that we have as people, collectively as a people, toward other Canadians who have moved to this land, and who have come to develop the resources that we shared with them. And so the question then becomes: "What is an Indian claim and how should it be handled?" Is it really a question of Indians looking for free handouts? Is it a question of just exactly what do we want in bringing these issues forth to the public at this time? I think the importance of Indian claims at this time lies in the fact that the wealth of this country is not distributed equally or adequately among all segments of society. One of the covenants − one of the basic premises of our treaties − is that the wealth of this country, the resources of this country, would be used for the development and growth of all Canadians − not more to some and less to others. Many young Indian people look at Indian claims, and, I think, perhaps erroneously, as an opportunity for them to get the resources that have been denied their people in terms of economic development, in terms of social development, and in terms of educational development. From our point of view here in Alberta, we look at the possibility of launching claims actions only when we are convinced that our people are going to be totally left out of a particular development or the benefits of a particular development. In another instance, we will be looking at launching claims if we feel that those areas of our province, which we hold sacred, are going to be destroyed by development or too disturbed by development. We look specifically to the Rocky Mountains.

One of the other areas that we have to begin seriously looking at in the next five or so years is the question of how our water is going to be utilized. Government and industry are increasingly looking at Alberta as a new Hamilton, a new Pittsburg, or a new San Diego. We have to begin being concerned as Indian people about the effect that type of development will have on our environment for which we share some responsibility. And I think our quest over the next five or ten years has got to be accomplished by sitting down with more and more groups, to try and discover the best methods we should use so that after we pass on, our children and our people will have at least as much of an environment as we had when we were born. One of the things we are going to be looking at is the direction of the thrust that our people will be taking in this province. I believe that others in other parts of the country will soon be following in that area.

In looking at our claims, we see this as an opportunity of talking about them — we should begin to update and perhaps to redefine the obligations of the larger society to us and our responsibility to this country. I do not feel that we can live indefinitely in a situation where we carry a mentality that says "you owe us this because we are equal". I think we have to look increasingly at our responsibility to this land and our responsibility in the partnership with the white people in the larger society.

There are different ways in which land claims will be handled. In the James Bay area, the hydroelectric companies were challenged by the Crees of northern Quebec in a court action that resulted in negotiations which have let to an interim settlement or a settlement to some degree. We have negotiations, greatly improved negotiations without legal action, which are taking place in the Yukon Territories, and they have now reached a stage where the federal government has come up with at least a working paper which may be the basis for agreement. I think that in our claims situation in Alberta we would not want to launch a massive claims action as they did in the James Bay area, or as they did in the Yukon area where negotiations are taking place lock, stock, and barrel. One of the things that we have to assess in a very close way is the manner in which the larger society, the manner in which our future partner, the sovereignty, has fulfilled its commitments to our people. We will have to examine in a very serious way the manner in which the federal government has handled this trusteeship responsibility, and based on that assessment, we will have to decide whether we can trust the Canadian sovereignty with further handling of our resources, further handling of our lands and our people. If we find in the process of our review that the record of the government, at the federal or provincial level, is such that it does not merit the trust of our people, then we have to take a serious look at the options for our people. We hope that we will be able to work out some agreement which will be of benefit not only to our people, but also to the larger society. We are mindful, as well, of the fact that there are other alternatives which we must seriously consider.

In looking at Indian claims in Alberta, it behooves us to understand why any group of Indian people would agree to surrender, for example, a territory as big as the Yukon, for what sounds like a phenomenal amount of 50 million or 100 million dollars, maybe for 5000 ha of land, when we know the value of the resources contained in this country and when we know the value of the land that we have. For example, if we were to launch a claims action in the province of Alberta, we could say that we want to claim our mineral resources. But we would be talking about something so valuable that it would be impossible to come up with a valid settlement. And I say this keeping in mind that just 16 000 ha of tar sands, at current market value, is worth something like 3, 5, or 6 billion dollars. If we were to launch an action based on mineral resources and someone was to offer the Indian people of this province 150 million dollars, we could end up in the long run being sold out in terms of getting an equitable settlement. So we have to look at some other way of approaching our claims process. We will be taking an approach that is considerably different from what we see going on in Quebec, British Columbia, or the Yukon.

With the help of our elders, we in Alberta are determined to gain an understanding of who we are and what we should be in this country. Perhaps even more important, through that process we are also determined to gain an understanding of who you are, what you are, and why you are. Because once we do that, perhaps we can talk on equal terms in trying to decide how we can work together to build this country in the future or whether your society is working on our trust again.

Problems and Priorities in the Canadian Environment

PIERRE DANSEREAU

Since this report is, in a sense, on the "state of the environment in Canada", it must identify the areas in which there is a crisis, an emergency, an urgency, or a problem, especially in the short-range view.

The perspective which I find most useful in drawing attention to actual points of crisis involves a grouping under four principal headings, inasmuch as the land mosaics that contain the productive ecosystems are the units that have to be protected or managed in some way.

In fact, it is on a world basis that one can recognize the four major components of all land mosaics: wild, rural, industrial, and urban.

They are briefly defined as follows:

1. Wild areas are virtually unaffected by direct human interference, although they may contain a highly scattered human population.

2. Rural lands are greatly transformed by man, but largely occupied by exploitable vegetation and animal life, whereas the total area of human construction and presence is quite small.

3. Industrial occupation implies purposeful harnessing of local or imported resources for redistribution or transformation by technological means.

4. Urban settlements conserve virtually nothing of the primeval mineral, vegetable, and animal resources, and consist of almost entirely built-on space or at least profoundly transformed landscape.

Looking at Canada as a whole, in the light of land use, we find that a very large amount of its territory is wild, but that the vast majority of its population is urban.

A sampling of land use across the country will show a great variety of mosaics, some with harmonious inter-fingering of wild and rural; of wild, rural, and urban; of rural-industrial; or rural-industrial-urban; but many also that are overwhelmingly of one or the other main types.

It will be pertinent to our present purpose to proceed in a thematic rather than regional way, since we cannot indulge much in an actual land inventory.

Wild Areas

The arguments for maintaining large tracts of the Canadian landscape in the wild state come under the following headings.
(a) The preservation of a complete repertoire of primeval kinds of rocks, streams, plants, and animals.
(b) The availability for recreational, esthetic, and other human-oriented activities, of essentially unmodified environments.
(c) The continuance of the wild way-of-life as an option.
(d) The need for further and continued scientific study of undisturbed ecosystems.

We must then ask ourselves what the prevalent land use is within each one of these major units and what the ownership and management regimes are. The principal questions are:
(a) How much public and how much private land?
(b) What is the variety of ecosystems within each major region?
(c) What is the distribution, size, and degree of protection afforded each major and minor grouping of regional ecosystems?
(d) What are the function and accessibility of wild lands in each region?

Parks

There is much to be proud of in Canada's federal park organization: its early start, its wise policy, its level of management, etc. As for the provinces, the record is not always so good, inasmuch as many provincial parks, to this day, do not exclude lumbering or even

● PIERRE DANSEREAU is at the Université du Québec à Montréal. This article is reprinted from the Canadian Environmental Advisory Council, *Annual Review*, 1973-74, pp. 10-19, with permission.

mining! No generalization can be made concerning municipal and private parks.

The park situation, as it stands in Canada today, reveals that there has been no visible plan or purpose to set aside a park within each and every one of the major natural regional units. Some of them, however, are fairly well sampled, notably the boreal forest, but not the prairie, the sub-arctic parkland, and least of all the eastern temperate forest!

There are no detailed surveys of the national or other parks. The systematic study, on a uniform methodological basis, of all parks (covering the gamut of Canada's major bioclimatic units) is not even underway.

The National Parks Act and its applications are much in need of revision, if only because it is based on an outmoded conservationist credo.

Water

Water is one of Canada's most abundant resources. Its capacity for electric power development, for irrigation of forest and farm land, for fisheries, and possibly for export as well, has long made its uses controversial, and we cannot be sure that all the right decisions were made, at the right time.

On the largest scale, the International Joint Commission (created by treaty between the USA and Canada in 1909 and formally set up in 1912) was able to settle management controversies between the two countries where waters flowed from the one into the other. But any long-term positive operation such as the St. Lawrence Seaway (very much desired by Canada from the 1920's onward and stoutly rejected by the USA until the late 1950's) was very slow in materializing.

The whole question of flood control (especially in the Great Lakes − St. Lawrence), so dramatically revived in the spring of 1974, and its effects on fishing, farming, industrial supply, and urban development, have been the object of much study. Nothing short of a bold synthesis of the known facts, a better quantitative geomorphological approach, and a tighter integration into long-range multi-purpose planning will put us in possession of a workable distribution of water.

Rural Landscapes

The varities of Canadian rural settlement and economy are, for the most part, coincident with the wild (or bioclimatic) divisions. Behind the present agriculturally-dominated landscapes lie one to three centuries of landscape/inscape interaction. In other words, the natural potential has been developed under the inspiration and limitations of cultural transfers and market tolerances.

As in wild lands, the rationale for maintaining a certain proportion of strategically located Canadian landscapes in the rural state can be summarized in a few simple propositions.
(a) Growing and producing our own food will always be a necessity, and is also an inalienable backdrop of our culture.
(b) The rural way-of-life is one of the main human options and must remain available to a large number of people.
(c) No considerable tract of landscape would seem to be harmonious without rural components.

An inventory of farm practices and of building and growing materials would give us the insight which we lack on Canadian rural settlements. But, again bypassing a study in depth and considering the outstanding aspects of the rural environment, we can get our bearings from the following issues.
(a) The rural-urban ratio of population distribution has reversed itself from 1871 (18.3 percent urban) through 1921 (47.4 percent urban), 1931 (52.5 percent urban), and 1966 (73.5 percent urban) to 90 percent predicted for the year 2000.
(b) Mechanization of agriculture has made many farms unviable below a certain area, especially in the mixed farming regions of eastern Canada.
(c) A certain conception of international balance-of-trade and of competitive markets has led to drastic reductions in farm production (especially of cereals in the West).
(d) The inroads of electronic and audi-visual information have resulted − together with (b) above − in the large-scale alienation of rural labour, and its migration to the Urban centres.
(e) The proximity of rural land to expanding urban areas has made it highly vulnerable to industrial, suburban, and urban development, often via the operations of real-estate speculation.

(f) A possible shift in the Canadian way-of-life towards greater self-reliance of the individual and of the family-unit would place higher value on small, but very productive, holdings otherwise in the urban matrix.

(g) Concern over too high a degree of dependence on imported food may well favour a directed shift in the rate of rural-urban transfer of population, and increased subsidy to agricultural production.

(h) Some alienated urban dwellers are moving to rural areas at the cost of a sometimes radical loss of income; and this may be important enough to determine new planning and legislation.

These eight issues probably dominate the rural scene in Canada at this time, and they call for an exhaustive study of rural-urban balance in all of its Canadian patterns, as well as for a comparison with other countries (most notably the Scandinavian states) where more deliberate experimentation has been carried out.

Some of the tasks that will permit a better utilization of the rural environment in Canada at this time, and that meet the issues mentioned above, would be as follows.

Land Use Potentials, as recorded and mapped by the Canada Land Inventory, show a number of conflicts (where high values are ascribed to more than one vocation) and a good number of mistakes (where present occupation is on a low-potential site for its vocation). These data are more than ready for the computers and we could engage in a large-scale evaluation, or at the very least in a judicious sampling of representative regional mosaics.

Agricultural Technology has operated through the substitution of new tools for old, the newer being presumably more efficient in that they require less time, less labour, and as often as not less personal skill. Balancing the gains and losses, not entirely in terms of economic yield but in terms of human satisfaction, could not the bulldozer and the plough co-exist in the same landscape (as they apparently do in contemporary China)? Could the artisanal-industrial dichotomy be a lure than cannot really be justified by "progress"?

The poisoning of air, water, and soil by industrially produced chemicals, applied in massive doses, has resulted in the crippling, death, and near-extinction of many forms of wildlife, and in the accumulation of toxic substances in otherwise luscious vegetables and healthy-looking poultry. Much is known, of course, on this subject, from the surprising amount of DDT in Antarctic penguins to the high doses of mercury in lake fish, to the killing-off of vast masses of plankton in the oceans. Lists are periodically produced of substances to be outlawed on the grounds of their noxiousness. It is not less well known, especially to those having attended UNESCO and UNEP conferences, that the weighing of positive versus negative effects (for instance of DDT) is not scaled in the same way in Brazil and in Canada, in India and in the United States of America.

The quality of rural life presents another set of problems. There is no sense in advocating a return to old farming habits and village life-styles. These are most likely to appeal to the unproductive retired community, and to present as much of a dead end as does total conservation of large wild areas. On the other hand, in these days of rampant social engineering, there is no reason to accept as incompatible scientific farming, well geared to market realities, and the leisure and amenities of a traditional rural ambience.

Improvement of the rural way-of-life through controlled markets, selected subsidies, and better transport and communication facilities should be the object of major national and provincial programs that can only be developed if the motivation and consensus are there. Such a way-of-life is a social objective; it is ecologically and technically well within our planning capacity; its long-run economy can easily be justified if speculation and un-planning are recognized as major evils; and its political implementation presents no major difficulties.

Land-banks and the *Regrouping of Farms* thus stand out in very sharp focus as indispensable, however our social values may shift, and even if the rural way-of-life mentioned above is the object of no special favour.

Many, if not most, of our Canadian cities are built upon agricultural land of high value, and continue to this day to proliferate at the expense of highly productive soil. This cannot be allowed to continue for any reason. The free play of the real-estate market must be checked. We are conscience-bound to know what we are doing, to face squarely the shortsighted trade-offs that we have accepted for several decades. It is true that a twenty-storey building brings forth more revenue than a cornfield over the same surface. But what are and what will be the respective needs and possible locations of agriculture and urbanized residence?

The sheer obedience to an automatic triggering of real-estate values and to their self-engendered growth has led to such widespread waste and to such irreversible spoilage that it cannot very well perpetuate itself without calamitous effects. Some of these are only too evident:
(a) farmers who can no longer afford to keep and to operate their farms;
(b) city and suburban dwellers who are forced into habitat patterns (e.g., high-rise) that are thoroughly uncongenial;
(c) destruction of entire landscapes that have picturesque value and some natural recreational facilities;
(d) destruction of historical sites, monuments, and buildings;
(e) reduction of the total agricultural production, sometimes of a very specialized crop.

The creation of land-banks by provincial governments is therefore imperative. A start has been made in British Columbia with the outlawing of construction on floodplain areas, and with various zoning laws or regulations which are also encountered in other parts of Canada. But no real pattern emerges at this time, and one cannot but feel that federal, provincial, and municipal governments in Canada are, in various degrees, lacking in purpose and have not even seriously studied the matter of land-banks.

Industrial Canada

Educational and cultural conditioning, political ties, wealth of heritage, and collective discipline seem as important to the perception of industrial impact as the indigenous resource base itself.

From an ecological point of view, it may be useful to look upon industrialization according to the level of exploitation which it most obviously affects, considering the nature of the raw materials (mineral, vegetable, animal) and how much of an energy input is required. The main distinctions concern the increasing energy input, on a scale that runs from mere extraction (mining, peat-cutting,

fisheries) of raw materials *in situ* to multi-level processing or manufacturing (refinery, distillery, leather goods), with two intermediary groups that ensure transport and provide power.

The questions that arise concerning industry in Canada are the following.
(a) How does location of industry relate to landscape as a whole?
(b) How efficient is the transport system for industrial purposes?
(c) What is the geography of industrial raw materials?
(d) What is the strategy of labour distribution in industrial work?
(e) In what ways and to what extent does industry spread pollution?
(f) What is the role of inter-industrial cycling and recycling?

These questions, like those posed above in the wild and in the rural environments, are best answered in the concrete vision of Canadian industries in the principal groups mentioned above.

Extractive Industries are numerous in Canada and cover the whole territory. Mining has resulted in a craterization only equalled by war. After many decades, Sudbury remains moonlike in aspect, with its burned vegetation and bared rock. Many other sites across the land are almost as ugly, although not always so productive of a wealth that fails to be re-invested in the rehabilitation of the landscape. Pollution of air (sulfur and silicon) may well be accompanied by pollution of streams (as in the salmon-bearing Matane River, in the Gaspé Provincial Park).

A very similar effect is achieved by quarrying and by borrow-pits, where gravel and sand have been extracted. In a radius of less than 160 km of Toronto and Montreal, the needs of road-building and urban construction have resulted in the razing of moraines, kames, and eskers, the ablation of terraces, and the digging of large pits more or less filled with water. Shining examples of the reclamation of such unsightly and unproductive areas are to be seen in British Columbia (Butchart's Gardens, Queen Elizabeth Gardens), but mostly there is no plan for the successional use of these lands, although Ontario has shown some preoccupation in this direction.

Lumbering of Canada's forests is by no means the worst in the world. But the myth of our inexhaustible woodlands is not all that far behind us, and the evidence of indiscriminate cutting is still very much in view. Unquestionably, more scientific and also wiser policies now prevail, although their application is somewhat ironic when a heavily lumbered area is turned into a national park (Parc National de la Mauricie).

Fisheries (the extraction of fish from fresh water and from the sea) is a world in itself, and Canada's role in covering this domain by scientific inquiry is a major one. It does not follow that we have managed our water resources all that well. What with damming, spraying of pesticides, or pollution of lakes and sea, all is not for the best. Moreover, the fishing force, up from the artisanal to the industrial stage, is not all that well disciplined and balanced. The artisanal-industrial ratio may well bear resetting.

Transport Industries range very widely, both physically *(a mari usque ad mare)* and psycho-socially (The Canadian Dream). The network of seaway, railroad, highway, road, street, walk, and path spreads its arterial-to-capillary pattern in ever-denser meshes that do not always relate harmoniously to other land uses.

At this time, transportation in Canada is afflicted with many uncertainties, all of them in some way related to the environmental crisis. Just to hit some of the high points:

(a) The ratio of public to private transportation in most cities is the cause of many breakdowns; unchecked, it leads to acute crises.

(b) The inefficient use of private transport (one person per car) is one of the main causes of excessive expenditure and clogging of streets and highways.

(c) An overwhelming concern with rapid transit has stamped out corresponding preoccupations with the shortening of travel, the protection of valuable land, the viewing of picturesque landscape, the emission of fumes and noise, etc.

(d) The prevalence of private transport has influenced the design and zoning of suburbs to the detriment of physical exercise and neighbourhood activity.

(e) The automobile is the principal agent of air pollution in urban areas.

(f) Noise-abatement laws and speed regulations are both insufficient and unenforced, especially where sports cars, motorboats, motorcycles, and snowmobiles are concerned, and so the dangers to physical and mental health constantly increase.

(g) Over Canada as a whole, and within each province, there is little coordination and planning of the competing services of steamship, plane, railroad, bus, truck, and private car.

(h) Walking is a lost art.

Energy-producing Industries probably thrust the greatest change upon the environment. The official document *An Energy Policy for Canada*, (Department of Energy, Mines and Resources, 1973), seems to be geared to an unconditional growth objective, although it presents itself primarily as a discussion paper.

It requires a very strong dose of optimism and a great deal of scientific and technical imagination to devise a plan (hopefully not too coercive) that will permit continued growth *and* the attainment of the intended economic goals.

Manufacturing is so diverse that lumping metallurgy, distilling, and textiles together may seem artificial. It is not so, I believe, inasmuch as the geographical location of most industrial plants (whether they use mineral, vegetable, or animal raw materials, supplied *in situ* or from afar) obeys very similar requirements. The manufacturing ecosystems need access to constantly renewed raw materials, they must have an uninterrupted amount of power, a reliable work force, and an assured system of transport and marketing. A refinery, a paper mill, or a shoe factory all have an ugly metabolism that has made them, since the beginning of the Industrial Revolution, the polluters par excellence, and rather more so than other kinds of industry considered above.

A picture of industrial Canada, if it could be mapped in some detail, would show the energy charge in terms of the number of successive processes involved in the elaboration of a product. The extraction of gravel, which is immediately spread on a nearby road, involves a small expenditure, whereas the delivery of a pair of shoes in a retail store is the last act of many, from the breeding, feeding, capture, and killing of the mammal; the stripping and tanning of hide; the cutting and sewing of leather; the designing and assemblage of a shoe; its ornamentation and polishing; its packaging, advertising display, and delivery to a wearer. If we can look at this sequence with a truly ecological eye, we are bound to ask ourselves what the hazards are in each of the several ecosystems where at least one stage of treatment occurs (pasture, stock yard, train, slaughterhouse, factory, warehouse, store).

Where does the manufacturing industry stand in Canada, as far as environmental adaptation is concerned? A brief rundown can be made in answer to the following crucial questions, which concern landscape, location, human ecology, zoning, legislation, and recycling.
(a) Where are the plants located – in wild, rural, industrial, or urban landscapes?
(b) Have the environmental conditions of soil, wind direction, stream proximity, natural vegetation, and nature of settlement been considered at the time of location? Afterwards?
(c) Has the human habitat been planned in the place-of-work ecosystem? The residential ecosystem? The recreational ecosystems? The transportation ecosystems?
(d) Does the plant fit a particular niche in municipal zoning?
(e) Are preventive, as well as corrective, measures being enacted? Does the management respond to environmentally designed planning?
(f) Are by-products and wastes the object of recycling practices?

Some partial answers to these six questions can be given which will help to set our sights.

Landscape

Oil refineries, tankers, cisterns, petroleum products, and distribution centres occupy wild areas (Normal Wells Mont-Louis, Gaspé Peninsula), rural areas (Thompson), industrial areas (Montreal East, Burlington), and urban areas (Saint John, Vancouver). A similarly wide spectrum is achieved by the pulp-and-paper installations and to a lesser degree by distilleries, breweries, textile mills, or leather goods plants. They tend to be in an urban setting, are sometimes in a rural one, but are seldom in a wild landscape. They are usually not large enough to create a properly industrial landscape. The food industries are, by and large, rural and urban.

Location

Meat-packing, fishmeal, and pulp factories are, as often as not, upwind from urban or village settlements. The decision-making on location and purchase of lots would seem to be economic-technical and to contain hardly any social or ecological elements, although an increasingly evident force is political (i.e., labour-management contests).

Human Ecology

Principles, foreshadowed by the endeavours of industrial social workers, may now be applied. The stock of information, however, is very low. It seems to have been no one's business to measure the responses of workers to their physical, biological, social, economic, and political environment in their place of work, their place of residence, and in the commuting spaces. The ecological background needs to be sharply etched as a framework for the study of physical and mental health and, by implication, of individual and collective fulfillment.

Zoning

Zoning has, for some time, restricted the free establishment of new industries, and has been instrumental in shutting out some of them. This has been increasingly obvious in the larger urban centres, but is not too apparent in rural areas, where huge storage tanks block the view of fine architectural and natural aspects. In fact, zoning is the latter part of planning (more about this in the section on Urban Canada).

Legislation

Legislation on environmental use has tended to be focussed on abuse, in the first years (from 1965?) of the "environmental crisis". Although Canada seems to have moved already into a more positive phase, where the planning of resource use as a whole and the management of land itself are in honour, the punitive measures are still with us. They are by-and-large rather paltry and inefficient. The cost of adapting cleaning devices to old machinery is alarming to the owners and producers who do not all enjoy the advantages of the planned obsolescence that graces the automobile industry. There are plenty of signs, nevertheless (and whatever the motivation may be), that industrial design has moved into the environmental phase. Canada has no spotless, esthetically pleasant, clean-smelling refinery comparable to Japan's major establishment of this kind at the gates of Tokyo, but it does have a number of good-looking "industrial parks"; some of our rivers have been cleaned of floating logs; the fumes of many industries have abated. But Sudbury, Murdochville, and many other industrial centres still wallow in a mixture of opulence and ugliness.

Recycling

Recycling is unfortunately not the order of the day. It is considered primarily an economic problem and is more often than not dismissed on the grounds of its costliness. It is inseparable from the problem of waste-disposal and is, of course, contingent upon our view of resource supply for the future and limits to growth. The Science Council's Report No. 14 (1971) emphasized recycling as one of the main issues in the Canadian environmental turnover. As we develop an index more satisfying than the GNP for the appraisal of a well-balanced society, and propose a more acceptable (and possibly more "realistic") accountancy for hidden costs and eroded benefits, a more austere "housekeeping" is likely to prevail. Many large companies advertise their exclusive use of recycled paper; some manufacturers of beer and soft drinks use returnable containers. Nevertheless, a shocking amount of potentially useful material is wasted.

The Canadian industrial landscape therefore suffers from a faulty integration into the various regional mosaics. It tends to disrupt the wild and the rural fabrics and it also perverts the urban tissues.

Urban Canada

The ecological searchlight on urban spaces is just beginning to reveal unsuspected aspects of the town as a human habitat. This is not the place to draw a historical perspective on the application of specifically ecological thinking and methodology to the urban milieu. A search for origins and ancestors would take us very far back.

Some of the realities most obvious on the urban scene today in Canada are the following:
(a) Cities are growing in size, area, and population.
(b) Most cities grew without a proper plan, and are obeying, almost exclusively, economic imperatives in their rate and direction of growth, and are not subjected to regional directives (and not too obviously to provincial ones).
(c) Cities are increasing their capacity to control wild, rural, and industrial landscapes.
(d) Pollution (chemical, physical, visual, acoustic) is rampant, too expensive to correct, impossible to prevent.
(e) Housing is inadequate in quantity and quality, and leaves little choice to the individual or to the family.
(f) Poverty and idleness have become the characteristics of most large cities, ever since the beginning of the Industrial Revolution.

Maybe I can attempt to sort out the positive and negative aspects of these six afflictions by regrouping them under six headings:
1. Urban/nonurban patterns and processes
2. Urban growth, size, and structure
3. Housing
4. Planning and zoning
5. Transport
6. Amenities and recreation

In considering the issues that arise under each of these items and in their interaction, the following principles may well serve as a guide.
(a) Diversity in a city consists in the interdigitation and compatibility of as many functions as possible within the neighbourhood cell (residence, commerce, services, recreational and cultural facilities, places of work).
(b) Health in a city depends upon the good repair of all facilities (including transportation and residential) and freedom from pollution (including noise), and easy access to care.
(c) Efficiency in a city is revealed in smoothness of operation, suitability of work-forces to tasks, average to high productivity.
(d) Amenity in a city results from relative freedom of choice in abode and occupations,

ease of access to essential and nonessential facilities, participation in neighbourhood (and city) decisions, pride in collective esthetic and social achievement.

One would like to designate a city, in Canada or elsewhere, in the present or in the past, to which very high points can be given on all four counts. But were not the amenities of Athens, Rome, Paris, or London achieved at the expense of a suffering minority — indeed a suffering majority? Injustice, poverty, and oppression have had historical (sometimes genetic) origins, but they have had, and still have, highly visible ecological consequences and dimensions. Pollution, malnutrition, frustration, and crime are rampant in the greatest centres of high efficiency and wealth, such as New York, Tokyo, or Berlin. We shall obviously have to deal with relative obedience to these four principles. Thus, Montreal rates much higher than New York, but not so high as Helsinki.

Urban/Nonurban Patterns and Processes

The landscape matrices within which our cities are set vary a good deal and afford contacts with wild, rural, and industrial land uses that are all too often chaotic and dysfunctional. The half-moon of Toronto opens on the vast expanse of Lake Ontario; Montreal straddles two islands and sprawls upon the south shore of the St. Lawrence; Vancouver stretches its tentacles along a marine inlet and stops abruptly against steep mountains; Winnipeg, Saskatoon, and Edmonton are each traversed by a more or less encased river and spread radially into adjacent flatland. Thus the possible contacts between the wild, rural, industrial, and urban landscapes (as defined above) are constrained in very different ways.

Urban Structure, Growth, and Size

Toronto, Montreal, and Vancouver all demonstrate the tendencies that dominate urban development in Canada and that will continue to do so in the foreseeable future.
(a) Structure will consist of a night/day — weekday/weekend pulsating core, and an amoebic urban tissue encroaching upon rural land in an ecologically indiscriminate way.
(b) Growth will follow the uncoordinated investments of private enterprise with no consideration for harmony.
(c) Size, whilst not unpredictable, is really no part of planning.

Servicing networks and schedules, should one compare Canadian cities with Stockholm or Copenhagen, are hopelessly empirical and lacking in foresight. This is due, in large part, to the land tenure system, and especially to the reluctance of Canadian municipalities to use their powers of expropriation and constraint.

A tallying of the various constraints permits the development of a grid that would show contrasts in the positive/negative forces in the concentration of populations in towns. The patterning thus obtained would be useful to the definition of urban/nonurban strategy, and would meet the following issues.
(a) How much urban development has occurred on high-grade agricultural land?
(b) What is the motivation of city-dwellers acquiring a second home in a rural or wild area?
(c) What is the true residential preference of suburbanites?
(d) How could land-banks be circumscribed and how can legislation deal with speculators?
(e) What is the real state of air and water pollution, and how is it perceived?

Housing

A study of the metabolism of towns is inseparable from an inquiry into the housing stock.

From an ecological point of view, one of the first things that comes to mind is the matter of building materials. From this point of view, Canadian cities (and even farmhouses) built before 1900 owe a great deal to the characteristics of the geographical region which they occupy: stone, brick, mortar, wood, shingle (and even metal and glass) are derived from local resources. Although these resources may still be available, contemporary houses throughout Canada are constructed with materials imported from great distances. Regional specialization and high industrial efficiency have promoted interregional and international exchange to the highest degree. This flow of building materials, accompanied by the acceptance of standardized architectural patterns, has tended to destroy any particular fitness or harmony of building and landscape. The "mobile home" is the final "all-purpose" structure that belongs nowhere, that has no proper ecological niche.

City-scapes have always shown evidence of historical addition, super-imposition, and succession. A single building, like the cathedral of Chartres, bears witness to the style of several centuries. The present variety of cities as a whole, and of their component neighbourhoods, is thus due to the structure and function of built-up masses. Entire quarters are homogeneous in height, breadth, size, texture, etc., and also belong to the same period. Some of the most attractive European cities display their history, ward by ward, street by street, in the form of buildings as functional now as they ever were. Fragments of such a pageant are visible in Victoria, Montreal, Quebec City, Fredericton, and St. John's, but mostly the juxtaposition of buildings suggests a shock of forms and functions where architectural recall clashes with textural anarchy and functional confusion. The resulting nonenvironment induces recoil and evasion, and makes anything like participation unlikely and unreal.

The human habitat, *sensu stricto*, or abode or dwelling-place, should meet the following requirements:
(a) Space enough to assure privacy and conviviality to all of its occupants.
(b) Peace, or freedom from noise, unsightliness, intrusion, and other nuisances.
(c) Comfort, or protection from excessive heat/cold, light/darkness, humidity/drought; adequate sanitation.
(d) Amenities in decoration, surroundings, storage, conservation, repair and display of possessions.
(e) Neighbourhood integration, or relative similarity of design combined with congenial variety of access to services.

Such are the functions of the individual's (or the family's) habitat. Continuity of occupation and psychological identity are ensured by familiar objects having significance for the inhabitants; fence, tree, door, furniture, books, furnishings, and works of art.

Just how well are Canadians, especially urban Canadians, housed as far as the five criteria mentioned are concerned? What is the range; what is the average in the wild, rural, industrial, and urban areas?

Wild: Canadian hunting and fishing populations (mostly Indian and Inuit) are reasonably well off, it would seem, in this respect, with benefit of nylon tents, portable stoves, etc.

Rural: Rural houses may be getting uglier with each succeeding generation, but are increasingly provided with comfort and household amenities.

Industrial: Typically industrial settlements (of the company-town type) are uniform and monotonous, to be sure, but are very comfortable and generally fairly spacious.

Urban: Urban dwellings show the largest range (from the mansion to the hovel) and reveal a real crisis.

Utilization (and nonutilization) of land is so obviously controlled by socio-economic factors that affect housing that they must at least be enumerated.
(a) Speculation maintains land in an unproductive state, whilst its value increases through public and private investments.
(b) Zoning laws are often obsolete and out-of-step with present requirements, and inhibit diversity.
(c) Building is a nonindustry, uncoordinated, suffering from undue restrictions (e.g., taxes) on the one hand and from lack of planning on the other.
(d) Residential development is completely dictated by economic and not by social benefit.
(e) Urban renewal has, by and large, been preferred to urban restoration.
(f) Taxation policy, as a rule, diverts land to higher density use.

The real-estate lobby, municipal bureaucracy, supply and labour troubles, acceptance of the economic imperative, and indifference to history all combine to allow the disruption of neighbourhoods, the uprooting of communities, and the "Hiltonization" of city-scapes (witness Quebec City).

Planning and Zoning

If zoning in Canadian cities is inadequate, obsolete, and unduly coercive, this may well be a result of administrative shortsightedness, compliance with vested interests, and rigidity of application. It is also due to the fact that alternative options are unknown to many citizens and neglected by planners, that prevalent architectural and social trends go unchallenged, and that the present economic order is assumed to be permanent. The perpetuation of legalized disorder and institutionalized injustice is therefore the standard. It is simply not possible to consider either zoning or planning in a purely technical and scientific way without reference to the underlying values.

Ecology and eco-planning are not likely to redress the Canadian urban prospect. Nor do eco-planners start as nearly from scratch as some of them assume. A submission of urban structures and functions to an ecological examination through the binoculars of natural and social science is nevertheless our principal hope. The presence of ecologists in virtually all phases of planning and implementation is the best possible guarantee of environmental protection and environmental management.

The principal questions that pose themselves to Canadian planners can be grouped under the following headings:
(a) Salvage operations: revitalization of roads, spaces, and buildings whose functions have lapsed or diminished (railroad tracks, empty lots, churches, warehouses).
(b) Renovation processes: destruction of slums, warehouses, commercial blocks, or unsanitary, unsafe, unattractive and nonsignificant buildings to be replaced by more functional ones.
(c) Growth control: patterning and zoning of all intended development, and re-patterning of already obsolete or undesirable servicing media.
(d) Transport and circulation flow: timing of traffic on existing throughways, and re-allocation of walking and driving areas, and of private and public vehicle use.
(e) Spacing and accessibility of amenities: increased ratio of green spaces and other recreational spots to area and population.

The environmental issues under each of these items are numerous. The principle of continuity, mentioned above, militates strongly against the destruction of churches whose congregations have dwindled, old railroad stations that cater to reduced traffic (or none at all), ornate Victorian houses with wastefully monumental staircases, etc. One may well turn to Moscow and Leningrad where some such buildings are turned into museums, and others have been dedicated to a new function (library, swimming pool, community centre), admittedly without much grace. The citizens of a city that forces them to live only in the present, with no reminders of their history (yes, even jails, slavery, colonialism, capitalism, . . .) are very impoverished indeed!

Renovation is justified mainly where a new dedication does not destroy a heritage which is capable of salvage and active use. In fact, an urban planner's program is geared to all six items under review in this section.

Transport

Some of the greatest adversities suffered by the urban population are to be encountered on the transport circuit. The nervous stress of the solitary commuter driving an over-powerful and wasteful vehicle at alternating high speeds and bumper-to-bumper jerking halts, presents a staggering expenditure of unproductive human effort. Harmful levels of chemical pollution and of noise are reached inside the city limits. Damage to vegetation, and to what is left of wildlife (and to human life), is thus rampant, and is increased in snowy Canada by the widespread use of brutal snow-removal machinery and its accompaniment of corrosive salt.

The work-residence-recreation movements, daily and yearly, on the intracity and intercity networks are badly scheduled, and are unaccompanied by the kind of relief which various forms of public or collective transport would offer. The deterioration of the public/private ratio is counter-historical in creating ever-increasing deadlocks.

The crisis in transportation stems from the following causes:
(a) too many vehicles and too much reliance on private transport,
(b) lack of residential facilities in densest working areas,
(c) absence of collective transport in suburbs,
(d) inadequacy of rapid surface and underground transit,
(e) narrow range of working hours,
(f) status attached to large, expensive cars,

(g) increasing cost of fuel,

(h) increasing rate of noise and pollution.

The occasional (and for some commuters, frequent) nightmare of traffic jams epitomizes all of these features.

Amenities and Recreation

In recent years, students of cities have attached much importance to perception of environment. Thus all bear witness to the regional and social diversity of the inner reflection that precedes the projected wish and the move to its implementation.

Perhaps I should start by saying that the five requirements under "Housing" above (space, peace, comfort, amenities, and neighbourhood integration) constitute a lower threshold of amenity, and that some reality of fulfillment is obtained only if an additional surplus is available.

The meaning, or meanings, of "amenity" and "recreation" are to some extent elusive, but even within the compass of a drifting definition it is not impossible to delineate the problems. The leading questions seem to boil down to the following:

(a) How is diversity of choice to be assured with respect to housing, transport, education, recreation, and other forms of sharing in the collective benefits?

(b) What mechanisms do we have for offering real alternatives and for quantifying the trade-offs?

(c) How do we measure efficiency or productivity at work, and value of service rendered?

(d) What is the final measure of health (physical and mental) and of community fitness?

(e) What is really "tolerable" in a polluted environment, and what price can a city ultimately pay for cleanliness?

(f) In what way do we really need green spaces?

(g) How can we assure a reasonable balance (although a shifting one) of privacy/togetherness, (architecturally, socially, environmentally)?

(h) What are the most satisfactory ways of ensuring consultation and also efficiency of decision-making?

In a strict sense, none of these questions are environmental. They are social and psychological and, of course, economic. But different answers have different ecological consequences.

Priorities

This cursory scanning of Canada's landscape with a roving ecological eye reveals a number of crises in the diverse environments. In a world perspective, Canada is very fortunate indeed, for it does not suffer from earthquakes like Costa Rica, from drought like the Sahelian lands, from erosion like Haiti, from hunger like India, or from crowding like Japan. That is not to say that the Canadian environment is not menaced. I have been at pains to point out many deteriorations, some of them irreversible, that scar so many landscapes. Good management has not always prevailed; even less, ecological wisdom.

Canada needs a program on environmental management. It may well be the Canadian Environmental Advisory Council's primary vocation to write such a program. The questions defined, considered, and investigated by the Council spread across a wide frame, essentially covering the questions outlined above. I shall not presume to outline such a program, but I will try to abstract from the foregoing survey what seem to me the inescapable focal points: austerity, science, planning, information, and consultation. Under each of these headings we may be able to gather the threads of sound principles to test the issues that have arisen in the face of various crises and, through study and research, to reach solutions that will allow better use of our resources.

1. Austerity requires a heightened consciousness of our whole environment and a willingness to spare and to re-use.

2. Science is the indispensable tool of investigation which will provide a true image of our ecosystems prior to decisions concerning their exploitation.

3. Planning is derived from moral, political, social, and scientific tests of the technically, economically, and ecologically allowable choices.

4. Information must be obtained in full freedom, and disseminated at various levels.

5. Consultation is of the essence if the people potentially affected by a decision concerning their environment are to participate; and, at the other end of the scale, if workable international agreements are going to be enacted.

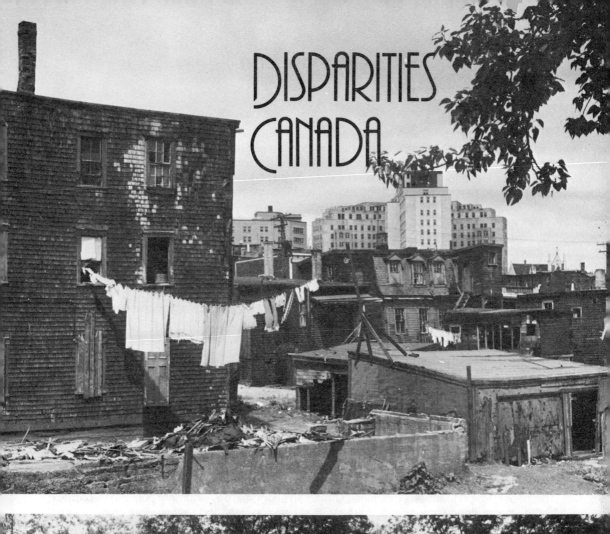

DISPARITIES
CANADA

Introduction

Canada is a diverse country — in resources, in physical attributes, in population, and in culture. Unfortunately, diversity, which is often thought of as a strength, has not led to widespread and more or less equally-shared economic development and opportunity. The problem of economic and social disparity is a long-standing one for Canadians and it continues to persist despite direct and indirect federal and provincial efforts to alleviate the imbalance.

In the introductory paper by the Economic Council of Canada, the social and economic anatomy of the disparity problem is unfolded. The Atlantic region suffers most, whereas Ontario, as in the past, continues to enjoy prosperity along with Alberta and British Columbia. The picture drawn by the Economic Council in its Second Annual Review, *Towards Sustained Economic Growth* (1965) has not changed substantially over the intervening thirteen years.

Income gaps and job opportunities continue to persist despite direct and indirect federal and provincial intervention to attract industry, create more employment opportunities, and upgrade skills and education in the lower income regions. One of the major federal bodies charged with reducing regional disparities is the Department of Regional Economic Expansion (DREE), created in 1969. The evaluation of DREE's performance in fulfilling its mandate is the subject of the second paper. Although the evidence is not conclusive, it does appear that DREE's activities, particularly the Regional Development Incentives Act program, have been at best only marginally successful in creating job opportunities.

Federal efforts in the regional economic development field have been subject to widespread criticism, largely because they are expensive, and in the view of some politicians, economists, and others, they have not operated effectively to reduce economic disparities. In fact some federal government ministers have been quoted as suggesting that DREE should be dismantled and its function assumed by other government departments. In his paper, Ralph Matthews examines Canadian economic development policy in the context of the dependency theory which argues that development and underdevelopment are not separate phenomena but are part of the same economic process, and that the primary cause of underdevelopment is not internal to those regions but external to them. He argues that Canadian economic development policy, since it is based on incentive grants to corporations, is trying to relieve regional underdevelopment by giving financial assistance to those corporations from developed regions (e.g. central Canada) which are essentially the cause of underdevelopment. He concludes that if this approach continues the hinterland regions of the country are doomed to perpetual stagnation.

In the final paper, Burke and Ireland demonstrate that in recent years it is no longer accurate to label all of the Atlantic Provinces as economically underdeveloped. The 1971-74 period witnessed the strengthening of the urban-industrial core region (Fredericton, Saint John, Moncton, Halifax-Dartmouth, and St. John's) to the point where this belt has acquired the economic characteristics of the more advanced parts of Canada. Unfortunately the regions peripheral to the developing urban core region have not experienced such favourable growth, and they continue to exhibit the characteristics of a lagging region. To alleviate the internal disparities (centre vs. periphery) in the Atlantic region they suggest continued federal government expenditures in the urban core followed by an increasing proportion of development support in the peripheral urban areas — a strategy that Ralph Matthews argues will continue to create dependency.

The Anatomy of Regional Disparities

ECONOMIC COUNCIL OF CANADA

The material well-being of a society depends not only on absolute wealth and purchasing power, but also on their distribution among individuals and families. Because there are returns to scale in family living, the same income per capita goes farther if a society is primarily comprised of families rather than of individuals living alone. There will also be less discontent in society if the wealth is not unduly concentrated in the hands of a few, or if the income distribution does not change too quickly. These factors differ from region to region. For example, in 1970 the average Newfoundland family had roughly one more child and about one-third less income than its Ontario counterpart and, apart from housing, it faced a higher cost of living; these factors help explain why Newfoundland is an area of net out-migration and Ontario is not.

In a work-oriented society, having a job is an important aspect of social well-being. Not only does it provide income, but it also brings respectability and a sense of self-importance. In addition, it frequently provides the worker with an agreeable social environment in which to spend a large part of his life. Unfortunately, there are large regional disparities in access to jobs. A worker in New Brunswick faces a higher probability of being unemployed than his Alberta counterpart; moreover, if he is unemployed, he can expect to remain that way for a longer time.

Since there are difficulties in measuring real income, it is also important to consider regional disparities in social areas such as housing, health, environment, and education, which provide some alternative indicators of individual well-being. These may in fact reveal that there are some nonmonetary advantages to living in a region where incomes are lower that may adequately compensate for differences in income. Furthermore, government may wish to intervene directly to provide greater equality between regions in the distribution of, for example, health and education services. Finally, these social indicators may provide some useful warnings of the dangerous side effects of economic activity. For example, the high incomes and rapid economic growth observed in Alberta, British Columbia, and the Yukon give the appearance of material happiness. One notes, however, that this economic success has been accompanied by abnormally high rates of suicide and marriage breakdown. Thus there may be serious complications in implementing policies designed to increase incomes and encourage people to move to areas where job opportunities are better; some migration policies may prove better than others.

Demographic Differences

It is useful to examine population figures for various regions because regional demographic differences influence certain economic variables, such as participation rates and per capita incomes, and are in turn determined partly by economic variables to the extent that these exert an influence on interregional migration.

For many years, the southern parts of Ontario and British Columbia have been the destinations preferred by foreign immigrants and Canadian migrants alike. Projections based on demographic trends show that by 1985 these are the only two regions where the working-age population will be increasing. In other regions, this growth will be almost zero.[1] Therefore, unless there are considerable increases in participation rates or major advances in productivity, these regions will experience very weak economic growth.

In the Atlantic region, Quebec, and the North, family sizes in 1971 exceeded the national average, and the smallest average families were to be found in British Columbia and Ontario (Table 1). However, current family size reflects past fertility rates, which have fallen dramatically since 1956. At that time, Canadian mothers could expect to bear an average of 3.9 children in their lifetime; by

● Reprinted from Economic Council of Canada, *Living Together: A Study of Regional Disparities* (Ottawa: Minister of Supply and Services, 1977), Ch. 4, pp. 31-60, with permission. The article has been slightly edited.

Table 1

DEMOGRAPHIC CHARACTERISTICS, CANADA, BY PROVINCE AND TERRITORY, SELECTED YEARS, 1961 TO 1976

	New-found-land	Prince Edward Island	Nova Scotia	New Bruns-wick	Quebec	Ontario	Mani-toba	Saskat-chewan	Alberta	British Colum-bia	Yukon	North-west Terri-tories	Canada
Population					*(Thousands)*								
Estimated total, January 1976	554	120	830	684	6224	8290	1023	929	1804	2481	21	38	22 998
Rate of:					*(Percent)*								
Population change, 1961-71	14.0	6.7	7.0	6.1	14.6	23.5	7.2	0.1	22.2	34.1		41.4*	18.3
Natural increase, 1961-71	23.7	13.1	13.2	14.8	14.2	13.8	13.0	13.7	18.0	11.8		35.8*	14.3
Net migration, 1961-71	-9.7	-6.4	-6.1	-8.7	0.4	9.7	-5.6	-13.6	4.2	22.3		5.6*	4.0
Interprovincial net migration, 1966-71[1]	-3.1	-0.9	-0.9	-2.3	-1.9	0.9	-3.1	-6.5	2.6	7.7		n.a.	n.a.
Interprovincial net migration, 1971-75[1]	-2.2	2.6	-0.2	0.6	-1.2	0.4	-3.6	-6.6	2.0	5.8		n.a.	n.a.
Urban population as a percentage of total population, 1971	57.2	38.3	56.7	56.9	80.6	82.4	69.5	53.0	73.5	75.7		n.a.	76.1
Family data					*(Average)*								
Persons per family, 1971	4.4	4.0	3.8	4.0	3.9	3.6	3.6	3.7	3.7	3.5		4.3*	3.7
Children per family, 1971	2.4	2.0	1.8	2.0	1.9	1.6	1.7	1.8	1.8	1.6		2.4*	1.7
Fertility rate, 1974[2]	n.a.	2.2	2.0	2.1	1.7	1.9	2.2	2.4	2.1	1.8	3.1	3.5	1.9
Dependency ratio													
Youth dependency ratio, 1971[3]	65.9	55.4	50.6	53.9	46.6	45.5	47.3	50.9	51.7	44.5	55.3	78.2	47.5
Old age dependency ratio, 1971[3]	10.9	19.3	15.2	14.5	10.8	13.3	15.8	17.2	11.9	15.0	4.4	3.9	13.0

n.a. – not available.

*Yukon and Northwest Territories combined.

[1]Estimates based on migration of children eligible for family allowance.

[2]Total fertility rate is the sum of the age-specific fertility rates. The age-specific fertility rate at age "a" is equal to:

$$\frac{\text{number of births among women at age ``a''}}{\text{total number of women at age ``a''}}$$

[3]Youth dependency ratio is 100 times the ratio of the number of people in the 0-14 age group to the number of people in the 16-64 age group. Old age dependency ratio is 100 times the ratio of people in the 65-and-over age group to people in the 15-64 age group.

Source: Estimates by the Economic Council of Canada, based on data from Statistics Canada.

1974 that number had dropped to less than 1.9. The most spectacular decline occurred in Quebec. Whereas the fertility rate in that province slightly exceeded the Canadian average in 1956, it had fallen to just below 1.7 by 1973.

As a result of past fertility and migration rates, the ratio of children to working-age people is high in Newfoundland, very high in the Northwest Territories, and low in Ontario and British Columbia. In Prince Edward Island, Saskatchewan, and Manitoba, the proportion of the provincial population over sixty-four years of age is higher than the Canadian average, partly because young people have tended to move out, leaving their parents behind. In contrast, the Yukon and Northwest Territories have very few elderly people, partly because a number of Whites leave the North when they reach retirement age and partly because native people, who have had a high mortality rate in the past, have a lower average age. Moreover, a number of young people have moved in from the South to work in the gas wells, mines, and defense establishments in the North, thus lowering the average age in the two territories.

Net interprovincial migration, which was substantial between 1961 and 1971, seems to have slowed down between 1971 and 1975, and the trend towards net out-migration seems to have reversed in Prince Edward Island and New Brunswick, which may be a sign that economic conditions in the Atlantic region are improving.[2] Saskatchewan stands out as the province losing the most people; its population has declined steadily from 960 000 in 1968 to 907 000 in 1974 — a loss of 5.5 percent. Most of that loss occurred in 1970 and 1971, when net farm income suffered badly. Since the province's unemployment rate did not rise precipitously during this period, it may be inferred that, when economic conditions deteriorate in Saskatchewan, the people simply leave to take jobs in the expanding economies of neighbouring Alberta and British Columbia.

Differences in Income

Data on personal income per capita — which have been available on a regional basis for many years — constitute a good starting point in measuring regional differences in income to show how these have changed over time. Note, however, that this indicator overstates regional income disparities because it ignores several important elements. First, progressive income taxes take a larger share of the incomes of wealthy regions, so that regional disparities in disposable income are not as large as differences in personal income. Moreover, personal income data neglect the fact that per capita income goes farther in lower-income regions where the population is generally organized into larger family units, and so disparities in income per family are not as large as income per capita. Finally, this indicator ignores the fact that the cost of living differs from region to region. It is also important to see how this income is distributed within a region — both among people and among geographic subregions.

Personal Income per Capita

There has been a persistent pattern of regional disparity in per capita incomes over the past half century, although there has been a very slow convergence towards the national average, especially since 1954 (Figure 1). Historical events have caused some notable variations in these patterns. Income per capita is highly variable in the Prairie Provinces because this region is dominated by competitive agricultural markets, notably wheat for export. The prices of agricultural products are subject to great fluctuations, because variable weather conditions cause crop yields to fluctuate against the price-inelastic world demand for food. The onset of the Great Depression in 1929, which caused a collapse of international grain prices, and the severe drought conditions in the 1930's, which destroyed crops and forced people to abandon certain parts of the Prairies, led to a dramatic drop in per capita income in this region. Relative prosperity returned during the Second World War and reached its peak during the Korean War, when the worldwide boom in commodity prices coincided with very large wheat crops and with rapid growth of the region's young petroleum and natural gas industries.

Per capita income grew rapidly in the Atlantic region during the Second World

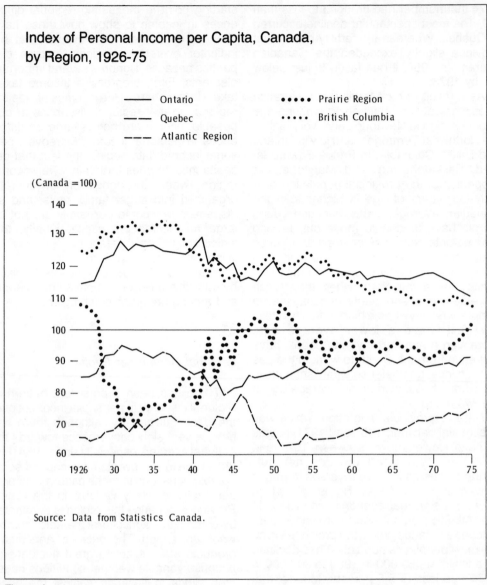

Index of Personal Income per Capita, Canada, by Region, 1926-75

—— Ontario	••••• Prairie Region
– – Quebec	••••• British Columbia
– – – Atlantic Region	

(Canada = 100)

Source: Data from Statistics Canada.

Figure 1

War, when Halifax became the point of departure for convoys of troops and supplies headed for Europe, but the end of the war brought relative economic decline. Since the early 1950's, per capita income in this region has grown a little more rapidly than the Canadian average, but extrapolation of current trends shows that it would still take about seventy years for incomes in the Atlantic region to reach the national average.

The components of personal income per capita – wages and salaries, transfer payments, dividends, and interest payments – have varied since 1954 (Table 2). It should

be noted that total farm income per capita has not grown at all in the Atlantic region or British Columbia; while income per farmer has risen, this increase has been offset by the decline in the number of farmers. Growth of farm income has been sluggish in Quebec and Ontario, but fast in the Prairies.[3] Since the end of the Second World War, agriculture has intensified in the Prairies, whereas a substantial amount of farmland in Quebec and the Atlantic Provinces is no longer used for agricultural production.

The fastest-growing sources of personal income have been dividends and interest in

GROWTH[1] OF COMPONENTS OF PERSONAL INCOME PER CAPITA, CANADA,
BY REGION, 1954-75

	Atlantic Region	Quebec	Ontario	Prairie Region	British Columbia	Canada
	Average annual rate of growth ($ current)					
	(Percent)					
Market income	7.1	6.8	6.4	6.9	6.1	6.7
Wages and salaries[2]	7.6	7.1	6.8	7.4	6.5	7.1
Farm income	0.0	1.8	1.8	5.1	0.5	3.2
Unincorporated nonfarm income	3.8	3.6	3.1	3.5	2.8	3.4
Dividends and interest	9.2	8.7	8.3	10.4	8.5	8.9
Government transfer income	10.2	9.4	8.9	8.2	7.2	8.9
Total income	7.6	7.1	6.6	7.0	6.3	6.9

[1]Estimated, using an unrestricted logarithmic regression.

[2]Including "other" income, which is mainly military pay and allowances.

Source: Estimates by the Economic Council of Canada, based on data from Statistics Canada.

Table 2

Figure 2

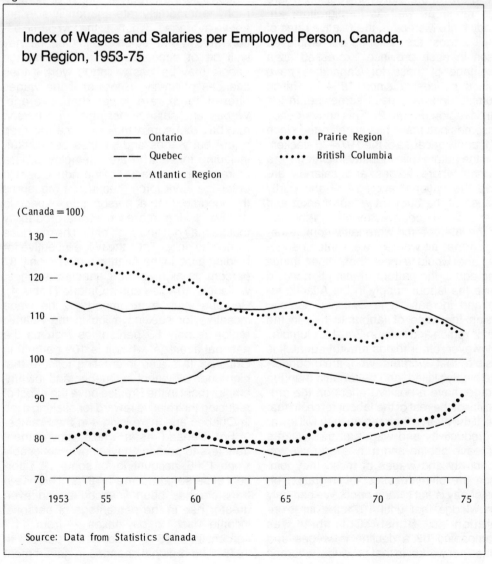

Index of Wages and Salaries per Employed Person, Canada, by Region, 1953-75

——— Ontario

—— Quebec

——— Atlantic Region

••••• Prairie Region

••••• British Columbia

(Canada = 100)

Source: Data from Statistics Canada

the Prairies and British Columbia, and government transfers[4] in the Atlantic region, Quebec, and Ontario. It is interesting to note that, in all regions, both these sources of personal income have grown much faster than unincorporated income and farm income. Everywhere, the corporate and government sectors have become more and more important, while the relative importance of individual proprietorship has declined.

The level of wages and salaries per capita is affected by factors such as rates of pay, migration, the proportion of the population that is of working age, and the proportion of the working-age population that is employed. Wages and salaries per capita have grown more slowly in regions where wage rates, as well as in-migration, are already high. Moreover, the relative rates of pay — wages, and salaries per employed person in each province expressed as a percentage of those for Canada — have declined noticeably since 1954 in British Columbia and have risen somewhat in the Atlantic region (Figure 2). This convergence of incomes nationally has accelerated since 1971, partly because of the 1974-75 slackening in the automobile industry in Ontario — a province where wages and salaries are above the national average — and partly because of the buoyant wheat market and the energy-related investments in Alberta.

If the labour force were fairly homogeneous, so that all workers were interchangeable, one would expect that, other things being equal, the pattern of migration would reduce the labour supply in the Atlantic region and increase it in British Columbia, causing the price of labour to rise in the Atlantic region and to fall in British Columbia.

However, other things are not equal. It is possible that workers who are younger, more highly skilled, and better paid than the average, have a positive effect on the productivity of the rest of the labour force. In that case, if they leave their region they will lower the productivity and wages of the workers they leave behind and may even raise the productivity and wages of those they join. Hence the ultimate effect of migration on rates of pay is not clear a priori. We can only acknowledge that until 1971 the large in-migration to British Columbia was accompanied by a decline in wages and salaries per worker in that province, whereas

the moderate in-migration to Ontario was not.

The proportion of population of working age grew most rapidly in Quebec and the Atlantic Provinces from 1954 to 1975 (Table 3). However, these demographic results were partly offset by the fact that the proportion of employed persons rose more rapidly in other regions. The fact that the Atlantic region had the lowest proportion of working-age people in its total population and the lowest proportion of employed people in its working-age population helps to explain why its personal income per capita was so far below the Canadian average.

Except for the effects of migration, the low proportion of working-age people in the Atlantic region is largely a matter of demography — its fertility rate and family size are above average — whereas the low participation rates in that region and in Quebec may well be of economic origin. Working-age people may be less willing to work if their chances of finding a job, or if the wages offered them, are lower than average. Wages and salaries per employed person may be a measure of the inducement to keep a job, but wages and salaries per worker, including those who are unemployed, is a more accurate measure of the inducement to enter the work force, because it measures the income that a person can expect to receive, taking into account the probability that he may not find a job or that he may lose it once he does. Thus the wage incentive for holding a job in the Atlantic region is only 87 percent as high as the Canadian average, while it is 107 percent in Ontario (Table 4). Perhaps even more important, the wage incentive for seeking a job in the Atlantic region is only 83 percent as high as the national average, while it is 109 percent in Ontario.[5] It is also interesting to note that high unemployment in Quebec and low unemployment in the Prairies have the effect of reducing the relative reward for seeking a job in Quebec and increasing it in the Prairies.

For Canada as a whole, government transfers to persons have grown noticeably since 1966, accounting for some 12.7 percent of personal income in 1975.[6] This rise in transfers has been financed by an even greater rise in the percentage of personal income used to pay taxes — from 11.7 percent in 1965 to almost 19.0 percent in 1975. The regional changes in government

EMPLOYMENT AND POPULATION RATIOS, CANADA, BY REGION, 1954-56 AND 1973-75

	Proportion of:		
	Population that is of Working Age	Working-age Population Employed	Total Population Employed
	(Percent)		
Atlantic Region			
1954-56	63.5	43.3	27.5
1973-75	70.5	46.1	32.5
Percentage change	11.0	6.5	18.2
Quebec			
1954-56	65.9	50.4	33.2
1973-75	75.4	52.2	39.3
Percentage change	14.4	3.6	18.4
Ontario			
1954-56	70.2	54.4	38.2
1973-75	74.8	57.7	43.1
Percentage change	6.6	6.1	12.8
Prairie Region			
1954-56	67.1	50.3	33.7
1973-75	71.4	57.0	40.7
Percentage change	6.4	13.3	20.8
British Columbia			
1954-56	70.2	49.0	34.4
1973-75	74.8	55.2	41.3
Percentage change	6.6	12.7	20.1
Canada			
1954-56	67.6	50.9	34.4
1973-75	73.8	54.7	40.4
Percentage change	9.2	7.5	17.4

Source: Based on data from Statistics Canada.

Table 3

Table 4

WAGES AND SALARIES, UNEMPLOYMENT, AND PARTICIPATION RATES, CANADA, BY REGION, 1975

	Atlantic Region	Quebec	Ontario	Prairie Region	British Colum- bia	Canada
Index of wages and salaries:						
Per employed person	87	95	107	91	106	100
Per worker in the labour force	83	93	109	95	105	100
	(Percent)					
Unemployment rate	11.6	8.8	6.0	3.4	8.3	7.1
	(9.9)	(8.1)	(6.3)	(3.9)	(8.5)	(6.9)
Participation rate of population aged 14 and over	51.9	57.2	61.3	59.3	60.2	58.8
	(53.5)	(58.5)	(64.2)	(62.8)	(61.3)	(61.1)

Source: Department of Finance, *Economic Review* (April 1976). The unemployment and participation rates in parentheses are based on Statistics Canada's Revised Labour Force Survey.

transfers per capita since 1953 have brought personal income per capita in British Columbia and in the Atlantic region closer to the national level (Figure 3). In the past, a large proportion of government transfers have gone to British Columbia, partly because its share of old age pensioners was larger than its share of total population, although this situation may now be changing. Transfers have particularly favoured the Atlantic region, because a relatively larger proportion of its population qualifies for family allowances, old age pensions, and unemployment insurance. The recent surge in per capita income transfers going to the Atlantic region and Quebec coincide with the revision of the Unemployment Insurance Act in July, 1971.

Regional Income Distributions

Ontario is doubly blessed compared with the Atlantic region; not only does it have higher income levels, but these are more evenly distributed within the population of the province (Figure 4). In Ontario, as well as in Alberta, British Columbia, and the Yukon,

Figure 3

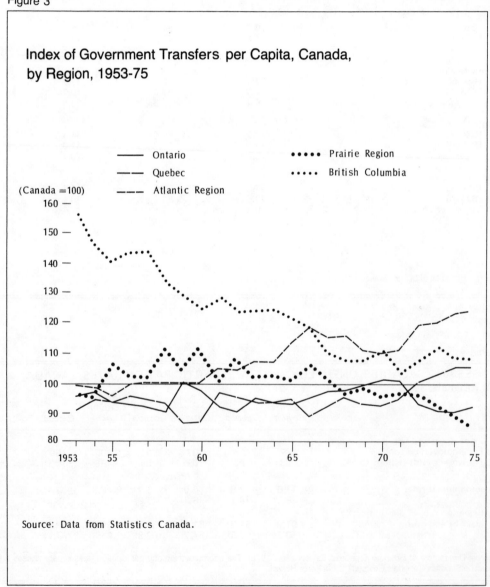

Index of Government Transfers per Capita, Canada, by Region, 1953-75

Source: Data from Statistics Canada.

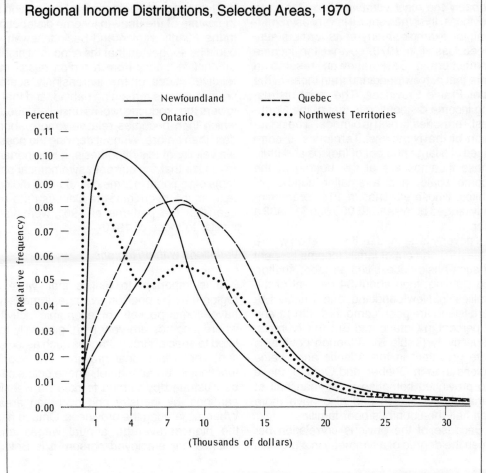

Regional Income Distributions, Selected Areas, 1970

Percent

——— Newfoundland   ———— Quebec
——— Ontario   •••••• Northwest Territories

(Relative frequency)

0.11 —
0.10 —
0.09 —
0.08 —
0.07 —
0.06 —
0.05 —
0.04 —
0.03 —
0.02 —
0.01 —
0.00 —

1   4   7   10   15   20   25   30

(Thousands of dollars)

Note. The family income distribution for Alberta, British Columbia, and the Yukon resembles that of Ontario. Prince Edward Island, New Brunswick, Nova Scotia and Saskatchewan have skewed distribution, which varies between that of Newfoundland and Quebec. Manitoba is similar to Quebec.

Source: Data from Statistics Canada.

Figure 4

the modal group — that is, the one with the greatest number of families — in 1970 comprised families receiving between $10 000 and $14 999 a year. The income distribution in this group of regions is fairly symmetrical and has a high mean and mode. It is interesting to note that the regions where mean income is highest and where the income distribution is most symmetrical are also those where net in-migration is greatest.

In 1970, the modal income group in Newfoundland — the province with the most net out-migration — consisted of families receiving between $1000 and $3999 a year. The

patterns of income distribution in Prince Edward Island, New Brunswick, and Nova Scotia were similar to that of Newfoundland, but they were somewhat more symmetrical. The modal group for these three provinces in 1970 consisted of families receiving between $4000 and $6999 a year. In Quebec, the modal income range (between $7000 and $9999) was lower than in Ontario. Quebec's net in-migration during 1961-71 period was almost zero. The income distribution that most closely resembles Quebec's is that of Manitoba, where net out-migration during this period was close to zero.

Saskatchewan's income distribution is probably the most variable of all Canadian provinces. In some years, it is very asymmetrical; for example, many of its farmers sustained losses in 1970, so that the income distribution in Saskatchewan resembled more that of Newfoundland than those of the other Prairie Provinces. The most distressing income distribution is that of the Northwest Territories, which is bimodal. The population of the Northwest Territories is composed of a large number of Indians and Inuit, whose incomes are at the bottom of the income scale, and a smaller number of Whites, whose earnings in 1970 predominantly ranged between $10 000 and $14 999 a year.

If poverty is defined as the situation where at least 70 percent of family income is spent on such basic necessities as food, shelter, and clothing, then about 34 percent of the families in Newfoundland and Prince Edward Island are poor, compared with 11 and 12 percent in Ontario and British Columbia, respectively (Table 5). Although poverty is more prevalent in the Atlantic and Prairie regions than in Quebec and Ontario, these two provinces comprise more than half of Canada's population and also have more than half the country's poor families.

Because of the obvious correlation between the degree of asymmetry in a region's income distribution and its net out-migration, it is tempting to hypothesize that migration is the cause of the uneven income distribution in the Atlantic region and Saskatchewan. It could be suggested that the exodus of highly productive labour from a region raises the relative wages of the increasingly scarce skilled workers remaining behind, and that it lowers the productivity of its unskilled labour, which then becomes relatively more abundant than before. Without denying the possible validity of this hypothesis, it is important to realize that the current asymmetrical patterns of regional income distributions did not take shape overnight; the Census data reveal that the patterns of 1961 were very similar to those of 1970.

Variations within Regions

It is important to realize that, while a region may be poor on average, it may still have some pockets of economic activity where incomes are very high, generally related to special circumstances such as isolation, very rapid local growth, or a poorly functioning labour market. These points can be illustrated by turning to data on weekly earnings, by industry and by urban area. Whereas, among the provinces, Ontario has the highest average annual wages and salaries per employed person, it is British

Table 5

REGIONAL NET MIGRATION, POVERTY, AND MEAN FAMILY INCOME, CANADA, BY PROVINCE AND TERRITORIES, SELECTED YEARS

	New found-land	Prince Edward Island	Nova Scotia	New Bruns-wick	Quebec	Ontario	Mani-toba	Saskat-chewan	Alberta	British Colum-bia	Yukon and North-west Terri-tories	Canada
	(Percent)											
Net migration rate, 1961-71	-9.7	-6.4	-6.1	-8.7	0.4	9.7	-5.6	-13.6	4.2	22.3	5.6	4.0
Low-income families as a percentage of all families, 1971	33.7	34.0	23.0	24.1	17.7	11.2	19.4	27.9	17.9	12.0	n.a.	15.9
Regional distribution of low-income families, 1971	4.3	1.0	5.0	4.1	29.9	26.1	5.7	7.4	8.5	8.0	n.a.	100
	(Dollars)											
Mean family income, 1970	6680	6989	7858	7479	9260	10 661	8646	7328	9475	10 019	11 194 (Yukon) 8 449 (N.W.T.)	9600

n.a. – not available.

Source: Data from Statistics Canada.

UNEMPLOYMENT AND AVERAGE WEEKLY EARNINGS, CANADA, BY REGION AND
MAJOR URBAN AREA, JUNE 1976

	Unemployment Rate (Seasonally Adjusted)	Weekly Earnings[1]
	(Percent)	(Dollars)
Atlantic Region	11.3	202
Newfoundland	12.4	224
St. John's		204
Corner Brook		238
Prince Edward Island	12.4	162
Charlottetown		178
Nova Scotia	10.5	194
Halifax		193
Sydney		215
Truro		164
New Brunswick	11.5	202
Saint John		219
Edmundston		210
Quebec	7.8	224
Sept Iles		326
Baie Comeau		286
Chicoutimi		248
Rouyn-Noranda		257
Montreal		226
Ontario	6.3	231
Sudbury		256
Cornwall		225
Oshawa		278
Toronto		230
Hamilton		237
St. Catharines		261
Sarnia		291
Prairie Region	4.2	224
Manitoba	4.5	207
Winnipeg		196
Saskatchewan	5.0	215
Regina		212
Alberta	3.8	238
Calgary		231
British Columbia	8.8	262
Vancouver		255
Prince George		271
Kamloops		238
Victoria		226
Yukon Territory	n.a.	316
Northwest Territories	n.a.	268
Canada	7.0	229

n.a. — not available.

[1]Industrial composite — annual average of larger firms.

Source: Data from Statistics Canada.

Table 6

Columbia that has the highest weekly earnings.[7]

Data for all industries combined show that, although the unemployment rate is highest in Newfoundland, the rate of remuneration in that province is higher than in five other provinces (Table 6). The industrial mix, of course, differs from province to province, but weekly earnings in Newfoundland are higher in several industries. In particular, weekly earnings in mineral ore mining and in pulp and paper are higher in that province than in Ontario or Quebec (Table 7). Newfoundland's labour market does not function well because many of its unemployed workers are fishermen scattered among the outports along the island coast, where it is not easy to recruit skilled labour for the commencement of large projects. The extent of unemployment and of labour force participation differs markedly within Newfoundland. In the very depressed North Shore region,

AVERAGE WEEKLY EARNINGS IN SELECTED INDUSTRIES, BY PROVINCE, JUNE 1976

	Average Weekly Earnings
	(Dollars)
Forestry	
Newfoundland	286
New Brunswick	169
Quebec	247
Ontario	308
British Columbia	342
Metal mining	
Newfoundland	347
Quebec	342
Ontario	297
Manitoba	275
British Columbia	340
Durable manufacturing	
Nova Scotia	219
New Brunswick	205
Quebec	238
Ontario	266
Manitoba	219
Saskatchewan	244
Alberta	254
British Columbia	294
Food and beverages	
Newfoundland	164
Nova Scotia	171
New Brunswick	159
Quebec	225
Ontario	237
Manitoba	225
Alberta	234
Saskatchewan	240
British Columbia	246
Pulp and paper	
Newfoundland	337
New Brunswick	306
Quebec	309
Ontario	319
British Columbia	335
Smelting and refining	
Quebec	306
Ontario	288
Chemicals	
Quebec	252
Ontario	267
Alberta	300
British Columbia	263

Source: Data from Statistics Canada.

Table 7

participation rates are apparently well below 50 percent, and of those relatively few persons seeking employment, perhaps one in five may be without a job.

Apart from Newfoundland's rather special circumstances, average weekly earnings are noticeably lower in the Atlantic region than in the other major regions. Within Quebec, average weekly earnings are noticeably higher in more remote areas such as Baie Comeau and Sept-Iles, but lower in older established areas such as Chicoutimi and Rouyn-Noranda. In Ontario, average earnings are higher in cities containing automobile assembly plants (Oshawa) or petrochemical plants (Sarnia), and in cities near a large U.S. industrial centre (St. Catharines). In the Prairies, wages are noticeably higher in Alberta than in Saskatchewan or Manitoba. In British Columbia, average earnings are higher in Vancouver and in northern interior cities, such as Prince

Table 8

ALTERNATIVE MEASURES OF REGIONAL INCOME DISPARITIES, BY PROVINCE AND TERRITORIES, 1970[1]

	New-found-land	Prince Edward Island	Nova Scotia	New Bruns-wick	Quebec	Ontario	Mani-toba	Saskat-chewan	Alberta	British Colum-bia	Yukon and North-west Terri-tories
Regional indexes:					*(Canada = 100)*						
Market income per capita[2]	55	60	75	68	88	120	92	70	100	109	101
	(56)	(57)	(75)	(68)	(89)	(114)	(96)	(96)	(104)	(110)	(101)
Personal income per capita	63	67	77	72	89	118	93	72	99	109	95
Personal disposable income per capita	68	72	79	75	90	116	94	75	100	109	93
Average family income	70	73	82	78	96	111	90	76	99	104	99
		(77)			(94)	(109)		(100)		(105)	(n.a.)
Average family disposable income[3]	74	79	84	81	98	109	91	79	99	104	97
Family disposable purchasing power											
Excluding housing cost differential[4]	70	76	82	78	98	110	94	80	101	101	69
Including housing cost differential	75	77	83	87	102	106	97	85	101	97	70

n.a. – not available.

[1] All figures in parentheses are for 1974.

[2] Personal income minus transfers to persons, divided by population as of June 1, 1970.

[3] Family income adjusted by the ratio of disposable income to personal income in each region.

[4] Family disposable income deflated by intercity partial consumer price index: Winnipeg = 100, May 1971. The price index for the Yukon and Northwest Territories was assumed to exceed that of Edmonton by 48 percent.

Source: Estimates by the Economic Council of Canada, based on data from Statistics Canada; and Gemini North Limited, *Social and Economic Impact of Proposed Arctic Gas Pipeline in Northern Canada,* Book 1, May 1974, p. 506.

George, than elsewhere. Average earnings are even higher in the Yukon and Northwest Territories because workers are reluctant to locate in the North and so must be offered higher wages as an attraction.

Alternative Measures of Regional Income Disparities

Regional income disparities appear largest if data on market income per capita are used (Table 8). This measure is interesting in that it shows the disparities in earning power.[8] As might be expected, Ontario leads the nation while, at the other end of the spectrum, the earning power of the average Newfoundlander, in 1970, was only 55 percent of the national average, and substantially less than half that of the average Ontarian. Market income data for both 1970 and 1974 illustrate how extremes in business conditions can affect regional earning power per capita. In particular, 1970 was the poorest year for farm income in the Prairies since the 1961 recession, whereas 1974 was the peak of the boom in grain exports generated by the recent world-wide food shortage, and was the year in which the sharp decline in automobile exports led Ontario into one of its worst recessions. While fluctuations in farm income have a substantial effect on the market incomes of all three Prairie Provinces, the most extreme variations occur in Saskatchewan, where aggregate net farm income rose from $185 million in 1970 to $1094 million in 1974. In a depressed year, earning power in Saskatchewan is similar to that in some of the Atlantic Provinces. Even in 1974, its earning power per capita did not reach the Canadian average.

It is important to consider how much of the regional income disparities remain if regional differences in income transfers, personal taxation, and living costs are taken into account, and if income per family is used rather than income per capita.

Transfers, which constitute the essential difference between market income and personal income, generally have the effect of reducing interregional income differences. In 1970, they narrowed the income gap between Newfoundland and Ontario residents from 65 percentage points for market income to 55 points for personal income (Table 8). Per capita transfers in 1970 ranged from a high of $433 in Newfoundland, where they constituted over 40 percent of personal income, to $333 in Ontario, $306 in Quebec, and only about $140 in the Yukon and Northwest Territories. While government transfers narrowed the disparities between the Atlantic region, Manitoba, Saskatchewan, and Ontario, they made little difference to the relative positions of personal incomes in Quebec, Alberta, or British Columbia.

Since per capita transfers to persons in the North in 1970 were well below half the Canadian average, the transfer programs had the effect of pulling personal income per capita in this region down to below the Canadian average. Northern government and company employees, most of whom live in the Yukon, receive subsidies, whereas the Inuit and Indians, most of whom live in the Northwest Territories, receive very little subsidy unless they are on welfare. Hence transfers increase income disparities among people within this region. The situation is changing, however; although transfer income per capita is still lower in the North than in any other region, transfers to persons quadrupled in the Yukon and Northwest Territories between 1970 and 1974, partly as a result of expanding the unemployment insurance program.

A comparison of disposable income per capita with personal income per capita shows that the progessive income tax system generally works to reduce regional income disparities. In 1970, more than 86 percent of personal income in Newfoundland and Prince Edward Island was disposable, whereas only 79.3 percent in Ontario and 80.9 percent in British Columbia was disposable. Hence, even before interprovincial equalization payments are made, it appears that high-income provinces are bearing the largest burden of personal income tax. Taxation apparently hits the Yukon and Northwest Territories fairly hard. Only 79.7 percent of their personal income was disposable, which puts the North second only to Ontario in the share it paid in taxes in 1970. This proportion was about 78 percent during the years 1971-75, partly because the highly paid mining and government employees in the North receive a very large share of those regions' incomes, and the progressive income tax system cuts deeply into their earnings.

The number of persons per family and the ratio of children to working-age people are higher in the low-income regions. Since there are some returns to scale in family living, and since children need less income than adults, the situation in these regions is not truly as bad as the figures on disposable income per capita would suggest. The range of income disparities between Newfoundland and Ontario residents, in comparison with the average for all Canadians, falls from 55 percentage points for personal income per capita to only 41 points for income per family (Table 8). If the proportion of family income that is disposable is the same as that of personal income generally, then the maximum range in average disposable family incomes between provinces is only 35 percentage points.

A dollar of disposable income in 1971 had greater purchasing power in cities like Winnipeg,[9] where the cost of living was low, than in cities like Toronto or Vancouver, where the cost of housing, for example, was some 40 to 50 percent above Prairie levels.[10] Except for housing, the cost of living in Montreal and in the capital cities from Ottawa to Edmonton differed by no more than 3 percentage points. However, the nonhousing costs of living in the capitals of the Atlantic region exceeded those of Winnipeg by 6 to 9 points whereas in Vancouver they exceeded them by about 7 points. Since these nonhousing cost differentials probably extend to the areas surrounding these cities, it is reasonable to surmise that, in the Atlantic region and in British Columbia, the cost of living, excluding housing, is 6 to 8 points higher than in the rest of the country. In the Atlantic region and in the interior of British Columbia, there are substantial transportation costs, which cause retail prices to be higher than in the more central regions. If regional family disposable income is deflated by the relevant city's nonhousing price index, the differential between Newfoundland and Ontario families widens to roughly 40 percentage points, and the average pruchasing power of families in British Columbia becomes comparable to that of families in Alberta and closer to that of Quebec families. If cost of living in the North exceeds that of southern regions by some 45 percent,[11] then average purchasing power per family is lower in the North than in Newfoundland.

Housing costs are much more difficult to assess regionally because they differ dramatically from one urban centre to another, and from one period to another. Our estimates show that, in 1971, housing costs in Toronto and Ottawa exceeded those in Winnipeg by 46 and 19 percent, respectively, but that they were not nearly as high in the rest of Ontario as they were in Toronto. In Regina and Saskatoon, housing was inexpensive during the depressed year 1970-71, but by 1976, improved economic conditions had increased those housing prices relative to those in Ontario, although the increase was less than in Alberta. The fact that housing costs were much lower in Saint John than in Halifax suggests that the average purchasing power of New Brunswick families was closer to that of Nova Scotia families than was indicated by per capita data. Given the fact that the proportion of disposable income is lower and the cost of living higher in British Columbia than in Quebec, and that the average Quebec family is larger, it is possible to argue that the purchasing power of the average family in British Columbia is lower than that of the average Quebec family.

Differences in Employment Opportunities

Unemployment

Giving a man a job solves a lot of problems. Unfortunately, the unemployment rates show that the success with which people find and hold jobs differs markedly from region to region (Figure 5). The pattern of regional disparities in unemployment is as persistent as that of differences in income per capita, and it favours the same regions, with only one exception; whereas income per capita is higher in British Columbia than in the Prairies, the reverse is true for unemployment. From 1953 to 1975, the unemployment rate in British Columbia averaged 6.0 percent, nearly twice the rate of 3.3 percent recorded in the Prairies. The unemployment rate in the Atlantic region was the worst, at 8.6 percent, while the second worst rate was in Quebec, at 7.0 percent. A point that is less obvious is that, when unemployment across Canada increases, it

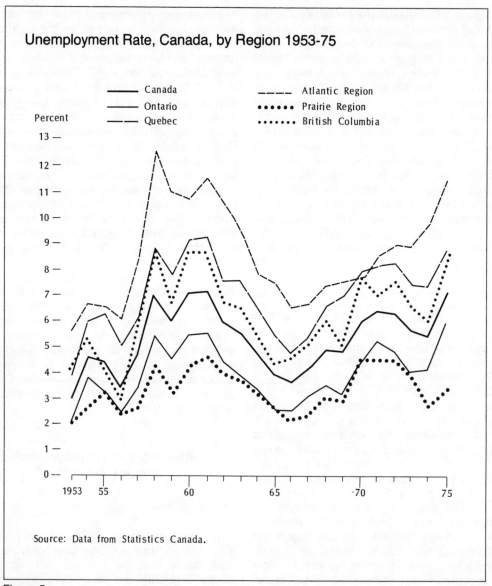

Unemployment Rate, Canada, by Region 1953-75

———— Canada   ————— Atlantic Region
———— Ontario   •••••• Prairie Region
——— Quebec   ••••••• British Columbia

Source: Data from Statistics Canada.

Figure 5

increases most in the regions where it is already the highest. Past experience shows that an increase of 2 percentage points in the Canadian unemployment rate is typically accompanied by an increase of roughly 3.7 points in the Atlantic region, 2.6 points in Quebec, 1.3 points in Ontario, 1.7 points in the Prairie Provinces, and 1.9 points in British Columbia.

The unemployment rate reflects two factors: the number of workers who experience some unemployment at any given time during the year, and the duration of the unemployment period experienced by individual workers. Precise and accurate data on duration, by province, are unavailable. However, data from Unemployment Insurance Commission records give a reasonable indication, though they pertain only to the duration of unemployment up to the time the monthly survey is taken by Statistics Canada rather than the duration of completed spells. Among the people receiving unemployment insurance benefits, the likely duration of unemployment differs markedly from one region to another (Figure 6). Unemployment insurance benefits, since mid-1971, have been payable for a greater number of weeks in regions like Newfoundland, where unemployment is higher than the Canadian average. The unemployment rates and the time distribution of unemployment indicate that the probability of finding a job is much lower in the Atlantic region than in Ontario or Alberta. However, recent evidence leaves room for hope, because the relationship between the level of unemployment in the Atlantic region and in the rest of Canada seems to have shifted downward since 1969. The actual levels of unemployment experienced in the Atlantic region from 1970 to 1975 are a little lower than would have been predicted by the relationships that existed from 1953 to 1969.

The Importance of Seasonal Unemployment

Differences in the scope of seasonal variations constitute one of the major causes of regional disparities in unemployment. Between one-quarter and one-third of Canadian unemployment is seasonal, as is at least one-third of Atlantic and Prairie unemployment. The problem arises not so much because one region's climate is more severe than that of another, but because a larger proportion of the economy of some

Figure 6

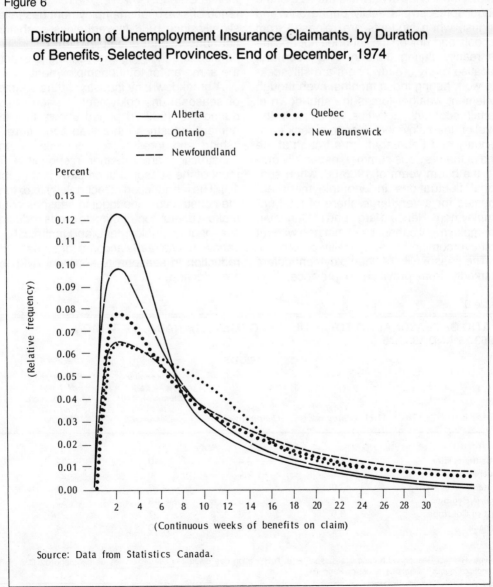

Distribution of Unemployment Insurance Claimants, by Duration of Benefits, Selected Provinces. End of December, 1974

Alberta
Ontario
Newfoundland

Quebec
New Brunswick

Percent

(Relative frequency)

0.13
0.12
0.11
0.10
0.09
0.08
0.07
0.06
0.05
0.04
0.03
0.02
0.01
0.00

2 4 6 8 10 12 14 16 18 20 22 24 26 28 30

(Continuous weeks of benefits on claim)

Source: Data from Statistics Canada.

regions – particularly the Atlantic and Prairie regions – consists of industries that are especially vulnerable to seasonal changes, notably fishing, farming, and forestry.

An examination of actual and seasonally adjusted time series of employment, unemployment, and labour force data gives a good illustration of this problem. For Canada as a whole, there are very regular fluctuations in employment, which reach their peak in July and August and their trough in January and February. The growth path of the labour force shows that the supply of labour also moves seasonally to accommodate some of the fluctuations in employment. However, while the size of the labour force swells easily during July and August, when students are released from schools and universities, it does not decline as readily during January and February, because many breadwinners are still seeking work during those months, even though inclement weather forces the shut-down of certain economic activities. As a result, the level of unemployment rises to a peak each January and February from a trough at the end of the previous summer, especially during the boom years of 1965-66, when seasonal fluctuations in unemployment accounted for a very large share of total unemployment. Hence a large part of Canadian unemployment cannot be eliminated without first overcoming the seasonality problem.

The seasonality of unemployment differs markedly from province to province. It is particularly severe in Prince Edward Island, whose economy is dominated by potato farming, tourism, and fishing – activities that are closely tied to the seasons. The situation is not much better in Newfoundland. Ontario has the fewest seasonal fluctuations in employment.

It is apparent that, except for the Prairies, the percentage of seasonal unemployment is largest in those regions with the highest toal unemployment (Table 9). In other words, it is partly the greater prevalence of seasonal unemployment in the Atlantic region, Quebec, and British Columbia that makes unemployment rates higher in those regions than in Ontario. However, the relative importance of seasonal unemployment has fallen slightly since 1966. Nevertheless, susbstantial nonseasonal unemployment disparities persist, and their regional ranking is exactly the same as for total unemployment.

A breakdown, by industry, of the sources of seasonal unemployment in each region during the 1966-75 period shows that the primary industries other than agriculture – fishing and forestry, for example – and construction are together responsible for most of the seasonal unemployment (Table 10). The introduction of technology to extend the construction and logging seasons could make a useful contribution towards reducing seasonal unemployment, and technical advances may already account for some of the reduction in seasonality apparent over the past decade.

Table 9

RATIO OF SEASONAL TO TOTAL UNEMPLOYMENT, CANADA, BY REGION, 1953-75 AND 1966-75

| | 1953-75 | | | 1966-75 |
	Total Unemployment	Seasonal Unemployment	Percentage of Total Unemployment That Was Seasonal	Percentage of Total Unemployment That Was Seasonal
	(Percent)			
Atlantic Region	8.6	3.4	40	33
Quebec	7.0	2.1	30	23
Ontario	3.9	1.0	26	21
Prairie Region	3.3	1.6	48	33
British Columbia	6.0	1.6	27	21
Canada	5.3	1.6	30	24

Source: Richard Beaudry, "Le chômage saisonnier et l'explication des disparités interrégionales de chômage au Canada", Economic Council of Canada Discussion Paper 84, 1977.

BREAKDOWN OF THE SOURCE OF SEASONAL UNEMPLOYMENT, BY INDUSTRY,
CANADA, BY REGION, 1966-75

	Atlantic Region	Quebec	Ontario	Prairie Region	British Columbia	Canada
	(Percent)					
Agriculture	4.5	4.3	6.2	6.3	6.7	5.3
Other primary industries	22.8	15.3	3.1	6.3	13.3	12.2
Manufacturing	18.2	23.9	31.2	12.5	20.0	22.9
Construction	27.3	28.2	34.4	31.2	26.7	29.8
Transportation, communication, and utilities	13.6	10.9	6.3	12.5	6.7	9.9
Trade	4.5	6;5	9.4	12.5	13.3	8.4
Services	9.1	10.9	9.4	18.7	13.3	11.5
Total	100.0	100.0	100.0	100.0	100.0	100.0

Source: Richard Beaudry, "Le chômage saisonnier et l'explication des disparités interrégionales de chômage au Canada", Economic Council of Canada Discussion Paper 84, 1977.

Table 10

Social Indicators and Other Social Measures

Economic and demographic variables do not tell the whole story about regional differences in the welfare of individuals. The analysis of certain social phenomena will, in effect, verify whether the conclusions drawn from the economic data are valid (Table 11). Social indicators – such as the crowding index in housing, infant mortality, and life expectancy – are direct measures of some aspects of human welfare. Other social measures, such as the number of telephones per 100 inhabitants or physicians per 100 000 inhabitants, suggest that there are regional differences in access to household gadgetry and medical services. These measures are not accepted as social indicators, however, because it has not yet been demonstrated that human welfare is closely related to them. In particular, there is doubt as to whether a person's state of health is directly related to the number of doctors. Also it has been demonstrated that teachers' training and access to good physical facilities are not as important as the students' peer group and socio-economic background in determining the effectiveness of education.[12] These social measures may nevertheless serve as warnings that some factors do differ regionally, although the implications of these differences are not clear.

The extent of crowding in housing and the distribution of telephones show that, while housing is more expensive in Ontario, British Columbia, and Alberta than in Quebec and the Atlantic region, it is also better, on average.

As for health standards, Ontario and the Prairies seem to be the most favoured regions. Infant mortality is lowest in Ontario, and life expectancy is greatest in the Prairies. Although British Columbia has a high infant mortality rate, life expectancy there is above average for persons beyond infancy. After the Northwest Territories, Newfoundland and Saskatchewan appear to have the poorest record on infant mortality. Note that, between 1971 and 1973, great progress was achieved in reducing the infant mortality rate, most notably in the North, the Atlantic region (except New Brunswick), and Alberta. Quebec and Nova Scotia have the lowest life expectancy at birth, although Quebecers and Nova Scotians spend an above-average share of their personal income on physicians' services. Newfoundland, Prince Edward Island, and New Brunswick have attracted (or retained) relatively fewer doctors than the more wealthy, urbanized provinces. Medical services appear to be far more scarce in Newfoundland than in other provinces.

With regard to education, Ontario has the largest proportion of young people in school at age sixteen, whereas Alberta has the largest proportion enrolled in postsecondary education.[13] Quebec had only 29 percent of its seventeen-year-old boys and 14 percent of its seventeen-year-old girls in school in 1961 – a much smaller proportion than in any other province. Following the changes in its education system during the 1960's, Quebec had the fourth largest percentage of eighteen to twenty-four-year-olds in postsecondary and university education by 1971-72, after Alberta, Ontario, and Nova Scotia.

Table 11

SOCIAL MEASURES, CANADA, BY PROVINCE AND TERRITORIES, SELECTED YEARS, 1966 TO 1976

	Newfoundland	Prince Edward Island	Nova Scotia	New Brunswick	Quebec	Ontario	Manitoba	Saskatchewan	Alberta	British Columbia	Yukon	North west Territories	Canada
Housing and telephones													
Crowding index, 1976 *(Persons per room)*	0.74	0.60	0.59	0.62	0.62	0.55	0.55	0.58	0.57	0.55			0.58
Telephones, 1975 *(Number per 100 inhabitants)*	36.1	41.0	47.5	46.9	55.0	61.4	55.9	51.3	62.4	59.9	50.4	45.0	57.2
Health													
Infant mortality rate, 1972-74 *(Number of deaths)[1]*	19.3	17.7	15.6	15.9	16.5	14.2	17.0	19.3	15.6	16.5	22.7	42.8	15.9
Average life expectancy, 1971 *(Years)*													
Males at birth	69.3	69.3	68.7	69.1	68.3	69.6	70.2	71.1	70.4	69.9			69.3
Males at age 20	51.9	52.0	51.0	51.6	50.7	51.6	52.7	53.8	52.9	52.3			51.7
Females at birth	75.7	77.4	76.0	76.4	75.3	76.8	76.9	77.3	77.3	76.7			76.4
Females at age 20	57.9	59.4	57.8	58.4	57.2	58.4	58.9	59.6	59.2	58.5			58.2
Active free-practice physicians, 1971 *(Number per 100 000 inhabitants)*	44.3	81.3	93.0	67.2	95.4	107.4	89.7	88.3	101.8	125.8			100.4
Index of physicians' fees ("price"), 1973 *(Ontario = 100)*	88	91	97	87	92	100	100	94	116	123			
Expenditures on physicians as a percentage of personal income, 1971 *(Percent)*	1.51	1.86	1.75	1.48	1.71	1.66	1.78	1.58	1.82	1.67			1.68
Education													
Percentage of population aged 16 attending elementary and secondary schools, 1974-75													
Male	74.9	62.1	76.9	75.9	81.5	88.7	84.4	82.6	85.3	86.5	74.5	63.8	84.3
Female	74.0	79.2	83.3	80.8	81.9	88.5	86.2	87.2	89.6	88.9	69.5	85.3	85.7
Percentage of 18-24 age group enrolled full-time in post-secondary institutions, 1971-72	13.0	17.6	19.4	16.3	18.7	19.5	17.3	16.6	20.5	15.9			18.5
Federal and provincial aid per student enrolled in post-secondary institutions, 1971-72 *(Dollars)*	1 099	740	778	712	378	473	480	502	641	288			475
Per capita expenditure on education, 1973	400	416	373	333	464	449	410	379	440	361			436
Average salary of university teachers, 1972-73	15 158	14 314	16 111	14 932	16 168	18 047	15 737	17 243	18 026	17 919			17 184
"Warning" statistics Divorce rate index, 1974 *(Index)[2]*	271	381	868	519	889	775	756	489	1 232	1 169	963[2]		855
Offence rate (both sexes), 1974 *(Number of offences)[3]*	6 296	8 001	7 520	6 054	5 329	8 882	9 372	10 050	12 471	12 732	33 331	31 929	8 459
Standardized suicide rate, 1966-68 *(Number per 100 000 inhabitants)*	2.57	8.30	8.19	4.81	7.13	10.39	10.82	8.72	10.06	13.25	27.38	10.63	9.31

[1]The infant mortality rate is the number of deaths of infants (under one year old) per 100 000 live births.

[2]The divorce rate index is $100\ 000 \times \dfrac{\text{number of divorces}}{\text{half the number of married persons}}$

Statistics Canada no longer gives separate divorce rates for the Territories; but, in 1971, it reported a divorce rate index of 1164 for the Yukon (the highest rate of any province or territory in Canada at that time) and only 80 for the Northwest Territories.

[3]The offence rate is the number of offences reported per 100 000 population, aged seven and over.

Source: Based on data from Statistics Canada; and Robert G. Evans, "Beyond the Medical Marketplace: Expenditure, Utilization and Pricing of Insured Health Care in Canada," in Spyros Andreopoulos, ed., National Health Insurance: Can We Learn from Canada? (new York: Wiley, 1975).

Alberta and British Columbia are the provinces where teachers have the highest incomes and where the highest percentage of them have university degrees; Ontario's university professors are paid the most. Among the Atlantic Provinces, Nova Scotia appears to have the teachers with the highest salaries, as well as the largest percentage of students taking higher education. Whereas per capita expenditures on education, rates of enrolment, and teachers' salaries and training seem to be below the national average in the Atlantic region, these provinces all have above-average financial aid per student enrolled in postsecondary institutions and universities. Quebec had relatively low per capita expenditures and rates of enrolment in 1960, but these have now increased. Hence measures have been taken to reduce regional disparities in education. Those born during the postwar baby boom, however, who will be dominating the labour force for the rest of this century, have now finished their studies, and it seems likely that those raised in Quebec and the Atlantic region have been disadvantaged to some extent by the regional disparities in education that existed in the two decades following the Second World War.

These measures of housing, health, and education standards reinforce the findings on regional disparities in income, and show that the standard of living is higher in Ontario and western Canada than in Quebec and the Atlantic region. Quebec may have been particularly disadvantaged in education in the past, but the situation has considerably improved over the last decade.

There is, however, a set of social measures that serves as a warning that rapid economic growth and large in-migration may involve some high social costs to be borne by a few unfortunate individuals. While the Atlantic region is a low-income area from which people are migrating, it seems to enjoy the lowest rate of social disruption and stress, to the extent that statistics on divorce, crime, and suicide, are indicative of such things.[14] The most troubled region, from this point of view, seems to be the Yukon, followed by the Northwest Territories, British Columbia, and Alberta. Since these are the regions with the greatest in-migration, it is possible, then, that migration and rapid economic growth may have higher social costs than is commonly believed. Perhaps the migration process can

be improved so that it has a less disruptive effect on society.

Conclusion

While regional differences in family purchasing power are not nearly as large as disparities in earning power per capita, our analysis shows that no amount of legitimate "tinkering" with average income figures will make interregional income disparities disappear. Ontario undeniably has the highest average real income, exceeding levels in the Atlantic region by some 20 to 30 percent. Economic expansion and population growth have been very rapid in British Columbia, Alberta, and Ontario; as a result, for the last several decades, their economic and political power base has been increasing relative to other regions. While incomes per capita are higher than average in these three provinces, Ontario apparently suffers less social disruption than Alberta or British Columbia. Moreover, Ontario is still clearly the leader where purchasing power is concerned. In contrast, British Columbia has higher average weekly earnings in some industries, but this advantage is offset by higher rates of unemployment and living costs than prevail in Ontario. In addition, Ontario, British Columbia, and Alberta have the most symmetrical income distributions in Canada. Other social indicators – in housing, health, and education – tend to confirm that they enjoy a higher standard of living than other provinces.

The Prairie region is rather heterogeneous. Saskatchewan's farm income is variable and large relative to the rest of its economy. In poor years, its income distribution has more in common with that of Newfoundland than with that of neighbouring Alberta or Manitoba. The centre of population is moving westward, in that Winnipeg, which dominated the Prairies during the expansionary period of the 1920's, has now been eclipsed by the burgeoning Edmonton-Calgary economic axis based on energy and petrochemicals. As a result, Alberta has the highest average real incomes in the Prairie region.

Quebec has a serious unemployment problem, although it is not as bad as that of the Atlantic region. While income per capita is lower in Quebec than in the Prairies and

British Columbia, its purchasing power per family is not very much less, and may even be higher, depending on whether housing costs are included in the price index used. There were more children per family in 1971 in Quebec than in Ontario and the West, but its reduced birth rate suggests that, in the near future, it will have fewer children per family and a smaller youth dependency ratio. As a result, its income per capita can be expected to converge towards that of the western provinces.

An important matter is Quebec's relatively slow rates of growth in population and employment. Quebec's share of the national population has declined about 2.5 percentage points since 1947 and is now back to what it was in 1921. Population projections indicate a further decline of some 1 to 4 points by the year 2001. Since the Francophone population is concentrated in Quebec, these figures have implications for the ability of French-speaking Canadians to make their voices heard in a federation whose population is slowly becoming more Anglophone.

Whether one looks at income per capita or purchasing power per family, the Atlantic region has the lowest incomes. Newfoundland and Prince Edward Island are in an unfavourable position relative to Nova Scotia and New Brunswick. This low standard of living is confirmed by social indicators on housing, health, and education. The Atlantic region also has the poorest (least symmetrical) income distributions, apart from that for the dual society in the Northwest Territories. Its wages and salaries per employed worker are the lowest, and its unemployment rates are the highest. In a period of recession, its unemployment rate rises the most, and its unemployed workers stay out of work the longest. Since its workers have a lower incentive than others to enter the labour force, it is not surprising that Atlantic participation rates are so low.

In sum, regional disparities in incomes and job opportunities are indeed substantial and remarkably persistent, in spite of the amount of labour migration that has taken place over the years.

NOTES AND REFERENCES

[1] Economic Council of Canada, *Twelfth Annual Review: Options for Growth* (Ottawa: Information Canada, 1975), p. 51. In the Prairie region, the anticipated growth in Alberta is offset by the anticipated losses in Saskatchewan and Manitoba.

[2] A thorough understanding of population movements between 1971 and 1976 must await analysis of the 1976 Census data. The observations noted here are based on a study of data pertaining to the movement of children eligible for family allowances. However, caution is necessary in dealing with these data. Young adults without children are over-represented among out-migrants, as are young adults with children among return migrants. Since the proportion of the population in these two groups has changed rapidly in recent years, thanks to the repercussions of the postwar baby boom, the comparability of 1961-71 migration data, which are not based on movements of children, with 1971-76 data is fairly questionable. Caution is therefore warranted, and our observations here are only tentative.

[3] The data in Table 2 are expressed in current dollars; the rates of growth therefore include the effects of inflation. In real terms, farm income per capita – that is, farm income per member of the entire population rather than per farmer – has fallen in Quebec and Ontario.

[4] Government transfers include family and youth allowances, unemployment insurance benefits, pensions to veterans and government employees, Canada and Quebec Pension Plan payments, adult retraining payments, and provincial and municipal welfare payments.

[5] These disparities were even larger in 1974 prior to the surge in Ontario unemployment caused by 1975 layoffs in the automotive industry.

[6] In particular, growth in unemployment insurance benefits has been very large in recent years, rising from $300 million in 1966 to approximately $3.1 billion in 1975 and $3.3 billion in 1976.

[7] The difference arises from the fact that these data are the result of different surveys by Statistics Canada. See *The Labour Force*, Cat. No. 71-001, for estimates of employed persons obtained from a survey of 30 000 households; *Estimates of Labour Income*, Cat. No. 72-005, for estimates of wages and salaries on a National Accounts basis; and *Employment, Earnings and Hours*, Cat. No. 72-002, for estimates of weekly earnings obtained through a mail survey of large business establishments.

[8] To the extent that lower participation in the labour force is voluntary, lower market income may partly reflect disparities in the willingness to earn such income rather than disparities in the power to earn it.

[9] In the discussion of regional differences in the cost of living, Winnipeg was chosen as the base point. See Statistics Canada, "Canadian Inter-City Retail Price Comparisons: A Study of Place-to-Place Relationships between Selected Urban Centres as of May 1971," *Prices and Price Indexes*, Cat. 62-002, 1972.

[10] However, by 1976, the burst of petrochemical-related investment in Alberta had driven house prices in Calgary and Edmonton up to levels comparable to those in Toronto.

[11] See Gemini North Ltd., *Social and Economic Impact of the Proposed Arctic Gas Pipeline in Canada*, Book 1, May 1974, p. 506.

[12] See, for example, Ivan Illich, *Medical Nemesis* (Toronto: McClelland and Stewart, 1975); and Vernon Henderson, Peter Mieszkowski, and Yvon Sauvageau, *Peer Group Effects and Educational Production Functions* (Ottawa: Economic Council of Canada, 1976).

[13] Secondary education ends at Grade 11 in Newfoundland and Quebec, at Grade 13 in Ontario, and at Grade 12 in other provinces.

[14] As symptoms of stress and social disruption, these statistics must be interpreted with care. They are influenced by many other factors as well. The standards of law enforcement, for example, differ regionally. Also divorce rates in predominantly Roman Catholic areas can be expected to differ from those in Protestant areas. Divorce rates are also influenced by marriage rates. Whereas the Northwest Territories have a low divorce rate, this situation probably reflects more on the prevalence of common-law marriages than on the state of marital bliss.

The Role of the Department of Regional Economic Expansion in Alleviating Regional Disparities

ECONOMIC COUNCIL OF CANADA

In the early 1960's growing anxiety concerning the problems associated with regional economic disparities led the federal government to play a more active role in determining both the level and the direction of economic activity in the lagging regions. This culminated in the formation of the Department of Regional Economic Expansion in 1969.

Prior to DREE, several programs had been introduced, the first being the Agricultural Rehabilitation and Development Act in 1961, which was intended to alleviate the high incidence of low income in rural (agricultural) areas. It emphasized soil and water conservation and land use conversion programs, and its financing was shared equally by the provinces and the federal government. With the introduction of the Agricultural and Rural Development Act (ARDA) in 1965, rural nonagricultural poverty was recognized, and the original program was appropriately expanded, while the land use programs of its predecessor were continued.

In 1966 the Fund for Rural Economic Development (FRED) was introduced to provide assistance to those areas requiring resources beyond those supplied by ARDA. The program covered land management, education, infrastructure investment, and industrial development, especially in the primary sector, although tourism and manufacturing were also assisted. Agreements covered all of Prince Edward Island and parts of New Brunswick, Quebec, and Manitoba.

In 1962 the Atlantic Development Board (ADB) was established as an advisory body on economic problems in the Atlantic region and, beginning in 1963, it administered a development fund that supported large infrastructure investment projects.

The Area Development Agency was set up in 1963 to provide incentives for firms to

● Reprinted from Economic Council of Canada, *Living Together: A Study of Regional Disparities* (Ottawa: Minister of Supply and Services, 1977), Ch. 8, pp. 145-172, with permission. The article has been edited slightly.

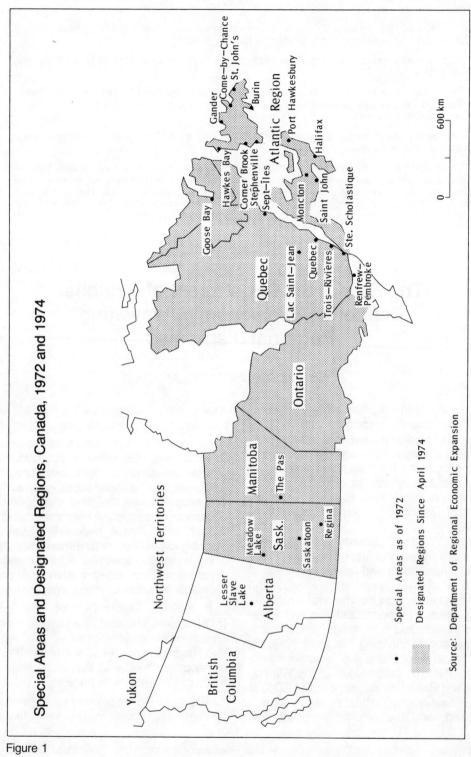

Special Areas and Designated Regions, Canada, 1972 and 1974

Figure 1

Yukon

British Columbia

Alberta

Lesser Slave Lake

Northwest Territories

Sask.

Meadow Lake

Saskatoon

Regina

Manitoba

The Pas

Ontario

Quebec

Goose Bay

Hawkes Bay

Corner Brook

Stephenville

Sept–Îles

Lac Saint–Jean

Quebec

Trois–Rivières

Renfrew–Pembroke

Ste. Scholastique

Saint John

Moncton

Halifax

Port Hawkesbury

Atlantic Region

Gander

Come–by–Chance

St. John's

Burin

600 km

0

• Special Areas as of 1972

    Designated Regions Since April 1974

Source: Department of Regional Economic Expansion

294

locate in designated areas of high unemployment and, until 1965, it provided tax concessions and accelerated depreciation allowances to such firms. In 1965, when the Area Development Incentives Act (ADIA) was introduced, the program of tax incentives was modified, and a system of capital grants was introduced.

The Department of Regional Economic Expansion

From this brief inventory[1] it can be seen that, prior to 1969, the federal government was involved in several regional development programs. It failed, however, to coordinate these programs and policies towards any specific goals for any region. At the same time, the individual provinces also instituted regional development policies of their own, with the result that there was a hodgepodge of development efforts that lacked any general overall direction. In the light of these circumstances, after the Economic Council of Canada having suggested it,[2] DREE was established in 1969.

The Department's goal is to ensure that economic growth is more widely dispersed across Canada and that employment and earning opportunities in the slow-growth regions are brought as close as possible to those in the rest of the country.

In defining the functions of the Minister of the Department of Regional Economic Expansion to extend to, and include, all matters pertaining to economic expansion and social adjustment (in areas requiring special measures to improve opportunities for productive employment) that were not already assigned to other federal departments, the Government Organization Act, which established DREE, charged the Minister with responsibility for federal regional development efforts. Thus authority for most of the existing regional development programs was transferred to DREE.

The Act also assigned to the Minister the responsibility for Special Areas (SA), in which, by reason of exceptionally inadequate opportunities for productive employment, special measures were needed. Special areas were designated from 1969 to 1972 (Figure 1); after 1972, DREE adopted a new policy strategy, which suspended the Special Area approach.

This earlier Special Area legislation was to be implemented by a series of agreements to be signed with the provinces, under which DREE could provide industrial incentives to firms to establish or expand industrial facilities, or it could provide assistance to establish or expand the infrastructure required for the economic expansion of the SA. Eighteen of the twenty-three SA's named were deisgnated to receive infrastructure assistance, including all of the areas in the Atlantic Provinces. Within this infrastructure assistance program, expenditures were heavily weighted in favour of transportation facilities such as highways and streets, with expenditures for municipal services, such as water and sewer systems, ranking second in importance.

A second piece of legislation, the Regional Development Incentives Act, adopted in 1969, replaced the ADIA legislation of 1965, although commitments under that Act were to be honoured. The objective of the RDIA program was to stimulate industrial expansion in Designated Regions (DR's) of Canada by providing grants to firms starting a new manufacturing or processing operation, or expanding or modernizing an existing one. These industrial incentives were to apply to most industries, with the exception of pulp and newsprint mills and oil refineries.

Table 1

MAXIMUM INCENTIVE GRANTS AVAILABLE UNDER RDIA

| | Type of project | |
	Modernization or Expansion	New Plant or New Product Expansion
Designated regions		
Atlantic	20% of eligible capital costs	25% of capital costs plus 30% of wage bill for one year[1]
All other	20% of eligible capital costs	25% of capital costs plus 15% of wage bill for one year[1]

[1]Based on average of projected second- and third-year wage bills.

Source: Department of Regional Economic Expansion.

The incentives have always varied in generosity according to the region concerned and the nature of the project, new facilities in the Atlantic region being the most favoured. In recent years, however, greater emphasis has been placed on the expansion and modernization of existing firms. The basis for calculating the grants has changed from time to time, with the common thread being that the amount available for new facilities or new products has always depended on both the capital investment and the expected employment creation, whereas it has depended on capital only for modernizations or expansions. Table 1 shows that capital-linked subsidies are presently between 20 percent and 25 percent, and employment-linked subsidies between 15 percent and 30 percent, of the cost of wages for one year.

The DR's originally encompassed about 30 percent of the country's population, compared with about 18 percent under the ADIA. An amendment to the Act added a third DR in 1971, bringing the proportion of the Canadian population covered by DR's to about 50 percent. Further changes have been made since 1973, which have excluded some areas previously included in Alberta, Ontario, and British Columbia. These Designated Regions are shown on Figure 1.

Following the policy review undertaken during 1972-73, the department introduced the concept of development opportunities, which, once discovered, were to be exploited through a series of subsidiary agreements under ten-year General Development Agreements (GDA's) signed with each province except Prince Edward Island, beginning in 1974. The RDIA program was retained, while other existing DREE programs were to continue for an interim period, but it was claimed that the development opportunities concept would become the central element of regional development policy. In fact, up to 1975, less than one-fifth of all expenditures were made under the GDA program, and the types of expenditure were quite similar to those made under other programs. There is, however, a considerably enhanced degree of participation by provinces, by federal departments other than DREE, and by the private sector in the subsidiary agreements signed within the GDA program.

Expenditures by DREE

Under both the old and new approaches, it is possible to classify DREE expenditures on grants and contributions into four main categories.

First, there is direct assistance to the private sector, consisting mainly of industrial incentives provided under the RDIA program but also under the GDA. Second, there is public sector nonrural assistance, which is mainly infrastructure designed to make certain locations more attractive to business. It includes the SA infrastructure assistance programs, the Special Highways Agreement, and the ADB infrastructure program in the Atlantic Provinces, as well as similar assistance under the GDA. Third, there is public sector rural assistance, provided through FRED, ARDA, and GDA, which was designed to improve infrastructure and methods of production in areas heavily dependent on the primary industries. Finally, there are manpower programs, studies, and research projects.

Expenditures under each of these four categories for Canada as a whole are shown in the bottom right corner of Figure 2. It may surprise many that the much publicized assistance to the private sector − the part that *is* DREE to most people − accounted for only 30.7 percent of total spending ($476 million from 1969-70 to 1974-75). Nonrural infrastructure − highways, streets, and water and sewer systems − accounted for 35.4 percent, while rural assistance accounted for 29.3 percent. In total, $1551 million was spent.

A breakdown of total expenditures, by province, may be of some interest. Quebec received just over one-third of the total. Its share was exceeded by the Atlantic region as a whole, which received just over 45 percent, with New Brunswick getting the largest individual share within the region, followed by Newfoundland, Nova Scotia, and Prince Edward Island in that order. Of the remaining 20 percent, the Prairies received 12 percent, Ontario 6 percent, and British Columbia 2 percent.

Table 2 presents expenditures more meaningfully on a per capita basis. Prince Edward Island benefited most over the six-year period from 1969-70 to 1974-75 ($807). All of the Atlantic Provinces received far more per person than Quebec or any of the

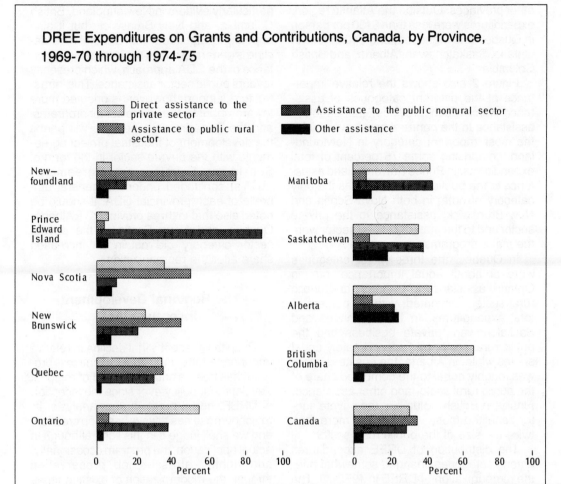

DREE Expenditures on Grants and Contributions, Canada, by Province, 1969-70 through 1974-75

Source: W. Irwin Gillespie and Richard Kerr, The Impact of Federal Regional Economic Expansion Policies on the Distribution of Income in Canada, Economic Council of Canada Discussion Paper 85, 1977; and various subsidiary agreements signed by the provinces and DREE.

Figure 2

Table 2

TOTAL AND PER CAPITA DREE EXPENDITURES ON GRANTS AND CONTRIBUTIONS, BY PROVINCE, FISCAL YEARS 1969-70 THROUGH 1974-75

	New-found-land	Prince Edward Island	Nova Scotia	New Bruns-wick	Quebec	Ontario	Mani-toba	Saskat-chewan	Alberta	British Colum-bia
				(Millions of dollars)						
Expenditures	185.9	91.2	182.0	247.3	522.7	97.0	89.9	35.2	65.8	33.5
				(Thousands)						
Population[1]	528.0	113.0	793.0	641.0	6048.5	7749.3	991.8	926.0	1639.0	2221.7
				(Dollars)						
Expenditures per capita	352	807	230	386	86	13	91	38	40	15

[1]Average, 1969-74.

Source: W. Irwin Gillespie and Richard Kerr, "The Impact of Federal Regional Economic Expansion Policies on the Distribution of Income in Canada", Economic Council of Canada Discussion Paper 85, 1977; and the CANSIM databank.

other provinces. Outside the Atlantic region, expenditures were less than $100 per person in Quebec and Manitoba, and $40 or less in Ontario, Saskatchewan, Alberta, and British Columbia.

Figure 2 also shows the relative importance of the different categories of assistance within each province. Infrastructure assistance to the nonrural sector was by far the most important category in Newfoundland, comprising some 75 percent of total expenditures. In Prince Edward Island assistance to the public rural sector was the only category of note; in both Nova Scotia and New Brunswick, assistance to the private sector and to the public nonrural sector were the major programs.

In Quebec, the three major categories were of about equal importance; but, in Ontario, assistance to business and public rural sector dominated, with 94 percent of total expenditures. In both Manitoba and Saskatchewan, private business and the public rural sector had approximately equal shares while, in Alberta, the business share was roughly equal to the combined share of the public rural sector and other assistance. Finally, in British Columbia, the private sector benefited most, with a share more than twice the size of the public rural sector.

The distribution of DREE expenditures among categories changed somewhat after the reorganization of DREE in 1972-73. The share devoted to the private sector declined in importance, dropping from about 33 percent to about 28 percent. This pattern is particularly evident in Newfoundland, British Columbia, and New Brunswick, but it also appears in Manitoba and Ontario. The decline is generally due to the increasing importance of the GDA approach, which presently favours public sector assistance. This infrastructure support, however, is oriented more towards industry than were earlier programs, and the Department hopes this will permit the development of industrial project agreements with the private sector in the form of Subsidiary Development Agreements (SDA's) concluded under the general umbrella of each provincial GDA. It should be noted also that in three provinces – Alberta, Quebec, and Nova Scotia – the private sector category did obtain an increased share after the reorganization.

The Regional Development Incentives Act

Despite its recent slight decline in relative importance, the industrial incentives program has been a critical element of regional development policy ever since the inception of DREE, and it merits further analysis. Its major purpose has always been job creation, and we shall judge it in this light, although in actual application, the program occasionally, and rightly, stresses job preservation through the modernization of existing firms. Table 3 gives the total number of new jobs claimed for the RDIA program for each province during the 1969-75 period. These new

Table 3

EMPLOYMENT ASSUMED TO BE ASSOCIATED WITH RDIA INCENTIVE GRANTS, BY PROVINCE, 1969-75

	1969	1970	1971	1972	1973	1974	1975	1969-75
			(Number of jobs)					
Newfoundland	189	167	625	539	923	478	687	3 608
Prince Edward Island	–	223	232	97	394	146	292	1 384
Nova Scotia	130	582	1 287	2 405	1 936	1 231	886	8 457
New Brunswick	229	1476	1 837	1 011	1 779	2 865	973	10 170
Quebec	327	4922	14 779	20 771	15 148	10 654	7 963	74 564
Ontario	25	189	2 276	3 201	1 615	2 249	829	10 384
Manitoba	246	870	1 918	1 853	1 867	1 898	1 649	10 301
Saskatchewan	–	573	730	403	1 482	660	1 001	4 849
Alberta	–	534	430	1 127	319	514	604	3 528
British Columbia	7	123	458	294	360	152	89	1 483
Total	1153	9659	24 572	31 701	25 823	20 847	14 973	128 728

Source: Department of Regional Economic Expansion, *Report on Regional Development Incentives,* various monthly issues.

jobs are associated with incentive grants accepted[3] for new facilities, for expansions and modernizations, and for new product expansions.

The table shows that, for the period as a whole, Quebec appears to have benefited the most, at least in an absolute sense, with 74 564 jobs claimed, almost 58 percent of the total. The Atlantic region has 18 percent of the total, with New Brunswick faring the best with 8 percent. Interestingly enough, over the whole period, both Ontario and Manitoba obtained more jobs than any of the individual provinces of the Atlantic region. These comments must, however, be viewed in proper context, and any evaluation of the relative impact or benefit in terms of employment must be made with reference to the average number of unemployed (or employed) in the province over some relevant time period. For example, although Quebec received approximately three times as many new jobs as the Atlantic region, it also had, in absolute terms, about three times the number of unemployed persons as the Atlantic region over the same period.

New jobs can be associated either with new plants or with the expansion and modernization of old ones. New plants are perhaps of special interest in their own right, and data on the number of new manufacturing or processing establishments assisted by DREE are reported in Table 4 for each province for the 1969-75 period. The pattern previously observed in Table 3 is repeated, with Quebec obtaining by far the largest single share – just over half – and the Atlantic region obtaining about 21 percent. New Brunswick had the highest number of new plants within the Atlantic region but, as with new jobs, it had fewer than Manitoba.

Both ARDA and FRED programs continued, following the policy review of 1972-73. Some of the projects undertaken under ARDA include improvements to the industrial water supply at Port-aux-Basques, in Newfoundland; assistance to the coal wash plant at Stellarton, Nova Scotia; and assistance to the Indians on the Manitou Rapids Reserve, in Ontario, with the production of wild rice. Under the FRED program, expansion of vocational training opportunities in Prince Edward Island was achieved through the establishment and development of Holland College, while housing units were provided in Bathurst, New Brunswick. In deriving Figure 2 both the ARDA and FRED programs were classified under the public rural sector category.

Some of the projects undertaken under the Special Areas legislation, which were classified under the public nonrural sector category in Figure 2, include the construction of a road linking Highway 4 to Port Hawkesbury Industrial Park, in Nova Scotia; the Marsh Creek sewage system in Saint John, New Brunswick; and the paving of the Slave Lake Airport, in Alberta.

Beginning in 1974, DREE signed ten-year GDA's with the provinces (all except Prince Edward Island, which is covered by a long-term FRED agreement) and, in turn, a series

Table 4

NUMBER OF RDIA INCENTIVE GRANTS FOR NEW FACILITIES, BY PROVINCE, 1969-75

	1969	1970	1971	1972	1973	1974	1975	Total
Newfoundland	—	2	10	7	11	8	14	52
Prince Edward Island	—	5	4	3	4	3	11	30
Nova Scotia	1	8	12	17	14	23	22	97
New Brunswick	3	12	19	19	27	41	24	145
Quebec	5	53	124	177	176	137	121	793
Ontario	—	5	25	20	32	28	23	133
Manitoba	2	12	29	29	19	26	45	162
Saskatchewan	—	9	9	4	10	15	31	78
Alberta	—	8	10	10	9	16	5	58
British Columbia	—	6	2	3	4	1	3	19
Total	11	120	244	289	306	298	299	1567

Source: Department of Regional Economic Expansion, *Report on Regional Development Incentives*, various monthly issues.

EXPENDITURES BY DREE UNDER GDA, BY PROVINCE, 1973-74 AND 1974-75

	Private Sector	Public Nonrural Sector	Public Rural Sector	Other	Total
			(Millions of dollars)		
Newfoundland	–	12.5	23.2	–	35.7
Nova Scotia	–	3.1	0.9	0.5	4.5
New Brunswick	–	10.8	14.7	0.9	26.4
Quebec	14.5	12.1	1.8	–	28.4
Ontario		4.0	0.6	–	4.6
Manitoba	–	–	7.7	–	7.7
Saskatchewan	0.3	–	0.2	–	0.5
Alberta	–	–	2.9		2.9
British Columbia	–	–	2.5	–	2.5
Total	14.8	42.5	54.5	1.5	113.2

Source: The various subsidiary agreements signed with the provinces and published by DREE

Table 5

of subsidiary agreements that specified the exploitation of "development opportunities" in each province. The DREE expenditures on these subsidiary agreements for fiscal years 1973-74 to 1974-75 are shown by category in Table 5. The emphasis on public sector assistance, both nonrural and rural, is evident. The total expenditures to date have been fairly small compared with the $635 million spent by DREE over the two-year period, with only Newfoundland, Quebec, and New Brunswick obtaining any substantial amounts. On the other hand, the agreements almost invariably involve provincial expenditures as well as federal, so they are more important than DREE spending alone would indicate. The provinces also play a substantial role in the planning and implementation of the agreements.

Examples of some of the actual subsidiary agreements that were signed include a forestry agreement in Newfoundland in 1974, which provided for the acquisition of forest land and the construction of access roads as a package deal; a one-year agreement with Nova Scotia in 1975, which provided for the development of the Halifax-Dartmouth waterfront, including the acquisition of land and the provision of basic facilities; and an agreement with Ontario for the construction of a new sewage line for Thunder Bay and the reconstruction of a portion of Highway 599.

The Effectiveness of DREE

Figure 3 shows two measures of the unemployment rate, in Quebec and the Atlantic region. The solid line shows the actual unemployment rate. The broken line shows what the unemployment rate *could have been expected to be on the basis of its historical relationship to the Canadian level.*

Figure 3 suggests that something may have recently improved the situation in the Atlantic region but that little has changed in Quebec. In the Atlantic region, the actual unemployment rate was lower than might have been expected on the basis of past experience for eight of the nine years from 1967 to 1975. More detailed calculations suggest that the average rate was half a percentage point closer to the national rate than in earlier years – a significant closing of the gap.[4] In Quebec, the actual unemployment shows no systematic tendency to diverge, in either direction, from what might have been expected on the basis of past experience. In British Columbia also, though not shown in the chart, there is no evidence of any improvement in the unemployment rate relative to the nation as a whole.

In terms of earned income levels per person employed, slow convergence has been occurring since the early 1950's. A very slight acceleration can be detected for the Atlantic region, but not elsewhere, during the

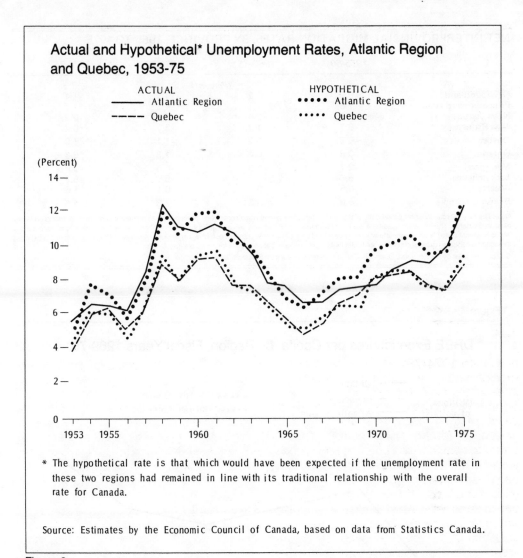

Actual and Hypothetical* Unemployment Rates, Atlantic Region
and Quebec, 1953-75

ACTUAL
——— Atlantic Region
– – – – Quebec

HYPOTHETICAL
•••• Atlantic Region
••••• Quebec

(Percent)

* The hypothetical rate is that which would have been expected if the unemployment rate in
these two regions had remained in line with its traditional relationship with the overall
rate for Canada.

Source: Estimates by the Economic Council of Canada, based on data from Statistics Canada.

Figure 3

years in which DREE and its predecessors have been operational.[5]

Another way in which the effects of DREE might show up is through a change in out-migration. Migration can be good for the leavers, but it may be a mixed blessing for the stayers. However one views it, a decrease in migration is likely to be one consequence of success in either creating jobs or in raising income levels in a given region. Table 6 shows net migration rates for the most recent period and compares them with rates in three earlier periods.

The figures in the table shows the amount of migration over five-year periods, with the gain or loss expressed as a percentage of the population at the start of the period. Thus, in Newfoundland, the number who moved out to other provinces (not abroad) from 1955 to 1960 exceeded the number who moved in from other provinces by an amount equivalent to 0.9 percent of the 1955 population. This figure is shown with a negative sign to indicate population loss through migration; when there is a net gain in population, as in British Columbia, the figures are positive.

Among the regions with heavy net out-migration – Atlantic, Saskatchewan, and Manitoba – there is some indication of a recent slowdown in out-migration in all provinces of the Atlantic region except Newfoundland, with Prince Edward Island and New Brunswick actually moving to a situation of net in-migration. No strong indication

NET INTERPROVINCIAL MIGRATION RATES, BY PROVINCE, 1955 TO 1975

| | Estimated Average | | | |
	1955-60	1960-65	1965-70	1970-75
	(Percent)			
Newfoundland	-0.9	-1.2	-4.0	-1.4
Prince Edward Island	-1.6	-1.1	-4.3	1.7
Nova Scotia	-3.3	-2.4	-2.8	-0.2
New Brunswick	-0.7	-1.7	-3.7	0.7
Quebec	-0.3	0.2	-1.1	-1.0
Ontario	0.8	0.4	1.8	0.3
Manitoba	-2.7	-1.3	-4.5	-2.9
Saskatchewan	-5.3	-3.5	-5.9	-5.3
Alberta	0.5	0.1	0.1	1.6
British Columbia	4.0	3.3	7.8	4.6

Note: The only interprovincial migration data available for intercensal periods on a continuous basis pertain to movements of children, based on family allowance records. The migration rates here are for movements of children as a percentage of the number of children at the beginning of each period, and they almost certainly underestimate what the rate would be for the whole population. It is highly unlikely, however, that *changes* in migration rates through time based on data for children would differ much from *changes* based on data for the whole population.

Source: Estimates by the Economic Council of Canada, based on data from Statistics Canada.

Table 6

Figure 4

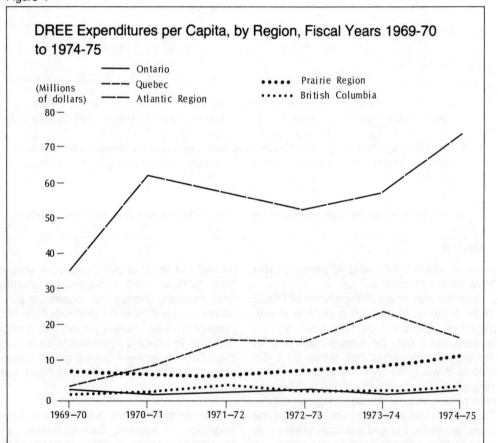

DREE Expenditures per Capita, by Region, Fiscal Years 1969-70 to 1974-75

Source: W. Irwin Gillespie and Richard Kerr, The Impact of Federal Regional Economic Expansion Policies on the Distribution of Income in Canada, Economic Council of Canada Discussion Paper 85, 1977; and CANSIM data bank.

of this kind of change appears in Saskatchewan or Manitoba.

These developments in unemployment, income, and migration suggest that job opportunities in the Atlantic region have improved over the last few years, although nothing much seems to have changed elsewhere. Can this partial improvement in the Atlantic region be attributed to DREE?

Certainly the timing of the change is about right. In a broad way, one would have expected programs that were instituted mainly in the early and middle 1960's, and which expanded in the late 1960's and early 1970's when they were gathered together under the DREE umbrella, to have begun to make a dent in the problems late in the 1960's and early in the 1970's. Moreover, the only changes that can be detected are in the Atlantic region, and it is only in that region that the DREE expenditures are at all significant on a per capita basis, as Figure 4 shows. Where DREE expenditures have been very small in per capita terms, as in Quebec, Saskatchewan, and elsewhere, no detectable changes in disparities occurred.

The correspondence between DREE's claims of job creation in the Atlantic region and the combined reduction in unemployment and out-migration is far from exact. But it is of the right order of magnitude, in the sense that much of the actual change in employment opportunities could in principle have been accounted for if DREE did actually create the number of jobs it claims.

Table 7 gives some details on this. In the first column are found the estimated number of extra workers needing jobs in the Atlantic region by reason of reduced out-migration between the 1965-70 period and the 1970-75 period. If out-migration had not declined, fewer people would have been in the region and needing work; our estimates indicate job requirements of 45 000 on this account. In addition, to keep the unemployment rate half a percentage point lower than in earlier years meant that extra jobs were needed — 4000 in the region as a whole. The third column gives the total extra jobs needed to accommodate both reduced out-migration and a lower unemployment rate. Finally, the last column shows the number of jobs that DREE claims to have created. It can be seen that the extra DREE jobs, if all were created as claimed, would have met about half the need in each of the three largest provinces, and in the region as a whole.

All of this evidence is suggestive, but far from conclusive. It is not impossible that the situation improved quite independently of DREE and that the latter's existence made no real difference. We do know, for example, that the rapid expansion of income maintenance programs in the late 1960's led to a

Table 7

JOBS NEEDED AND CLAIMED TO BE CREATED BY DREE, ATLANTIC REGION, 1970-75

| | Extra jobs required | | | |
	To Accommodate Reduced Out-migration between 1965-70 and 1970-75	To Reduce Unemployment Rate by Half a Percent, 1975	Total Extra Jobs Required, 1970-75	Extra Jobs Claimed by DREE, 1970-75
Newfoundland	7 374	960	8 334	3 419
Prince Edward Island	4 536	234	4 770	1 384
Nova Scotia	14 208	1515	15 723	8 327
New Brunswick	18 630	1300	19 930	9 941
Atlantic Region	44 748	4009	48 757	23 071

Note: The number of adults who migrated during each of the above time periods was estimated by assuming equal proportions of adult and child migrants (see note, Table 6). This procedure probably leads to an understatement of the number of adult migrants, but there is no satisfactory way to correct for this.

Source: Estimates by the Economic Council of Canada, based on Table 6 above and on data from Statistics Canada and the Department of Regional Economic Expansion.

massive and continuing net transfer of purchasing power to the Atlantic region — "net" in the sense of after taking account of the taxes levied on Atlantic residents themselves as part of the taxes needed to pay for the programs.[6] Moreover, the size of the transfers would have been sufficient on its own to reduce the unemployment rate by the amount observed, though not to reduce out-migration as well. We also know that transportation subsidies to Atlantic trade have been increased in recent years. Nor is it impossible for the myriad of other influences on a region's economic destiny to have changed recently in the Atlantic region so as to favour it more than hitherto — whether permanently or transiently, no one can tell.

Although we have stressed the effects on broad aggregative measures of regional disparities, such as the unemployment rate and the rate of employment growth in the Atlantic region, it is certainly possible that DREE has had beneficial consequences in particular areas of other provinces, where the general level of DREE expenditures is so low (because these provinces are generally prosperous) that no detectable effect on aggregate measures could ever be expected. In such cases, success could be ferretted out only by detailed cost-benefit studies. In British Columbia and Alberta, for example, total DREE assistance has been very small; nevertheless, it may have had a high pay-off, per dollar spent, for people in particular areas, such as Kelowna in British Columbia and Medicine Hat in Alberta, either directly or through aid in establishing innovative firms, which others then follow.[7]

Thus more analysis is needed, going beyond a *post hoc ergo propter hoc* level of reasoning that uses simple inspection of changes, or their absence in broad aggregate measures, in order to reach valid conclusions on the question of DREE's effectiveness. This analysis can conveniently be centred around the question of whether DREE really does encourage the formation of new enterprises or whether it simply ends up giving money to firms who would have set up plants in the region in any event. In technical jargon, are the DREE jobs "incremental" or not? In presenting this analysis, we confine ourselves to the Atlantic region on the grounds, already mentioned, that DREE's expenditures elsewhere have been small on a per capita basis.

The Question of Incremental Growth

In theory, DREE gives a firm only as much money as is required to make it locate where DREE wants it to be rather than somewhere else. If a firm would have located where DREE wanted it even if there had been no grant, it is not eligible to receive any money. The same is true of grants for expansion of existing facilities.

In practice, it is not possible to screen grants as carefully as this, and some establishments will be subsidized unnecessarily. In such cases, it is commonly said that the establishments, and the employment associated with them, are not "incremental". Conversely, establishments for which the subsidy was necessary (and the associated employment) are called incremental.[8]

There have been a few empirical investigations into the question of incremental growth. The major ones have been by the Atlantic Provinces Economic Council, by Springate, by the Atlantic Development Council, and by DREE itself.[9] Their estimates of the proportion of DREE-subsidized investment or employment that is incremental range from a low of 30 percent (Springate) to a high of 80 percent (Atlantic Provinces Economic Council).

We have examined this question ourselves, but only for the Atlantic region. Elsewhere, DREE expenditures are so low, relative to the size of the economy, that we doubt that their effects could be detected, given the crudity of available econometric methods and published statistics. Our basic technique is to look at how many establishments were newly formed or "born", on average each year, in a given industry in the Atlantic region during the period before the federal grants were available. Then we ask whether there has been any increase in the number of "births" in recent years. If there has, is the increase as large as is implied by the DREE claims?

Our method allowed us to classify DREE-supported establishments into three broad groups: those whose existence seemed improbable without DREE, those whose existence seemed probable without DREE, and those for which it was impossible to make any reasonably reliable judgement one way or the other.

For the Atlantic region as a whole, we established that about 25 percent of the DREE-supported establishments were incremental; 41 percent were not; and no reasonably reliable judgement could be made for 34 percent. Thus the incremental growth ratio for establishments, according to these tests, appears to lie between 25 percent and 59 percent. This corresponds to an incremental growth ratio for employment of between 39 percent and 68 percent.

The Crowding-out Effect

It would be quite possible for all DREE-subsidized establishments and jobs to be incremental and yet have no net effect on total investment and employment in the region. This would happen if the competition from subsidized firms for scarce types of labour, for particularly favourable production sites, or for the other scarce resources needed for production, together, made it sufficiently more difficult for other firms to produce or survive in the area. To take a hypothetical example: if a subsidized company required a number of middle-management personnel and such personnel were in short supply in the region, a plant in some other type of industry might fail to be born; existing plants might be hindered from expanding as rapidly as they otherwise could; or establishments in other industries might fold from their inability to hold key labour or other productive resources. More generally, the influx of subsidized capital could cause a fall in the rate of return to nonsubsidized capital investment in the region. Subsidized firms might "crowd out" unsubsidized firms.

One way to approach the crowding-out effect (if any) quantitatively would be to examine births of establishments in manufacturing as a whole, instead of industry by industry, and to look at the total number of firms going out of existence. If there was a significant crowding-out effect, the extra DREE births – the 25 percent to 59 percent of DREE claims that we have concluded to be incremental – should have been offset by either an above-normal increase in the number of other nonsubsidized establishments going out of existence, by a decline in births of such establishments, or by some combination of both. Complete crowding-out would mean that the annual net increase in

establishments in manufacturing as a whole – newly formed firms, less firms going out of existence – would show no change relative to past experience. No crowding-out would mean that the annual net increase[10] would rise by the number of incremental DREE-subsidized births – i.e., by a total of between 25 and 58 establishments over the 1970-72 period.[11]

From 1962 to 1968, in the Atlantic region as a whole, 671 manufacturing establishments were born[12] and 784 went out of existence, for a net increase of -16 a year. From 1970 to 1972, 344 establishments were born and 283 disappeared, for a net increase of +20 a year.[13] The rise in the annual net increase is 36, which is within our range of 25 to 59, towards the lower end. One should be cautious in interpreting these numbers, as data for individual years are very erratic;[14] but they do suggest that, if there is a crowding-out effect, it is not very strong or it occurs through inhibiting expansions in existing establishments.[15] Certainly it would seem safe to suggest that the lower end of our range for incremental growth, in terms of establishments – 25 percent – is a fairly conservative estimate of DREE's success rate. This corresponds to an incremental growth ratio for employment of 39 percent, if all employment in incremental establishments is itself incremental. Most of it probably is, since the unemployment rate in practically all occupational groups is high in the Atlantic region and thus 39 percent is not likely to be a serious overestimate.

Other DREE Activities

The success or failure of other DREE programs, mainly infrastructure-spending and spending on rural assistance, but also RDIA outside the Atlantic region, is hard to assess. No published quantitative work on this question exists, to our knowledge.

The Department of Regional Economic Expansion also plays a role in the co-ordination of spending by federal government departments, with a view to keeping the distribution of federal expenditures as equitably spread among the regions as possible. No information is released on this coordinating activity, and it is therefore impossible to judge its value or to know whether shifts in federal expenditures induced in this way make it significantly more costly to deliver federal government services. It may be as costly to relocate federal "establishments" to outlying regions as it is to relocate private establishments. In the latter case, inefficiency can be measured by the subsidy required; gain, by the value of employment generated. In principle, the extra costs of relocating federal activities could also be calculated and compared with the value of employment gains; but if this is being done, the facts have not yet been made publicly available.

National Output and Income Distribution

From the point of view of the private businessman, projects undertaken as a result of DREE support would generally be losers without that support. This is true not only for direct support under ADIA and RDIA[16] but also for indirect support through infrastructure partially or wholly financed by DREE funds. We say "generally losers", because there will be cases where the businessman with a grant would have gone ahead without a grant, and there will be cases where the businessman's judgment, before starting out, was that a grant was necessary but in fact, in practice, it was not. To counterbalance the latter, there are also cases where the government grant turns out to be insufficient and an establishment folds.

Since most of the projects would have taken place somewhere in Canada if they had not been influenced to locate in particular areas by DREE, there appears to be a loss of real production to the economy as a whole, roughly equal to the amount of DREE money expended, through grants, infrastructure assistance, and other devices designed to attract firms. This is the loss that results from locating industry in places that are inefficient from a private point of view.

However, from the point of view of the citizenry as a whole, the DREE projects may be considered efficient if enough of the jobs they create are for people who would otherwise be unemployed. As we have seen, not all the jobs in projects subsidized by DREE are incremental jobs, and so they do not always lessen unemployment. Moreover, those DREE firms that are incremental or those induced by government infrastructure

may employ labour that would have been employed anyway in firms not crowded out by government assistance to other firms, or in firms that would have been larger but for the subsidzed competition. Even if this were not the case, some of the unemployed labour might have migrated to other regions where jobs would have been available. On the other hand, to the extent that jobs are in fact created in firms subsidized by DREE or attracted by DREE-assisted infrastructure, there will be a multiplier effect that creates other jobs and incomes in addition to those created directly.

Now, if jobs are obtained by people who would otherwise have been unemployed, output is higher than it would have been without DREE. From the social point of view, the value of that output is a credit that may offset, or more than offset, the loss of output resulting from industry locating in places that are inefficient from the private point of view.

To put some perspective on this, it would be desirable to compare the DREE budget for grants and infrastructure with the value of output attributable to labour that would otherwise have been unemployed. It is difficult, however, to be dogmatic about what proportion of jobs under the RDIA and ADIA programs is incremental, let alone to add to these jobs those created as a result of increasing the attractiveness of certain locations by infrastructure-spending. What we can do, though, is compare DREE spending on RDIA grants alone with the present value of output attributable to formerly unemployed labour, under various assumptions about the *incrementality* of these jobs.[17] Table 8 does this, assuming that 40 percent, 25 percent, and 10 percent of the jobs are incremental. Jobs are assumed to last five years and to pay a wage of $5500 a year.[8] Their present value is calculated using a discount rate of 12 percent. Assuming a job life of more than five years or a lower discount rate[19] would increase the values of the jobs beyond the figures shown in the table.

Table 8 shows that the RDIA program would be extremely profitable, from a social point of view, if as many as 40 percent of the jobs associated with DREE grants were in fact incremental. Even if only one job in four were incremental (25 percent), there would be a gain of $82 million at the conservative job-lifetime and discount rates chosen. If, however, fewer than 18 percent of the jobs

NET EXPECTED INCREMENT TO NATIONAL OUTPUT FLOWING FROM RDIA
AGREEMENTS MADE DURING 1970-72, UNDER VARIOUS ASSUMPTIONS ABOUT
GROWTH

	Incremental growth ratio (%)		
	40	25	10
	(Thousands)		
Jobs created	26.4	16.5	6.6
	($ Million)		
Value added from jobs created	554	346	138
DREE-spending on grants	264	264	264
Net output increase (+) or decrease (−)	+290	+ 82	-126

Note: Some analysts argue that the value of (forced) leisure to workers formerly unemployed should be subtracted from the numbers in this row. This was not done. If, for example, leisure was worth $2000 a year, no output increase would occur until the incremental growth ratio was 28 percent.

Source: Table 3; and W. Irwin Gillespie and Richard Kerr, "The Impact of Federal Regional Economic Expansion Policies on the Distribution of Income in Canada", Economic Council of Canada, Discussion Paper 85, 1977.

Table 8

created were truly incremental, the program would be a waste of money, from a social point of view.

Our evidence suggests that the incremental growth ratio in the Atlantic region is such as to put the program into the profitable range, with at least 25 percent of establishments, and perhaps as much as 39 percent of employment, being incremental. If the same rate of success applies in other regions, it can be concluded that DREE is a success, as far as the proportion of its spending devoted to assisting private industry (approximately one-third) is concerned. As yet, no evidence exists to judge the value of the remaining spending.

The effects of DREE on national output, whatever they be, should be sharply distinguished from its effects on the distribution of national output, whether it be among provinces and regions or among the rich, the poor, and the in-betweens. To illustrate, if tax money spent on DREE does create jobs in the Atlantic region, unemployment insurance and welfare payments will be less there. Since the latter are partially paid for by federal taxes, the total federal taxes needed to finance DREE will be less than total expenditures on DREE. Conceivably total

taxes could even be lower with DREE than without it and, even going beyond that, they could be lower for taxpayers outside the Atlantic region if the saving to them of their share in federally financed unemployment insurance payments and welfare outweighed their share in the cost of DREE.

There are other effects from DREE on income distribution. If DREE changes the return to capital invested in certain regions, it will affect the distribution of income between wages and profits, both in the favoured regions and, to the extent that capital is owned by nonresidents of the favoured regions, outside them. Since profit recipients tend to be higher up the income scale than those who have wage income alone, the size distribution of income will be affected. Spending by DREE on rural assistance programs also affects the distribution of income, by size and location, in Canada. For these and other reasons, the full effects of DREE on income distribution are extremely complex.

In a study done for the Economic Council of Canada by the Regional and Urban Policy Analysis Centre at Carleton University, Professor Gillespie has tried to analyse all these effects. His major findings concern the distribution of income among major regions and

Table 9

EFFECTS OF DREE ON THE DISTRIBUTION OF INCOME, CANADA, BY REGION, 1969-75

	Proportion of Expenditure that Benefits Region	Share of Taxes Paid	Net Gain or Loss
	(Percent)		
Atlantic Region	31	6	25
Quebec	24	25	-1
Ontario	19	42	-23
Prairie Region	17	16	1
British Columbia	9	12	-2
Canada	100	100	0

Source: W. Irwin Gillespie and Richard Kerr, "The Impact of Federal Regional Economic Expansion Policies on the Distribution of Income in Canada", Economic Council of Canada, Discussion Paper 85, 1977.

by size classes within regions. Table 9 shows the effects, by region.

The first column of Table 9 shows the percentage of DREE expenditures that accrued as benefits to the residents of each region. The second column shows how much of the tax needed for DREE was paid by people in the region. Thus, for every $100 spent by DREE, residents of the Atlantic region eventually got $31 and they paid $6, resulting in a net average gain of $25. The zero at the bottom of the final column indicates no net gain or loss to the nation as a whole — an assumption made by Gillespie in order to isolate the pure distributional effects of DREE. As we have seen, there is probably a net gain nationally, and adding it in would change the final column, making the positive numbers larger and the negative ones less negative.

Gillespie's own conclusions from the table are that:

The Atlantic and Prairie regions are net gainers whereas Ontario, British Columbia, and Quebec are net contributors. The Atlantic region is the big gainer with almost 25 percent of DREE expenditures showing up as a net gain. Ontario's net contribution is almost 22 (sic) percent of DREE expenditures. Given the possible errors in a study of this nature one can reasonably conclude that Quebec, the Prairies, and British Columbia are neither net gainers from, nor net contributors to, DREE programs.[20]

These conclusions would remain substantially unchanged if the net gain to the nation was regionally distributed, except that Quebec would probably show a small net gain rather than a small net loss.

One can similarly analyse the burdens of financing DREE and the distribution of benefits from it, by income size, within each region. This is done in Table 10.[21]

The bottom row of Table 10 shows the same information as the final column of Table 9 — that the Atlantic region gained,

Table 10

NET GAINS (+) OR LOSSES (-) AS A PERCENTAGE OF TOTAL DREE EXPENDITURES, BY FAMILY INCOME CLASS, CANADA, BY REGION, 1969-75

	Atlantic Region	Quebec	Ontario	Prairie Region	British Columbia	Canada
Family income class:	(Percent)					
Under $2000	1.3	0.5	0.1	1.0	—	2.9
$2 000 − 2 999	2.0	0.4	—	0.9	-0.1	3.2
$3 000 − 3 999	2.7	0.4	-0.2	1.0	0.1	4.0
$4 000 − 4 999	2.3	0.4	-0.3	0.6	—	2.9
$5 000 − 5 999	2.3	—	-1.4	—	-0.2	0.8
$6 000 − 6 999	2.3	—	-1.3	-0.4	-0.1	0.5
$7 000 − 9 999	5.3	-1.3	-6.8	-1.4	-2.3	-6.6
$10 000 − 14 999	4.2	-1.3	-9.3	-2.0	-2.1	-10.4
$15 000 and over	2.1	—	-3.4	1.3	2.5	2.6
Total	24.7	-0.8	-22.6	0.9	-2.2	0.0

(−) Less than 0.1 in absolute value.

Source: W. Irwin Gillespie and Richard Kerr, "The Impact of Federal Regional Economic Expansion Policies on the Distribution of Income in Canada", Economic Council of Canada, Discussion Paper 85, 1977.

that Ontario lost, and that the net effects elsewhere were negligible. The other rows show some interesting and important details of how this gain or loss was spread among income size classes. Within Ontario, the biggest losers were the two upper/middle family income classes, between $7000 and $15000. The top and lower income groups lost the least. In British Columbia, the top group actually gained, while the next two highest groups lost a little. These statements remain true on a family basis — i.e., even after allowing for the fact that the number of families within each income class is not the same. All groups gained in the Atlantic region — the rich, the poor, and the in-betweens. Gains and losses elsewhere were all rather small.

Thus, while rich Ontarians were subsidizing poor Atlanticans, poor Ontarians were also subsidizing rich Atlanticans, though the former effect outweighed the latter. These are interesting distributional side effects of a program that tries to redistribute activity rather than simply redistribute income.

Other Government Assistance to the Private Sector

The Department of Regional Economic Expansion is not the only government body with programs that have a direct impact on industrial location. All levels of governments are involved to some degree in programs that attempt — through grants and subsidies; tax or capital depreciation incentives; research, development, or other technical aid; public sector lending institutions; or assistance in the procurement and training of the

required labour force — to increase the level of productive activity and thereby influence the most profitable location for business enterprise. Needless to say, a complete analysis of the regional impact of such programs would constitute a monumental task; in fact, a recent listing of industrial assistance programs and regulations alone comprised nearly three hundred pages.[22] While we have been unable to make an evaluation of all such programs, we do offer a few figures to indicate the level of DREE-type government assistance that is undertaken with other economic or social goals in mind. These programs are presently not screened for their regional impact, as they would be in France, for example; it is, however, important that they be looked at from the regional viewpoint in order to gauge the overall regional impact of government policy. This is necessary also in order to judge whether DREE's efforts to redirect economic activity are undertaken in a context of offsetting policy initiatives, and whether the magnitude of DREE efforts are significant in the industrial system.

Federal transfers to business have increased steadily from 2 or 3 percent of gross federal expenditures in the early fifties to 8.6 percent or $2.6 billion in 1974-75. We saw earlier that nearly one-third of DREE's expenditures went directly to the private sector in the form of capital assistance and incentive grants. The $69.7 million that DREE transferred to private business in 1974-75 amounted to less than 3 percent of all subsidies and capital assistance payments from the federal government in that fiscal year. This figure is astonishingly low and would remain so even if oil and gasoline payments

Table 11

FEDERAL TRANSFERS TO BUSINESS, 1974-75

	($ Million)	(Percent)
Natural resources	1181	44.7
(of which oil and gas)	(1162)	(44.0)
Agriculture	485	18.3
Transportation and communications*	415	15.7
Canadian Broadcasting Corporation	299	11.3
Trade and industry	183	6.9
Housing	43	1.6
Labour, employment, and immigration	37	1.4
Total	2644	

Note: The categories employed in this table are those used in the source publication and do not exactly conform to the ones on which the regional distribution of assistance discussed in the text are based. Thus, for example, DREE plus IT&C industrial assistance more than exhausts the trade and industry category shown above. The correspondence is close enough, however, to convey the general point being made regarding where transfers to business are going.

*Mainly Canadian National Railways and Canadian Pacific Railways ($399 million).

Source: Data from Statistics Canada.

309

– 44 percent of the total – were removed.[23] An additional 11.3 percent is accounted for by the Canadian Broadcasting Corporation. The remaining large components of transfers to business are: agriculture, 18.3 percent; transportation and communications, 15.7 percent; and trade and industry, 6.9. percent.

In principle, all of these transfers may affect the location of production and employment by encouraging the expansion of existing firms on site; by making payment directly dependent on location, as in the case of DREE; or by affecting transportation and input costs so that new or expanded production is more likely to be undertaken in one place rather than another. While re-emphasizing the somewhat piecemeal development of industrial policy in Canada and its consequent myriad of programs fashioned as industrial needs have been perceived to emerge, we are able to present data, on a regional basis, not only for DREE expenditures but also for thirteen industrial assistance programs under the Department of Industry, Trade and Commerce (IT&C) for four agriculture programs, and for the Department of Manpower and Immigration's Industry Training Program (Table 12).

The figures show, for example, that the Department of Industry, Trade and Commerce, under its various industrial assistance programs, transferred $159 million to the private sector, more than twice as much as did DREE in 1974-75. Moreover, the Industry, Trade and Commerce transfers are highest on a per capita basis in British Columbia and Ontario, precisely those areas which DREE would consider low-priority areas for assistance in economic development.

In agriculture, public subsidy and regulation have become a topic of increasing discussion and controversy. One of the issues in the debate surrounding marketing boards and supply management is the implication that government involvement may influence the profitability and hence investment in, and development of, agricultural resources differently in different regions. Of the $357.1 million paid out in agricultural subsidies in 1974-75,[24] Quebec and Ontario received $116.4 million and $101.0 million, respectively; together they account for more than 60 percent of the expenditures. The Atlantic region received $7.3 million or just slightly over 2 percent. The Prairie region received the largest share, on both an absolute and per capita basis; Quebec also received more than the Canadian per capita average.

Table 12 also shows payments to employers for industrial training; hence the regional distribution is in line with the severity of unemployment in the various regions.

In addition to federal programs that affect the regional distribution of economic activity in Canada, the provinces and territories have their own industrial and economic development policies. Conceptually, provincial development efforts may be grouped into two categories: those which seek to increase output and incomes by encouraging the use of unemployed resources or improving the efficiency of a combination of resources, and those which seek to influence the location decisions of firms. Policies of the latter type may bring the provinces into competition with each other or with DREE for a given stock of new or mobile enterprise; thus, while one province or region may come out ahead of another, it is not clear that any gain in

Table 12

GRANTS, CONTRIBUTIONS, AND SUBSIDIES TO BUSINESS, CANADA, BY REGION, 1974-75

	Atlantic Region	Quebec	Ontario	Prairie Region	British Columbia	Canada	Total assistance
	(Dollars per capita)						($ Million)
IT&C industrial assistance programs	8	6	9	1	11	7	158.7
DREE	8	6	1	3	–	3	69.7
Agriculture	3	19	12	35	3	16	357.1
M&I industry training program	3	2	1	1	2	2	36.9
Total	23	33	23	40	17	28	622.4

Source: Estimates by the Economic Council of Canada.

national output will be achieved or that disadvantaged regions will benefit relative to others.

The provinces offer industrial incentive grants and loans, as well as a variety of special business services such as technical information, research, and management consultation. Three of the Atlantic Provinces and Alberta also provide assistance to development corporations for the formation of industrial estates along the lines of the British model where government assembles land and provides infrastructure and basic buildings for secondary industry. In 1973-74, provincial programs involved a transfer to business of $116 million for agriculture, trade and industry, and tourism projects, and an additional $23 million for natural resources and the supervision and development of regions and localities.

On a regional basis, these transfers were highest in the Atlantic region at $11 per capita; Quebec followed at $9; the Prairie region and Ontario spend $8 and $6, respectively; and British Columbia spent $3 per capita.

It is clear that governments in general, and the federal government in particular, are funneling massive amounts of assistance to business. This is true even if one excludes the large oil import compensation program on the grounds that this is more properly considered of assistance to consumers, with private business as the distributing agency, than as assistance to business per se. The regional distribution of total payments under all programs other than oil compensation is not fully known, but the data we have been able to present suggest that it is different from the distribution of DREE expenditures, being much less favourable to the Atlantic region and Quebec. A clear distinction should be drawn, however, between the distribution of the monetary benefits of these programs and the effects of this distribution on the amount of economic activity in each region. The Prairies do well from assistance to agriculture, for example, but it is not likely that the amount of employment there, in agriculture or in total, is as much influenced by this assistance, dollar for dollar, as it is by assistance under DREE. Similarly, IT&C dollars of industrial aid go heavily to Ontario and British Columbia, but that is mainly because firms already in these provinces make more intensive use of IT&C help than firms elsewhere; it is not that firms are attracted to go there by the greater availability of IT&C money.

In short, though DREE expenditures are small in relation to total spending for industrial assistance, their effects on industrial location are proportionately more important. The RDIA part of DREE expenditures does seem to work, in the sense that enough firms are encouraged by the grants to relocate to cause national output to be higher than otherwise, as a result of making use of labour that would otherwise be unemployed. The effect of the remaining DREE expenditures, notably the heavy spending for infrastructure, remains a matter for further research.

NOTES AND REFERENCES

[1] A number of rather more specialized initiatives have also been taken from time to time, such as the Prairie Farm Rehabilitation Act, the Maritime Marshland Rehabilitation Act, and the formation of the Cape Breton Development Corporation.

[2] Economic Council of Canada, *Fifth Annual Review: The Challenge of Growth and Change* (Ottawa: Queen's Printer, 1968), p. 180.

[3] These grants are net of revisions and declines or withdrawals.

[4] Also statistically significant at 1 percent, on a formal test. On the other hand, one could not rule out the hypothesis that by 1973-75 the previous historical relationship had been restored — i.e., that there was temporary improvement only in the years 1970-72.

[5] See Economic Council of Canada, *Living Together: A Study of Regional Disparities* (Ottawa: Minister of Supply and Services Canada, 1977), Ch. 3; and Economic Council of Canada, *Twelfth Annual Review: Options for Growth* (Ottawa: Information Canada, 1975), Ch. 2.

[6] N. Swan, P. MacRae, and C. Steinberg, *Income Maintenance Programs: Their Effect on Labour Supply and Aggregate Demand in the Maritimes*, A Joint Report by the Council of Canada (Ottawa: Minister of Supply and Services Canada, 1976).

[7] The question of where DREE should put its money, whether it should put any into a region as generally prosperous as Alberta, or whether the decision to spread economic activity within a prosperous province should be a provincial rather than a federal financial responsibility, is conceptually separable from the

question of whether the money, once spent, achieves its purpose. It is the latter question that we are examining here. On the former question, the amounts of DREE money going to prosperous provinces is, in any case, so small as to make the matter rather academic.

[8] For those familiar with the literature on these matters, we should stress that we are using "incremental" in the narrow sense that the subsidized establishment really needed the subsidy. The word is also used in the literature in a broader sense, to indicate not only that the establishment really needed the subsidy, but also that no other establishments were competed out of existence by the subsidized one. We call this latter effect, if it exists, the "crowding-out" effect, and it is dealt with later in this paper.

[9] Atlantic Provinces Economic Council, *The Atlantic Economy*, Fifth Annual Review (Halifax: APEC, October 1971); David J. V. Springate, "Regional Development Incentive Grants and Private Investment in Canada: A Case Study of the Effect of Regional Development Incentives on the Investment Decisions of Manufacturing Firms", Ph.D. thesis, Harvard University, Graduate School of Business Administration, 1972; Atlantic Development Council, *Regional Development Incentives Program: Atlantic Region* (St. John's: ADC, 1976); and Department of Regional Economics Expansion, "Assessment of the Regional Development Incentives Program", a staff paper, April 1973.

[10] Or decrease, if negative.

[11] The longest, recent period when DREE was active, for which complete data are available and during which DREE *claimed* to be responsible for the births of 98 establishments in the Atlantic region.

[12] These data were obtained from special computer runs done by Statistics Canada on our behalf, and they exclude bakeries, fish-processing plants, and sawmills. These particular industries were omitted because of their tendency to frequently appear, disappear, and reappear in the collected statistics on births, consequently making them of dubious validity.

[13] We excluded 1969 as being neither before nor after DREE became really active.

[14] Data for individual years were as follows:

	1962	1963	1964	1965	1966	1967	1968	1969	1970	1971	1972
Going out of existence	99	142	86	89	92	161	115	133	101	92	90
Births	89	83	146	82	91	69	111	128	125	119	100
Net increase	−10	−59	+60	−7	−1	−92	−4	−5	+24	+27	+10

The change in the net increase from 1962-68 to 1970-72 is "significant" at 10 percent, but not at 5 percent.

[15] Data on the growth of total employment in manufacturing could, in principle, be used to check this. In practice, they are much more subject to cyclical variations than those on establishments, and this problem prevents their being used for our purposes.

[16] Indeed, a firm cannot legally get help unless it can show that the proposed establishment needs a subsidy; i.e., it has to show that it would be a loser without DREE money.

[17] Assuming no crowding-out (or subsuming crowding-out effects into a lower incremental growth ratio), assuming no multiplier effects (or subsuming them also), and assuming that no portion of other DREE expenditures were necessary preconditions for RDIA grants to work.

[18] This was a reasonable wage for the 1970-72 period, to which the data in Table 8 relate.

[19] Normal procedure is to discount at twice the private rate, in order to allow for the 50 percent corporation tax. We have assumed a private rate of 6 percent, which may perhaps be a little high. Allowance was made for that proportion of agreements made during 1970-72 which later lapsed due to withdrawal of the offer by DREE or rejection of the offer by the firm.

[20] W. Irwin Gillespie and Richard Kerr, "The Impact of Federal Regional Economic Expansion Policies on the Distribution of Income in Canada", Economic Council of Canada, Discussion Paper 85, 1977.

[21] To calculate the distributional effects, it was assumed that the money to finance DREE policies comes from all federal taxes in proportion to their importance in total tax collections, whose incidence, by region and family income class, is known. On the expenditure side, each of the major items (capital incentive grants, highways, sewers, etc.) was treated separately.

[22] P. E. McQuillan and G.H.R. Goldsmith, *Industrial Assistance Programs in Canada*, Fourth Edition (Ottawa: CCH Canadian Ltd., 1976).

[23] The major component of oil and gasoline payments is transfers under the Oil Import Compensation Program. This program is designed to adjust the Canadian oil and gas market to new international conditions and to achieve regional equalization in the final price of oil and gas products.

[24] We have included here payments under four agencies or programs: Canadian Dairy Commission, Agricultural Stabilization Board, Two-Price Wheat Payments, and Prairie Farm Emergency Fund.

An Urban/Economic Development Strategy for the Atlantic Region

CHRIS D. BURKE AND DEREK J. IRELAND

Past and Present Trends

General

1. Economic Development

It is convenient to look at the Atlantic region in terms of a simplified export base approach. This is not to imply that exports are the most important element, but this model offers a convenient categorization in terms of the present situation.

Simply stated, the economy can be seen as having three components. First, exports; second, backward linkages providing inputs to the export industries; and third, local multipliers, which represent the cumulative effect of the respending of the income generated in the first two sectors.

In the Maritimes, as is normal in a relatively underdeveloped region, exports and their backward linkages have a strong resource orientation. Pulp and paper, fish and fish products, and minerals dominate the export picture.

In the past, manufacturing industries based on exports from the region serving national and international markets have typically found the distance to market a major restraining factor. Export-oriented manufacturing industries have traditionally been based either on resources or on low wages. Michelin is the exception rather than the rule. In a shift-share analysis we undertook of the manufacturing sector of the Nova Scotian economy through the 1960's, it was most striking that the "Province Effect" almost always had an opposite sign from the "Industry Effect". In other words, Nova Scotia was doing well in those industries which were in decline, and doing badly in those which were booming. The former could no longer afford high wages and high land costs in the central markets; the latter could not afford to leave the centre, where their skilled labour and service needs could be readily met.

These factors remain true, to some extent, and yet there has been considerable economic growth in the region in the past five years.

This growth has arisen from four general thrusts in the economy, and these thrusts will likely continue to generate growth for the next decade.

The first two thrusts are not export-oriented, but rather serve to increase the multiplier. They represent import substitution and constitute most of net new employment, as the export potential is somewhat limited, although far from nonexistent.

The most important in terms of number of jobs is the expansion of the service sector selling to local and Atlantic region markets. The local market is being served in smaller urban centres, and the regional market is primarily being served out of Halifax-Dartmouth, with some warehousing and distribution through other centres, notably Moncton. This sort of growth will probably continue to account for the major part of regional employment growth.

The second multiplier thrust is the dominant element in goods production, and is the growth of manufacturing industries to serve the Atlantic region market. Such industries as modular and mobile homes, plastic fabrication, automobile parts, truck body manufacturing, and furniture manufacturing have been established. Again, this is expected to be a major factor in manufacturing growth.

Of the export thrusts, the first is that based on regional resources. The region has developed two important new resources in recent years: first, changes in shipping technology have made deep water harbours a major resource, for bulk commodities and for container movements; second, mineral exploration is turning up new resources such

---

● CHRIS BURKE is with Environment Canada, Ottawa, and DEREK IRELAND is Director of Research and Programme Development, Saskatchewan Department of Industry and Commerce, Regina. This article is Chapter 3 in *An Urban/Economic Development Strategy for the Atlantic Region* (Ottawa: Information Canada, 1976), pp 17-48, and is reprinted with permission.

as offshore oil and gas, and lead-zinc deposits. In terms of future growth, these resources will be more important than traditional resource sectors. Agriculture and fishing look to declining employment, and forestry has expanded close to the limits of the resource. In these sectors, opportunities exist in relation to better utilization of the resource and further processing, but major expansions in employment are not anticipated.

Finally, there is room for development of industries based on imported components for further processing before distributing to central Canada and other U.S. and Canadian markets. These industries are significantly assisted by the development of the container service to Halifax and Saint John.

In addition to these four thrusts, there are still some industries which are locating in the region because of low wages, cheap land, and generous incentives. As wages rise and incentives are more tightly administered, these industries are becoming scarcer, although there will remain some wage differentials which will be very important to some industries in some parts of the region.

2. *Urban Implications*

The major economic thrusts have strong urban biases. Except for resource-based activities, all manufacturing and service activities tend to prefer urban locations because of the availability of support services and a pool of available, generally skilled labour.

The regional service sector growth has the strongest urban bias, and a particularly strong propensity towards the Halifax-Dartmouth area, which is emerging as the major service centre for the entire Atlantic region, particularly for "higher order" services. On a smaller scale, St. John's is emerging as the service centre for Newfoundland. Other service activities are tending to locate in the smaller urban centres, typically in those of 5000 people or more. Warehousing and distribution has tended to locate in Moncton or Halifax, with some in Truro and Amherst. The provincial capitals have shown particularly strong growth, the result of the expansion of the public sector.

Manufacturing for the regional market is typically constrained by distribution factors and the available labour force. It has tended to locate in the Truro and Amherst areas of Nova Scotia, and in Saint John and Moncton

in New Brunswick, with some activity in New Glasgow, Bridgewater, and Halifax-Dartmouth. Other manufacturing has located either close to the resources or in the central area between Saint John and Halifax, particularly if container shipping and ready access to central markets are required.

3. *Spatial Implications*

Growth is not only occurring primarily in urban areas: it is also very much concentrated in a central corridor, approximately from Halifax to Fredericton.

This is creating a sharp dualism in the regional economy, with a prospering area, successfully attracting private investment, and a lagging area, with little or no employment growth. As far as private investment is concerned, it is our opinion that, without government policy interventions, this spatial pattern will continue for the next decade, with the central corridor getting most of the employment growth. The only major exceptions to this will be in the two growing resource sectors, namely minerals and deep water harbours. (Particular recent interest has focussed on the concept of a massive steel complex at Gabarus Bay, close to Industrial Cape Breton.)

The major growth centre is Halifax-Dartmouth, which is increasingly taking on the role of service centre for the region. Because it is growing, it is a major factor in strengthening the reality of the regional market; a concentrated market is much easier to serve than a dispersed market, and the growth of the regional market is as much a reflection of trade concentration as of increases in the volume of sales. The metropolitan area is also serving as a "broker" for the region, as the intermediary between the major cities of Canada and the urban centres of the region. It is linked to the major cities of other regions for exchange of ideas and services, and it is linked to its hinterland centres to bring them business and personal services. It is our opinion that the growth of the Halifax-Dartmouth area has only just begun, and that it will be growing at a very rapid rate for the next decade at least. In fact, the indications available to us at present show that in retail sales per capita, in construction activity, and employment growth, the metropolitan area has been growing at close to the same rate as Calgary and Edmonton, the "boom" cities of Canada. Income per capita and family income figures show Halifax-Dartmouth now on a par with Montreal and Quebec City, having already surpassed Winnipeg, Regina, and Victoria.

Saint John, New Brunswick, is also growing very rapidly, based in part on manufacturing growth together with growth in the Saint John port. It is the second largest centre in the Maritimes, but does not seem likely to develop as a focal point of the region in the same way as Halifax-Dartmouth. Rather, the comparative advantage of Saint John appears to lie in other activities that are more or less complementary to those of the Halifax-Dartmouth metropolitan area, but which hold a similar potential for generating new employment and high incomes.

The rest of the central corridor is growing, primarily around its urban centres. Some of the smaller-sized centres, such as Truro, are posing new problems, and we have very little literature to tell us what to do with the small-sized growth centre. One major problem is that, although industrial development has provided steady employment at sufficient wages to reduce out-migration, wages are not high enough to encourage in-migration to a small town with limited amenities.

In terms of the rest of the region, a capsulized description at this stage should establish the framework for an understanding of the more detailed quantitative analysis, and confirm our urban biases.

Northern New Brunswick lacks a major urban centre, and so is extremely susceptible to variations in markets for resources. It consistently records extremely high unemployment rates, has an extremely low female participation rate, and in development terms behaves very much as a classic hinterland economy.

Western Nova Scotia has similar problems, but they are much less intense, and the settlement patterns are far more traditional and established. Declining employment in agriculture and fisheries poses continuing problems for this area, and it is noticeable that, in recent years, the only areas showing significant prosperity are in the vicinity of Yarmouth, which is the largest urban centre, and of Kentville, which is the centre of agricultural processing and a farily substantial service centre in its own right.

Cape Breton Island does not suffer from the lack of an urban centre, but rather from a declining economic base for the existing urban centre. Industrial Cape Breton has been in decline because of the decline of the coal and steel industries. The coal industry has recently shown signs of defying the pundits who condemned it to death and, if the international steel complex can be established at Gabarus Bay, then the steel industry in Cape Breton will also be placed on a sound footing. However, the urban situation is critical to the full development of this area, assuming that the industrial revitalization occurs, as it is in the urban centre that the life-style and multiplier impact will be felt.

Prince Edward Island is implementing a development plan which is based largely on development of tourism, fishing, and a small amount of forestry. It seems likely that the future employment generation in Prince Edward Island will be urban-focussed, even though physical output will continue to be concentrated in the farming sector, which is, by definition, rural. The nature of growth and development in Charlottetown and Summerside is, therefore, a critical factor.

The Avalon Peninsula area, centred on St. John's, Newfoundland, is doing relatively well because of the development of the service sector, led by government activities. Indications are that activity will be increasingly focussed on this area, so prosperity should continue.

The rest of Newfoundland, particularly those areas dependent upon the fishery (and in Newfoundland that means much of the rest of Newfoundland), face a much more difficult time. Resource depletion has hit the Newfoundland fishery to the same extent, if not more so, than other provincial fisheries. Attempts at forced urbanization through the resettlement program have had mixed success. Alternative employment opportunities are few, outside of the existing centres, such as Corner Brook, and the mining and forest industries. Any new employment opportunities that do occur will likely be urban in nature.

In summary, urban factors are critical to the development of most areas within the Atlantic region. Areas of successful, concentrated urban development, such as the central corridor of Nova Scotia-New Brunswick and the Avalon Peninsula area of Newfoundland, are prospering. Their continued prosperity seems to rest very much on their urbanization and upon the metropolitanization of the Halifax-Dartmouth area. The other areas, facing various kinds of economic problems, cannot compensate for declining resource employment by development of new manufacturing and service employment, because these activities require an urban setting and a geographically central place. Rapid urbanization will improve that situation.

Quantitative Findings

1. The Region as a Whole

The following will attempt to "flesh out" the above arguments with a few statistics, and to place the urban profile of the Atlantic in the context of recent trends in the regional economy.[1] As economists generally do, we shall begin with looking backward.[2]

From the Second World War to about 1971, the Atlantic region experienced the least economic expansion of any part of Canada, and possessed all the usual characteristics of a lagging region. These were:

- high unemployment rates and depressed growth in employment;
- high unemployment rates and depressed participation rates, the result of the slow growth in employment;
- high dependence on primary and other slow growth industries, an important factor behind the lack of employment growth in the region;
- well below average incomes, reflecting both depressed labour force utilization rates and an industrial structure skewed towards lower-paying industries;
- low population growth, resulting from heavy and almost continuous out-migration; as usual, the out-migration was concentrated in the youngest and most skilled age groups, lending a further downward bias to educational levels, and technical, managerial, and entrepreneurial skills in the region.

The lack of growth in the Atlantic economy can perhaps best be summarized by two simple comparisons. From 1951 to 1971, the Canadian population recorded an annual average growth rate of 2.2 percent, or nearly double the Atlantic rate of 1.2 percent. Corresponding to this, employment in the country as a whole expanded at an annual average rate of 2.4 percent from 1951 to 1971 — exactly double the Atlantic rate. In short, the two decades from 1951 to 1971 were a period of adjustment rather than expansion for the Atlantic economy.[3]

More detailed research suggests, however, that the Atlantic economy possessed more underlying strength during the sixties than the aggregate statistics point out. This strength was masked, though, by a number of unfavourable developments specific to the region and the decade. In order to better identify these 1961-71 developments and their quantitative impacts, shift-share and location quotient analysis was conducted on sectoral trends in employment for the total Atlantic region, its four provinces, and sixteen subprovincial regions. The analysis was further extended to the larger urban centres (generally 10 000 population and above) in the region. The results of this research are fully documented in a background study to this report.[4]

The most important finding was that the region, as a whole, and a number of its component parts were seriously held back during the sixties by a number of nonrecurring events, events that are either no longer relevant or whose impacts have been reversed. Many of the Atlantic's largest centres were particularly affected by these developments, thus preventing these centres from playing the "growth centre" role generally ascribed to them by regional economic theory (and, to varying extents, by government planners).

These nonrecurring events can be separated into five categories:

(a) Military cutbacks (both American and Canadian), which had a much sharper impact on the Atlantic than elsewhere in Canada, reflecting in part the military's much greater importance to the region at the outset of the decade. These reductions were a serious constraint to growth in a number of areas, but most particularly in the region's largest centre, Halifax-Dartmouth, which suffered a direct job loss of about 5000. Moreover, the military is a high-wage activity in the context of the Atlantic, and thus the effects on the residentiary (largely service) industries were particularly severe.

This subject should, perhaps, be pursued a little further. Viewed from this perspective, federal government policy in the region takes on a dichotomous and confusing tone. On one hand, the sixties was a period of marked acceleration in the federal government's regional development efforts, through such programs as ADB, ADIA, FRED, ARDA, and the other members of the "alphabet soup". On the other hand, the senior level of government removed close to 10 000 high-paying jobs from the region, with little attempt at direct job replacement. The growing interest in the decentralization of federal government employment is a highly favourable development, but comes a little late for many residents of the Atlantic region.

(b) Mine cutbacks and closures, particularly involving Cape Breton coal and Bell Island (Newfoundland) iron ore. These offset substantial new mining development elsewhere in the region, especially in northeastern New Brunswick and Labrador. Moreover, the indirect (i.e., multiplier) effects associated with the new developments — particularly those in Labrador — are probably significantly smaller than those of the mines that shut down, reflecting the remoteness of the new mines. On balance then, the region probably lost from this redistribution of mining activity.

(c) Massive rationalization and employment adjustment in the primary sectors, particularly agriculture and forestry. According to the 1971 Census, the reduction in agricultural employment was from 36 to 23 000, while for forestry the decline was from 23 to 13 000 between 1961 and 1971. Thus, the adjustments in the resource sectors, which occurred earlier in other parts of Canada, were concentrated in the sixties in the Atlantic. These adjustments can be viewed as "nonrecurring" in the sense that, in many parts of the Atlantic — and likely for the Atlantic as a whole — these reductions were of an absolute magnitude that is unlikely to occur again.

(d) Rationalization trends in the manufacturing sectors of many of the region's larger centres, especially Halifax-Dartmouth and Saint John. These acted as a constraint to growth but left these centres with a much stronger manufacturing base.

(e) Little or no growth in transportation, particularly port employment, reflecting the growing — and, as it turns out, highly subsidized — competition from the St. Lawrence Seaway.

These nonrecurring events tended to offset some of the highly favourable developments that were occurring in the Atlantic both in manufacturing and services, and were most likely to be the major factors preventing the region from keeping pace with the rest of the country during the 1960's.

The disappointing performance of the sixties is very familiar. What is less familiar is that a dramatic reversal has occurred in the economic fortunes of the Atlantic region over the last three years. In almost all of the key indicators (employment, output, population, rates of labour force utilization, income per capita), the performance of the Atlantic economy has equalled or bettered trends in the rest of the country from 1971 to 1974. The most dramatic outcome is that the Atlantic region has experienced some in-migration since 1971, a sharp reversal from the heavy out-migration of the preceding twenty years. Moreover, preliminary evidence suggests that large amounts of the migration flow into the region are accounted for by the younger and more skilled segments of the population.

Five main factors lie behind the recent surge in the Atlantic economy, some of which were alluded to earlier. These are:

— improved international markets for the region's natural resources and semi-processed goods, including pulp, paper and other wood products; mineral products (both metallic and nonmetallic); agricultural goods; and, until 1974, fish products;

— the continuing rationalization of the Atlantic economy from lower to higher productivity industries, with secondary manufacturing coming to play a lead role for perhaps the first time;

— the continuing growth of the Atlantic region market, which is now of sufficient size to support a wide range of manufacturing and services oriented towards the local market;

— the further broadening of the region's service sector base, leading to significantly increased economic multipliers;

— changes in transportation and other technologies, which are favouring the Atlantic region for perhaps the first time in 100 years; included here are the growing importance of deep water harbours and containerization in transportation; other factors of some significance now, but holding even greater potential for the future, are the increasing emphases being given to ocean science and technology, and the growing trade links between Canada and Europe.

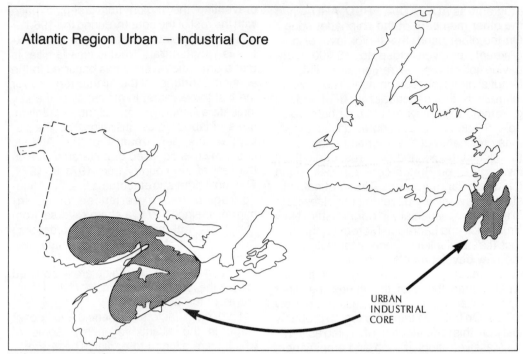

Figure 1

2. *Subregional Trends*

In many respects then, the economic adjustments of the previous two decades laid the foundations for the strong economic growth experienced by the Atlantic economy from 1971 to 1974. Perhaps most important, the adjustment period left the Maritimes with an urban/industrial core region, the area stretching from Saint John and Fredericton through to Moncton and Halifax-Dartmouth (Figure 1), which holds many economic characteristics in common with the more advanced parts of Canada and with the so-called "national average". The urban core region is defined to include the cities of Halifax-Dartmouth, Saint John, Moncton, and Fredericton (and the areas surrounding them), plus northern Nova Scotia and Lunenburg County (including the mainly manufacturing centres of Amherst, Truro, New Glasgow, and Bridgewater).

As of 1971, this urban core contained a population of 811 000, or two-fifths of the Atlantic region total. Nearly two-thirds of the urban core population was urbanized, only 10 percent below the national average. Moreover, the small distances in the core region allow considerable commuting from rural farm to urban job; thus, the urbanization ratio, when viewed in terms of economic

function, was probably much closer to the Canadian level.[5] Further to this, nearly 50 percent of the urban core population resides in centres with 30 000 or more population, only a few percentage points below the national average. Turning to labour force utilization, 52 percent of the urban core's population was gainfully employed in 1971, not markedly below the national rate of 53.4 percent.

Moreover, the structural adjustments of the sixties left the urban core region with a fairly diversified, well-balanced industrial structure, one which is not substantially different from the industry mix in Canada as a whole. The urban core region has less representation in the primary industries and manufacturing, but this is compensated for by higher representation in public administration and defence, transportation, and trade. The largest and perhaps most critical divergence occurs in manufacturing, which accounts for a much high proportion of activity in Canada than the Atlantic.

Finally, compared with a shortfall of 28 percent for the Atlantic region as a whole, personal income per capita in the urban core was estimated to be about 14 percent behind the Canadian average in 1971. When account is taken of the diversity of life-styles, the easy access to recreational areas, and the lack of pollution offered by most parts of the urban core, its overall "quality of life" is probably not significantly below the "more prosperous" parts of Canada.

In stark contrast, the outlying or peripheral areas viewed as a single entity continue to possess the economic characteristics that, up to now, have generally been associated with the Atlantic region as a whole. As of 1971, less than one-half of its population was residing in urban centres, and only 11 percent in centres of more than 30 000 population. Only 43 percent of its working-age population was employed, and the share of employment accounted for by the primary industries was three-fifths above the national average. In contrast, manufacturing (particularly secondary manufacturing) and the more modern service industries are significantly under-represented. Finally, and most important, per capita income in the peripheral areas was nearly two-fifths below the national standard. In other words, all of the usual characteristics of a lagging region are in evidence. Certain subregions among the peripheral areas are particularly disadvantaged. For example, per capita income in northern New Brunswick and in Newfoundland outside of the Avalon Peninsula comes to slightly over 50 percent of the Canadian average.

For reasons alluded to earlier, the urban core region failed to play a leadership role in the Atlantic economy prior to 1971, notwithstanding its generally favourable economic structure. This is exemplified by the fact that its 1961-71 population growth rate, at 1 percent, was only marginally above the Atlantic rate of 0.8 percent. Of the subregions making up the urban core, only the Halifax-Dartmouth area was significantly above that level; but still, Halifax-Dartmouth showed the least growth of any metropolitan centre in Canada over the period. On balance then, the urban core was not a major destination point for migrants from other parts of the Atlantic region prior to the last few years.

3. *The Four Years, 1971-1974*

The picture changes dramatically when the 1971-74 data are examined.

In contrast to earlier trends, all the available evidence suggests a clear leadership role for the urban core region of the Maritimes, and, in particular, its two largest centres – Halifax-Dartmouth and Saint John – during the recent upturn in the Atlantic economy.

Turning first to trends in employment, the growth in large establishments employment[6] in the industrial/urban core was well above the trend for the total region – and country – over the past few years. For example, from 1971 to 1974 the advance in the urban core was 15.4 percent, or well above the increase of 9.2 percent in the peripheral areas of the region.[7]

Not all of the outlying areas performed badly. Perhaps most significantly, the Avalon Peninsula in Newfoundland – the most urbanized of the peripheral areas – recorded a very strong employment advance. Reflecting the dominance of St. John's, the Avalon Peninsula holds more in common with the industrial/urban core than with the older peripheral areas. Prince Edward Island also moved strongly upward, the consequence of the high construction activity under the Comprehensive Development Plan.

In sharp contrast, northeastern New Brunswick, southwestern Nova Scotia, Cape Breton, the Strait of Canso, and most subregions of Newfoundland, outside of the Avalon Peninsula, all experienced well below average growth in large establishments employment from 1971 to 1974. It is significant that all four of these subregions rank among the least prosperous parts of the Atlantic region.

Moreover, current evidence suggests that the divergence in performance between the urban core and the peripheral areas tended to widen through the period under review. For example, from 1973 to 1974 the growth in large establishments employment in the urban core was four times that of the peripheral areas taken together. And the Halifax-Dartmouth and Saint John areas combined, accounted for nearly three-fifths of the growth from 1973 to 1974, compared with their one-third share of the 1973 "level" or employment.

An analysis of employment trends by urban centre shows a similar pattern. Most of the urban centres within the urban/industrial core experienced employment gains above the Atlantic average from 1971 to 1974, with Halifax-Dartmouth and Saint John again leading the way. Other centres in this category are New Glasgow, Amherst, Truro, and Sackville. In contrast, with a few notable exceptions particularly St. John's, most of the centres in the outlying areas experienced below average employment gains from 1971 to 1974.

Trends in labour force utilization were in accord with developments in large establishments employment. In the industrial/urban core region of the Maritimes, the unemployment rate fell in line with national trends from 1971 to 1974, going from 6.1 to 5.6 percent. In marked contrast, the unemployment rate in the remainder of the Atlantic moved upward from 10.7 to 13.1 percent over the same period. Unemployment is particularly high in Newfoundland outside of the Avalon Peninsula, and in northeastern New Brunswick. In both cases, the rate of unemployment averaged close to 20 percent through 1974, representing increases of about a half over their 1971 rates, while through the first seven months of 1975, their averages have been around 25 percent.

Trends in labour force participation rates generally followed those for unemployment, as the urban core experienced much sharper gains than the peripheral areas. The upshot is that the urban core now approaches the national average in terms of the overall rate of labour force utilization (as measured by employment over the working-age population). In sharp contrast, the outlying areas have fallen further behind the all-Canadian level. Similar to the trends in employment, not all of the peripheral areas fared badly; but again, generally the most disadvantaged areas showed the least improvement, and in some cases, a decline, including northeastern New Brunswick, southwestern Nova Scotia, Cape Breton Island, and the areas of Newfoundland outside of the Avalon Peninsula.

As a consequence, income differentials *within* the three largest provinces in the region increased dramatically over the last three years. Incomes in southern New Brunswick are now 50 percent above those in the northern parts of the province, compared with 43 percent three years earlier. Similarly, the income differential between the central corridor of Nova Scotia (i.e. Halifax-Dartmouth and northern Nova Scotia) and the remainder advanced from 31

to 37 percent from 1971 to 1974. Perhaps most remarkable is that our estimates suggest that the income differential between the Avalon Peninsula and the remainder of the province close to doubled, and is now approaching the 50 percent mark. Growing income differentials within provinces or broad regions are often viewed as an inevitable consequence of a successful development process. Even so, the continuation of these diverging trends could cause serious economic and social dislocations. These, in the final analysis, could act as serious constraints to sustained economic growth in the Atlantic region over the longer term.

Before closing this section, a closer look should be given to economic trends in Halifax-Dartmouth, St. John's, and Saint John, in light of their pivotal roles in the recent upturn in the Atlantic economy. After a very sluggish performance during the sixties, the Halifax-Dartmouth economy has shown considerable strength in recent years. The recent economic upsurge has been very broadly based, with nearly all key sectors making contributions. Manufacturing employment, after stagnating for over a decade, has moved upward, based, in particular, on expansion in electronics, auto assembly, shipbuilding, and activities oriented towards the Atlantic region market. Growth in the container port has also been a major factor, as evidenced in part by an 18 percent increase in transportation and communications employment from 1971. Construction activity has moved up sharply, with an employment rise of 12 percent in 1974 alone, based especially on business and commercial development in the central business districts of both Halifax and Dartmouth.

Finally, and perhaps most important, is that service activities have been expanding rapidly over the past few years, as Halifax-Dartmouth is now emerging as the prime service centre for the Atlantic region. Of particular significance is the marked expansion in the finance, insurance, and business service categories. Many national firms have established regional offices in Halifax-Dartmouth, as they now find it easier and more economical to service the Maritime and Atlantic markets from the Metro area than from Montreal and Toronto. Also, employment in the trade category (including wholesale) has moved sharply upward, reflecting in part the growing role of Halifax-Dartmouth as a major distribution centre in the Atlantic economy.

In consequence, employment in the trade, finance, insurance, real estate, and personal and business services categories, taken together, rose by nearly 30 percent from 1971 to 1974. It should be emphasized that much of the growth in services involves an important type of import substitution, and is not just "passive" or "induced" growth based on expansion in the goods-producing industries. Accordingly, the current upward trend in services is resulting in a significant expansion and diversification in the economic base of the Halifax-Dartmouth area.

These forces combined to provide Halifax-Dartmouth with a 20 percent increase in "total" large establishments employment from 1971 to 1974, nearly double the rate of increase for the Atlantic region or the country as a whole. This has led to substantial improvements in the rate of labour force utilization. The rate of unemployment in the Metro area averaged 3.7 percent during 1974 (compared with 5.4 percent nationally), while the labour force participation rate averaged 61.7 percent, or three percentage points above the national standard. In both cases, the improvements from 1971 are more favourable than those for the country as a whole.

In short, Halifax-Dartmouth not only remains the most prosperous part of the Atlantic region but, more important, it is now coming to play a leadership role within the Atlantic economy. The renewed growth of the port, the development of Halifax-Dartmouth as the prime service centre for the entire Atlantic region, and the steady development of manufacturing have transformed the metropolitan area from a slow growth area to a major growth centre, with increasing links to the major metropolitan centres in the rest of Canada and the Eastern Seaboard of the United States.

St. John's has also expanded rapidly in recent years – continuing an upward trend which began in the early sixties – and has apparently been affected only marginally by the difficulties experienced by the rest of the province in 1974 and 1975. Similar to Halifax, St. John's development has been very broadly based, with manufacturing, construction, and services all making major

contributions. Of particular significance is the 35 percent increase in construction employment from 1971 to 1974, related in part to the redevelopment of St. John's central business district. The upshot was a 28 percent increase in "total" large establishments employment in the St. John's CMA from 1971 to 1974, or slightly above the Halifax-Dartmouth increase. Together with the similarity in growth performance, St. John's now has much in common with Halifax-Dartmouth, as it is coming to perform many of the same functions for Newfoundland (albeit on a smaller scale) that Halifax-Dartmouth performs for the Maritimes.

While nearly as impressive in aggregate terms, the growth in the Saint John economy has been less broadly based and has been more concentrated. From December, 1973, to December, 1974, alone, large establishments employment in Saint John rose by over 14 percent. The result of this concentrated growth has been serious labour shortages, particularly in the construction trades, and a housing shortage which has impeded the permanent attraction of labour to the Saint John area. For most of 1974, Saint John has effectively had "overfull" employment, with the number of vacancies often exceeding the number of registered clients without employment in the records of the local Canada Manpower Centre Office.

The main engines of growth have been transportation (related to the expansion of the Saint John port), manufacturing (metal fabricating, and for a while at least, the Bricklin), and in particular, construction with several major projects underway in the Saint John vicinity. Construction activity is expected to remain at very high levels for the next few years at least, the major project being the nuclear power plant at Point Lepreau. Accordingly (relative to Halifax-Dartmouth and St. John's), industrial activities are expected to play a more pivotal role in the future development of Saint John, although this city will likely remain the financial centre for New Brunswick.

Because of data constraints, it is not possible to update this analysis to take full account of 1975 developments. The available evidence suggests, however, that the Atlantic economy has fared much better during the 1975 recession than in previous Canadian downturns, providing a further illustration of the growing economic strength of the region. In addition, the indications are that the spatial pattern has been consistent with that of the 1971-74 period, although some softening is apparent in the labour markets of Moncton and Halifax-Dartmouth. For the latter, however, this softening may be as much the consequence of an influx of migrants looking for work, as a downturn in employment. Consistent with this, large establishments employment in the Halifax subregion was up 1.7 percent in the first three quarters of 1975, compared with a similar period in 1974 – an increase that compares favourably with most other parts of the region.

4. *Broad Implications of Quantative Trend*
Certain broad conclusions emerge from this analysis. These should help set the stage for the future projections to follow.
(a) Considerable growth has occurred in the Atlantic region over the past number of years, but most of this growth has been concentrated in the most urbanized subregions. Thus, in terms of both the level of growth and its spatial distribution, recent economic trends in the Atlantic have been very close to the typical Canadian pattern, a situation which did not prevail during the 1960's. While this urban concentration provides support for the authors' "urban bias", it could be a major cause for concern if it should continue indefinitely.
(b) The spread of "beneficial" effects from growth in Halifax-Dartmouth and Saint John, while not unimportant, has been limited both in magnitude and geographical extent. Most spread effects have been concentrated in the smaller urban centres within the central corridor. As evidence of this, many of these communities have realized substantial increases in manufacturing and other activities serving the growing regional market.
(c) The limited nature of these spread effects is not surprising in light of the limited duration to date of the economic upturn in the larger centres. Analysis of the development process in other parts of Canada suggests that, as the urban economies continue to expand, beneficial effects for the outlying areas can be expected to emerge.
To cite a few examples:

– Further improvements in the transportation system of the urban core, particularly the port facilities at Halifax-Dartmouth and Saint John, will greatly enhance the international competitiveness of the primary products and semi-processed goods of the outlying areas.
– Sustained increases in income and population in the urban core region will allow the agricultural economy of Prince Edward Island to move away from a single export staple – potatoes – towards mixed farming serving the concentrated high income market provided by the urban core. In a similar vein, a growing urban core will offer the forest products industries of the peripheral areas an expanding local market for further processed goods, thus increasing value added and lessening dependence on volatile export markets.

These arguments will be expanded in the next section (Policy Framework).
(d) At the same time, the evidence to date (as well as recent literature) strongly suggests that the spread effect from the central core will be slow to develop, and will likely not be sufficient to generate sustained economic growth in the outlying areas of the Atlantic economy. Direct policy intervention in these subregions is clearly required.
(e) Adding support to the fourth point is the evidence that the main centres in the region received perhaps as much benefit from the recent improvements in international commodity markets as the outlying areas themselves – the areas that possess these resources. This reflects the low economic multipliers in these subregions, related in turn to their lack of urbanization.
(f) Thus, the recent performance of the Atlantic economy, while based on the growing strength of its largest centres, also serves to point out serious weaknesses in the total urban system of the region. This suggests that urban initiatives should not be limited to the three metropolitan areas, but should be expanded to include the second-order centres, particularly those in the peripheral areas. This judgement will play a key role in the projections and policy prescriptions offered in later sections.

Policy Framework

The analysis has stressed the importance of intraregional differences within the Atlantic region. Although the Atlantic region is becoming more integrated as a market, it is becoming more disparate in terms of economic opportunities and economic well-being.

This poses a particular set of problems, in terms of establishment priorities. It is no longer valid in development efforts in the region to adopt the attitude of "taking anything we can get", or to follow the simple slogan of "jobs at any price". As far as a significant portion of the region is concerned, it is now possible to be more selective.

The first major piece of priority setting has to be in terms of apportioning priorities between the more prosperous areas, which are growing, and the less prosperous areas, which are lagging and which may be considered "needy". With limited development resources, to what extent should efforts be directed towards the developing areas, and to what extent should they be directed towards the needy, in terms of the spatial orientation of development expenditures? Within this question, there are a whole series of technical priorities to determine what specifically should be done to make the developing areas more viable, and what should be done to make the needy areas less needy.

There are a number of broad guidelines which seem to us to emerge from our analysis and our experience.

Centre vs. Periphery

The prospering areas, particularly the "central corridor", must be allowed to grow, and special inputs must be made to accommodate that growth, particularly in the form of assistance to municipal infrastructure and other services which support residential growth, including education and health services. The growth of the central corridor and of the Avalon Peninsula makes good, competitive, economic sense, and so relatively little will be needed in the way of industrial incentives to accommodate such growth. If that growth is deterred from occurring in those areas, it is our considered opinion that it will *not* spin off into other parts of the region: rather, it will not occur in the Atlantic region at all. Many services and manufactured goods will continue to be supplied from the major industrial areas of Ontario and Quebec. The overall effect will be to substantially dampen growth of the region. This emphasis is the only way that overall employment targets will be met for the region as a whole, given limited development resources.

However, where choice exists (for example, in agriculture and fishing), all other things being equal, priority should be given to developing these industries in the areas where the need is greatest and where alternatives are least. It is not possible to lay down general rules, but consideration should be given to applying this approach even where things are slightly less than equal. This may involve some special subsidization — for example, for the movement of agricultural produce, or for the provision of fish landing and processing facilities. In this approach, based on our present knowledge, we would advocate that Industrial Cape Breton be considered the prime location for a steel complex; that northern New Brunswick be given special consideration as a future nuclear power plant location; that southern Newfoundland and southwestern Nova Scotia be given priority in terms of the offshore fishery; that efforts be directed towards maintaining the inshore fishery in northeastern Newfoundland and northeastern New Brunswick; and that Prince Edward Island, northwestern New Brunswick, and the Annapolis Valley of Nova Scotia be accorded priority in agriculture.

Activities can thus be viewed as a continuum with respect to their location options. At the top are activities that are highly location-specific, and that, in most cases, are closely tied to the "centre" and, more specifically, the region's largest centres (although this quartile would also include activities tied to a particular resource such as mining). Attempting to shift these activities out of the centre would be self-defeating and would lead to their flight from the region. The bottom quartile would be composed of activities that are totally "footloose", a condition better known in fiction than in fact. To the extent that they exist, these activities should be attracted towards the periphery.

In the middle are activities that have some, but not unlimited, location options. For example, the second quartile may be composed of economic functions which may have a preference for the metro areas in the centre, but which could be induced to locate

323

in the centre's smaller communities. Governments may often wish to provide such inducements to prevent over-heating of the metro area economies. In a similar vein, the third quartile may be made up of activities with a preference for the centre's smaller communities, but which could be encouraged to locate in the urban centres in the periphery. In most cases, governments will likely offer this encouragement, in order to stabilize and diversify the economic base of the periphery.

The above provides, of course, a highly simplified approach to industrial location and incentives, and begs two key questions: what activities are located in each quartile, and what are the most appropriate forms of government incentives? Still, this approach should provide a useful starting point for an in-depth analysis of the location characteristics of different activities, and the development opportunities of different subregions.

Finally, the centre-periphery question can be extended by looking at the relative development efforts that might be appropriate at the present time in each of the two broad subregions, and how these efforts might change over time. The following provides only rough orders of magnitude, but can help to further illustrate the authors' basic position. We would argue that, for the next seven years or so, the centre, with 40 percent of the Atlantic population, should, perhaps, receive 50 percent or better of the government expenditures of all federal departments directed towards development in the region, [8] for it is in the centre that the federal government will receive the best return from its investments. This ratio is probably not out of line with current expenditure allocation. However, as the centre's economy continues to build up momentum, this ratio should shift in favour of the periphery in the next decade, perhaps in the direction of 40 percent or below for the centre. While crude, these numbers provide some indication of the authors' position on evolving development priorities over time.

Regional Integration

A very high priority should be given to activities which integrate the Atlantic regional economy. This is mutually advantageous to both the viable and the needy areas, as it confirms and strengthens the regional market, makes metropolitan and urban services available to "have-not" areas, enables some of the demands of the growing areas to be met by supplies from the poorer areas, and in particular, provides employment opportunities for the labour force of the "have-not" areas within the region, rather than forcing them to leave the region to seek fame and fortune in central Canada. The major program activities which would emerge from this aspect would be in the development of the air passenger services within the region, perhaps utilizing STOL air services, and also the continued improvement of the all-weather highway system and the ferry services to Newfoundland and Prince Edward Island.

Urbanization is an important element in this, as communication occurs much more readily between urban centres than between a distant urban centre and a sparsely settled region. For example, linkages between northeastern New Brunswick and Moncton or Halifax would be much better if Bathurst developed as a major centre.

The Human Dimension

Programs and plans must be geared towards meeting the aspirations of people. This is not only an idealistic stance, it is also a recognition of the fundamental right of individuals to choose where they live. Therefore, population redistribution is the result of a set of individual decisions, which are not exclusively motivated by the prospects of economic gain.

It is important that the range of social amenities available to the population matches the economic prospects. With limited fiscal capacities, local governments will not be able to provide these in great profusion. It may, therefore, be necessary for senior governments to place particular emphasis on such items as recreation, entertainment, and similar amenities.

Particularly in areas of limited opportunity, social services such as health and education must be stressed. These are not only a matter of human rights, they are also important in equipping the youth of these areas with the basic skills necessary for successful relocation into other areas. Without such efforts, the poor get poorer, and real disparity increases.

Fiscal Limits

Resources are scarce, and it is not possible to do everything. This approach involves a conscious trade-off between the allocation of public funds to viable areas, where a high income and employment return can be earned, and to lagging areas, where massive injections of public funds may only provide temporary relief for a few individuals.

Fundamentally, the approach taken is the "development opportunities" approach adopted by DREE. Government intervention in economic development should be directed towards the creation of viable, long-term employment.

As a corollary to this, urban development efforts should be focussed on urban areas where real economic opportunities exist, and in particular, focussed on major growth poles in each subregion, in order to maximize employment multipliers.

The Quality of Urban Environment

It is not enough to merely supply viable economic opportunities (although that alone is a monstrous task for which ready solutions are not typically available). Beyond this, people are very much affected by the level of urban amenities. The consumption of public goods is a major part of the total consumption package.

If populations are to remain in the region (and even more, if they are to be attracted to the region), the level of urban amenities must be at least at a comparable level to urban centres in the rest of Canada. With limited fiscal capacities, and with the extra cost pressures resulting from growth, this must become a matter of particular concern to senior governments.

A recurring problem throughout the Atlantic region is urban sprawl. This is not only costly in terms of public services (sewer, water, roads, schools) and private costs (primarily transportation), it also limits economic multipliers by dispersing demand, takes recreational and agricultural land out of use, and is typically aesthetically unpleasant.

Major Metropolitan Centres

A particularly important role has to be played by the metropolitan centres. Not only must they grow, they must also be substantially transformed in order to take on their key economic roles.

Halifax-Dartmouth must emerge not only as the largest centre, but also as a true regional centre, and as a part of the network of major Canadian cities. It must be the focal point of the Atlantic region, in touch with the outside, and in contact with the region. Recent trends indicate that it is rapidly taking on such a function. Links with other Canadian and foreign centres are improving rapidly. (A glance at the Air Canada schedule shows four direct flights a day to Toronto, four to Montreal, three to Boston, one to New York, three to St. John's, and two to Ottawa, with four a week to London and a weekly flight to Calgary and Vancouver.) The transformation of the downtown area to house commercial services, etc., has already taken place, and is continuing. In the next few years, the transformation should see the development of high-technology industry, the development of Halifax as a nationally significant cultural and entertainment centre (adding to Neptune Theatre and the Atlantic Symphony Orchestra, and perhaps including recording studios) with local amenities to match, including a full-scale transit system, regional parks, and convention centres.

St. John's must be ready to take on a similar role, on a smaller scale. The geographical isolation of Newfoundland is real — Halifax is as remote as Toronto in many respects. Also, Newfoundland has a distinctive culture to which St. John's should give expression. This will involve an increasing emphasis on the downtown area as the location of the service industries, which will provide the bulk of new employment. It is also important for St. John's to extend its linkages to the outside and to the rest of Newfoundland, particularly Labrador. This means upgrading air service (perhaps with fog dispersal for Tor Bay!) and, perhaps, routing ship traffic more frequently to St. John's and Argentia from Halifax and Montreal, rather than using the expensive and cumbersome North Sydney to Port-aux-Basques routing.

Saint John must pass through a similar transformation. It is seen today as primarily a heavy industry town and a seaport. These activities will probably prosper, but the city must move into white collar areas (with the downtown as the focal point), establishing itself more firmly as the commercial centre of New Brunswick. Just building the downtown area, however, is not enough. The external links must be looked after, and again, air service is critically important. (And it should be noted that air service is more than the airport. The ground transportation service needs good buses, plentiful taxis, and a well-paved road!) Overall, the city needs to change its self-image, in order to recognize the value and importance of the commercial activity. Furthermore, if Saint John is to serve as the major growth centre in New Brunswick, it must be recognized that a large proportion of its new residents will be Francophones. It is important that the city be consciously designed to accommodate such an infusion, especially in cultural and entertainment facilities, retail trade, and public administration.

Urban Development in Lagging Areas

Urban centres have an important role to play in lagging areas of the region. Economically they are critical in providing "multiplier"

jobs, in particular, service jobs based on consumer demands. Socially they are critical in providing public amenities, in particular, recreation, entertainment, higher levels of education, health services, and the like. In the service sector, urban centres offer particular opportunities for female employment, increasing family incomes.

Resource sectors do not offer major employment growth prospects, and the effective development of lagging regions demands concentrated urban development. In many cases, this entails a conscious choice of a particular town as the prime focal point. Thus, in northeastern New Brunswick, Bathurst should take priority over Cambellton and other smaller centres. In western Nova Scotia, Yarmouth should take priority over Shelburne and Digby. Other important focal points in the region are Corner Brook, Grand Falls, Edmundston, Charlottetown, and Sydney. Charlottetown has particular potential as a government centre. Sydney is an existing urban centre which should be enhanced as the focal point of Industrial Cape Breton.

NOTES

[1] A statistical package is included in D. J. Ireland, "The Atlantic Region to 2001: A Detailed Forecast".

[2] But, unlike many practitioners of our trade, we promise not to stop there.

[3] To be more precise, the decade of the fifties was largely a period of stagnation, while the sixties was a period of adjustment for the Atlantic economy.

[4] See D. J. Ireland, "Structural Changes in the Atlantic Economy".

[5] A further factor is the high cost of housing in the region's larger centres relative to its rural areas, which has strongly dampened the stimulus to permanently migrate.

[6] The only employment indicator available between Census years. This includes firms with twenty or more employees and excludes some sectors, particularly agriculture, forestry, and government.

[7] The advance for the total Atlantic was 12.4 percent which compares favourably with the national trend.

[8] Including infrastructure assistance in support of development.

Canadian Regional Development Strategy: A Dependency Theory Perspective

RALPH MATTHEWS

This article will argue that there are fundamental biases and weaknesses in Canada's regional development strategy when it is examined from the point of view of a sociologist. More specifically, the federal approach to eliminating regional disparity and regional underdevelopment in Canada will be examined and an attempt made to demonstrate:

1. that Canadian regional planning has undergone a major shift in value orientation during the past fifteen years;

2. that the present value orientation of Canadian regional planning biases it in favor of economic considerations to the point where it fails to take into account important aspects of the social vitality of Canada's underdeveloped areas;

3. that the biases of current Canadian regional development policy are most likely to result in failure to reduce Canadian regional disparity significantly; instead it is likely to exacerbate class differences within the region and to institutionalize disparities to the point where they become permanent;

4. that, over the long term, such policies are actually likely to create the further dependency of Canada's underdeveloped regions, rather than assist in their development; and

5. that there is empirical evidence which suggests that such a process is already underway.

The Changing Value Orientation of Canadian Regional Development Policy

Any policy is, by definition, a goal-value system. It is an attempt to achieve a goal which is deemed legitimate in terms of a set of values. Usually the statement of policy spells out in detail the *means* whereby this goal is to be achieved. These means are also governed by values which determine those considered legitimate and those considered illegitimate. As I have argued elsewhere,[1] the goals and values incorporated in many types of "plan" are frequently those of the planners who design them, and often have little relationship to the goals and values of those people who are most likely to be affected by the plan.

As far as regional planning is concerned, Canada has a comparatively long history when compared with other countries. Efforts in this area date from the 1930's when Canada passed the Prairie Farm Rehabilitation Act (PFRA), which was a deliberate attempt to revitalize small rural farms which had suffered from several successive years of drought and crop failure.[2] A similar program to aid east coast farmers was introduced under the Maritime Marshlands Rehabilitation Act (MMRA) in the 1940's. In terms of their value orientation, both programs were *rural development* oriented.

A similar value orientation was incorporated into the early programs begun under the Agricultural Rehabilitation and Development Act (ARDA) of the early 1960's. ARDA was an attempt to do nationally what the earlier programs had done for selected regions. It focused on "land use" projects designed to "salvage lands abandoned as agriculture retreated from marginal areas".[3]

However, the mid-1960's saw a major change in the value orientation of Canadian regional development planning. One of the major forces in bringing about this change was a report on Canadian regional development programs by Helen Buckley and Eva Tihanyi for the Economic Council of Canada. It was highly critical of the land development

● RALPH MATTHEWS is Associate Professor of Sociology at McMaster University. The article reprinted from *Plan Canada*, Vol. 17, No. 2, 1977, pp. 131-143 with permission.

strategy of ARDA and earlier programs, arguing that they were "unlikely to have had any appreciable impact on the problem of low-income farming".[4] The new direction advocated by Buckley and Tihanyi was away from programs of rural development toward measures which they described as "adjustment".[5] These were aimed at moving large numbers of rural people out of rural areas so that their farms could be consolidated into larger farming units. Obviously "adjustment" was their way of arguing for rural depopulation. Thus, Buckley's and Tihanyi's work marked the beginning of a major change in the value orientation of Canadian regional development programs. Economic considerations began to take precedence.

Buckley's and Tihanyi's formulations appear to have had a considerable impact on the next major phase of Canadian regional development planning. The FRED program (Fund for Rural Economic Development), begun in 1965, was designed to provide comprehensive development planning for selected regions of Canada. Most of the plans for these regions embodied programs of population movement.[6] However, few of these comprehensive development plans were ever implemented. As part of his campaign for leadership of the Liberal Party and the position of Prime Minister of Canada, Pierre Elliot Trudeau turned regional development into a national issue. He argued that regional disparity had the same devisive potential as the more publicized French-English cleavage,[7] and promised that, if elected, he would make an effort to rectify regional economic differences. After his victory, he undertook to make good on this promise by establishing in 1969 a new Department of Regional Economic Expansion (DREE).

DREE marked a further change in the value orientation of Canadian regional development policy toward the direction encouraged by Buckley and Tihanyi. The new department encouraged regional development in Canada through three interrelated programs: (1) industrial incentive grants to large industries to encourage them to establish in areas of regional economic disparity; (2) infrastructure assistance grants to provincial governments and municipalities to encourage them to build those amenities which "industries require"; and (3) programs of "social adjustment".[8]

Thus, with DREE, Canadian regional development planning has shifted its value orientation 180 degrees. From a concern with the importance of rural life and with maintaining the family farm, the emphasis has now shifted to building up selected urban areas as "growth centres". The dominant concern of DREE is with encouraging the large scale industrialization of these centres. It appears to be an underlying assumption of DREE that economic growth will not take place in underdeveloped areas unless industries establish there. Finally, DREE, as its name suggests, is very much a department of *economic* expansion. Its dominant consideration is economic development and there is relatively little effort spent in examining the social vitality of the areas it is affecting.

The Value Orientation of the Growth Centre Strategy of Development

The growth centre strategy of development has its origin in the growth pole concept of French economist-geographer François Perroux[9] and in the works of American economist A.O. Hirschman.[10] Perroux's concern was with "economic space as a field of forces",[11] in which all economic forces were seen as centred around certain "poles" or foci. Hirschman argued that regional underdevelopment could be overcome by developing particular growth poles. The key to such development was the introduction of "master industries" which have the ability to alter the whole pattern of economic relationships within the region. They were able to do this through their ability to develop strong *forward and backward linkages.* Backward linkages were essentially those to the suppliers of raw materials which the master industry processed. Forward linkages were those outlets which received the product of the industry. A master industry was, by definition, one which could bring changes to the area around it by encouraging new companies to come and supply it with the materials which it needed (backward linkages) and which could also develop external markets for its products so as to bring needed cash into the area (forward linkages).

The basic way in which a master industry could accomplish this was through cost reduction. By producing its product at a lower cost than was available elsewhere, it was able to attract these other industries, and market its product. The two major factors which would enable it to reduce costs were its more modern technology and its large size. The former would lower labour costs and the latter would produce economies of scale.

There were, of course, obvious problems with this theory. It hinged almost entirely on the ability of a master industry to produce at a cheaper rate than industries located elsewhere, thereby overcoming the transportation costs involved in shipping its product to market. However, immediately an industry of similar size was built closer to the metropolis, or even when an older industry modernized its technology, the master industry lost much of its advantage. Moreover, by employing more modern technology, it tended to employ relatively few workers, thereby defeating some of the reason for its existence. However, perhaps the major flaw of this model is that it is formulated virtually exclusively in economic terms without any consideration of its social impact. As we shall see, it is the social impact of this model which has created major problems and which undermines its usefulness as a strategy of regional development.

Economic vs. Social Considerations in Canadian Regional Development

In order to have a growth pole, one must have a large concentration of people, i.e., an urban area. This is implied in the model, for a large industrial operation must have a labour supply upon which it can draw. Moreover, a large industrial complex needs infrastructural services generally unavailable in nonurban areas. For Canadian regional planners, this presented few problems as their plans were already leading them toward an emphasis on urban areas. The only question was whether to provide the urban area before or after the industry had been established.

For a number of reasons "growth centres" tended to appear long before there was sufficient industry to justify calling them "growth poles". First, with little attention being paid to developing jobs in rural areas, many farm workers now had little choice but to move out, providing a pool of unemployed, unskilled workers searching for employment. Second, the proposed industrial developments in urban areas received widespread publicity and their proposed sites became meccas for migrant workers. Third, as the roads, sewers, schools, hospitals, and airports were being built to lure such industries to selected locations, they provided a large number of temporary construction jobs at incomes considerably higher than those of most rural residents. Fourth, in some cases there were actual government policies aimed directly at phasing out rural communities. For example, the Newfoundland resettlement program was originally designed to move 70 000 rural Newfoundlanders to more urban areas.[12] Fifth, there seems to have been a general feeling among Canadian planners that a large unemployed labour force was an attractive incentive to prospective industry, for the industry was thereby assured a continuing labour supply of willing workers.[13] The result was a massive movement of rural population to urban centres throughout the 1960's.

One of the major social consequences of Canada's growth pole strategy was its failure to recognize and develop the potential of Canada's rural communities. Almost invariably small communities were judged by economic criteria and rarely or never by social criteria. Thus the lack of *economic viability* was accepted as evidence of a community's failure without ever considering attributes of *social vitality*. Too often highly vital patterns of living were disrupted and even wiped out without any real attempt to salvage them. It is impossible in the few pages provided here to establish that this was the case and that, for many rural communities, there might have been alternatives. To do so requires intensive community case studies.[14]

However, the weaknesses in Canada's present strategy of regional development, while certainly sufficient to impair its effectiveness and wreak havoc on the social lives of many Canadians, are unlikely to doom it to failure. To understand this claim, it is necessary first of all to have a basic understanding of a theory of underdevelopment generally known as "dependency theory".

Dependency Theory and Underdevelopment

Traditional approaches to underdevelopment explain the plight of underdeveloped areas as either caused by their isolated location or as the result of some internal weakness. Underdeveloped regions are said to suffer from being too far from major trading and industrial centres. Mention is also frequently made of a lack of resources within the region, its lack of capital, the low level of training of its population, or the lack of administrative and entrepreneurial skills amongst its political and economic leaders. Occasionally such explanations are also couched in social and socio-psychological terms. Thus, we are told by some theorists that residents of underdeveloped areas lack "achievement motivation", or that they suffer from overly stringent ties to a traditional culture that is essentially opposed to modern development. The obvious solution to such problems is to change the character of the underdeveloped area through training programs, and through grants for economic development.

However, in recent years, there has appeared a rival theory of underdevelopment, generally known as dependency theory, which argues that development and underdevelopment are not separate phenomena but are actually manifestations of the same economic process.[15] Furthermore, these theorists argue that the primary cause of underdevelopment does not rest within the underdeveloped regions but is actually external to it. From this perspective, underdevelopment is the product of an historical pattern of exploitation of the impoverished areas by more wealthy ones. The wealth of these latter areas is seen as largely a result of their ability to harness the resources of underdeveloped areas for their own gain.

From this perspective, the world economic system is an interlocking set of metropolis-hinterland relationships. Thus, developed countries are the economic metropoli and underdeveloped countries are their hinterlands, both bound together in economic relationships which serve to maintain the economic status of each. However, within each nation there are also metropolis-hinterland relationships between richer and poorer regions.[16]

In this relationship the underdeveloped areas perform the function of providing the scarce natural resources needed to operate industries controlled by central economic interests. Usually the resources are sent outside the underdeveloped region for refining, ensuring that little of the economic gains remain within the underdeveloped region. However, even if resources are refined locally, economic control of the manufacturing industry usually rests outside the area, and the bulk of the surplus profit is still likely to leave the underdeveloped region.

The underdeveloped region is also exploited for its labour. The surplus of unemployed local labour usually ensures that the natural resources of the region can be obtained at minimal labour cost. Moreover, this labour pool is also of benefit to manufacturing industries located outside the region. It can be drawn on in times of economic expansion, and it simply returns home when times get tight.

Quite frequently, the underdeveloped area even provides the economic capital for its own exploitation. Political and military leaders usually want goods and equipment manufactured in the economic centres. In order to finance them, the area must export even more of its scarce (and frequently nonreplenishable) resources, or it must borrow money from the metropolis. This ensures its economic dependence for long years to come. The same process occurs at the level of the individual consumer who uses up his surplus income in his attempt to acquire expensive consumer goods generally produced elsewhere. Such actions ensure that there is no local fund of surplus capital available to finance local economic growth. However, for purposes here, the key portions of this theory are those which explain how the more developed areas are capable of economically exploiting their dependent satellites without encountering extensive opposition. Adapting Galtung's argument, this is possible because both the metropolitan and hinterland areas have a small and powerful "centre" group and a large undeveloped and generally powerless "periphery" group in their population.[17] Galtung argues that the economically powerful groups within both the metropolis and the periphery have "interests" in common. As a result, the powerful unit in the metropolis is able to develop an alliance with the controlling groups in the hinterland. Galtung refers to this alliance as a "bridge-head".[18] Galtung further argues that there is a disharmony of interest between the centre and periphery groups inside each of the two areas, and there is a conflict of interest as well between the two large periphery groups.[19]

The consequences of the alliance between the two dominant centre groups can be inferred from the work of Sunkel. Sunkel argues that a "simultaneous process of dual polarization" occurs in both the developed and the underdeveloped nations.[20] The result is a growing division of the societies into advanced and backward regions. Industrial activity, supported by outside investment, creates large urban concentrations. However, this is frequently accompanied by the stagnation of the traditional sector.

Dos Santos gives us greater insight into the implications of this type of analysis for the dependent region.[21] He theorizes that the establishment of a metropolis-hinterland "dependency" relationship is only possible through the formation of "a certain type of internal structure"[22] in the hinterland region. "Domination is practicable only when it finds support among those local groups which profit by it."[23] Dos Santos thus parallels Galtung by suggesting that there is a group within the peripheral area which actually benefits from its dependency, and he argues that certain internal groups are "compromised" into supporting their own domination.

The preceding synopsis should, at least, make clear that the dependency theory explains underdevelopment in terms of class interests. Galtung, Sunkel, and Dos Santos all imply that a segment of the population of an underdeveloped region becomes the mediator of external economic interests, and owes its position of economic power primarily to these external ties.

The main instrument facilitating this process is seen by Sunkel to be the multinational corporation. The multinational corporation is the economic agent of the dominant area operating within the dependent one.

The multinational corporation is a medium for the intrusion of the laws, politics, foreign policy, and culture of one country into another . . . Multinational corporations reduce the ability of the government to control the economy.[24]

Dos Santos likewise suggests that such external corporations are particularly detrimental to the development of locally owned industry serving local needs, for they move much of the surplus capital out of the underdeveloped area.[25] Both writers argue that the domination of these external industries places severe restrictions on the development of local entrepreneurial activity. On the one hand, the multinational corporation tends to centralize its decision-making in the home country or region,[26] thus allowing its local agents little scope to exercise entrepreneurial skills. On the other hand, many of the residents of the underdeveloped region who show independent entrepreneurial skill are ultimately bought out by foreign capital.[27]

The Implications of Dependency Theory for Canadian Society

The dependency framework presented above was developed by South American economists and sociologists to explain the interrelationship between their countries and those more economically advanced. Thus, it may be difficult at first glance to see the relevance of this framework to regional underdevelopment in Canada. However, it will be argued that dependency theory provides an alternative to the traditional modernization approach to underdevelopment, and also implies an alternative to the growth centre strategy of regional development discussed earlier.

From a dependency perspective, Canada is a series of metropolis-hinterland relationships in which the relatively rich central region dominates and exploits the hinterlands of the east, west, and north. At the same time, Canada is itself a hinterland of economic interests elsewhere, primarily those in the United States. It is possible to point to already published work which verifies that such a perspective has empirical validity.

A good place to start is the work of Canada's best known social scientist, economist Harold Innis. Innis argued that the Canadian economy can best be considered a "staples economy" in which economic growth and development is based on the exportation of raw materials.[28] He claimed that new countries such as Canada were able to develop a high economic standard of living simply because there was an overabundance of raw materials and a favorable man/land ratio. Though conditions have changed since Innis wrote his theory in the 1930's, the structure of the Canadian economy has not altered appreciably. Much of Canada's export income still comes from the exportation of raw materials rather than from manufactured products, though fish and fur have now been replaced by oil, gas, and hydroelectric power. In this respect, Canada has been lucky, for the greatest threat to a staples economy is that it falls into what Watkins has called a "staples trap".[29] Either it runs out of staples for which there is world demand, or it stagnates to the point of being unable to develop new staples as they are needed.

Innis' work emphasizes the highly *dependent* nature of the Canadian economy. As Watkins has also noted, "The essential characteristics of Canada as a capitalist country are that it is rich *and* dependent."[30] Any needs have been alleviated primarily by selling more raw materials. However, Canada's resource dependency goes further than that. It has generally used very little of its own capital to develop its resources. Instead it has offered incentives to foreign-owned, corporations to come and develop Canada's raw materials for their own use. The result, as Porter,[31] Mathias,[32] Watkins,[33] and a host of other studies have demonstrated, is that Canadians do not own the majority of resource-based corporations operating in this country. Of course, there has been *some* industrial development, particularly in central Canada. However, as Levitt[34] and others document, most of this industrial development has also been financed by outside interests which maintain control over it. Thus, Canada closely fits the model of a dependent or "satellite" economy described by the dependency theorists.

The same model can also be applied to the relationships among the regions within Canada. One of the first to recognize this was A. K. Davis, who argued that Canada consisted of a series of metropolitan-hinterland relationships in which "the metropolis continuously dominates and exploits hinterlands whether in regional, national or class terms".[35] He emphasized the inherent "oppositions" between Canadian regions.[36]

More concrete evidence of this basic organization within Canadian society is to be found in the work of a number of economic historians. Naylor has shown how Canada's banks and major business enterprises slowly moved out of the hinterland regions into the central areas of Ontario and Quebec.[37] This process was hastened by the National Policy of 1876, which was a systematic attempt to centralize Canadian wealth. Acheson has also traced the failure and decline of Maritime metropoli in the latter part of the nineteenth century.[38] More recently, Laxer has documented the continuation of this process as Toronto has come to replace Montreal as a centre of metropolitan dominance in Canada.[39]

Lest this be thought a particularly radical interpretation, it should be noted that the leader of the Progressive Conservative Party, the Honourable Joe Clark, has recently made the following similar statement.

But it is, in fact, quite justified for the West and for Atlantic and Northern Canada to note that national policy, from the construction of seaways, through to most transportation and tariff policy, has encouraged growth to concentrate in Canada rather than to disperse in Canada.[40]

American multinational corporations have also been major contributors to the centralization of Canadian wealth. They have tended to locate their branch plants as close to the major American industrial centres as possible. For example, Ray calculated that 45 percent of American-controlled manufacturing is within 160 km of Toronto, and that if American-controlled manufacturing employment had the same distribution as Canadian-controlled manufacturing employment, there would be an approximately 20 percent increase in employment in the Atlantic Provinces.[41] This foreign control of industry and resources has made it difficult for Canada to develop her own economic elite in these areas.[42] Instead most elite Canadian businessmen are in trade, transportation, and communications.

That Canada's dependent economic status has implications for our social structure is readily apparent in the work of Wallace Clement, who has shown how, particularly in the 1960's, "U.S. economic elites have penetrated the Canadian power structure".[43] His work demonstrates that Canada now has three elite groups.

First is the indigenous elite, closely associated with dominant Canadian-controlled financial, utilities, and transportation corporations, with smaller respresentation in the manufacturing and resource extraction sectors. Second, is the comprador elite, the senior management and directors of dominant foreign-controlled branch plants, mainly in manufacturing and the resource sectors. This group is subservient to the third group, the parasite elites, who control major multinational corporations which dominate important sectors of the Canadian economy through branch plants. They focus on resource and manufacturing sectors and operate their enterprises as part of an integrated organization.[44]

The picture of Canada's social structure presented by such researchers epitomizes that described by dependency theorists such as Dos Santos, Sunkel, and Galtung. The comprador elite are the "compromisers", Canadians who act as the agents of external economic interests. In Canada they are joined by a body of non-Canadians who act directly for these external organizations. This is particularly true in the areas of resource exploitation and manufacturing.

The Implications of Dependency Theory for Canada's Regional Development Strategy

In the preceding two sections the approach to underdevelopment known as dependency theory has been examined and the attempt made to demonstrate that, in many key respects, Canadian society and Canadian regional underdevelopment fit the model of a dependent economy and society.[45]

Throughout the remainder of this article, the way in which Canadian regional dependency is affected by Canada's present strategy of regional development will be examined. Hodge has suggested that Canadian regional planning is essentially "colonialism", and is done with an eye to exploiting rural and hinterland areas to serve the needs of metropolitan ones.[46] While I am unwilling to attribute such Machiavellian motives to Canadian development planners, nevertheless I would agree that this is most

likely one of the consequences of their plans. *I would contend that Canadian development policy, to the extent that it is based on incentive grants to corporations, is trying to combat regional underdevelopment by giving money to those economic interests which are essentially the cause of underdevelopment.* Such an approach can lead only to ultimate failure.[47]

In tracing the dependency framework, several reasons why this is so have already been suggested. If the corporations which receive incentives happen to come from outside the region, their basic aim will be to extract wealth from the area rather than to increase the wealth and benefit the living conditions of those who live there. Moreover, most major decisions will be made by metropolitan "head-office", and the head of the local branch plant will essentially be only an administrative figure charged with carrying out the day-to-day implementation of such decisions.

A significant subgroup of these industries presents still further problems. When they first come to the hinterland area, such industries are usually welcomed as providing much needed jobs. However, many such industries are attracted to the area because of the incentives, because wage levels are low, and because local workers are unorganized. Their continued operation is often conditional on their continuing ability to pay lower wages than they would have to pay elsewhere. Thus, instead of eliminating regional disparity and underdevelopment, such industries are more likely to "institutionalize" it to the point where regional wage differentials become permanent.

Before this happens, the workers in such an operation frequently organize in order to protect themselves from easy dismissal and to receive wages on a par with those paid elsewhere. Only in this way can it be explained why some of the most militant labor union locals in the country at the present time are located in Newfoundland and in the western provinces.[48] As it is unable to meet these workers' demands and still remain in operation, the company's only recourse is to appeal to government. Unless still further "incentives" are forthcoming, the company will have to be closed. Given the company's size, this will have devastating economic impact, to say nothing of the impact on the careers of politicians who have loudly proclaimed their own role in attracting such concerns. As a consequence, ways are frequently devised to continue assistance on a permanent basis, often concealed as tax concessions or subsidized resources and power. When this happens, scarce resources are being exploited even further while the industry has now become a permanent subsidized dependent. If this scenario appears far-fetched, it should be noted that it fits the circumstances surrounding Newfoundland's defunct oil-refinery, New Brunswick's Bricklin plant, and the Churchill Industries operation in Manitoba. These have come to light because the respective governments finally refused to keep up ever increasing requests for assistance. Many more such operations may still be continuing.

Of course, not all incentive grants go to firms with head offices outside the underdeveloped region. A number of local entrepreneurs also benefit from such assistance. On the surface, this would seem to have considerable benefit for it allows local entrepreneurs to develop their enterprises and it builds up capital within the region. Certainly this has advantages over the giving of funds to outside corporations. However, it also has its pitfalls. Many such firms are engaged solely in resource exploitation and thus they are simply being subsidized to export the scarce resources rather than to develop them locally.

One can look more benignly on those local firms which receive incentive grants to further the secondary manufacturing of local resources in the region before they are exported. Even though the resource is being taken from the region, a considerable portion of the labour values is being added there, and local workers are benefiting. One can even smile pleasantly upon those locally owned firms which get incentive grants to develop local resources and manufacture local products for consumption within the region itself. However, one still must be aware of the apparent illogic involved in an approach to alleviating regional underdevelopment and worker poverty which begins by giving large sums of money to those who are already well-off and who, if successful, will succeed in increasing their fortunes even more. However, lest this point be carried too far, it may be noted that one of the significant features of Canada's industrial incentives programs has been its difficulty in getting local entrepreneurs to take advantage of the program. It would seem that most such persons are involved in commercial activities and are unwilling to become heavily engaged in manufacturing.

Such reflections move us further into a consideration of the social class implications of Canada's regional development policy. Already noted is the possibility of increased labour strife and the likelihood that the local elite will get richer. Regretfully, further explication must be essentially speculative for we do not have any studies of the regional class structure of Canadian society paralleling the major work on national elites of Porter and Clement.[49] However, the work of Galtung, Sunkel, and Dos Santos leads us to expect certain patterns to emerge.

First, much greater involvement could be expected of multinational and central Canadian manufacturing firms in this region and, with them, a substantial increase in the number of people who fit the categories which Clement has labelled the comprador and parasitic elite. As these people will have important economic ties to outside sources of capital, they will tend to replace the traditional economic elite of the region in terms of power and prestige. As the parasitic elite is essentially a transient group, the social elite of the region is likely to have a greater representation of compradors, the native-born agents of outside economic interests. However, we would also expect to see a significant change in the characteristics of the merchant elite as well. Whereas the old commercial elite was likely to be comprised of "self-made men", the new merchant elite is more likely to be composed of the agents of external economic interests. They will be what we would label a "franchise elite", the owners of franchised branches of national and international chain stores, rather than independent businessmen.

It could be suggested that the effect which DREE is likely to have on the hinterland region has been overdramatized. At least one study which interviewed executives of some businesses receiving assistance discovered that two-thirds of the firms represented would have made their investment without the grant which they received.[50] A similar study undertaken by the Atlantic Provinces Economic Council indicated that this was certainly the case for at least one-third of their sample.[51] One might therefore conclude that claims that DREE subsidies lead to further dependency and exacerbate class differences are highly exaggerated.

However, such an argument overemphasizes economic factors and fails to consider properly the likely social consequences of DREE assistance. As Usher notes, the fundamental fact remains that the DREE incentive program is a transfer of income "from all Canadians to that group of Canadians rich enough to be owners or stockholders in the subsidized firms".[52] He further notes that the "principle of need", whereby applicants must prove that they need subsidies and that the designated investments would not be made without them, in fact favors outside entrepreneurs over local ones. Firms already operating in the region would find it much more difficult to prove that they would not make the proposed investment without governmental assistance.[53] Woodward has also shown that DREE subsidies are biased in favor of capital, for they reduce the price of plant machinery and equipment proportionally more than the cost of other inputs, such as labor.[65] As a result, new installations are more likely to be technologically efficient rather than labour intensive. Woodward contends that this is "inconsistent with the department's primary objective – employment".[55]

The social implication of all this becomes apparent when it is noted that the studies by Springate and the Atlantic Provinces Economic Council cited previously quite emphatically indicate that between one-third and two-thirds of all grants are virtually windfall profits for the companies involved. Moreover, the people who are structurally most advantaged are entrepreneurs from outside the underdeveloped area. Thus the available evidence appears to support the theoretically derived conclusion that DREE subsidies are likely to enhance regional dependency and exacerbate class differences within the region.

The DREE incentives and development program is also likely to have another important impact on the social structure of Canada's underdeveloped regions. If Galtung and Sunkel have any validity, there is likely to be a widening "gap" between developed urban and industrialized areas, and the underdeveloped and increasingly impoverished peripheral areas within the regions themselves. In short, on this point as on many others, the dependency theory is in direct contradiction to the growth centre theory. While the growth centre approach envisages a spread effect throughout the whole regional economy as a result of backward linkages of the master industry, the dependency approach envisages a cleavage and increasing duality in the economy and society between the developed and underdeveloped segments. Galtung explains this in terms of different class interests between the newly developed central area and the still dependent peripheral area within the region. Sunkel explains it in terms of the tendency of the new urban areas to produce consumer goods for the urban population with little ties to the traditional economy. But perhaps the best explanation can be built out of the Innis staple theory. It is the argument that, as much of the investment goes for increased resource exploitation, there is little involvement and integration of the local rural economy beyond a few openings for labourers.

Whatever the reason, there is now evidence that this dualism is occurring in at least one of the major Canadian hinterland regions. In a recent study, Burke and Ireland document that, since the start of the DREE incentive programs, there has been a growing division in Atlantic society between the industrialized urban core and the rural hinterland. They describe the area as having "a sharp dualism in the regional economy, with a prospering area, successfully attracting private investment, and with a lagging area with little or no employment growth".[56] Indeed, they even describe one subregion as "very much a classic hinterland colonial economy".[57]

Burke's and Ireland's findings are highly significant, for they suggest that there has been an improvement in the economic well-being of the urban core in the Atlantic region since the inauguration of the DREE incentive program. However, I would suggest that their findings help demonstrate that there has been little change in the hinterland status of the region. The federal incentive grants have enabled some major industries to develop in the region, but they have either been re-source exploiting (e.g., mining), or they have done so without drawing heavily on local resources (e.g., petroleum refining and automobile assembly). Indeed, the most significant factor in propelling the economic growth of the region in a positive way has been the development of the major ports as transhipment centres, perhaps reflecting Watkins' long standing claim that transportation is the key factor in generating economic spread effects.[58] While the urban areas of the Atlantic region may be flourishing, they thus remain highly dependent on the economic fluctuations of other countries and other regions of Canada. Given this, the economic future may not be as rosy as Burke and Ireland suggest, for their prosperity depends on factors totally beyond their control.[59] In the meantime, the duality in the local economy and society continues to increase.

Towards a Better Way

It would take considerably more space than is available to develop an alternative strategy of regional development which would overcome some of the negative consequences of the growth centre strategy without falling into the pit of dependency. However, it does not seem right to end an article as critical as this one has been without some constructive suggestions. One place to start such a consideration is with an examination of the alternatives suggested by others.

Burke and Ireland argue for a continuation of the incentives program, only this time with a focus on smaller centres. Such a strategy would not overcome some of the long term problems of providing incentives. If incentive grants were made to outsiders, the region would still remain economically dependent. If the grants were made to existing local entrepreneurs, the major class divisions would become greater as the rich got richer. All that is likely to happen is that the next level of semi-urbanized centres becomes drawn into the vortex of dependency. Moreover, there are major obstacles to the development of local entrepreneurial activity. I have already indicated how they can become transformed easily into comprador and parasitic elites.

The dependency theorists, virtually uniformly, all agree on a solution. That solution is socialism; government ownership of new industrial development. That solution certainly has considerable attraction for it ensures that profits remain in the region and are likely to be plowed back into further economic expansion. It also ensures that the "owners" have an overriding interest in having local resources refined and manufactured within the region. The biggest problem here is a decided lack of interest among the people of Canada's hinterland region in electing governments committed to government-sponsored resource exploitation and government ownership of manufacturing, though there are some signs of changing attitudes in western Canada.[60] The other problem is that, within Canada, government-run organizations have tended to become bureaucratic preserves and have not displayed a strong record of economic efficiency or imagination. Indeed, if past experience is any indication, Canada's provincial governments seem even more likely to end up in a "staple trap" than is private industry.

However, the one thing which does seem to be clear from the preceding analysis is that every effort should be taken to develop to the fullest extent possible, the regional economy as an independent entity. The primary goal should be the development of local resources for local consumption, rather than external consumption. In this way, the local residents get the fullest value for their resources, the local economy remains integrated, and local entrepreneurs (be they private individuals or governments) are not supplanted by outside interests.

Thus, I would argue that the dependency theory incorporates the precepts of the position generally espoused by those who claim that "small is beautiful"[61] and who argue for "the limits of growth". The goal is self-sufficiency, not industrialization and urbanization. If our planners fail to recognize this, they are likely to doom our hinterland regions to perpetual underdevelopment.

NOTES AND REFERENCES

[1] Ralph Matthews, "Ethical Issues in Policy Research", *Canadian Public Policy-Analyse de Politiques*, vol. I, no. 2., 1975, pp. 204-11.

[2] Helen Buckley and Eva Tihanyi, *Canadian Policies for Rural Adjustment: A Study of the Economic Impact of ARDA, PFRA, and MMRA.* Special Study no. 7. Ottawa: Economic Council of Canada, 1967, p. ii.

[3] *Ibid., p. 18.*

[4] *Ibid., p. 17.*

[5] *Loc. cit.*

[6] L. E. Poetschke, "Regional Planning for Depressed Rural Areas: The Canadian Experience", in John Harp and John Hofley, (eds.), *Poverty in Canada.* Scarborough: Prentice Hall of Canada, 1971, pp. 274-280.

[7] R. W. Phidd, "Regional Development Policy", in G. Bruce Doarn and Seymour Wilson, (eds.), *Issues in Canadian Public Policy.* Toronto: MacMillan of Canada, 1974, p. 174.

[8] J. P. Francis and N. G. Pillai, *Regional Development and Regional Policy: Some Issues and Recent Canadian Experience.* Ottawa: Department of Regional Economic Expansion, 1972, p. 46.

[9] François Perroux, "Notes sur la notion de 'pole de croissance", *Economic Appliquée,* janvier-juin, 1955, pp. 307-320. Translated as 'Notes on the Concept of "Growth Poles", in David McKee *et al,* (eds.), *Regional Economics.* New York: Free Press, 1970, pp. 93-103.

[10] A. O. Hirschman, *Strategy of Economic Development.* New Haven, Connecticut: Yale University Press, 1958.

[11] Morgan D. Thomas, "Growth Pole Theory: An Examination of Some of its Basic Concepts", in Niles M. Hansen (ed.), *Growth Centers in Regional Development.* New York: Free Press, 1972, pp. 50-81.

[12] Noel Iverson and Ralph Matthews, *Communities in Decline: An Examination of Household Resettlement in Newfoundland.* Newfoundland Social and Economic Studies no. 6, Institute of Social and Economic Research, St. John's: Memorial University of Newfoundland, 1968. See also Noel Iverson and Ralph Matthews, 'Anderson's Cove,' in W. E. Mann (ed.), *Poverty and Social Policy in Canada.* Toronto: Copp Clark, 1973, pp. 233-240.

[13] I would hasten to point out that I do not accept the argument that a large labour pool is necessarily attractive to industry. Most of the people who make up this labour pool in most of Canada's so-called growth centres are unskilled labourers who are generally unsuited for employment in the technologically advanced master industries. However, such a labour pool of unskilled workers is attractive to those industries which employ unskilled workers and/or which survive by paying low wages. The competition for scarce jobs among such workers may mean that the industry can pay lower wages than if workers were in short supply, and may also serve to cut down labour unrest. Thus a worker who causes trouble for the company can quickly be replaced.

[14] See Ralph Matthews, *There's No Better Place Than Here: Social Change in Three Newfoundland Communities.* Toronto: Peter Martin Associates, 1976. Also Ralph Matthews, "Economic Viability versus Social Vitality in Regional Development", in H. Guindon, D. Glenday, and A. Turowitz, (eds.), *Modernization and the Canadian State.* Toronto: Macmillan of Canada, 1977 (forth-coming).

[15] One of the founders of the dependency position is Latin American sociologist Andre Gunder Frank whose books, *Capitalism and Underdevelopment in Latin America.* New York: Monthly Review Press, 1967, and *Latin America: Underdevelopment or Revolution.* New York: Monthly Review Press, 1970, are key works in the area. In presenting the Frankian position I have taken pains to omit the frequently accompanying rhetoric. In this way, I hope to show the potential of the framework for explaining

underdevelopment whether or not one accepts the Marxist assumptions. Certainly one does not have to be a Marxist to make the argument presented here.

16 Frank, *op. cit.,* pp. 146-147.

17 Johan Galtung, "A Structural Theory of Imperialism", *Journal of Peace Research,* vol. 8, 1971, pp. 81-114.

18 *Ibid.,* p. 83.

19 *Ibid.,* p. 84.

20 Osvaldo Sunkel, "Transnational Capitalism and National Disintegration in Latin American", *Social and Economic Studies,* vol. 2, 1973. p. 140.

21 Theotonio Dos Santos, "The Crisis of Development Theory and the Problem of Dependence in Latin America", in Henry Bernstein, (ed.), *Underdevelopment and Development: The Third World Today.* Middlesex: Penguin Books, 1973, pp. 57-80.

22 *Ibid.,* p. 76.

23 *Ibid.,* p. 78.

24 Sunkel, *op. cit.,* p. 166.

25 Theotonio Dos Santos, "The Structure of Dependence", *American Economic Review: Papers and Proceedings of the 82nd Annual Meeting of the American Economic Association,* vol. 60, no. 2, 1970, pp. 235.

26 Sunkel, *op. cit.,* p. 167.

27 Dos Santos, "Crisis of Development Theory", in Bernstein, *op. cit.,* p. 71.

28 See Harold A. Innis, *The Fur Trade in Canada: Introduction to Canadian Economic History.* New Haven: Yale University Press, 1930; revised edition, Toronto: University of Toronto Press, 1956. *The Cod Fisheries: The History of an International Trade.* New Haven: Yale University Press, 1940; revised edition, Toronto: University of Toronto Press, 1954. *Political Economy in the Modern State.* Toronto: The Ryerson Press, 1946.

29 Melville Watkins, "A Staples Theory of Economic Growth", *Canadian Journal of Economics and Political Science,* May, 1963. Reprinted in W. T. Easterbrook and M. Watkins, (eds.), *Essays in Canadian Economic History.* Toronto: McClelland and Stewart, 1967. See particularly pp. 62-63.

30 Melville Watkins, "Economic Development in Canada", in Immanuel Wallerstein, (ed.), *World Inequality.* Montreal: Black Rose Books, 1975, p. 73.

31 John Porter, *The Vertical Mosaic.* Toronto: University of Toronto Press, 1965.

32 Philip Mathias, *Forced Growth: Five Studies of Government Involvement in the Development of Canada.* Toronto: James, Lewis & Samuel, 1971.

33 Melville Watkins, "Resources and Underdevelopment", in Robert M. Laxer, (ed.), *Canada Ltd: The Political Economy of Dependency.* Toronto: McClelland & Stewart, 1973, pp. 107-126.

34 Kari Levitt, *Silent Surrender: The Multinational Corporation in Canada.* Toronto: Macmillan of Canada, 1970.

35 Arthur K. Davis, "Canadian Society and History as Hinterland versus Metropolis", in Richard Ossenberg, (ed.), *Canadian Society: Pluralism, Change and Conflict.* Toronto: Prentice Hall of Canada, Ltd., 1971, pp. 6-32. Particularly p. 13.

[36] Arthur K. Davis, "Metropolis/Overclass, Hinterland/Underclass: A New Sociology", *Canadian Dimension,* vol. 8, March/April 1972, pp. 36-43 and 49-50. See particularly p. 36.

[37] Tom Naylor, *The History of Canadian Business, 1867-1914.* 2 volumes. Toronto: James Lorimer & Co., 1975.

[38] T. W. Acheson, "The National Policy and the Industrialization of the Maritimes", *Acadiensis,* vol. 1, no. 2, Spring, 1972, pp. 3-28.

[39] Gord Laxer, "American and British Influences on Metropolitan Development in Canada 1878-1913", paper presented to the Annual Meeting of the Canadian Sociology and Anthropology Association, Frederiction, New Brunswick, June 12, 1977.

[40] House of Commons *Debates,* February 18, 1977.

[41] Cited in Watkins, "Staples Theory", in Easterbrook and Watkins, (eds.), *op. cit.* p. 119.

[42] See Watkins, "Economic Development", in Wallerstein, (ed.), *op. cit.,* pp. 78-79 and Naylor, *op. cit.*

[43] Wallace Clement, *The Canadian Corporate Elite: An Analysis of Economic Power.* Toronto: McClelland & Stewart, 1975, p. 117.

[44] *Loc. cit.*

[45] I am acutely aware that, in the confines of this article, I have not been able to "prove" that Canada is a dependent society. What I have tried to do is indicate how several major studies of Canadian society can be used to demonstrate that this is indeed the case.

[46] Gerald Hodge, 'Regional Planning: Where It's At,' *Plan Canada,* vol. 15, no. 2, 1975, pp. 87-94.

[47] One of the anonymous reviewers of this article has argued that the industrial incentives program is but "a minor part of DREE expenditure" and that I am therefore attaching too much importance to its social and economic effects. In reply, I would point out that, while the industrial incentives grants may not be a major part of DREE's budget, they nevertheless represented a very considerable sum. Between 1965 and 1972 they amounted to nearly two and a half billion dollars. (See Dan Usher, "Some Questions About the Regional Development Incentives Act", *Canadian Public Policy-Analyse de Politiques,* vol. I, No. 4, p. 561). Second, I would note that many of the other DREE expenditures are also indirect grants and subsidies to industry. Thus money spent on infrastructural development may go for large finger piers required by new oil refineries, for new roads to paper mills, for water and sewage systems which will particularly benefit new factories, and for trade and technical schools to train the workers required in all these enterprises.

[48] This may also be a partial explanation for the militancy of labour unions in Quebec. However, the Quebec situation is obviously also complicated by ethnic and cultural factors which make generalization more difficult.

[49] I regard it as significant in itself that Porter's and Clement's studies of Canada's national elite really provide us with a picture of the lifestyle of a handful of very powerful individuals living almost exclusively in southern Ontario, Montreal, Vancouver, and Winnipeg. Virtually no one from the hinterland centres is included. This, in itself, demonstrates the differences in economic power between metropolis and hinterland in Canada.

[50] David J. U. Springate, *Regional Development Incentive Grants and Private Investment in Canada,* Ph.D. dissertation, Cambridge, Massachusetts: Harvard University, Graduate School of Business Administration, 1972.

[51] Atlantic Provinces Economic Council, *Fifth Annual Review,* October, 1971.

[52] Usher, *op. cit.,* p. 569.

[53] *Ibid.,* p. 562.

[54] Robert S. Woodward, "The Capital Bias of DREE Incentives", *Canadian Journal of Economics,* vol. 7, no. 2, May, 1974, pp. 162 and 163.

[55] *Ibid.,* p. 173.

[56] C. D. Burke and D. J. Ireland, *An Urban Economic Development Strategy for the Atlantic Region.* Ottawa: Ministry of State for Urban Affairs, 1976, See p. 20.

[57] *Ibid.,* p. 22.

[58] Watkins, "A Staples Theory of Economic Growth", in Easterbrook and Watkins, (eds.), *op. cit.,* p. 55.

[59] For a more detailed analysis of the implications of Burke's and Ireland's work see Ralph Matthews, "Growth and no growth in Atlantic Canada", *City Magazine,* vol. 2, no. 8, June, 1977, pp. 45-47.

[60] There, the tendency seems more toward state-capitalism than state-socialism.

[61] E. F. Schumacher, *Small Is Beautiful: Economics as if People Mattered.* N.Y., Harper & Row, 1975.

FUTURE CANADA

Introduction

". . . all futurists agree on the object of the exercise. It is to improve one's perception of change, to become more sensitive in judging how the society or the community is changing.
All important decisions concern the future, anticipate it, spread into it, change it. To make a decision is to look into the future, and yet we have managed to arrange things so that most of the people who make the most important decisions are the people most strongly attached to the past, the people who find the greatest difficulty in looking to the future. That is why futurism has such a uniquely important function in modern life."

(John Kettle, *Hindsight on the Future,* 1976).

"You can never plan the future by the past."

(Edmund Burke, *A Letter to a Member of the National Assembly,* 1728-1797).

Hindsight on the Future

JOHN KETTLE

"How did it happen?" moaned Aldice, shaking his head and jabbing the printop as the printop flashed back with graphic clarity:

FINAL FORECAST FROM ONTARIO RIDING #128

JANE KINGSTOWN	(ODP)	20 511
PAUL ALDICE	(PL)	19 275
GABE PARICOLO	(CIC)	3 481
BRANT SHNER	(MLA)	762

Reading it, Aldice could only cry again "How did we get to this? How in hell did we get here?"

It was one of the sillier questions of the year.

For at least two years everyone in that room had known what was happening, had watched with a kind of rapt dismay the inevitable unfolding of trends established decades before, had shuddered as the daily estimates from Demographics Canada etched the emerging pattern in sharp detail. Some of DemoCan's futurists had been painting the picture fairly accurately since the late 1980's. But politicians – and Aldice was nothing if not political – have always disliked long-range forecasts of trouble. Something would turn up: a brilliant policy would turn the trend, a computer-input error would be discovered, even a new fad might come to the rescue. Why borrow trouble from tomorrow? And anyway, some other poor jerk will be holding the stick when the time comes, not me, I'm not going to get through three more elections. . . .

So the years rolled by without interference: trends that could have been stopped or diverted were left to reach full flower, and consequences that could not have been avoided – but might have been prepared for – allowed to strike with full force.

By one of those familiar, clumsy miscalculations, 2001 is an election year as well as Census year. Of course Census has been a basic DemoCan benchmark for a thousand studies, but it still has its historic functions: turning up the numbers for redistribution of the 330 seats in the House of Commons and, since 1981, resetting the culture-group quotas in government and industry. But the election came before the politicians could agree to authorize the new electoral boundaries the computer produced at the end of Census Day. Really, there was nothing to argue about. The machine's redistribution program took account of historic voting and ethnic blocs, natural geography, old boundaries; it looped the gerrymander loop thousands of times for the one best answer that meant fewest boundary changes, and least political disruptions. But the members were fighting to retain prerogatives, and this was one they hated to give up. So they argued, on and off, for weeks. When the president gave the statutory five days' notice of the election, the committee ran for the telehustings as one. With no one to authorize them, the new boundaries would have to await the next election.

And that was why Aldice, on that September evening, was staring at his big desk printop watching his right-wing opponent take the riding – even polls that had been his for years were sliding over to that snappy little . . . lady with the gleaming white hair and the Rarotonga tan and all that line of talk about "the spirit of '85" and the old-age bonus and the seniors sabbatical grants, and on and on.

"How in hell, how in hell," Aldice wailed, and no one answered. Because they all knew, Aldice as well as any of them, when he could bring himself to step off his pink campaign cloud and face reality. A lot had happened to Canada in their lifetimes, some of it, in the last five or ten years, a fairly sharp shock, even to the astute. It had been endlessly analyzed in the printops and played to death in topshows. It had monopolized lunchroom chatter, parties, hotel talk. They

● JOHN KETTLE is a consulting futurist and writer, John Kettle Incorporated, Toronto, Ontario. This article is extracted from John Kettle, *Hindsight on the Future* (Ottawa: Ministry of State for Urban Affairs, 1976), pp. 7-36. The article has been slightly edited and is reproduced with permission.

all knew — in principle, in broad outline, and down into a considerable amount of factual detail. And still it hurt a bit to see a good minister like Aldice — well good*ish* — go under.

What *had* happened? How had they got there, to the Canada of 2001?

Canada was shifting as early as the 1970's. For huge numbers of young people, the Zero Population Growth arguments coincided with their own disinclination to be parents, and by 1980 the total fertility rate had dropped well below two, and many fewer were being born than dying each year. The

Figure 1

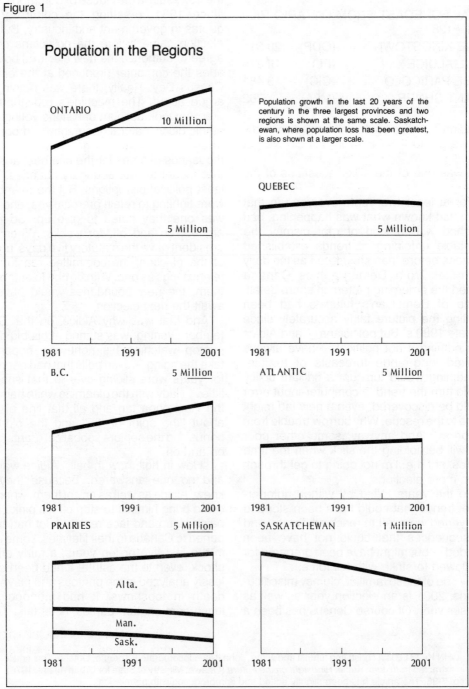

Population in the Regions

ONTARIO

10 Million

5 Million

1981   1991   2001

Population growth in the last 20 years of the century in the three largest provinces and two regions is shown at the same scale. Saskatchewan, where population loss has been greatest, is also shown at a larger scale.

QUEBEC

5 Million

1981   1991   2001

B.C.   5 Million

1981   1991   2001

ATLANTIC   5 Million

1981   1991   2001

PRAIRIES   5 Million

Alta.

Man.

Sask.

1981   1991   2001

SASKATCHEWAN   1 Million

1981   1991   2001

country's population was still growing but it took tens of thousands of immigrants to keep the number positive rather than negative — on average, something like 60 000 more immigrants than emigrants each year.

The population continued to shift westward, out of some of the Atlantic Provinces, out of Quebec, out of the Prairies, until today half of the twelve provinces are losing people, and the population is becoming increasingly concentrated. Rationalization and mechanization of the resource industries was the first cause of the exodus, coupled with some dismal failures of new factories that tried to staff their assembly lines with displaced loggers, fishers, miners, and farmers. People still go West for fun, fame, and fortune. In the case of Quebec, racial antagonism also caused people to leave. Quebec's population peaked in 1993 at just under 6 500 000; it is now smaller by 100 000. Nova Scotia and New Brunswick both saw their population totals shrink by the middle of the 1990's, and the number of people in Prince Edward Island began to decline four years ago. The drop in Manitoba's population began in 1989, not long after the province topped the million mark (it is now below 975 000). But the largest drop has been in Saskatchewan, of course, whose population peaked way back in 1968, just short of a million. Since then it has lost over 40 percent of its people (Figure 1).

The beneficiaries of these migrations have been Ontario and Alberta, both of which have increased by nearly a third in the last twenty years, and British Columbia, which has grown by over 40 percent. These three provinces now have 64 percent of the country's population, compared with 57 percent as recently as the Census of 1981. To a lesser extent the provinces of Newfoundland, Nunavut,* and Yukon have also benefited, but the numbers are much smaller.

Outside Saskatchewan, none of this is as startling as the changes that have occurred in the age composition of the Canadian people. Those who remember the 1970's, or before, tend to get wild-eyed when they compare today's population with what they knew a generation ago, but everyone agrees it is remarkably different. People used to say that the population was aging — which was right in a way — but what we say now is that it

has been middle-aging. The most notable development is the drop in kids under 15 — from 34 percent of the total population to 20 percent. The most startling effect, psychologically, has been that very few people now grow up knowing what it is to have a brother or sister. What it means in practical terms is that the people over fifteen, and especially the people aged fifteen to sixty-four — call them the "working age adults" if you like — are a much more dominant part of the population. Where forty years ago every hundred working-age adults had seventy-one old and young people depending on them, that number is now down to forty-six. In some provinces the shift is even more marked: and in Saskatchewan, because of the exodus, very large numbers of old people are left — thirty-four per hundred working adults, twice as many as the average province.

This change in the shape of the population pyramid has drastically altered the pattern of family and household spending, shifted government spending, changed the style of housing, moved tastes and fashions away from their former youth orientation. The median age has risen fairly steadily, from twenty-eight in 1981 to thirty-six today. And of course the twenty-eight was higher than anything earlier generations knew. Food markets, hotels, clothing, vacation packages, art, architecture, topshows, all reflect middle-aged tastes, particularly the tastes of the people born in the period 1950-1965.

That is the Big Generation (Figure 2), what used to be called the baby-boom children. Even when they were being born — nearly 500 000 a year at the peak just before 1960 — people knew they were going to have a large effect on future demographics. But the effect has been even more profound because of the drop in the birth rate ever since. Inevitably they caused their own mini-boom as they reach child-bearing age; but fertility rates, however, had already dropped quite far by that time and the peak was smaller — scarcely more than 400 000 — in 1985. Births have been fewer and fewer each year since then. So the Big Generation has made waves all the way through — their music, their stimulants, their schooling . . . . They nearly wrecked the universities in the

*Called the Northwest Teritories before it became a province.

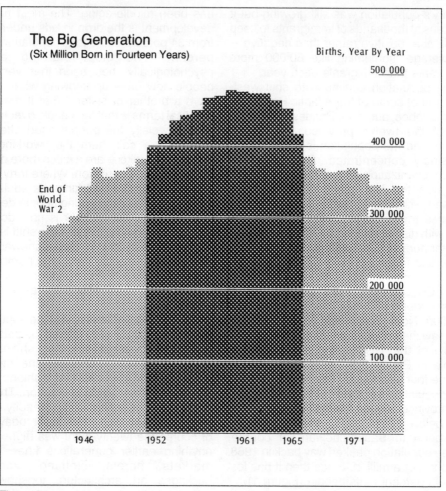

The Big Generation
(Six Million Born in Fourteen Years)

Births, Year By Year

500 000

400 000

End of
World
War 2

300 000

200 000

100 000

1946   1952   1961   1965   1971

Figure 2

1970's, as they first poured in, then held back, dropped out, or stopped and started: huge numbers of them vacillating between work, study, and welfare, quite undecided about what they meant to do with their lives (they must have had premonitions even then), yet already a group with a kind of independence and confidence stemming from their numbers, and the awareness that they were indeed a peak, to be followed inevitably by a decline. Today, with the advantage of hindsight, we tend to see them like a large ship ploughing through the air, building a shock wave just in front, sweeping aside whatever they move into, and leaving turbulence and a slight vacuum behind them. Yet when the Big Generation first made a substantial impression on society, when they entered public school, people could not

know how isolated and therefore important a phenomenon they were; they just looked like a new pattern that might or might not be repeated in the future, a pattern that ought not to be ignored in alternative scenarios. Thus, for years the universities justified their large plant on the grounds that other peaks might follow. That's why the plant for the intensive adult education of the late 1980's, and particularly the 1990's, was available; and why we have been so easily able to take the people on early retirement back into university since the Senior Sabbatical Act of 1992. In fact some of the sabbatical seniors are early retired members of the Big Generation that we have to thank for the existence of the university facilities in the first place. Funny the way some of those mistakes worked out. . . .

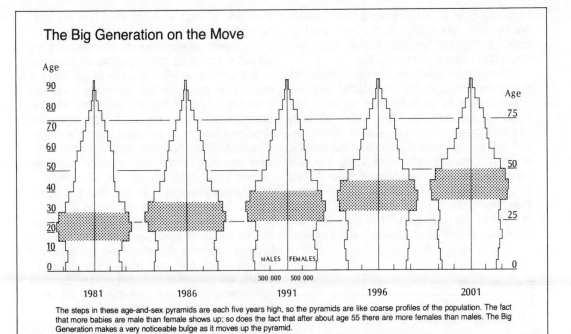

The Big Generation on the Move

The steps in these age-and-sex pyramids are each five years high, so the pyramids are like coarse profiles of the population. The fact that more babies are male than female shows up; so does the fact that after about age 55 there are more females than males. The Big Generation makes a very noticeable bulge as it moves up the pyramid.

Figure 3

Figure 4

One other thing has helped make the population distribution what it is today: changes in mortality rates. Not as spectacular as the declines in death rates earlier in the century, these changes nevertheless have had a substantial effect. Not only are there more old people – one in eight of the total population is now over sixty-five – but on average they are also living longer. This increase in old people has become very noticeable in the last five years – it is what sank Aldice, as you know. The figure we all knew from the Census, but still couldn't quite believe, was the median age of voters. It was thirty-eight in 1981, and increased quite slowly until the last few years, and now today it is forty-six. Half the voters are older than forty-six! The trends have surged powerfully. And it goes without saying that this trend at any rate can only continue upward.

I should give some detail of another thing that is helping to make the old people a still more important factor. That is earlier retirement. When the old age pension was first introduced in 1927 it was paid at age seventy. That was reduced to sixty-five in 1970 and since 1979, in a sort of indexing of the official date of seniority, it has been moved down until now it is sixty-one. As a result

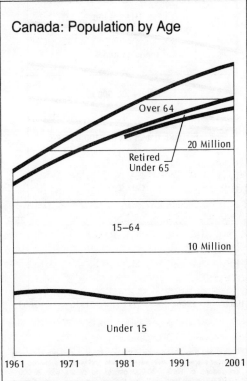

Canada: Population by Age

This simple time chart emphasizes the changes in the main age groups, young, working-age adults, and old. While total population has grown about 50 percent in 50 years, there are now actually fewer under-15's. There are more older people, especially more retired people as retirement comes earlier and earlier. But the startling increase is in people between youth and retirement.

there are over 800 000 people under sixty-five but officially retired today – a total of about 4 200 000 retirees rather than the 3 300 000 people over sixty-five.

Urbanization has been another massive movement, though not as solidly into the biggest cities as people seem to have expected in the old days. The highest growth rate has been in the cities with 30 000 to 100 000 populations (Figure 5). I suppose it is worth remarking that now there are half again as many Census Metropolitan Areas as there were in 1971 – three dozen rather than two dozen. The old CMA's now hold 60 percent of the population, but with the new ones, which are all in Quebec and Ontario, the metropolitan population is now 67 percent of the total. The towns and smaller cities have grown rapidly, though all together they do not yet hold as many people as Toronto, and the villages and little towns still have a sizeable population. It is the rural population that has shrunk: the people who left the places where the only work was hard and dirty and low-paying – the farming and logging and fishing communities – are the ones who brought about this change as they moved into bigger places, as often in their own home provinces as not. But the countryside is emptying rapidly, left to agribusiness, arbobusiness, and piscibusiness, left to the vacation packagers, and to the few brave souls who can stand the solitude and the risks. The city, as it has for thousands of years, still means civilization, and civilization still means protection from famine, disease, poverty, fire, and robbers. (Did you see the topshow last month on that gang of toughs that had been terrorizing Cape Breton? Hardly deserved to be called women. . . .)

It is not hard to see why the people living in the little villages and the raw countryside are now a mere 7 percent of the national total, and mostly in the resource or vacpac businesses. Rural population was once Canada's backbone (still 16 percent of the population in 1981), but it has declined by nearly half in the last twenty years, and soon those wide open spaces will be down to a hard core of maybe a million people or less.

The pattern varies from region to region, of course. Ontario has had a net loss of population in places with populations under 100 000, and now nearly four out of five live in the metropolises. In Quebec, the places with populations over 30 000 have grown, though in the last few years the metropolises, even Montreal, have actually begun to lose population. And that's a trend that is being felt on the west coast, too. Victoria is losing population and Vancouver's growth is slackening noticeably. For the most part, that is because they continue to keep the lower mainland cities separate and resist metropolizing them, even though that is where most of the growth has been. Today over 80 percent of the people live in places with populations of 30 000 or more.

On the Prairies there has been a huge decline in rural population, a sharp increase in the small towns, a decline in medium-size

Figure 5

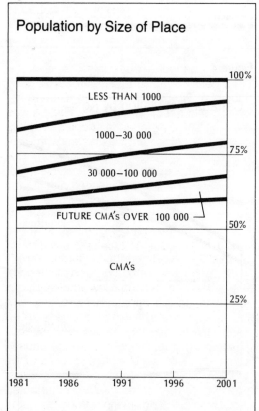

This shows how the balance of the population is continually moving to larger places – from the countryside and villages to places with populations of 1000 - 30 000 people, to places with 30 000 - 100 000 population, and to the metropolises (the Census Metropolitan Areas or CMA's, including some places that only graduated to CMA status in the last quarter century).

cities, and growth in the CMA's — but that is localized, principally in Calgary and Edmonton. Regina and Saskatoon have both lost population.

The Atlantic remains the least urbanized region. Three out of ten are still living in the countryside, and the nation-wide decline in rural populations has had least effect there. The greatest population growth has been in the small towns, places of 1000 to 30 000 people. Halifax started to lose population in the late 1980's, Saint John, N.B., in the early 1990's.

For a time the population of the countryside was markedly older, because old people hated to leave the places and friends they grew up with. But that is no longer true — not surprisingly, under the circumstances. It is the small urbanized places, to which old people have moved, that now have the oldest populations in Canada. The fast-growing demand for urban services began to shift the emphasis of government spending even in the 1970's, of course, but this latest development has brought on one of the worst crises yet. The metropolises, with their more balanced distribution of age groups, are coping quite well with the demand since they got the income and wealth taxes, but today there are 9 300 000 people living in places with populations of fewer than 100 000. Over 2 000 000 are in hamlets and countryside, and the resource companies and vacpackers look after them well enough. It is the 7 200 000 in the smaller towns, with their high proportion of old people, who are experiencing the worst difficulties. No doubt this will cause a further shift toward the bigger cities in the next ten or twenty years. Meanwhile, it is a problem we could have foreseen, but somehow didn't.

There are now nine cities in Canada with more than half a million people, of which Toronto is by far the largest. At 4 400 000, it has 16 percent of the country's total population, compared with 12 percent in 1971. It is followed by Montreal and Vancouver. These three metropolises shelter almost exactly a third of the Canadian population, and their influence on Ottawa is enormous. Next, though some way back, comes Calgary, which is now bigger than Ottawa-Hull, Hamilton, and Edmonton — all of which were larger than Calgary was twenty years ago, and some as recently as the mid-1990's.

Of the old metropolitan centres, eight, or more than a third, are now losing population — principally to other, faster-growing provinces. Montreal is the biggest of these; its population peaked about 1996 at 3 200 000 and is now down by 20 000. Saskatoon is down to 92 000 from its old peak of 135 000. The troubles in these minus-growth cities are very bad — the worst in Canada, much worse than the rural problem. It was Chicoutimi that set the pattern that has now spread from coast to coast. About 1984 it first came to national notice that there were rival gangs of squatters in empty houses — and even in a large apartment building abandoned by its landlord when he moved to Montreal. Old-time residents were frightened away from some of the streets where there were a lot of squatters, and in time whole sections of the town were taken over. What started with looting and some vandalism soon turned to gangsterism. Those involved were long-term unemployed — people pushed out of their resource industry jobs, people who could not survive in small villages that had depopulated below a viable level. The fierce gang war that wrecked downtown Cichoutimi occurred in the spring of 1985. Regina was the next city to experience troubles, and this time the development was faster, with disillusioned youngsters from Winnipeg, Edmonton, and even Toronto, flocking in for excitement. Halifax experienced particularly violent troubles, or maybe they only seem special because they were more recent. At the present moment the worst situation exists in two old Anglophone sections of Montreal — Westmount and Hampstead.

So far the troubles have been confined to depopulating cities and a few smaller places that were losing people quite rapidly. But many small towns have had to raise taxes for much larger security forces and special squatter squads, which in turn has caused some residents to move out and so make the problem worse. In retrospect, it is clear that nation-wide fear of the troubles contributed more than anything else to making the home scanner acceptable, despite all the cries of "Invasion of privacy!" and the resistance organized by the Orwell Brigade. Millions of household in effect agreed that loss of privacy was not too high a price to pay for security. The Orwell Brigade said there was no guarantee the scanner would only feed into the alterter-computer, and no way to tell whether human supervisors were looking in. But after a few rapists and a lot of burglars were caught in the act by scanner security, resistance dropped. (The Domestic Security Surveillance Act of 1993 put an end to the argument, anyway.)

In the countryside, of course, it's everyone for oneself: you get your corporation to put in floodlights, hot wire, sirens, and outside scanners; keep your gun handy; and pray scanner security will come in time if there is an attack. But the aggression is less organized than the city troubles – bums looking for food mostly (Cape Breton is an exception), and not the sadistic squatter gangs of the cities. The death rate has been relatively low.

The troubles come at the top of the list of worries for the millions who live in depopulating metropolises. In the others, the crime level is quite low, thanks to efficient policing and intelligence, both overt and covert. The medical toll is another matter. Blame it on diet and lack of exercise and someone will tell you the problem is psychological, the tension of crowding, the night work, the man-made environment. But blame it on psychological stress and someone else will tell you it has never been easier to move out to a smaller place where there is still interest, variety, good jobs, good income, and almost anything else you might want, plus lots of outdoor leisuresports. Who's right? We still don't know. Something draws people to the metropolises, and holds most of them, yet the level of psymedicare continues to climb.

Households are smaller throughout the country, on average about three-quarters of what they were twenty years ago: smallest in Quebec and largest in the Atlantic region, but all well below three people per household on average. The number of households has grown much faster than the population as a whole – needless to say, a principal reason for the terrific boom in housing. Since 1981 the number of households has jumped by more than half, while the population has grown by less than 20 percent. The family can no longer be thought of as the typical unit, though of course there are still more family than nonfamily households in the 12 600 000 total. Between 1 and 2 percent of existing marriages end in divorce each year. Lots of people still get through their lives on one marriage, but quite a number now get into three or more. The mode is two. There are reliable estimates that a third of the cohabiting couples under thirty are unmarried.

Our experience with the labour force in the last third of the twentieth century shows how wrong people were a generation ago when they worried about the "flood of women" starting to look for work. For one thing, it was not a new development but something that had been going on since World War 1, they just started to notice more women in the lists for jobs and Unemployment Insurance. For

Figure 6

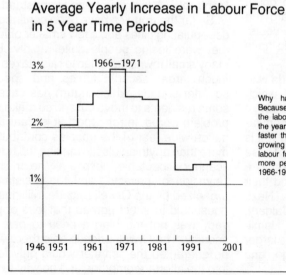

Average Yearly Increase in Labour Force in 5 Year Time Periods

Why has the unemployment problem disappeared? Because the chief cause of it was the high growth rate of the labour force between 1956 and 1981, particularly in the years 1966-1971. The labour force was growing much faster than the economy. Now it is the economy that is growing faster. The actual peak in numbers added to the labour force came in 1976-1981, an average of 260 000 more people a year, compared with 240 000 a year in 1966-1971.

another, the growth of the labour force, despite the "flood", was slackening even in the 1970's, after hitting its post-war peak in the second half of the 1960's (Figure 6). In other words, the flood of females was a figment of male fears; in fact, without the women the economy would have been in real trouble by the 1980's. Why?

Because what had always seemed like an unstoppable spring of new recruits to the labour force abruptly dried up. The Big Generation had passed through the entranceway and the inevitable vacuum followed it. Female participation rates continued their climb, and we were bringing ever more women into the labour force, some for the first time at forty and fifty, by every conceivable inducement − money, comfort, status, and so on. And not only women, of course, as you know from your own experience if you were born after 1965 or 1970. We went all out for kids who should have been in university, and by the late 1980's we were even going after high school kids: "Come work for us and we'll give you a new threedee printop every year", "Work part-time and we'll put you through college!" (" . . . later", we murmured under our breath) − that sort of thing. And of course it worked. In 1991 the participation rate for boys aged fifteen to nineteen went over the 50 percent mark (though it dropped again in a couple of years), and we were getting over 40 percent of teenage girls. In the same way we got the male twenty to twenty-four participation rate back up to 94 percent in the mid-1990's, and the female rate − well, it's 89 percent and climbing.

Older people thought we were heartless, and even stupid. But their thinking was still dominated by the old patterns. For instance, when they thought about the population they still expected there would be more people aged zero to twenty than people aged twenty to forty, more twenty to forty than forty to sixty, and so on. They had a nice orderly picture of the population as a kind of pyramidal layer cake. In 1980 DemoCan pointed out that there were more people in the twenty to forty age bracket than in the zero to twenty age bracket. People of my age had seen it coming, knew what kind of trouble it meant, and were ready to drop old ideas for new ones. (Next year, by the way, there will be more people aged forty to sixty than people aged twenty to forty or zero to twenty − and that really is a shock.)

The other side of the coin is, of course, darker.

The Big Generation began to feel the pressure in the mid-1980's. We all remember that spectacular and tragic suicide of Ben Winton in 1986. Winton was born in 1957, the last son of a World War II veteran on his second marriage; had been brought up on the Beatles, Viet Nam, the liberation movements, the automated office, and so on. He'd done very well at university, graduated in the top ten in his year, gone on to post-grad work, got a reasonable job with General Electronics, was living with a nice kid − the typical boy-next-door. He took out half the 43rd floor of Place Riel, where Gelec's Ottawa branch was located, when he blew himself up: "only" seventeen others killled with him. The note in his apartment was unarguable. No promotion this year, no promotion next year, and not much in the way of prospects. Not his fault, the supervisor said. Just too many others of his age group. Too many of the Big Generation fighting for too few spots. Winton was twenty-nine. He'd spent eight years getting a Ph.D. in fluid-state physics, and he'd grown up believing that that was as good as a ticket to the top. But it wasn't, not by then, and it wasn't going to be. He had had a sudden clear vision of what the rest of his life was going to be like − the endless struggle, the back-stabbing, the office politics, the pressure of competition in his thirties, his forties, his fifties, his sixties. . . .

Of course it got a lot of time in the top-shows and got a lot of ink, as they used to say, in the newspapers (yes, they were still publishing then). Somehow Winton's death crystallized something the Big Generation had known all along, but hadn't quite faced. Suddenly there were a lot of drop-outs. By 1993, which by any reckoning was a bad year, the labour force participation rate for men aged twenty-five to forty-four was down to 86 percent. That meant more than a million dropouts in the prime of life, counting both men and women, for even though we once thought there were going to be as many women as men in that age group in the labour force, it hasn't been the case with the Big Generation, and who can tell if it ever will be.

These two developments wrenched th business world into a new shape. Look at the situation in 1993. At the bottom of the

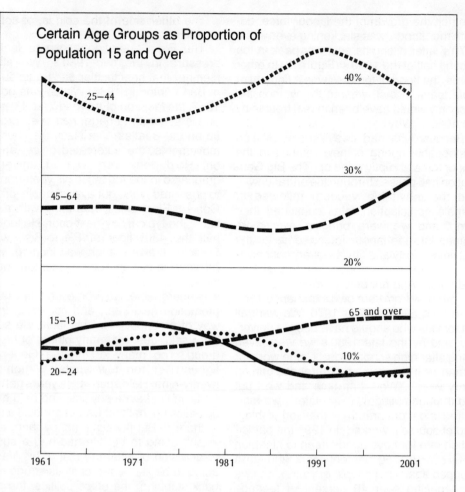

Certain Age Groups as Proportion of Population 15 and Over

40%

25–44

30%

45–64

20%

65 and over

15–19

10%

20–24

1961   1971   1981   1991   2001

Five important age groups are shown here, each represented as a fraction of the total population over 14, during the last 50 years. The most noteworthy thing is the huge gap that developed around 1985-1990 between the 25-44-year-olds, who have traditionally been the lively, innovative middle management people, and the 45-64-year-olds, the more conservative top managers. It had two principal effects: strong competition between the ambitious in middle life, and powerful pressure from below on those who had moved into their last, top jobs. The recent shortage of younger people is also very obvious.

Figure 7

pyramid there was an uncomfortable gap where there should have been new workers coming in. Those who did come were pampered, overpaid, too conscious of their scarcity value, not keen to do the dirty work, go for stim, run maintenance shifts. But there were thousands of more experienced employees ten years older whose advancement was blocked, many of whom had got used to doing the chores in their time, and who were ready to take out their frustration on the youngsters. And the people at the top — they could see, could almost *feel*, the Big Generation coming.

Now, eight years later, we can see Big Generation toughs in many of the top jobs — forty-year-olds running big corporations, both public and private — and a lot of fifty- and sixty-year-olds have taken early retirement. The pressure is building; more of the Big Generation are dropping out. It is harder to see the shape of the economy right now than to describe it as it was five or ten years ago, and I suspect that a different trend is developing as the Big Generation takes charge. Certainly that looks as though it is the case, politically, and it seems to be in the corporations too. The children of the Big

Generation are just beginning to make an impression on the labour force, and I think it is subtly changing a lot of things, attitudes to work, to not working; to support, to not supporting. We should know by 2010.

Since 1980 the overwhelmingly largest proportion of new jobs has been in business, community, and personal services, which now employ 7 000 000 people, or half of the work force. By contrast, there are fewer than 500 000 in the old primary sectors — agriculture, fishing, forestry, mining, and so on — which we now lump together as one sector: the resource industry. It is highly automated. In the last twenty years so many farms have been taken over by agribusiness corporations that the statistics have become confused: some take-overs were mergers, some outright purchases, some lease-purchases, and there were other variations, with the result that by the end of the 1980's DemoCan data showed fewer than one "farmer" per farm on average ("farmer" by DemoCan's definition is an agribusiness person whose principal work is with resources, nonadministrative, nonmanagerial). A better figure is the number of hectares per farmer, which crept up from 186 ha in 1981 to a staggering 672 ha per farmer today. Obviously DemoCan is going to have to fix up the definitions and insist on clearer reporting from agribusiness. Farmers are highly paid: in the case of some smaller operations, better paid than the owners, whose reward will come when they sell out to the big corporations (if they haven't already borrowed too heavily against the sale). More than half the farms are now in the Prairies region; farm area has dropped sharply in the East, increased slightly in the West. Perhaps it is not surprising the agricorporations had to move in and automate, with trends like that. But 20 percent of the work force is still in secondary industry — manufacturing and construction (it was 25 percent in 1981). The number of public servants has actually dropped, which people did not expect a generation ago, but so many government services are now being contracted out or optionalized — "secularized", in the Progressive Liberal Party's campaign phrase — that the public service is boiling down to an elite of managers and professionals, particularly social scientists.

Figure 8

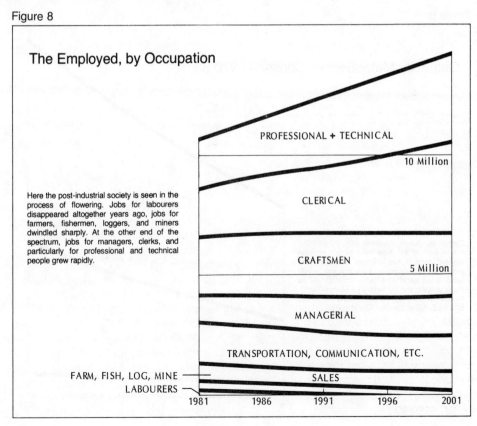

The Employed, by Occupation

Here the post-industrial society is seen in the process of flowering. Jobs for labourers disappeared altogether years ago, jobs for farmers, fishermen, loggers, and miners dwindled sharply. At the other end of the spectrum, jobs for managers, clerks, and particularly for professional and technical people grew rapidly.

PROFESSIONAL + TECHNICAL

10 Million

CLERICAL

CRAFTSMEN

5 Million

MANAGERIAL

TRANSPORTATION, COMMUNICATION, ETC.

FARM, FISH, LOG, MINE

LABOURERS

SALES

1981   1986   1991   1996   2001

The change in occupations has been more spectacular. Over half the work force, or 7 500 000 people, are professional, technical, or clerical. Another 12 percent are managerial. These are, to make an old-fashioned point, essentially clean, dry, sit-down, inside, protected occupations. Of course, a sizable proportion of managers have the title more by courtesy than by responsibility: workers in chief's clothing, as Sandy Mukherjee put it on his topshow a couple of months ago. Virtually all of these honorary bosses are Big Generation people, for whom someone has softened the blow.

Half the people in the resource industry, too, are managerial, professional, technical, or clerical people. They scarcely know what the sea looks like, have never been down a mine, and couldn't tell one end of a cow from the other. The vice-president, in charge of data processing in an agribusiness, could hardly be expected to plant soybeans! Fewer than half the people in trade are sales clerks. But most of those in secondary industry are skilled craftsmen. They call it blue collar work, for reasons only an academic could explain now. There have been hardly any labourers since the 1980's; the labourers' union quietly folded up last November, a mere nine years after the last labourer retired (or, maybe, was edugraded).

Automation has wrought startling changes in productivity. In the resource industry, where more capital has been invested in automatic, semi-automatic, and robot equipment than in any other, output per employee more than doubled in twenty years (constant dollars). In the secondary industries it did not quite double, but still increased substantially. In the tertiary industries the increase was a mere 20 percent or so. The nature of most tertiary work is personal and resists automation, of course, but the bulge of the Big Generation made us unwilling to automate away any more jobs than we had to — had to, of course, because we couldn't get anyone to do them.

In economic terms, these developments have caused surprisingly little change. The resource industry still produces 7 percent of the gross national product, and a mere three percentage points have shifted between the

Figure 9

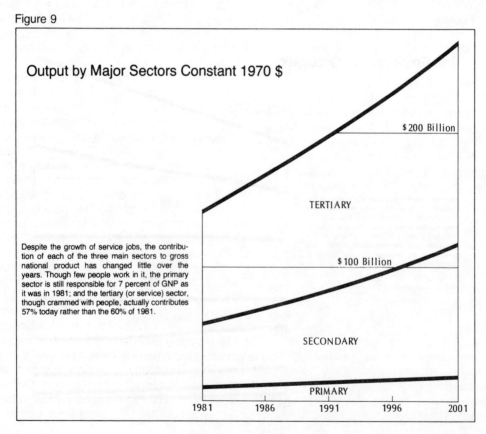

Output by Major Sectors Constant 1970 $

$ 200 Billion

TERTIARY

$ 100 Billion

Despite the growth of service jobs, the contribution of each of the three main sectors to gross national product has changed little over the years. Though few people work in it, the primary sector is still responsible for 7 percent of GNP as it was in 1981; and the tertiary (or service) sector, though crammed with people, actually contributes 57% today rather than the 60% of 1981.

SECONDARY

PRIMARY

1981   1986   1991   1996   2001

other two: secondary has gone from 34 percent in 1981 to 37 percent today, and tertiary has gone from 60 to 57 percent. Output per employee is up 42 percent, about 1.8 percent a year on average, though it is slowing down now, and averaged only 1.6 percent in the last five years — not exactly zero growth, of course, but quite a change from our economic heyday. The largest change, compared with the middle of the last century, say, is the near-disappearance of small business. It had been declining, of course, but the decline has turned to a rout. Eight percent of national income went to small businesses, including unincorporated farmers, in 1981, but this year the proportion is down to 3 percent. But of course there are hardly any real individual owners left in *any* size of corporation. In most large corporations, government is the largest shareholder, and taxes have made it useless for an individual to hold more than the standard share allowance (equal to one year's salary). Most of the rest of the national income is split between the corporations and the employees, and there has been a shift there: labour income is up from 78 percent to 83 percent and net corporate profits are down from 14 to 13 percent. In real terms, both have about doubled; small business income, on the other hand, has actually declined by a third.

Old-line economists still say that slower economic growth from an unprecedentedly large work force is unhealthy. But the key figure to look at is income. Since the typical household is more likely now than ever before to contain more than one worker, yet likely to be smaller than ever before, with fewer dependents, slower growth of output per employee makes perfectly good sense. This is so particularly in some of the regions that were lagging in the last century, where household income has made large gains because of transfer payments and larger households as well as productivity. In fact the Atlantic region, though still at the bottom of the heap with the Prairie region when the measure is per capita personal income, is second only to Ontario in household income.

There has been a huge demand for housing. The increase in nonfamily households has been the chief contributor to the surge of home-building, but in recent years the replacement of older housing has accelerated, and this year the Canada Housing Corporation says more than one dwelling unit in four simply replaced one that had to be demolished. In the last twelve months, for the first time ever, more than 100 000 old units were demolished, while about 390 000 new ones were built. In the last five years the corporation has built (or had built) 1 750 000 units and demolished 435 000. Demolition is still concentrated on single-family houses, but apartment buildings, condominiums, semidetached homes, and other forms are a growing part of the mix going under; a batch of housing built just after World War II has now just about gone, and quite a lot of the half-million units built in the second half of the 1950's are looking increasingly tacky and are disappearing fast (that was where the Big Generation started!). The wreckers are even beginning to work on some of the million units built in the first half of the 1970's, though most are still occupied.

It took a long time, but in 1995 Canada officially became an "apartment" country, where less than half the population lives in single-family detached houses. (I put "apartment" in quotes because the term includes all rental housing, in fact everything except the single house.) That is something that happened decades ago in Quebec and in the early 1990's on the west coast, and it is just about to happen in Ontario and the Prairies. The Atlantic region is still house country by better than 2:1, even though more than half of what has been built there in the last twenty years has been apartment-type accommodation. Old anticipations, that most Canadians would soon be living in apartments, seem to have been based on two fallacies: that urbanizing meant moving everyone into the centre of a metropolis (whereas nearly 10 000 000 people today are living in places with populations under 100 000) and that the cost of serviced land would prevent the building of anything but multiple housing (whereas the domestic-wastes recycling technology innovated in the 1980's radically reduced land costs and brought the middle class back into the market for houses). All by itself, the plumberator has probably put two or three million people in houses who would otherwise have been in multihousing today, including a significant number of nonfamily households. On the other hand, it is also the second most important cause of the government service crisis in

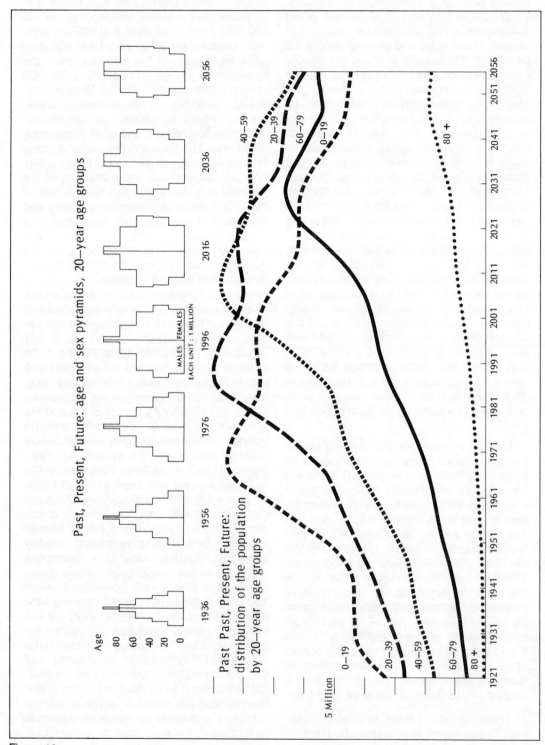

Past, Present, Future: age and sex pyramids, 20-year age groups

EACH UNIT : 1 MILLION

MALES | FEMALES

2056

2036

2016

1996

1976

1956

1936

Age
80
60
40
20
0

40–59
20–39
60–79
0–19
80 +

Past Past, Present, Future:
distribution of the population
by 20-year age groups

0–19
20–39
40–59
60–79
80 +

5 Million

1921 1931 1941 1951 1961 1971 1981 1991 2001 2011 2021 2031 2041 2051 2056

Figure 10

356

the smaller towns, and for the same reason: it has allowed many more people to live in houses in small towns than would otherwise have been able to afford to, including many refugees from the wild and woolly countryside. Prefabricated housing units, or factory homes, as they are commonly called, are now home to about one in ten households. By far the largest number are in places with a population under 100 000, but some ingenious assemblies of factory homes have been seen in bigger cities. The corporation is now producing well over 50 000 a year. It is a cliché, I suppose, to remark how little they resemble the prototype "mobile" homes (which never went anywhere) of mid-twentieth century.

Housing technology could also be thanked for cutting back the number of automobiles, though there was ample indication years ago that the market was moving rapidly toward saturation. This year there are 12 200 000 cars on the road, actually fewer than in any year since 1995. The peak year was 1997. The statistics show forty-three cars per hundred people of all ages, fifty-nine per hundred people aged twenty or older (the driving age was moved up to twenty in 1988 after three provincial auto insurance corporations folded in one year). No doubt the size of the major metropolises has also contributed to the cutback. Meanwhile, the auto dealers are still selling about 1 250 000 cars a year, as they have been ever since the early 1980's: up a little some years, and down others. Nowadays more autos are scrapped than are sold each year, of course, but the difference is still small.

Factory homes are taking over a few of the auto plant assembly lines, however, and no doubt will make up a much larger part of the auto corporations' sales in fifteen or twenty years.

The primary schools are once more suffering withdrawal symptoms as the Big Generation's children move on to secondary school, and in a couple of years the bulge will be felt in postsecondary — the universities, the comcolls, and the corporations' new management institutes. We learned from the mess the Big Generation made of education a generation ago, and we were ready with a whole lot of uses for emptying classrooms, chiefly the edugrading program for workers who had to shift up a job scale or two in mid-career — or in many cases across rather than up — but also, and increasingly today, the senior sabbatical program. When you have seen the rural coachlines picking up their loads of seniors on a bright fall morning and whistling off to regional schools you've witnessed the signs of a big improvement in the educational establishment. The people who pay the school taxes began to make a lot more noise when they became the recipients of the education!

And what of the future?

We can foresee the population distribution settling at last in subsiding undulations as the Big Generation's grandchildren and great-grandchildren and great-great-grandchildren make their successively smaller waves. We can foresee, this time, a truly aging population. Today 16 percent of the population is over sixty but in a quarter of a century the proportion will be 25 percent and by 2051 it will be 30 percent or more; the population pyramid is rapidly turning into a column. We can foresee a further slowing of the economy and we know that any phenomenon approaching equilibrium is likely to become quite unstable, with larger swings up and down: something Bank Canada is already modeling.

But of course all of us have something of the politician in us. We don't like borrowing tomorrow's trouble. We have some fine brains behind the printops and a smart system to hunt deviations and put up warning signals in good time.